WITHDRAWN
UTSA LIBRARIES

RENEWALS

Philosophical Problems of SCIENCE & TECHNOLOGY

PHILOSOPHICAL PROBLEMS OF *SCIENCE & TECHNOLOGY*

Edited by
ALEX C. MICHALOS

University of Guelph
Guelph, Ontario, Canada

ALLYN & BACON, INC. ◆ BOSTON

© Copyright 1974 by Allyn & Bacon, Inc.,
470 Atlantic Avenue, Boston.

All rights reserved. No part of the material protected by this copyright notice may be reproduced or utilized in any form or by any means, electronic or mechanical, including photocopying, recording, or by any informational storage and retrieval system, without written permission from the copyright owner.

Library of Congress Catalog Card Number: 73–84907

Printed in the United States of America.

LIBRARY
University of Texas
At San Antonio

To the memory of
Daniel L. Pucci

and others like him,
who lost their lives in
wars that were not of
their own making.

CONTENTS

PREFACE

This is a collection of essays on philosophical problems of science and technology intended for college students at the introductory level. The readings do not presuppose any particular competence in philosophy, science or technology, although here as elsewhere, the more one brings to a subject, the more one is likely to carry away.

The collection differs from others in the field in three ways. First, more of the selections tend to be *introductory* in their style than is usually the case for readers in this field. One of the biggest problems teachers and students have had in this area is that the material available in texts has typically been too difficult for beginning students. Hopefully, the mixture of articles assembled here will solve this problem. Second, most of the selections in this set are *comprehensive* in their treatment of problems. For example, Ziman considers several proposed definitions of 'science'; Mehlberg, Jobe and Alexander examine several approaches to laws; Rapoport and Achinstein review several approaches to theories, and so on. My aim here has been to expose a number of alternatives to students in order to reduce their learning task to that of appraisal and, possibly, selection. Rather than ask them to create new views at this stage, I would ask them to think critically about the more or less plausible views that are currently in the literature. Third, the *range* of issues considered in this set of essays is well beyond that of all other collections in the field. In fact, the range has been expanded in two directions. On the one hand, essays have been included on the *nature* of technology and its *impact* on society at large. On the other, essays have been included on the social *responsibilities* of scientists and technologists, and on the problems of designing *policies* for the growth and development of science and technology.

No other collection of readings in the philosophy of science has taken notice of problems related to the social responsibilities of scientists and technologists *as* scientists and technologists, and no other collection has addressed itself to the problems involved in the creation of a national or international science policy. However, it is precisely in these areas where philosophy or humanistic concerns generally are likely to have the greatest impact on the future development of science and technology. It is with these areas that a non-scientist is most likely to come in contact. For example, ordinary citizens (as well as professional scientists and technologists) are often and increasingly called upon to sit on boards or councils to determine the allocation of public funds to various scientific and technological enterprises. Hopefully, the readings offered here will provide some insights into the sorts of problems and possible solutions that an ordinary citizen or a citizen-scientist/technologist will confront later.

Each of the sections is preceded by a brief introduction in which the student is given some hints as to what is coming, what to look for and what questions to ask. The sections are followed by suggestions for further reading. Since many of the essays deal with issues of a peripheral nature while grappling with their central problems, the Index can be used to identify various treatments of topics covered in this book. The order of presentation used is about as good as can be chosen for an introduction, but instructors and other interested persons can alter that order to suit their own purposes.

Finally, I wish to express my thanks to the other members of the philosophy department at Guelph who have shared their ideas, headaches, heartburn, joys and problems of introducing the scientific enterprise to nonscience students. They are Hugh Lehman, John Leslie, Ken Montague, Helier Robinson, Michael Ruse and Tom Settle.

A.C.M.

SECTION I

The Nature of Science and Technology

The primary aim of the first article in this section is to distinguish science or the scientific enterprise from virtually everything else. Hence, what is required is a principle of distinction or demarcation. What is it that distinguishes science from disciplines like art, law, religion, technology and so on? Are there several distinctive characteristics or only one?

Ziman presents four possible demarcation principles, gives reasons for regarding them as inadequate, and then introduces an allegedly adequate account. Let us consider briefly the first account that he rejects, namely, that "science is the mastery of man's environment." We ask first: Is the view reasonably clear? How is it elucidated? Evidently Ziman means to say that in this view science is identified with its products, and he cites penicillin as an example of a scientific product. So, in this view science is identified with products like penicillin. (At this point you should be wondering, "How can anyone become penicillin or even like penicillin, in the way that one can become a scientist or like a scientist?")

According to Ziman, this first account involves two mistakes. One of those mistakes is that it confuses science with technology. How? By emphasizing scientific output or products without ever mentioning "intellectual procedures." He seems to assume that the rejected account of science might well be an adequate account of technology. Should we grant that assumption? Furthermore, he is assuming that an adequate characterization of science must have some reference to "intellectual procedures." Is that necessarily true?

After carefully examining each of the views Ziman rejects in the light of his objections, you should then try to determine exactly what his allegedly adequate account is and whether or not it is really adequate. When he states, for example, that "Science is unique in striving for, and insisting on, a consensus" you should raise questions like: Isn't there *any* other activity in which a consensus is sought? Isn't there any enterprise in which consensus is equally sought, however much it is sought? Is the neighborhood bully being scientific when he forces everyone to agree with him? Is Molly Milktoast being scientific when she wins agreement from everyone by only holding vague, unclear, tentative and easily adjusted views? Is your local politician being scientific when he plays Molly Milktoast's game? In short, does Ziman's proposal turn too many or too few activities into science? Is it too narrow or too broad, or is it just right in its scope?

The objective of the second article is to distinguish technology from science by a detailed comparison of key features in both of these activities. For example, Bunge explains how a scientific theory can be "employed successfully in applied research (technological investigation)" and yet be false, and how it can be nearly true but a "practical failure." Insofar as these explanations are adequate, science and technology are distinguishable on the basis of what it means to have a *good* theory in either sphere.

Let us consider briefly the first account he offers as to how a false theory might be very efficient. He points out, quite simply, that "a theory may contain just a grain of truth, and this grain alone is employed in the theory's applications." He assumes, then, as he says more or less explicitly in the next sentence, that a scientific theory is a set of hypotheses all of which do not stand or fall together for technological purposes. On the other hand, he seems to assume that for purposes of pure science, all of the hypotheses in a set constituting a theory *do* stand or fall together. Moreover, if one thinks of a theory as at least a *conjunction* of hypotheses, it is easy to see why one might hold Bunge's view. A conjunction is false if one of its conjuncts is false; so a theory might be false if it contained only one false hypothesis besides lots of true ones.

Since there is a section on theories later on in this book, we do not want to worry about their nature too much now. What concerns us here is the question of whether or not science and technology may be distinguished on the basis of the different criteria of goodness employed by scientists and technologists. Scientists want true theories, while technologists want practicably useful theories. Does this distinction commit one to the view that true theories may be useless? Bunge has an answer to this question. Could a true theory be useless in *every* sense of the word "useless?" If so, then a blanket commitment to the pursuit of truth—*any* old truth—might turn out to be little more than a commitment to the wasting of one's time. But if a true theory cannot be useless in every sense, then it would seem that there is something practical about the pursuit of truth for its own sake. Or would it?

READING 1

What Is Science?

John Ziman

To answer the question 'What is Science?' is almost as presumptuous as to try to state the meaning of Life itself. Science has become a major part of the stock of our minds; its products are the furniture of our surroundings. We must accept it, as the god lady of the fable is said to have agreed to accept the Universe.

Yet the question is puzzling rather than mysterious. Science is very clearly a conscious artifact of mankind, with well-documented historical origins, with a definable scope and content, and with recognizable professional practitioners and exponents. The task of defining Poetry, say, whose subject matter is by common consent ineffable, must be self-defeating. Poetry has no rules, no method, no graduate schools, no logic: the bards are self-anointed and their spirit bloweth where it listeth. Science, by contrast, is rigorous, methodical, academic, logical and practical. The very facility that it gives us, of clear understanding, of seeing things sharply in focus, makes us feel that the instrument itself is very real and hard and definite. Surely we can state, in a few words, its essential nature.

It is not difficult to state the order of being to which Science belongs. It is one of the categories of the intellectual commentary that Man makes on his World. Amongst its kith and kin we would put Religion, Art, Poetry, Law, Philosophy, Technology, etc.—the familiar divisions or 'Faculties' of the Academy or the Multiversity.

At this stage I do not mean to analyse the precise relationship that exists between Science and each of these cognate modes of thought; I am merely asserting that they are on all fours with one another. It makes some sort of sense (though it may not always be stating a truth) to substitute these words for one another, in phrases like 'Science teaches us . . .' or 'The Spirit of *Law* is . . .' or '*Technology* benefits mankind by . . .' or 'He is a student of

Reprinted by permission of the author and publisher from Ziman, J.M.: *Science Is Public Knowledge,* © 1968. Cambridge University Press.

Philosophy'. The famous 'conflict between Science and Religion' was truly a battle between combatants of the same species—between David and Goliath if you will—and not, say, between the Philistine army and a Dryad, or between a point of order and a postage stamp.

Science is obviously like Religion, Law, Philosophy, etc. in being a more or less coherent set of ideas. In its own technical language, Science is information; it does not act directly on the body; it speaks to the mind. Religion and Poetry, we may concede, speak also to the emotions, and the statements of Art can seldom be written or expressed verbally—but they all belong in the non-material realm.

But in what ways are these forms of knowledge *unlike* one another? What are the special attributes of Science? What is the criterion for drawing lines of demarcation about it, to distinguish it from Philosophy, or from Technology, or from Poetry?

This question has long been debated. Famous books have been devoted to it. It has been the theme of whole schools of philosophy. To give an account of all the answers, with all their variations, would require a history of Western thought. It is a daunting subject. Nevertheless, the types of definition with which we are familiar can be stated crudely.

Science Is the Mastery of Man's Environment. This is, I think, the vulgar conception. It identifies Science with its products. It points to penicillin or to an artificial satellite and tells us of all the wonderful further powers that man will soon acquire by the same agency.

This definition enshrines two separate errors. In the first place it confounds Science with Technology. It puts all its emphasis on the applications of scientific knowledge and gives no hint as to the intellectual procedures by which that knowledge may be successfully obtained. It does not really discriminate between Science and Magic, and gives us no reason for studies such as Cosmology and Pure Mathematics, which seem entirely remote from practical use.

It also confuses ideas with things. Penicillin is not Science, any more than a cathedral is Religion or a witness box is Law. The material manifestations and powers of Science, however beneficial, awe-inspiring, monstrous, or beautiful, are not even symbolic; they belong in a different logical realm, just as a building is not equivalent to or symbolic of the architect's blueprints. A meal is not the same as a recipe.

Science Is the Study of the Material World. This sort of definition is also very familiar in popular thought. It derives, I guess, from the great debate between Science and Religion, whose outcome was a treaty of partition in which Religion was left with the realm of the Spirit whilst Science was allowed full sway in the territory of Matter.

Now it is true that one of the aims of Science is to provide us with a Philosophy of Nature, and it is also true that many questions of a moral or spiritual kind cannot be answered at all within a scientific framework. But the dichotomy between Matter and Spirit is an obsolete philosophical notion which does not stand up very well to careful critical analysis. If we stick to this definition we may end up in a circular argument in which Matter is only recognizable as the subject matter of Science. Even then, we shall have stretched the meaning of words a long way in order to accommodate Psychology, or Sociology, within the Scientific stable.

This definition would also exclude Pure Mathematics. Surely this is wrong. Mathematical thinking is so deeply entangled with the physical sciences that one cannot draw a line between them. Modern mathematicians think of themselves as exploring the logical consequences (the 'theorems') of different sets of hypotheses or 'axioms', and do not claim absolute truth, in a material sense, for their results. Theoretical physicists and applied mathematicians try to confine their explorations to systems of hypotheses that they believe to reflect properties of the 'real' world, but they often have no license for this belief. It would be absurd to have to say that Newton's *Principia,* and all the work that was built upon it, was not now Science, just because we now suppose that the inverse square law of gravitation is not perfectly true in an Einsteinian universe. I suspect that the exclusion of the 'Queen of the Sciences' from her throne is a relic of some ancient academic arrangement, such as the combination of classical literary studies with mathematics in the Cambridge Tripos, and has no better justification than that Euclid and Archimedes wrote in Greek.

Science Is the Experimental Method. The recognition of the importance of experiment was the key event in the history of Science. The Baconian thesis was sound; we can often do no better today than to follow it.

Yet this definition is incomplete in several respects. It arbitrarily excludes Pure Mathematics, and needs to be supplemented to take cognisance of those perfectly respectable sciences such as Astronomy or Geology where we can only observe the consequences of events and circumstances over which we have no control. It also fails to give due credit to the strong theoretical and logical sinews that are needed to hold the results of experiments and observations together and give them force. Scientists do not in fact work in the way that operationalists suggest; they tend to look for, and find, in Nature little more than they believe to be there, and yet they construct airier theoretical systems than their actual observations warrant. Experiment distinguishes Science from the older, more speculative ways to knowledge but it does not fully characterize the scientific method.

Science Arrives at Truth by Logical Inferences from Empirical Observations. This is the standard type of definition favoured by most serious philosophers. It is usually based upon the principle of induction—that what has been seen to happen a great many times is almost sure to happen invariably and may be treated as a basic fact or Law upon which a firm structure of theory can be erected.

There is no doubt that this is the official philosophy by which most practical scientists work. From it one can deduce a number of practical procedures, such as the testing of theory by 'predictions' of the results of future observations, and their subsequent confirmation. The importance of speculative thinking is recognized, provided that it is curbed by conformity to facts. There is no restriction of a metaphysical kind upon the subject matter of Science, except that it must be amenable to observations and inference.

But the attempt to make these principles logically watertight does not seem to have succeeded. What may be called the positivist programme, which would assign the label 'True' to statements that satisfy these criteria, is plausible but not finally compelling. Many philosophers have now sadly come to the conclusion that there is no ultimate procedure which will wring the last drops of uncertainty from what scientists call their knowledge.

And although working scientists would probably state that this is the Rule of their Order, and the only safe principle upon which their discoveries may be based, they do not always obey it in practice. We often find complex theories—quite good theories—that really depend on very few observations. It is extraordinary, for example, how long and complicated the chains of inference are in the physics of elementary particles; a few clicks per month in an enormous assembly of glass tubes, magnet fields, scintillator fluids and electronic circuits becomes a new 'particle', which in its turn provokes a flurry of theoretical papers and ingenious interpretations. I do not mean to say that the physicists are not correct; but no one can say that all the possible alternative schemes of explanation are carefully checked by innumerable experiments before the discovery is acclaimed and becomes part of the scientific canon. There is far more faith, and reliance upon personal experience and intellectual authority, than the official doctrine will allow.

A simple way of putting it is that the logico-inductive scheme does not leave enough room for genuine scientific error. It is too black and white. Our experience, both as individual scientists and historically, is that we only arrive at partial and incomplete truths; we never achieve the precision and finality that seem required by the definition. Thus, nothing we do in the laboratory or study is 'really' scientific, however honestly we may aspire to the ideal. Surely, it is going too far to have to say, for example, that it was 'unscientific' to continue to believe in Newtonian dynamics as soon as it had been observed and calculated that the rotation of the perihelion of Mercury did not conform to its predictions.

This summary of the various conceptions of science obviously fails to do justice to the vast and subtle literature on the subject. If I have emphasized the objections to each point of view, this is merely to indicate that none of the definitions is entirely satisfactory. Most practicing scientists, and most people generally, take up one or other of the attitudes that I have sketched, according to the degree of their intellectual sophistication—but without fervour. One can be zealous for Science, and a splendidly successful research worker, without pretending to a clear and certain notion of what Science really is. In practice it does not seem to matter.

Perhaps this is healthy. A deep interest in theology is not welcome in the average churchgoer, and the ordinary taxpayer should not really concern himself about the nature of sovereignty or the merits of bicameral legislatures. Even though Church and State depend, in the end, upon such abstract matters, we may reasonably leave them to the experts if all goes smoothly. The average scientist will say that he knows from experience and common sense what he is doing, and so long as he is not striking too deeply into the foundations of knowledge he is content to leave the highly technical discussion of the nature of Science to those self-appointed authorities the Philosophers of Science. A rough and ready conventional wisdom will see him through.

Yet in a way this neglect of—even scorn for—the Philosophy of Science by professional scientists is strange. They are, after all, engaged in a very difficult, rather abstract, highly intellectual activity and need all the guidance they can from general theory. We may agree that the general principles may not in practice be very helpful, but we might have thought that at least they would be taught to young scientists in training, just as medical students are taught Physiology and budding administrators were once encouraged to acquaint themselves with Plato's *Republic.* When the student graduates and goes into a laboratory, how will he know what to do to make scientific discoveries if he has not been taught the distinction between a scientific theory and a non-scientific one? Making all allowances for the initial prejudice of scientists against speculative philosophy, and for the outmoded assumption that certain general ideas would communicate themselves to the educated and cultured man without specific instruction, I find this an odd and significant phenomenon.

The fact is that scientific investigation, as distinct from the theoretical *content* of any given branch of science, is a practical art. It is not learnt out of books, but by imitation and experience. Research workers are trained by apprenticeship, by working for their Ph.D.'s under the supervision of more experienced scholars, not by attending courses in the metaphysics of physics. The graduate student is given his 'problem': 'You might have a look at the effect of pressure on the band structure of the III–V compounds; I don't think it has been done yet, and it would be interesting to see whether it fits into the pseudopotential theory'. Then, with considerable help, encouragement and criticism, he sets up his apparatus, makes his measurements, performs his calculations, etc. and in due course writes a thesis and is accounted a qualified professional. But notice that he will not at any time have been made to study formal logic, nor will he be expected to defend his thesis in a step by step deductive procedure. His examiners may ask him why he had made some particular assertion in the course of his argument, or they may enquire as to the reliability of some particular measurement. They may even ask him to assess the value of the 'contribution' he has made to the subject as a whole. But they will not ask him to give any opinion as to whether Physics is ultimately *true,* or whether he is justified now in believing in an external world, or in what sense a theory is verified by the observation of favourable instances. The examiners will assume that the candidate shares with them the common language and principles of their discipline. No scientist really doubts that theories are verified by observation, any more than a Common Law judge hesitates to rule that hearsay evidence is inadmissible.

What one finds in practice is that scientific argument, written or spoken, is not very complex or logically precise. The terms and concepts that are used may be extremely subtle and technical, but they are put together in quite simple logical forms, with expressed or implied relations as the machinery of deduction. It is very seldom that one uses the more sophisticated types of proof used in Mathematics, such as asserting a proposition by proving that its negation implies a contradiction. Of course actual mathematical or numerical analysis of data may carry the deduction through many steps, but the symbolic machinery of algebra and the electronic circuits of the computer are then relied on to keep the argument straight.* In my own experience, one more often detects elementary *non sequiturs* in the verbal reasoning than actual mathematical mistakes in the calculations that accompany them. This is not said to disparage the intellectual powers of scientists; I mean simply that the reasoning used in scientific papers is not very different from what we should use in an everyday careful discussion of an everyday problem.

This is a point to which we shall return in a later chapter. It is made here to emphasize the inadequacy of the 'logico-inductive' metaphysic of Science. How can this be correct, when few scientists are interested in or understand it, and none ever uses it explicitly in his work? But then if Science is distinguished from other intellectual disciplines neither by a particular style or argument nor by a definable subject matter, what is it?

The answer proposed in this essay is suggested by its title: *Science Is Public Knowledge.* This is, of course, a very cryptic definition, with almost the suggestion of a play upon words.† What I mean is something along the following lines. Science is not merely *published* knowledge or information. Anyone may make an observation, or conceive a hypothesis, and, if he has the financial means, get it printed and distributed for other persons to read. Scientific knowledge is more than this. Its facts and theories must survive a period of critical study and testing by other competent and disinterested individuals, and must have been found so persuasive that they are almost universally accepted. The objective of Science is not just to acquire information nor to utter all noncontradictory notions; its goal is a *consensus* of rational opinion over the widest possible field.

In a sense, this is so obvious and well-known that it scarcely needs saying. Most educated and informed people agree that Science is true, and therefore impossible to gainsay. But I assert my definition much more positively; this is the basic principle upon which Science is founded. It is not a subsidiary consequence of the 'Scientific Method'; it *is* the scientific method itself.

The defect of the conventional philosophical approach to Science is that it considers only two terms in the equation. The scientist is seen as an individual, pursuing a somewhat one-sided dialogue

* This point I owe to Professor Körner.

† There is also, unfortunately, the hint of an antithesis to *Personal Knowledge,* the title of Polanyi's book to which I have already referred. No antagonism is meant. Polanyi goes a long way along the path I follow, and is one of the few writers on Science who have seen the social relations between scientists as a key factor in its nature.

with taciturn Nature. *He* observes phenomena, notices regularities, arrives at generalizations, deduces consequences, etc. and eventually, Hey Presto! a Law of Nature springs into being. But it is not like that at all. The scientific enterprise is corporate. It is not merely, in Newton's incomparable phrase, that one stands on the shoulders of giants, and hence can see a little farther. Every scientist sees through his own eyes—and also through the eyes of his predecessors and colleagues. It is never one individual that goes through all the steps in the logico-inductive chain; it is a group of individuals, dividing their labour but continuously and jealously checking each other's contributions. The cliché of scientific prose betrays itself 'Hence we arrive at the conclusion that . . .' The audience to which scientific publications are addressed is not passive; by its cheering or booing, its bouquets or brickbats, it actively controls the substance of the communications that it receives.

In other words, scientific research is a social activity. Technology, Art and Religion are perhaps possible for Robinson Crusoe, but Law and Science are not. To understand the nature of Science, we must look at the way in which scientists behave towards one another, how they are organized and how information passes between them. The young scientist does not study formal logic, but he learns by imitation and experience a number of conventions that embody strong social relationships. In the language of Sociology, he learns to play his *role* in a system by which knowledge is acquired, sifted and eventually made public property.

It has, of course, long been recognized that Science is peculiar in its origins to the civilization of Western Europe. The question of the social basis of Science, and its relations to other organizations and institutions of our way of life, is much debated. Is it a consequence of the 'Bourgeois Revolution', or of Protestantism—or what? Does it exist despite the Church and the Universities, or because of them? Why did China, with its immense technological and intellectual resources, not develop the same system? What should be the status of the scientific worker in an advanced society; should he be a paid employee, with a prescribed field of study, or an aristocratic dilettante? How should decisions be taken about expenditure on research? And so on.

These problems, profoundly sociological, historical and political though they may be, are not quite what I have in mind. Only too often the element in the argument that gets the least analysis is the actual institution about which the whole discussion hinges—scientific activity itself. To give a contemporary example, there is much talk nowadays about the importance of creating more effective systems for storing and indexing scientific literature, so that every scientist can very quickly become aware of the relevant work of every other scientist in his field. This recognizes that publication is important, but the discussion usually betrays an absence of careful thought about the part that conventional systems of scientific communication play in sifting and sorting the material that they handle. Or again, the problem of why Greek Science never finally took off from its brilliant taxying runs is discussed in terms of, say, the aristocratic citizen despising the servile labour of practical experiment, when it might have been due to the absence of just such a communications system between scholars as was provided in the Renaissance by alphabetic printing. The internal sociological analysis of Science itself is a necessary preliminary to the study of the Sociology of Knowledge in the secular world.

The present essay cannot pretend to deal with all such questions. The 'Science of Science' is a vast topic, with many aspects. The very core of so many difficulties is suggested by my present argument—that Science stands in the region where the intellectual, the psychological and the sociological coordinate axes intersect. It is knowledge, therefore intellectual, conceptual and abstract. It is inevitably created by individual men and women, and therefore has a strong psychological aspect. It is public, and therefore moulded and determined by the social relations between individuals. To keep all these aspects in view simultaneously, and to appreciate their hidden connections, is not at all easy.

It has been put to me that one should in fact distinguish carefully between Science as a body of knowledge, Science as what scientists do, and Science as a social institution. This is precisely the sort of distinction that one must *not* make; in the language of geometry, a solid object cannot be reconstructed from its projections upon the separate cartesian planes. By assigning the intellectual aspects of Science to the professional philosophers we make of it an arid exercise in logic; by allowing the psychologists to take possession of the personal dimension we overemphasize the mysteries of 'creativity' at the expense of rationality and the critical power of well-ordered argument; if the social aspects are handed over to the sociologists, we get a description of research as an N-person game, with prestige points for stakes and priority claims as trumps. The problem has been to discover a unifying principle for Science in all its aspects. The recognition that scientific knowledge must be public and *consensible* (to coin a necessary word) allows one to trace out the complex inner relationships between its various facets. Before one can distinguish and discuss separately the philosophical, psychological or sociological dimension of Science, one must somehow have succeeded in characterizing it as a whole.*

* 'Hence a true philosophy of science must be a philosophy of scientists and laboratories as well as one of waves, particles and symbols.' Patrick Meredith in *Instruments of Communication,* p. 40.

In an ordinary work of Science one does well not to dwell too long on the hypothesis that is being tested, trying to define and describe it in advance of reporting the results of the experiments or calculations that are supposed to verify or negate it. The results themselves indicate the nature of the hypothesis, its scope and limitations. The present essay is organized in the same manner. Having sketched a point of view in this chapter, I propose to turn the discussion to a number of particular topics that I think can be better understood when seen from this new angle. To give a semblance of order to the argument, the various subjects have been arranged according to whether they are primarily *intellectual*—as, for example, some attempt to discriminate between scientific and non-scientific disciplines; *psychological*—e.g. the role of education, the significance of scientific creativity; *sociological*—the structure of the scientific community and the institutions by which it maintains scientific standards and procedures. Beyond this classification, the succession of topics is likely to be pretty haphazard; or, as the good lady said, 'How do I know what I think until I have heard what I have to say?'

The subject is indeed endless. As pointed out in the Preface, the present brief essay is meant only as an exposition of a general theory, which will be applied to a variety of more specific instances in a larger work. The topics discussed here are chosen, therefore, solely to exemplify the main argument, and are not meant to comprehend the whole field. In many cases, also, the discussion has been kept abstract and schematic, to avoid great marshlands of detail. The reader is begged, once more, to forgive the inaccuracies and imprecisions inevitable in such an account, and to concentrate his critical attention upon the validity of the general principle and its power of explaining how things really are.

SCIENCE AND NON-SCIENCE

In this chapter Science will be considered mainly in its intellectual aspects, as a system of ideas, as a compilation of abstract knowledge. The first question to be answered has already been posed in the introductory chapter: what distinguishes Science from its sister 'Faculties'—Law, Philosophy, Technology, etc.? The argument is that Science is unique in striving for, and insisting on, a consensus.

Take Law, for example. We all feel that legal thought is quite different from scientific thought—but what is the basis of this intuition? There are many ways in which legal argument is very close to Science. There is undoubtedly an attempt to make every judgement follow logically on statutes and precedents. Every lawyer seeks to clarify a path of implications through successive stages to validate his case. The judge reasons it out, on the basis of universal principles of equity, in the effort to arrive at a decision that will command the assent of all just and learned men.

The kinship of Law with the mathematical sciences is emphasized by the interesting suggestion that legal decisions might be arrived at automatically by a computer, into which all the conditions and precedents of the case would be fed and a purely mechanical process of logical reduction would produce exactly the correct judgement.* Although perhaps the idea is somewhat fanciful, if this procedure were technically feasible it would provide decisions that could not but command the assent of all lawyers—just as a table of values of a mathematical function printed out by a computer commands the assent of all mathematicians. To the extent, therefore, that the Law is strictly logical, it can be made 'scientific'.

Again, in the concept of 'evidence' there is close similarity. This is too primary and basic an idea to be defined readily, but, roughly speaking, it means 'any information that is relevant to a disputed hypothesis'. In Science, as in Law, we are almost always dealing with theories that are disputable, and that can only be challenged by an appeal to evidence for and against them. It is the duty of scientists, as of lawyers, to bring out this evidence, on both sides, to the full.

In the end, the case may hang upon some very minor item of information—was the man who got off the 3.57 at Little Puddlecome on Monday, 27 May, wearing a black hat? A scientific theory also may be validated by some tiny fact—for example, the almost imperceptible changes in the orbit of the planet Mercury. The question of the *credibilty* of evidence can become very important. We may find everyone in full agreement that, if a fact is as stated by a witness, it has vital logical implications for the hypothesis under consideration; yet the court may be completely undecided as to whether this evidence is true or not. The existence of honest error has to be allowed for. This sort of thing happens in Science too, though it does not usually get remembered in the conventional histories. For example, many scientists will recall the interest that was aroused by the publication of evidence for organic compounds in meteorites —probably an erroneous interpretation of a complex observation, but of the most profound significance if it proved to be true. In such cases there may even be questions about the relative reliability, in general, of two different observers—an assessment, perhaps, based upon their scientific standing and expert authority—just as the rela-

* I am indebted to Professor Julius Stone for sending me his fascinating critical essay on this subject.

tive veracity of conflicting witnesses may become the key issue in a legal case.

But, of course, in Science, when the evidence is conflicting, we withhold our assent or dissent, and do the experiment again. This cannot be done in legal disputes, which must be terminated yea or nay. If we are forced to a premature opinion on a scientific question, we are bound to give the Scottish verdict *Not Proven,* or say that the jury have disagreed, and a new trial is needed. In Criminal Law, where the case for the prosecution must be proved up to the hilt, or the accused acquitted, this is well enough; but Civil Law demands a decision, however difficult the case.

The Law is thus unscientific because it *must* decide upon matters which are not at all amenable to a consensus of opinion. Indeed, legal argument is concerned with the conflict between various principles, statutes, precedents, etc.; if there were not an area of uncertainty and contradiction, then there would be no need to go to law about it and get the verdict of the learned judge. In Science, too, we are necessarily interested in those questions that are not automatically resolved by the known 'Laws of Nature' (the analogy here with man-made Laws is only of historical interest) but we agree to work and wait until we can arrive at an interpretation or explanation that is satisfactory to all parties.

There are other elements in the Law that are quite outside science—normative principles and moral issues that underlie any notion of justice. As is so often said, Science cannot tell us what *ought* to be done; it can only chart the consequences of what *might* be done.

Normative and moral principles cannot, by definition, be embraced in a consensus; to assert that one *ought* to do so and so is to admit that some people, at least, will not freely recognize the absolute necessity of not doing otherwise. Legal principles and norms are neither eternal nor universal; they are attached to the local, ephemeral situation of this country here and now; their arbitrariness can never be mended by any amount of further logical manipulation. Thus, there are components of legal argument that are necessarily refractory to the achievement of free and general agreement and these quite clearly discriminate between Law and Science as academic Faculties.

To the ordinary Natural Scientist this discussion may perhaps have seemed quite unnecessary—Law, he would say, is a man-made set of social conventions, whilst Science deals with material, objective, eternal verities. But to the Social Scientist this distinction is by no means so clear. He may, for example, find it impossible to disentangle such legal concepts as personal responsibility from his scientific understanding of the power of social determination in a pattern of delinquency. The criterion of consensibility might temper some of the scientific arrogance of the expert witness—'Would *every* criminologist agree with you on this point, Dr. X?'—whilst at the same time throwing the full weight of personal decision and responsibility upon the judge, who should never be allowed to shelter behind the cruel and mechanical absolutes of 'Legal Science'. The intellectual authority of Science is such that it must not be wielded incautiously or irresponsibly.

At first sight, one would not suppose that much need be said about what distinguishes Science from those disciplines and activities that belong to the Arts and Humanities—Literature, Music, Fine Arts, etc. Our modern view of Poetry, say, is that it is an expression of a private personal opinion. By his skill the poet may strike unsuspected chords of emotion in a vast number of other men, but this is not necessarily his major intention. A poem that is immediately acceptable and agreeable to everyone must be banal in the extreme.

But, of course Arts dons do not write Poetry: they write about it. Literary and artistic critics do sometimes pretend that their judgements are so convincing that it is wilful to oppose them. An imperious temper demands that we accept their every utterance of interpretation and valuation. Fortunately, we have the right of dissent and if our heart and mind carry us along a different path we have no need to be frightened by their shrill cries of contempt.

The point here is that there are genuine differences of taste and feeling, just as there are genuine differences of moral principle. At the back of our definition of Science itself is the assumption that men are free to express their true feelings; without this condition, the notion of a consensus loses meaning. Under a dictatorship we might be constrained to pay lip service to a uniform standard of style or taste, but this is the death of criticism.

There are, of course, periods of 'classicism' and 'academicism' when some style or techique is overwhelmingly praised and practised, but no one supposes that this is in obedience to the commands of absolute necessity. The attempts of the stupider sort of academic critic to rationalize the taste of his age by rules of 'harmony' or 'dramatic unity' are invariably by-products of the fashion whose dominance he is seeking to justify, not its determining factors. No sooner are such rules formulated than a great artist cannot resist the temptation to break them, and a new fashion sweeps the land. By their very nature, the Arts are not consensible, and hence are quite distinct from Science as I conceive it.

Science is not immune from fashion—a sure sign of its socio-psychological nature. We shall return to some of the symptoms of this disease later. But what, abstractly, *is* fashion? It means doing what other people do for no better reason than that that is what is done. If everyone were to follow only fashionable lines of thought, there would be a false impression of a consensus; the inhibition of the critical imagination by such a conformist sentiment is the antithesis of the scientific attitude. It is also, of course, another way of death for true Poetry and Art.

But the products and producers of Literature, Art and Music may be studied in more factual aspects than for their emotional or spiritual message. For example, they are the outcome of, or participants in, historical events.

The place of *History* in this analysis is very significant; it seems to be truly one of the borderlands marching between scientific and non-scientific pursuits. Suppose that we are investigating such a problem as the date and place of birth of a writer or statesman. We search in libraries and other collections of material documents for written evidence. From various oblique references we might build up an argument in favour of some particular hypothesis—an argument to persuade our colleagues by its invincible logic that no other interpretation is tenable. This procedure seems quite as scientific as the research of a palaeontologist, who might reconstruct the anatomy of an extinct animal by piecing together fragments of fossil bone. Our aim is the same—to make a thoroughly convincing case which no reasonable person can refute. If, unfortunately, we cannot find sufficient evidence to clinch the case, we do not cling to our hypothesis and abuse our opponents for not accepting it; we quietly concede that the matter is uncertain, and return once more to the search. On such material points, the mood of historical scholarship is perfectly scientific.

The other mood in History is much more akin to Literature or Theology; it is the attempt to understand human history imaginatively and to 'explain' it. Having ascertained the 'facts', the historian tries to uncover the hidden motives and forces at work, just as the scientist goes behind the phenomena to the laws of their being.

The trouble is that the complex events of history can seldom be explained convincingly in the language of elementary cause and effect. To ascribe the English Civil War, for example, to the 'Rise of the Gentry' may be a brilliant and fruitful hypothesis, but it is almost impossible to prove. Even though one may feel that this is the essence of the matter, and though one may marshal factual evidence forcefully in its favour, the case can be no more than circumstantial and hedged with vagueness and provisos. It will go into the canon of interesting historical theories, but experience tells us that it will not, as would a valid scientific theory, be so generally acceptable as to eliminate all competitors.

The rule in Science is not to attempt explanations of such complex phenomena at all, or at least to postpone this enterprise until answers are capable of being agreed upon. Imagination in the search for such problems is essential, but speculation is always kept rigidly under control. Even in such disciplines as Cosmology, where it sometimes seems as if a new theory of the Universe is promulgated each week, the range of discussion is limited quite narrowly to model systems whose mathematical properties are calculable and can be critically assessed by other scholars.

History does not impose such restrictions upon its pronouncements. It is felt, quite naturally, that the larger questions, although more difficult, are very important and must be discussed, even if they cannot be answered with precision. To restrict oneself to decidable propositions would be to miss the lessons that the strange sad story has for mankind. A history of 'facts', of dates and kings and queens, although acceptable to the consensus, would be banal and trivial. In other words, History also has to provide other spiritual values, and to satisfy other normative principles, than scientific accuracy.

There are, of course, historians who have claimed universal 'scientific' validity for their larger schemes and 'Laws'. It is not inconceivable that historical events do follow discernible patterns, and that there are, indeed, hidden forces—the class struggle, say, or the Protestant ethic—which largely determine the outcome of human affairs. It would not be necessary for such a theory to be absolute and mathematically rigorous for it to acquire scientific validity, any more than the proof that smoking causes lung cancer requires every smoker to die at the age of 50. It is not inherently absurd to search for historical laws, any more than it was absurd, 200 years ago, to search for the laws governing smallpox. Seemingly haphazard events often turn out to have their pattern, and to be capable of rational explanation.

All I am saying is that no substantial general principles of historical explanation have yet won universal acceptance. There have been fashionable doctrines, and dogmas backed by naked force, but never the sort of consensus of free and well-informed scholars that we ordinarily find in the Natural Sciences. Many historians assert that historical events are the outcome of such a variety of chance causes that they could never be subsumed to simpler, more general laws. Others say that the number of instances of exactly similar situations is always too small to provide sufficient statistical evidence to support an abstract theoretical analysis.

Whatever their reasons, historians do not agree on the general theoretical foundations or methodology of their studies. Instead of establishing, by mutual criticism and tacit cooperation, a limited common basis of acceptable theory, from which to build upwards and outwards, they often feel bound to set up antagonistic 'schools' of interpretation, like so many independent walled cities.

They are not to be blamed for such behaviour; it only shows that this is a field where a scientific consensus is not the main objective. If you insisted that historians should work more closely together, they would object that the knowledge that they have in common is too dull, too trivial, too distant from the interesting problems of History, to circumscribe the thought of a serious scholar. To write about the Civil War without asking why the whole extraordinary thing happened is to compose a mere chronicle. For that reason, much of historical scholarship is not essentially scientific.

It would be wrong, on the other hand, to give the approving label 'Science' only to the new techniques of historical research derived from the physical and biological sciences and technologies—carbon dating, aerial photography, demographic statistics, chemical analysis of ink and parchment. Such techniques are often powerful, but they are not more 'scientific' than the traditional scholarly exercises of editing texts, verifying references and making rational deductions from the written words of documents. There is no reason at all why marks on paper in comprehensible language should be treated as inherently less evidential than the pointer readings of instruments or the print-out from a computer. In German the word *Wissenschaft,* which we translate as *Science,* includes quite generally all the branches of scholarship, including literary and historical studies.

To maintain, therefore, an impassable divide between Science and the Humanities is to perpetrate a gross misunderstanding that springs in the British case solely from a peculiarity of educational curricula. The Story, the Arts, the Poetry of Mankind are worthy both of spiritual contemplation and scholarly study, whether by laymen in general education or by experts as their life career. In many aspects this study is perfectly akin to the scientific study of electrons, molecules, cells, organisms or social systems: consensible knowledge may be acquired whether as isolated facts or as generally valid explanations. But to confine oneself, in education as in scholarship, to such aspects would impoverish the imagination, and even restrict the scope of possible further advance. Without general concepts as a guide—however uncertain, personal and provisional—we simply could not see any larger patterns in the picture. Historical and literary scholarship cannot therefore pretend to be scientific through and through, but that does not prevent their making progress towards a closer definition of the truth. In the end, bold speculative generalization and unverifiable psychological insight may go further in establishing a convincing narrative than a rigid insistence on precise minutiae.

It scarcely needs to be said that *Religion,* as we nowadays study and practise it, is also quite distinct from Science. This seems so obvious in our enlightened age that one wonders how there could have been any conflict and confusion between them. But was not Religion primitive Science—the corpus of generally accepted public knowledge? Should we not see Science as growing out of, and eventually severing itself from, this parent body—or perhaps as a process of differentiation and specialization within the unity of the medieval *Summum*? Just because many religious beliefs are now seen to be wrong, it does not follow that they were not seriously, freely and rationally accepted in their time. Conventional science too can be wrong at times.

Let me give an example. In the late eighteenth and early nineteenth centuries, prehistoric remains were found that we now see as pointing to the great antiquity of Man. But many scholars stood out against this interpretation because it did not square with the Biblical chronology of the past. Is it fair to treat this as a conflict between scientific rationality and religious prejudice? Would it not be more just to say that a widely accepted theory was being ousted by a better one as new evidence came to light?

The point is that this debate was open and free. The participants on one side may have been blinkered by their upbringing, but their beliefs were honestly held and rationally maintained. They may often have used poor arguments to defend their case—but they did not call in the secular arm or the secret police. In the end, they lost; and since then the appeal to Divine Scripture has ceased to be an acceptable element in scientific discussion.

What I am arguing is that there is a progressive improvement in the techniques and criteria of such discussion, and that the use of abstract theological principles was once respectable but is now discredited, just as the absolute justification of Euclidean geometry from the Parallelism Axiom is now discredited. The 'Scientific Revolution' of the seventeenth century is not a complete break with the past. The idea of presenting a rational non-contradictory account of the universe is perhaps a legacy of Greece, but it is very strong in medieval Philosophy and Theology. It may be that the very existence of a dogmatic system of metaphysics, implying a rational order of things and fiercely debated in detail, was the prerequisite for the development of an alternative system, using some of the same logical techniques but based upon different principles and more extensive evidence.* The doctrinaire consensus of the Church may have been prolonged beyond its acceptability to free men by the power of the Holy Office, but it had originally provided an example of a generally agreed picture of the world. These are subtle and deep questions which I am not competent to discuss, but I wonder whether the failure of Science to grow in China and India was due as much to the general doctrinal permissiveness of their religious systems as to any other cause. Toleration of deviation, and the lack of a very sharp tradition of logical debate may have made the very idea of a consensus of opinion on the Philosophy of Nature as absurd to them as the idea of absolute agreement on ethical principles would be to us.

The relationship between Science and *Philosophy* is altogether more complex and confused. In a sense, all of modern science is the Philosophy of Nature, as distinct from, say, Moral or Political Philosophy. But this terminology is somewhat old-fashioned, and we try to make a distinction between Physics and Metaphysics, between the Philosophy *of* Science and Philosophy *as* Science. Some philosophers attempt to limit themselves to statements as precise and verifiable as those of scientists, and confine their arguments to the rigid categories of symbolic logic. The consensus criterion would be acceptable to them, for they would hold that by a continuous process of analysis and criticism they would make progress towards creating a generally agreed set of principles governing the use of words and the establishment of valid truths. Others hold that such a hope can never be realized and that by limiting philosophical discourse in this way they would only allow themselves to make trivial statements, however unexceptionable. For this school of Philosophy it is important to be free to comment on grander topics, even though such comments will only reveal the variety and contradictory character of the views of different philosophers.

As with History we can only say that if Philosophy is what academic philosophers write in their books, then some of it is not very different from Science. But generally the motivation is nonscientific, by our definition, and the multiplicity of viewpoints indicates that there is no dominant urge to find maximum regions of agreement. Whatever their claims, the proponents of 'scientific' philosophical systems do not convince the majority of their colleagues that theirs is the only way to truth.

* This point is made in *Science in the Modern World* by A. N. Whitehead (New York: Macmillan, 1931).

* * *

Let us now consider *Technology*—Engineering, Medicine, etc. For the multitude, Science is almost synonymous with its applications, whereas scientists themselves are very careful to stress the distinction between 'pure' knowledge, studied 'for its own sake', and technological knowledge applied to human ends.

The trouble is that this distinction is very difficult to make in practice. Suppose, for example, that we are researching on the phenomenon of 'fatigue' in metals. We are almost forced into the position of saying that on Monday, Wednesday and Friday we are just honest seekers after truth, adding to our understanding of the natural world, etc., whilst on Tuesday, Thursday and Saturday we are practical chaps trying to stop aeroplanes from falling to pieces, advancing the material welfare of mankind and so on. Or we may have to make snobbish distinctions between Box, a pure scientist working in a University, and Cox, a technologist, doing the same research but employed by an aircraft manufacturer. There was once a time when Science was academic and useless and Technology was a practical art, but now they are so interfused that one is not surprised that the multitude cannot tell them apart.

Here again, a definition in terms of the scientific consensus can be really effective. The technologist has to fulfil a need; he must provide the means to do a definite job—bridge this river, cure this disease, make better beer. He must do the best he can with the knowledge available. That knowledge is almost always inadequate for him to calculate the ideal solution to his problem—and he cannot wait while all the research is done to obtain it. The bridge must be built this year; the patient must be saved today; the brewery will go bankrupt if its product is not improved.

So there will be a large element of the incalculable, of sheer art, in what he does. A different engineer would come out with quite a different design; a different doctor would prescribe quite different treatment. These might be better or worse, in their results—but nobody quite knows. Each situation is so complex, and has so many unassessable factors, that the only sensible policy for the client is to choose his engineer or doctor carefully and then rely upon his skill and experience. To look for a solution acceptable to all the professional experts is a familiar recipe for disaster—'Design by Committee.'

The technologist's prime responsibility is towards his employer, his customer or his patient, not to his professional peers. His task is to solve the problem in hand, not to address himself to the opinions of the other experts. If his proposed solution is successful, then it may well establish a lead, and eventually add to the 'Science' of his Technology; but that should not be in his mind at the outset.

What we find, of course, is that a corpus of generally accepted principles develops in every technical field. Modern Technology is deliberately scientific, in that there is continuous formal study and empirical investigation of aspects of technique, in addition to the mere accumulation of experience from successfully accomplished tasks. The aim of such research is not to solve immediate specific problems, but to acquire knowledge for the use of the experts in their professional work. It is directed, therefore, at the mind of the profession, as a potential contribution to the consensus opinion. This sort of work is thus genuinely scientific, however trivial and limited its scope may be.

The abstract distinction here being made between a 'scientific Technology' and 'technological Science' has its psychological counterpart. It is a commonplace in the literature on the Management of Industrial Research that applied scientists often suffer from divided loyalties. On the one hand, they owe their living to the company that employs them, and that expects its return in the profitable solution of immediate problems. On the other hand, they give their intellectual allegiance to their scientific profession—to Colloid Chemistry, or Applied Mathematics, or whatever it is—where they look for scholarly recognition. Although the rewards for 'technological' work are greater and more direct, they very often prefer to stick to their 'scientific' research.

This preference seems almost incomprehensible to management experts, because they fail to see that the scientific loyalty is not just towards a prestigious professional group but to an ideology. The young scientist is trained to make contributions to public knowledge. All the habits and practices of his years of apprenticeship emphasize the importance of making them convincing, and thus making them part of the common pool. Being a successful scientist is not just winning prizes; it is having other scientists cite your work. To give this up is worse than losing caste; it is to give up one's faith and be made to worship foreign idols.

Nevertheless, one must agree that Science and Technology are now so intimately mingled that the distinction can become rather pedantic. Take, for example, a typical Consumers Association report on a motor car. Some of the tests, such as the measure of petrol consumption, may be perfectly scientific in that their validity would be universally acceptable. Other tests, such as whether the springing was comfortable, would not satisfy this criterion, although it would be one of the important skills of the designer to attend to just such 'subjective' and 'qualitative' features. For this reason, to say that a car has been 'scientifically designed' is merely to assert that it has been well designed by competent engineers. Yet an account, by the designer, of the rationale behind various technical features of the model could rank as a serious contribution to the Science of Automobile Engineering by adding to the body of agreed principles at the basis of that mysterious art.

All that I can claim is that these distinctions, although subtle and perhaps pedantic, are not entirely arbitrary or unreal. We do not need to look far ahead to some conceivable remote application of the knowledge in question, nor do we need to examine the hidden, perhaps unconscious, motives of those who produce it. We do not need to decide whether some particular laboratory is 'technological' or 'scientific', and then attach the appropriate label to its products. The criterion is in the work itself, in the form in which it is presented, and in the audience to which it is addressed.

What are we to make of the so-called 'Social Sciences' in the light of this discussion? It is obvious that such a subject as Politics is very close to History and to Philosophy in its goals and achievements; to stick to ascertainable public 'facts' is to limit the discourse to the banal. To give this discipline the name of Political Science is unfortunate; it offers more than it can deliver and debases its ethical message.

On the other hand, *Economics* is a very technological subject; the experts are always being asked to diagnose the ills of the nation and to propose specific cures, long before they have sufficient scientific understanding to make a valid analysis. Yet the totally quantitative material medium—money—allows of convincing proofs, statistical or algebraic, of precise hypotheses, so that a body of agreed principles is gradually emerging. Leading economists may debate in public, and seem to be at loggerheads, but behind the scenes they teach much the same things to their students. It is typical of the tacit cooperation between scholars in a scientific subject that American and Soviet economists respect and learn from each other's work on Input-Output Analysis, however much they may disagree on more speculative issues of general social policy.

These are the neighbours of the new discipline of *Sociology* which is an attempt to escape from the 'unscientific' traditions of History and Politics, and to make the study of social systems, and of man in society, at least as scientific as Economics.

That is the reason why so much sociological research is by questionnaire and statistical analysis; the aim is to provide the necessary factual basis for firmly scientific theories. To the extent that observations of this sort are verifiable by repetition, and capable of being made quite convincing to a critical public, this attitude is sound. But the intractability of the subject must be reckoned with. Vast quantities of information do not add up to much serious knowledge without theories to give it meaning. Moreover, even to accept 'facts' of this sort may imply the acceptance of dubious hidden theories. Suppose, for example, that our car-testing organization decided to assess the comfort of various models by asking a hundred people to give their views, and reported that car A 'rated' 87 per cent and car B only 63 per cent. This is objective information, which might well be 'verifiable'. But there is the implication that it 'measures' something—'comfort'—which may not exist at all. This is a crude case, but sociological research is full of more subtle examples of the same difficulty.

Some sociologists have taken quite a different line. They deal in abstract categories, which they manipulate logically into various hypothetical relations in a sort of formal calculus. This approach also strives towards the creation of a consensus, in that the structure of the argument can be purged of contradictions and hence made unexceptionable and theoretically acceptable. But without much more rigorous connection of these abstractions with real systems and actual phenomena it is vacuous knowledge, without the power to persuade us that thus and thus is the world of men.

Nevertheless, Sociology is often genuinely scientific in spirit, although it has turned out to be an exceedingly difficult science whose positive achievements do not always match the effort expended on it. The 'methodological problem' has not been surmounted; there is not yet a reliable procedure for building up interesting hypotheses that can be made sufficiently plausible to a sufficient number of other scholars by well-devised observations, experiments or rational deductions. It was the sort of problem facing Physics before Galileo began seriously to apply mathematical reasoning and numerical measurement to the subject. The ideal of a consensus is there, but the intellectual techniques by which it might be created and enlarged seem elusive.

* * *

This survey of the Faculties has necessarily been brief and schematic. Why should we even want to decide whether a particular discipline is scientific or not? The answer is, simply, that, *when it is available,* scientific knowledge is more reliable, on the whole, than non-scientific. When there are conflicts of authority, when Sociolgy tells us to go one way and History another, we need to weigh their respective claims to validity. Our general argument here is that in a discipline where there is a scientific consensus the amount of *certain* knowledge may be limited, but it will be honestly labelled: 'Trust your neck to this', or 'This ladder was built by a famous scholar, but no one else has been able to climb it'.

In the end, the best way to decide whether a particular body of knowledge is scientific or not is often to study the attitudes of its professional practitioners to one another's work. A sure symptom of non-science is personal abuse and intolerance of the views of one scholar by another. The existence of irreconcilable 'schools' of thought is familiar in such academic realms as Theology, Philosophy, Literature and History. When we find them in a 'scientific' discipline, we should be on our guard.

This is the reason why for example we should be very suspicious of the claims of Psychoanalysis. The history of this subject is a continuous series of bitter conflicts between persons, schools and theories. Freud himself had the most honest and sincere desire to create a thoroughly respectable scientific discipline, but for some reason he failed to understand this key point—the need to move slowly forward, step by step, from a basis of generally accepted ideas. Perhaps the struggle to get anyone to listen at all was too bitter, or perhaps his mind was too active and impatient to endure continuous critical assessment of each new theory or interpretation. Whatever the reason, the mood of Psychoanalysis in its formative period was antagonistic to the covert cooperative spirit of true Science. Its clinical successes were only of technological significance, and did not scientifically validate the theories on which they were said to be based.

I have given this example, not out of prejudice against psychoanalytic ideas (one or other of the contending schools may well be right: we shall see) but to show that the principle of the consensus is a powerful criterion, with something definite to say on this vexed topic. To some people the words 'scientific' and 'unscientific' have come to mean no more than 'true' and 'false', or 'rational' and 'irrational'. In this chapter I have tried to show, by reference to other organized bodies of knowledge, that this usage is quite improper, and grossly unfair to those scholars who seek rationality and truth in bolder ways than by microscopic dissection of minutiae.

READING 2

Towards a Philosophy of Technology

Mario Bunge

The application of the scientific method and of scientific theories to the attainment of practical goals poses interesting philosophical problems, such as the nature of technological knowledge, the alleged validating power of action, the relation of technological rule to scientific law, and the effects of technological forecast on human behavior. These problems have been neglected by most philosophers, probably because the peculiarities of modern technology, and particularly the differences between it and pure science, are realized infrequently and cannot be realized as long as technologies are mistaken for crafts and regarded as theory-free. The present paper deals with those problems and is therefore an essay in the nearly nonexistent philosophy of technology.

SCIENCE: PURE AND APPLIED

The terms "technology" and "applied science" will be taken as synonymous, although neither is adequate: in fact, "technology" suggests the study of practical arts rather than a scientific discipline, and "applied science" suggests the application of scientific ideas rather than that of the scientific method. Since "technique" is ambiguous and "epitechique" unborn, we shall adopt the current lack of respect for etymology and go over to more serious matters.

The method and the theories of science can be applied either to increasing our knowledge of the external and the internal reality or to enhancing our welfare and power. If the goal is purely cognitive, pure science is obtained; if primarily practical, applied science. Thus, whereas cytology is a branch of pure science, cancer research is one of applied research. The chief divisions of contemporary applied science are the physical technologies (e.g., mechanical engineering), the biological technologies (e.g., pharmacology), the social technologies (e.g., operations research), and the thought technologies (e.g., computer science). In many cases technology succeeds a craft: it solves some of the latter's problems by approaching them scienctically. In other cases, particularly those of the social and thought technologies, there is no antecedent prescientific skill because the problems themselves are new. But in every case a distinction must be made between pure research, applied research, and the applications of either to action.

The division of science into pure and applied is often challenged on the ground that all research is ultimately oriented toward satisfying needs of some sort or other. But the line must be drawn if we want to account for the differences in outlook and motivation between the investigator who searches for a new law of nature and the investigator who applies known laws to the design of a useful gadget: whereas the former wants to understand things better, the latter wishes to improve our mastery over them. At other times the difference is acknowledged, but it is claimed that applied science is the source of pure science rather than the other way around. Clearly, though, there must be some knowledge before it can be applied, unless it happens to be a skill or know-how rather than conceptual knowledge.

What is true is that action—industry, government, warfare, education, etc.—often *poses problems* that can be solved only by pure science. And if such problems are worked out in the free and lofty spirit of pure science, the solutions to them eventually may be applied to the attainment of practical goals. In short, practice is one of the sources of scientific problems, the other being sheer intellectual curiosity. But giving birth is not rearing. A whole cycle must be performed before anything comes out from practice: Practice → Scientific Problem → Scientific Research (statement and checking of hypotheses) → Rational Action. Even so, this is far from being the sole way in which scientific research and action mingle. Ever since theoretical mechanics began, in the eighteenth century, to shape industrial machinery, scientific ideas have been the main motor and technology their beneficiary. Since then, intellectual curiosity has been the source of most, and certainly of all important, scientific problems; technology has often followed in the wake of pure research, with a decreasing time lag between the two.

This is not to debase applied science but to recall how rich its conceptual background is. In applied science a theory is not only the summit of a research cycle and a guide to further research: it is also the basis of a system of rules prescribing the course of optimal practical action. On the other hand, in the arts and crafts, theories are either absent or instruments of action alone. In past epochs a man was regarded as practical if, in acting, he paid little or no attention to theory or if he relied on worn-out theories and common knowledge. Nowadays a practical man is one who acts in obedience to decisions taken in the light of the best technological knowledge—not pure scientific knowledge, because this is mostly remote from or even irrelevant to practice. And such a technological knowledge, made up of theories, grounded rules, and data, is in turn an outcome of the application of the method of science to practical problems. Since technology is as theory laden as pure science, and since this either is overlooked or explicitly denied by most philosophers, we must take a closer look at technological theories and their application.

TECHNOLOGICAL THEORIES: SUBSTANTIVE AND OPERATIVE

A theory may have a bearing on action either because it provides knowledge regarding the objects of action, for example, machines, or because it is concerned with action itself, for example, with the decisions that precede and steer the manufacture or use of machines. A theory of flight is of the former kind, whereas a theory concerning the optimal decisions regarding the distribution of aircraft over a territory is of the latter kind. Both are *technological theories* but, whereas the theories of the first kind are *substantive,* those of the second kind are, in a sense, *operative.* Substantive technological theories are essentially applications, to nearly real situations, of scientific theories; thus, a theory of flight is essentially an application of fluid dynamics. Substantive technological theories are always preceded by scientific theories, whereas operative theories are born in applied research and may have little if anything to do with substantive theories, this being why mathematicians and logicians with no previous scientific training can make important contributions to them. A few examples will make the substantive-operative distinction clearer.

The relativistic theory of gravitation might be applied to the design of generators of antigravity fields (i.e., local fields counteracting the terrestrial gravitational field), which in turn might be used to facilitate the launching of spaceships. But, of course, relativity theory is not particularly concerned with either field generators or astronautics: it just provides some of the knowledge relevant to the design and manufacture of antigravity generators. Palaeontology is used by the applied geologist engaged in oil prospecting, and the latter's findings are a basis for making decisions concerning drillings; but neither palaeontology nor geology is particularly concerned with the oil industry. Psychology can be used by the industrial psychologist in the interests of production, but it is not basically concerned with production. All three are examples of the application of scientific (or semiscientific, as the case may be) theories to problems that arise in action.

On the other hand the theories of value, decision, games, and operations research deal directly with valuation, decision-making, planning, and doing; they even may be applied to scientific research regarded as a kind of action, with the optimistic hope of optimizing its output. (These theories could not tell how to replace talent but how best to exploit it). These are operative theories, and they make little if any use of the substantive knowledge provided by the physical, biological, or social sciences: ordinary knowledge, special but non-scientific knowledge (of, e.g., inventory practices), and formal science are usually sufficient for them. Just think of strategical kinematics applied to combat or of queuing models: they are not applications of pure scientific theories but theories on their own.

What these operative or non-substantive theories employ is not substantive scientific knowledge but the *method* of science. They may be regarded, in fact, as scientific theories concerning action, in short, as theories of action. These theories are technological in respect of aim, which is practical rather than cognitive, but apart from this they do not differ markedly from the theories of science. In fact, every good operative theory will have at least the following traits characteristic of scientific theories: (1) they do not refer directly to chunks of reality but to more or less idealized models of them (e.g., entirely rational and perfectly informed contenders or continuous demands and deliveries); (2) as a consequence they employ theoretical concepts (e.g., "probability"); (3) they can absorb empirical information and in turn can enrich experience by providing predictions or retrodictions: and (4) consequently they are empirically testable, though not as toughly as scientific theories.

Looked at from a practical angle, technological theories are richer than the theories of science in that, far from being limited to accounting for what may or does, did or will *happen* regardless of what the decision-maker does, they are concerned with finding out *what ought to be done* in order to bring about, prevent, or just change the pace of events or their course, in a preassigned way. In a conceptual sense, the theories of technology are definitely poorer than those of pure science; they are invariably *less deep,* and this because the practical man, for whom they are intended, is chiefly interested in net effects that occur and are controllable on the human scale: he wants to know how things within *his* reach can be made to work *for him,* rather than how things of any kind really are. Thus, the electronics expert need not worry about the difficulties that plague the quantum electron theories; and the researcher in utility theory, who is concerned with comparing people's preferences, need not burrow into the origins of preference patterns—a problem for psychologists.

Consequently, whenever possible the applied researcher will attempt to schematize his system a *black box;* he will deal preferably with external variables (input and output), will regard all others as at best handy interverting variables with no ontological import, and will ignore the adjoining levels. This is why his oversimplifications and mistakes are not more often harmful—because his hypotheses are superficial. (Only the exportation of this externalist approach to science may be harmful.) Occasionally, though, the technologist will be forced to take up a deeper, representational viewpoint. Thus, the molecular engineer who designs new materials to order, that is, substances with prescribed macroproperties, will have to use certain fragments of atomic and molecular theory. But he will neglect all those microproperties that do not show up appreciably at the macroscopic level: after all, he uses atomic and molecular theories as tools—which has misled some philosophers into thinking that scientific theories are *nothing but* tools.

The conceptual impoverishment undergone by scientific theory when used as a means for practical ends can be frightful. Thus, an applied physicist engaged in designing an optical instrument will use almost only ray optics, that is, essentially what was known about light toward the middle of the seventeenth century. He will take wave optics into account for the explanation in outline, not in detail, of some effects, mostly undesirable, such as the appearance of colors near the edge of a lens; but he will seldom, if ever, apply any of the various wave theories of light to the computation of such effects. He can afford to ignore these theories in most of his professional practice because of two reasons. First, the chief traits of the optical facts relevant to the manufacture of optical instruments are adequately accounted for by ray optics; those few facts that are not so explainable require only the hypotheses (but not the whole theory) that light is made up of waves and that these waves can superpose. Second, it is extremely difficult to solve the wave equations of the deeper theories save in elementary cases, which are mostly of a purely academic interest (i.e., which serve essentially the purpose of illustrating and testing the theory). Just think of the enterprise of solving a wave equation with time-dependent boundary conditions such as those representing the moving shutter of a camera. Wave optics is scientifically important because it is nearly true; but for most present-day technology it is less important than ray optics, and its detailed application to practical problems in optical industry would be quixotic. The same argument can be carried over to the rest of pure science in relation to technology. The moral is that, if scientific research had sheepishly applied itself to the immediate needs of production, we would have no pure science, hence no applied science either.

DOES PRACTICE VALIDATE THEORY?

A theory, if true, can be employed successfully in applied research (technological investigation) and in practice itself—as long as the theory is relevant to either. (Fundamental theories are not so applicable because they deal with problems much too remote from practical problems. Just think of applying the quantum theory of scattering to car collisions.) But the converse is not true, that is, the practical success or failure of a scientific theory is no objective index of its truth value. In fact, a theory can be both successful and false; or, conversely, it can be a practical failure and nearly true. The efficiency of a false theory may be due to either of the following reasons. First, a theory may contain just a grain of truth, and this grain alone is employed in the theory's applications. In fact, a theory is a system of hypotheses, and it is enough for a few of them to be true or nearly so in order to be able to entail adequate consequences if the false ingredients are not used in the deduction or if they are practically innocuous. Thus, it is possible to manufacture excellent steel by combining magical exorcisms with the operations prescribed by the craft—as was done until the beginning of the nineteenth century. And it is possible to improve the condition of neurotics by means of shamanism, psychoanalysis, and other practices as long as effective means, such as suggestion, conditioning, tranquilizers, and above all time are combined with them.

A second reason for the possible practical success of a false theory may be that the accuracy requirements in applied science and in practice are far below those prevailing in pure research, so that a rough and simple theory supplying quick correct estimates of orders of magnitude very often will suffice in practice. Safety coefficients will mask the finer details predicted by an accurate and deep theory anyway, and such coefficients are characteristic of technological theory because this must adapt itself to conditions that can vary within ample bounds. Think of the variable loads a bridge can be subjected to or of the varying individuals that may consume a drug. The engineer and the physician are interested in safe and wide intervals centered in typical values rather than in exact values. A greater accuracy would be pointless since it is not a question of testing. Moreover, such a greater accuracy could be confusing because it would complicate things to such an extent that the target—on which action is to be focused—would be lost in a mass of detail. Extreme accuracy, a goal of scientific research, not only is pointless or even encumbering in practice in most cases but can be an obstacle to research itself in its early stages. For the two reasons given above—use of only a part of the premises and low accuracy requirement—infinitely many possible rival theories can yield "practically the same results." The technologist, and particularly the technician, are justified in preferring the simplest of them: after all, they are interested primarily in efficiency rather than in truth, in getting things done rather than in gaining a deep understanding of them. For the same reason, deep and accurate theories may be impractical; to use them would be like killing bugs with nuclear bombs. It would be as preposterous—though not nearly as dangerous—as advocating simplicity and efficiency in pure science.

A third reason why most fundamental scientific theories are of no practical avail is not related to the handiness and sturdiness required by practice but has a deep ontological root. The practical transactions of man occur mostly on his own level; and this level, like others, is rooted to the lower levels but enjoys a certain autonomy with respect to them, in the sense that not every change occurring in the lower levels has appreciable effects on the higher ones. This is what enables us to deal with most things on their own level, resorting at most to the immediately adjacent levels. In short, levels are to some extent stable: there is a certain amount of play between level and level, and this is a root of both chance (randomness due to independence) and freedom (self-motion in certain respects). One-level theories will suffice, therefore, for many practical purposes. It is only when a knowledge of the relations among the various levels is required in order to implement a "remote-control" treatment, that many-level theories must be tried. The most exciting achievements in this respect are those of psychochemistry, the goal of which is, precisely, the control of behavior by manipulating variables in the underlying biochemical level.

A fourth reason for the irrelevance of practice to the validation of theories—even to operative theories dealing with action—is that, in real situations, the relevant variables are seldom adequately known and precisely controlled. Real situations are much too complex for this, and effective action is much too strongly urged to permit a detailed study—a study that would begin by isolating variables and tying some of them into a theoretical model. The desideratum being maximal efficiency, and not at all truth, a number of practical measures will usually be attempted at the same time: the strategist will counsel the simultaneous use of weapons of several kinds, the physician will prescribe a number of supposedly concurrent treatments, and the politician may combine promise and threats. If the outcome is satisfactory, how will the practitioner know which of the rules was efficient, hence which of the underlying hypotheses was true? If unsatisfactory, how will he be able to weed out the inefficient rules and the false underlying hypotheses?

A careful discrimination and control of the relevant variables and a critical evaluation of the hypotheses concerning the relations among such variables is not done while killing, curing, or persuading people, not even while making things, but in leisurely, planned, and critically alert scientific theorizing and experimentation. Only while theorizing or experimenting do we *discriminate* among variables and *weigh* their relative importance, do we *control* them either by manipulation or by measurement, and do we *check* our hypotheses and inferences. This is why factual theories, whether scientific or technological, substantive or operative, are empirically tested in the laboratory and not in the battlefield, the consulting office, or the market place. ("Laboratory" is understood here in a wide sense, to include any situation which, like the military maneuver, permits a reasonable control of the relevant variables.) This is, also, why the efficiency of the rules employed in the factory, the hospital, or the social institution can be determined only in artificially controlled circumstances.

In short, practice has no validating force; pure and applied research alone can estimate the truth value of theories and the efficiency of technological rules. What the technician and the practical man do, by contrast to the scientist, is not to *test* theories but to *use* them with noncognitive aims. (The practitioner does not even test *things,* such as tools or drugs, save in extreme cases; he just uses them, and their properties and their efficiency again must be determined in the laboratory by the applied scientist.) The doctrine that practice is the touchstone of theory rests on a confusion between practice and experiment and an associated confusion between rule and theory. The question "Does it work?"—pertinent as it is with regard to things and rules—is impertinent in respect of theories.

Yet it might be argued that a man who knows how to do something is thereby showing that he knows that something. Let us consider the three possible versions of this idea. The first can be summed up in the schema "If *x* knows how to do (or make) *y*, then *x* knows *y*." Yet we are unable to make them, and we know part of one million years, man has known how to make children without having the remotest idea about the reproduction process. The second thesis is the converse conditional, namely, "If *x* knows *y*, then *x* knows how to do (or make) *y*." Counterexamples: we know something about stars, yet we are unable to make them, and we know part of the past, but we cannot even spoil it. The two conditionals being false, the biconditional "*x* knows *y* if and only if *x* knows how to do (or make) *y*" is false, too. In short, it is false that knowledge is identical with knowing how to do, or know-how. What is true is rather this: knowledge considerably *improves* the chances of correct doing, and doing *may* lead to knowing more (now that we have learned that knowledge pays), not because action is knowledge, but because, in inquisitive minds, action may trigger questioning.

It is only by distinguishing scientific knowledge from instrumental knowledge, or know-how, that we can hope to account for the coexistence of practical knowledge with theoretical ignorance and the coexistence of theoretical knowledge with practical ignorance. Were it not for this the following combinations hardly would have occurred in history: (1) science without the corresponding technology (e.g., Greek physics); (2) arts and crafts without an underlying science (e.g., Roman engineering and contemporary intelligence testing). The distinction must be kept, also, in order to explain the cross-fertilizations of science, technology, and the arts and crafts, as well as to explain the gradual charcter of the cognitive process. If, in order to exhaust the knowledge of a thing, it were sufficient to produce or reproduce it, then certain technological achievements would put an end to the respective chapters of applied research: the production of synthetic rubber, plastic materials, and synthetic fibres would exhaust polymer chemistry; the experimental induction of cancer should have stopped cancer research; and the experimental production of neuroses and psychoses should have brought psychiatry to a halt. As a matter of fact, we continue doing many things without understanding how, and we know many processes (such as the fusion of helium out of hydrogen) which we are not yet able to control for useful purposes (partly because we are too eager to attain the goal without a further development of the means). At the same time it is true that the barriers between scientific and practical knowledge, pure and applied research, are melting. But this does not eliminate their differences, and the process is but the outcome of an increasingly scientific approach to practical problems, that is, of a diffusion of the scientific method.

The identification of knowledge and practice stems not only from a failure to analyze either but also from a legitimate wish to avoid the two extremes of speculative theory and blind action. But the testability of theories and the possibility of improving the rationality of action are not best defended by blurring the differences between theorizing and doing, or by asserting that action is the test of theory, because both of these are false and no program is defensible if it rests on plain falsity. The interaction between theory and practice and the integration of the arts and crafts with technology and science are not achieved by proclaiming their unity but by multiplying their contacts and by helping the process whereby the crafts are given a technological basis and technology is entirely converted into applied science. This involves the conversion of the rules of thumb peculiar to the crafts into grounded rules, that is, rules based on laws. Let us approach this problem next.

SCIENTIFIC LAW AND TECHNOLOGICAL RULE

Just as pure science focuses on objective patterns or laws, action–oriented research aims at establishing stable norms of successful human behavior, that is, rules. The study of rules–the grounded rules of applied science–is therefore central to the philosophy of technology.

A rule *prescribes* a course of action; it indicates how one should proceed in order to achieve a predetermined goal. More explicitly, a rule is an instruction to perform a finite number of acts in a given order and with a given aim. The skeleton of a rule can be symbolized as a string of signs, such as 1-2-3- . . . -n, where every number stands for a corresponding act; the last act, n, is the only thing that separates the operator who has executed every operation, save n, from the goal. In contrast to law formulas, which say what the shape of possible events is, rules are norms. The field of law is assumed to be the whole of reality, including rule-makers; the field of rule is but mankind; men, not stars, can obey rules and violate them, invent and perfect them. Law statements are descriptive and interpretive, whereas rules are normative. Consequently, while law statements can be more or less true, rules can be only more or less effective.

We may distinguish the following genera of rules: (1) *rules of conduct* (social, moral, and legal rules); (2) *rules of prescientific work* (rules of thumb in the arts and crafts and in production); (3) *rules of sign* (syntactical and semantical rules); (4) *rules of science and technology* (grounded rules of research and action). Rules of conduct make social life possible (and hard). The rules of prescientific work dominate the region of practical knowledge which is not yet under technological control. The rules of sign direct us how to handle symbols—how to generate, transform, and interpret signs. And the rules of science and technology are those norms that summarize the special techniques of research in pure and applied science (e.g., random sampling techniques) and the special techniques of advanced modern production (e.g., the technique of melting with infrared rays).

Many rules of conduct, work, and sign are *conventional*, in the sense that they are adopted with no definite reasons and might be exchanged for alternative rules with little or no concomitant change in the desired result. They are not altogether arbitrary, since their formation and adoption should be explainable in terms of psychological and sociological laws, but they are not necessary either; the differences among cultures are largely differences among systems of rules of that kind. We are not interested in such groundless or conventional rules but rather in founded rules, that is, in norms satisfying the following *definition:* A rule is *grounded* if and only if it is based on a set of law formulas capable of accounting for its effectiveness. The rule that commands taking off the hat when greeting a lady is groundless in the sense that is is based on no scientific law but is conventionally adopted. On the other hand, the rule that commands greasing cars periodically is based on the law that lubricators decrease the wearing out of parts by friction; this is neither a convention nor a rule of thumb like those of cooking and politicking—it is a well-grounded rule.

To decide that a rule is effective it is necessary, though insufficient, to show that it has been successful in a high percentage of cases. But these cases might be just coincidences, such as those that may have consecrated the magic rituals that accompanied the huntings of primitive man. Before adopting an empirically effective rule we ought to know *why* it is effective; we ought to take it apart and reach an understanding of its *modus operandi.* This requirement of rule foundation marks the transition between the prescientific arts and crafts and contemporary technology. Now, the sole valid foundation of a rule is a system of law formulas, because these alone can be expected to correctly explain facts,—for example, the fact that a given rule works. This is not to say that the effectiveness of a rule depends on whether it is founded or groundless but only that, in order to be able to *judge* whether a rule has any chance of being effective, as well as in order to *improve* the rule and eventually *replace* it by a more effective one, we must disclose the underlying law statements, if any. We may take a step ahead and claim that the blind application of rules of thumb has never paid and, second, to try to transform some law formulas into effective technological rules. The birth and development of modern technology is the result of these two movements.

But it is easier to preach the foundation of rules than to say exactly what the foundation of rules consists in. Let us try to make an inroad into this unexplored territory—the core of the philosophy of technology. As usual when approaching a new subject, it will be convenient to begin by analyzing a typical case. Take the law statement "Magnetism disappears above the Curie temperature (770° C for iron)." For purposes of analysis it will be convenient to restate our law as an explicit conditional: "If the temperature of a magnetized body exceeds its Curie point, then it becomes demagnetized." (This is, of course, an oversimplification, as is every other ordinary-language rendering of a scientific law: the Curie point is not the temperature at which all magnetism disappears but, rather, the point of conversion of ferromagnetism into paramagnetism or conversely. But this is a refinement irrelevant to most technological purposes.) Our nomological statement provides the basis for the nomopragmatic statement "If a magnetized body is heated above its Curie point, then it is demagnetized." (The pragmatic predicate is, of course, "is heated.") This nomopragmatic statement is, in turn, the ground for two different rules, namely, *R*1: "In order to demagnetize a body heat it above its Curie point", and *R*2: "To prevent demagnetizing a body, do not heat it above its Curie point." Both rules have the same foundation, that is, the same underlying nomopragmatic statement, which in turn is supported by a law statement assumed to represent an objective pattern. Moreover, the two rules are equiefficient, though not under the same circumstances (changed goals, changed means).

Notice, first, that unlike a law statement a rule is neither true nor false; as a compensation it can be effective or ineffective. Second, a law is consistent with more than one rule. Third, the truth of a law statement does not insure the efficiency of the associated rules; in fact, the former refers to idealized situations which are not met with in practice. Fourth, whereas given a law we may try out the corresponding rules, given a rule we are unable to trace the laws presupposed by it; in fact, a rule of the form "In order to attain the goal *G*, employ the means *M*" is consistent with the laws "If *M*, then G", "*M* or *G*", and infinitely many others.

The above has important consequences for the methodology

of rules and the interrelations between pure and applied science. We see there is no single road from practice to knowledge, from success to truth; success warrants no inference from rule to law but poses the problem of explaining the apparent efficiency of the rule. In other words, the roads from success to truth are infinitely many and consequently theoretically useless or nearly so, that is, no bunch of effective rules suggests a true theory. On the other hand, the roads from truth to success are limited in number, hence feasible. This is one of the reasons why practical success, whether of a medical treatment or of a government measure, is not a truth criterion for the underlying hypotheses. This is also why technology—in contrast to the prescientific arts and crafts—does not start with rules and ends up with theories but proceeds the other way around. This is, in brief, why technology is applied science whereas science is not purified technology.

Scientists and technologists work out rules on the basis of theories containing law statements and auxiliary assumptions, and technicians apply such rules jointly with groundless (prescientific) rules. In either case, specific hypotheses accompany the application of rules, namely, hypotheses to the effect that the case under consideration is one where the rule is in point because such and such variables—related by the rule—are in fact present. In science such hypotheses can be tested; this is true of both pure and applied research. But in the practice of technology there may not be time to test them in any way other than by applying the rules around which such hypotheses cluster—and this is a poor test indeed, because the failure may be blamed either on the hypothesis or on the rule or on the uncertain conditions of application.

SCIENTIFIC PREDICTION AND TECHNOLOGICAL FORECAST

For technology knowledge is chiefly a means to be applied to the achievement of certain practical ends. The goal of technology is successful action rather than pure knowledge, and accordingly the whole attitude of the technologist while applying his technological knowledge is active in the sense that, far from being an inquisitive onlooker or a diligent burrower, he is an active participant in events. This difference of attitude between the technologist in action and the researcher—whether pure or applied—introduces certain differences between technological forecast and scientific prediction.

In the first place, whereas scientific prediction says what will or may happen if certain circumstances obtain, technological forecast suggests how to influence circumstances so that certain events may be brought about, or prevented, that would not normally happen; it is one thing to predict the trajectory of a comet, quite another to plan and foresee the orbit of an artificial satellite. The latter presupposes a choice among possible goals, and such a choice presupposes a certain forecasting of possibilities and their evaluation in the light of a set of desiderata. In fact, the technologist will make his forecast on his (or his employer's) estimate of what the future *should* be like if certain desiderata are to be fulfilled; contrary to the pure scientist, the technologist is hardly interested in what would happen anyway; and what for the scientist is just the final state of a process becomes for the technologist a valuable (or disvaluable) end to be achieved (or to be avoided). A typical scientific prediction has the form "If x occurs at time t, then y will occur at time t' with probability p." By contrast, a typical technological forecast is of the form "If y is to be achieved at time t' with probability p, then x should be done at time t". Given the goal, the technologist indicates the adequate means, and his forecast states a means-end relationship rather than a relation between an initial state and a final state. Furthermore, such means are implemented by a specified set of actions, among them the technologist's own actions.

This leads us to a second peculiarity of technological forecast: whereas the scientist's success depends on his ability to separate his object from himself (particularly so when his object happens to be a psychological subject)—that is, on his capacity of detachment—the technologist's ability consists in placing himself, within the system concerned, at the head of it. This does not involve *subjectivity,* since after all the technologist draws on the objective knowledge provided by science; but it does involve partiality, a *parti pris* unknown to the pure researcher. The engineer is part of a man-machine complex, the industrial psychologist is part of an organization, and both are bound to devise and implement the optimal means for achieving desiderata which are not usually chosen by themselves; they are decision-makers, not policy-makers.

The forecast of an event or process that is not under control will not alter the event or process itself. Thus, for example, no matter how accurately an astronomer predicts the collision of two stars, the event will occur in due course. But if an applied geologist can forecast a landslide, then some of its consequences can be prevented. Moreover, by designing and supervising the appropriate defense works the engineer may prevent the landslide itself; he may devise the sequence of actions that will refute the original forecast.

Similarly, an industrial concern may forecast sales for the near future on the (shaky) assumption that a given state of the economy, say prosperity, will continue during that lapse. But if this assumption is falsified by a recession, and the enterprise had accumulated a large stock which it must get rid of, then instead of making a new sales forecast (as a pure scientist would be inclined to do), the management will try to *force* the original forecast to come true by increasing advertisement, lowering sale prices, and so on. As in the case of vital processes, a diversity of means will alternatively or jointly be tried to attain a fixed goal. In order to achieve this goal any number of initial hypotheses may have to be sacrificed: in the case of the landslide, the assumption that no external forces would interfere with the process and, in the case of the sales, that prosperity would continue. Consequently, whether the initial forecast is *forcefully falsified* (as in the case of the landslide) or *forcefully confirmed* (as in the case of the sales forecast), this fact cannot count as a *test* of the truth of the hypotheses involved; it will count only as an efficiency test of the rules that have been applied. The pure scientist, on the other hand, need not worry about altering the means for achieving a preset goal, because pure science *has* no goals external to it.

Technological forecast, in sum, cannot be used for controlling things or men by changing the course of events perhaps to the point of stopping them altogether, or for forcing the predicted course even if unpredictable events should interfere with it. This is true of the forecasts made in engineering, medicine, economics, applied sociology, political science, and other technologies: the sole formulation of a forecast (prognosis, lax prediction, or prediction proper), if made known to the decision-makers, can be seized upon by them to steer the course of events, thus bringing about results different from those originally forecasted. This change, triggered by the issuance of the forecast, may contribute either to the latter's confirmation (self-fulling forecast) or to its refutation (self-defeating forecast). This trait of technological forecast stems from no logical property of it; it is a pattern of social action involving the knowledge of forecasts and consequently is conspicuous in modern society. Therefore, rather than analyzing the logic of causally effective forecast, we should start by distinguishing three levels in it: (1) the conceptual level, on which the prediction p stands; (2) the psychological level—the knowledge of p and the reactions triggered by this knowledge; and (3) the social level—the actions actually performed on the basis of the knowledge of p and in the service of extra-scientific goals. This third level is peculiar to technological forecast.

This feature of technological forecast sets civilized man apart from every other system. A nonpredicting system, be it a jukebox or a frog, when fed with information it can digest will process it and convert it into action at some later time. But such a system does not purposely produce most of the information, and it does not issue projections capable of altering its own future behavior. A predictor—a rational man, a team of technologists, or a sufficiently evolved automation—can behave in an entirely different way. When fed with relevant information I^t at time t, it can process this information with the help of the knowledge (or the instructions) available to it, eventually issuing a prediction P^t, at a later t'. This prediction is fed back into the system and compared with the present goal that controls the whole process (without either causing it or supplying it with energy). If the two are reasonably close, the system takes a decision that eventually leads it to act so as to take advantage of the course of events. If, on the other hand, the prediction differs significantly from the goal, this difference will again trigger the theoretical mechanism, which will elaborate a new strategy: a new prediction, $P^{t''}$, will eventually be issued at time t'', a forecast including a reference to the system's own participation in the events. The new prediction is fed back into the system and, if it still disagrees with the goal, a new correction cycle is triggered, and so on until the difference between the prediction and the goal becomes negligible, in which case the system's predicting mechanism comes to rest. Henceforth the system will gather new information regarding the present state of affairs and will act so as to conform to the strategy it has elaborated. This strategy may have required not only new information regarding the external world (including the attitudes and capabilities of the people concerned) but also new hypotheses or even theories which had not been present in the instruction chart originally received by the predictor. If the latter fails to realize it or to obtain and utilize such additional knowledge, his or its action is bound to be ineffective. Moral: the more brains the better.

TECHNOLOGICAL FORECAST AND EXPERT PROGNOSIS

The preceding account of technological forecast is based on the assumption that it relies on some theory, or rather theories, whether substantive or operative. This assumption may be found wanting by anyone knowing that the forecasts issued by experts in medicine, finance, or politics are often successful and yet involve no great deal of theorizing. True, most often *expert prognosis* relies on inductive (empirical) generalizations of the form "*A* and *B* occur jointly with the observed frequency *f*," or even just "*A* and *B* occur jointly in most cases," or "Usually, whenever *A* then *B*." The observation that a given individual, say a human subject or an economic state of affairs, has the property *A* is then used to prognose that it has, or will acquire, the property *B*. In daily life such prognoses are all we do, and the same applies to most expert prognoses. Occasionally such prognoses made with either ordinary knowledge or specialized but nonscientific knowledge are more successful than predictions made with full-fledged but false or rough theories; in many fields, however, the frequency of hits is not better than the one obtained by flipping a coin. The point, though, is that expert forecast using no scientic theory is not a scientific activity—just by definition of "scientific prediction."

Yet it would be wrong to think that experts make no use of *specialized knowledge* whenever they do not employ scientific theories; they always judge on the basis of some such knowledge. Only, expert knowledge is not always explicit and articulate and, for this reason, it is not readily controllable: it does not learn readily from failures, and it is hard to test. For the progress of science, the failure of a scientific prediction is by far preferable to the success of an expert prognosis, because the scientific failure can be fed back into the theory responsible for it, thereby giving us a chance to improve it, whereas in the case of expert knowledge there is no theory to feed failure into. It is only for immediate practical purposes that expert prognoses made with shallow but well-confirmed generalizations are preferable to risky scientific predictions.

Another difference between expert prognosis and technological forecast proper would seem to be this: the former relies more heavily on *intuition* than does scientific prediction. Yet the difference is one of degree rather than of kind. Diagnosis and forecast, whether in pure science, in applied science, or in arts and crafts, involve intuitions of a number of kinds: the quick identification of a thing, event, or sign; the clear but not necessarily deep grasp of the meaning and/or the mutual relations of a set of signs (text, table, diagram, etc.); the ability to interpret symbols; the ability to form space models; skill in realizing analogies; creative imagination; catalytic inference, that is, quick passage from some premises to other formulas by skipping intermediate steps; power of synthesis or synoptic grasp; common sense (or rather controlled craziness), and sound judgment. These abilities intertwine with specialized knowledge, whether scientific or not, and are reinforced with practice. Without them theories could neither be invented nor applied—but, of course, they are not suprarational powers. Intuition is all right as long as it controlled by reason and experiment; only the replacement of theorizing and experimenting by intuition must be feared.

A related danger is that of *pseudo-scientific projection tools,* so common in applied psychology and sociology. A number of techniques have been devised to forecast the performance of personnel, students, and even psychologists themselves. A few tests, the objective ones, are somewhat reliable; this holds for intelligence and skill tests. But most tests, particularly the subjective ones, (the "global evaluation" of personality by means of interviews, the thematic apperception test, the Rorschach, etc.) are in the best of cases inefficient and in the worst of cases misleading. When they have been subjected to the test of prediction—that is, when their results have been checked with the actual performance of the subjects—they have failed. The failure of most individual psychological tests, and particularly of the subjective ones, is not a failure of psychological testing in general; what is responsible for such failures is either the total absence or the falsity of the underlying psychological theories. Testing for human abilities without first establishing *laws* relating objective indices of abilities or personality traits is as thoughtless as asking a tribesman to test an aircraft. As long as no theoretical foundations of psychological tests are secured, their employment as predictive instruments is not better than crystal gazing or coinflipping: they are practically inefficient and, even if they succeeded, they would not contribute to psychological theory, because they are unrelated to theory. The limited success of psychological testing has led many to despair of the possibility of finding a scientific approach to human behavior, but the right inference is that such an attempt has been tried only after a large number of alleged tests invaded the market. What is wrong with most of "applied" (educational, industrial, etc.) psychology is that it does *not* consist in the application of scientific psychology at all. The moral is that practical wants—such as personnel training and selection—should not be allowed to force the construction of "technologies" without an underlying science.

Technological forecast should be maximally reliable. This condition excludes from technological practice—not, however, from technological research—insufficiently tested theory. In other words, technology will ultimately prefer the old theory that has rendered distinguished service in a limited domain and with a known inaccuracy to the bold new theory that promises unheard-of forecasts but is probably more complex and therefore partly less well tested. It would be irresponsible for an expert to apply a new idea in practice without having tested it under controlled conditions. (Yet this is still done in pharmacy: recall the affair of the mutagenic drugs in the early 1960's.) Practice, and even technology, are bound to be more conservative than science. Consequently, the effects of a close association of pure research with applied research, and of the latter with production, are not all of them beneficial: while it is true that technology challenges science with new problems and supplies it with new equipment for data-gathering and data-processing, it is no less true that technology, by its very insistence on reliability, standardization (routinization), and speed at the expense of depth, range, accuracy, and serendipity, can slow down the advancement of science.

OTHER PROBLEMS

We have looked into a few problems of the philosophy of technology. Many other challenging problems have been left out, for example, the logic of technological rules; the test of technological theories; the patterns of technological invention; the reason textile, aircraft, and other industries are still largely based on crafts; and the power of technology to bring together previously separate fields (cases of cybernetics, nuclear engineering, computer science, space science, and bioengineering). These and many other problems are waiting to be discovered and worked out by philosophers attentive to their own times. Why should the waiting time be so long?

SUGGESTIONS FOR FURTHER READING

Campbell, N. *What Is Science?* New York: Dover Publications, Inc., 1952, pp. 1–36.

Feigl, H., and Brodbeck, M. (ed.) *Readings in the Philosophy of Science.* New York: Appleton-Century-Crofts, Inc., 1953, pp. 3–20.

Kemeny, J. G. *A Philosopher Looks at Science.* Princeton, N.J.: D. Van Nostrand Co., Inc., 1959, pp. 174–186.

Krimerman, L. I. (ed.) *The Nature and Scope of Social Science.* New York: Appleton-Century-Crofts, Inc., 1969, pp. 7–40, 139–204.

Kuhn, T. S. *The Structure of Scientific Revolutions.* Chicago: University of Chicago Press, 1962.

Martin, M. *Concepts of Science Education.* Glenview, Ill.: Scott, Foresman and Co., 1972, pp. 6–43.

Nagel, E. *The Structure of Science.* New York: Harcourt, Brace and World, Inc., 1961, pp. 1–15.

Popper, K. R. *Conjectures and Refutations,* 4th ed., London: Routledge and Kegan Paul, 1972, pp. 253–293.

Ravetz, J. R. *Scientific Knowledge and Its Social Problems.* Oxford: Clarendon Press, 1971, pp. 7–241.

Storer, N. W. *The Social System of Science.* New York: Holt, Rinehart & Winston, Inc., 1966.

SECTION II

The Logic of Discovery and Growth

The primary aim of the first article in this section is to show that "the proposal of a hypothesis can be a reasonable affair," although the reasons for *proposing* a hypothesis are in significant respects unlike the reasons for *accepting* a hypothesis. It is easy to see why one might doubt that there is a difference in kind between reasons for accepting and reasons for proposing a hypothesis. After all, one might think, we propose hypotheses because we suspect that they are true and the more this suspicion tends to be borne out, the more we tend to accept the proposed hypotheses. In other words, accepting-reasons are of the *same sort* as proposing-reasons, only of a *different degree*. We don't have, if you like, a question of buckets and bananas, but only of bigger and smaller bananas.

Hanson resists the above line of reasoning roughly on the ground that one might propose a hypothesis which is almost certainly (or believed to be) false while, at the same time, it is believed to be somewhere near the truth. For example, he considers Kepler's conjecture that the orbit of Mars is elliptical. Without imagining for a moment that he would hit upon the precise formula for describing the elliptical orbit of Mars, Kepler could nevertheless make a reasonable conjecture. He could make a plausible beginning in the presence of a firm conviction that his first shots almost had to miss their mark. And the "beginning" would not be plausible because it is probably or believed to be true. Rather, it would be plausible merely because it is believed to be "in the right ball park," i.e., it is believed to be of the right *type*.

Hanson's critic, Feyerabend, seems prepared to accept the view outlined in the preceding paragraph. He is more disturbed by another line taken by Hanson, namely, that because the so-called *hypothetico-deductive model* of scientific explanation cannot account for the alleged distinction between proposing- and accepting-reasons, that model is "incomplete." There is a section on this model of explanation below (Section III), but the bare bones of the model may be explained briefly. H-D theorists begin with the assumption that hypotheses are tested by deriving predictions from them and checking to see whether or not events have occurred as they were predicted. If they have, that is a mark in favor of the hypothesis; if not, that is a mark against it. What bothers Hanson about this model is that it does not seem to provide any place for his distinction between proposing- and accepting-reasons. It seems to begin with hypotheses as *given,* whereas Hanson wants to provide some account of why, say, this hypothesis rather than that one is proposed and how the logic of proposing (or discovery in his terms) differs from the logic of accepting (or justification). He does not object to talking about justification in much the same may that the H-D theorists talk about it. He just wants everyone to know that the H-D story cannot be the whole story. As a peacemaker, Feyerabend tries to explain how the H-D model might accommodate Hanson's distinction.

What you should try to determine is, first, whether or not the distinction Hanson wants to draw is legitimate. Would anyone propose a hypothesis if he knew or strongly believed it to be false? Should anyone propose such a hypothesis? Is there a difference between proposing a hypothesis because it seems to be the best shot at the truth that one can take, *and* proposing a hypothesis because it might well be acceptable? Second, you should try to determine whether or not Feyerabend's suggested accommodation does not turn the H-D model into something quite different from what any H-D theorist would be willing to accept.

The purpose of the second major article in this section is, as Popper asserts "to stress the significance of one particular aspect of science—its need to grow, or, if you like, its need to progress." From this fundamental assumption a most extraordinary view of scientific methodology emerges. Let us reconstruct the rudiments of the view slowly.

It is apparent that two meaningful independent assertions

about the world cannot tell us less than either one of them alone. In other words, the empirical content of the conjunction of such propositions must be greater than or equal to the content of either one of them by itself. Moreover, if all other things are equal, the more content, the better. The more we can say about the world, the better. On the other hand, the more we say about the world, the less chance there is that *everything* we say will be true. It is generally easier to get one item straight than ten, easier to get ten straight than a hundred, and so on. Hence, the probability that the conjunction of two meaningful independent assertions about the world is true must be lower than or equal to the probability that either of them alone is true, i.e., the probability of such propositions cannot be greater than the probability of either one of them by itself.

Now, if you recall Popper's assumption that growth is essential for science, you see that, contrary to popular opinion, a high probability is *not* the sign of a good hypothesis. If scientists want to say more rather than less about the world and if all other things are equal, then they ought to prefer hypotheses with low probability to hypotheses with high probability, because the lower the probability, the greater the empirical content. We don't want hypotheses that are probably true. We want hypotheses that are probably false!

Such is the beginning of Popper's contribution to the philosophy of science. You will have to lay aside your preconceptions about probability as necessarily a good thing long enough to follow Popper's line of reasoning. Furthermore, you should consider the possibility that he may be guilty of special pleading. You will of course want to know more about the idea of probability which is assumed in the article, and that is elucidated in the section on probability and induction (Section VII).

READING 3

Is There a Logic of Scientific Discovery?

Norwood Russell Hanson

Is there a logic of scientific discovery? The approved answer to this is "No." Thus Popper argues: [1] "The initial stage, the act of conceiving or inventing a theory, seems to me neither to call for logical analysis nor to be susceptible of it." Again, "There is no such thing as a logical method of having new ideas, or a logical reconstruction of this process." Reichenbach writes that philosophy of science "cannot be concerned with [reasons for suggesting hypotheses], but only with [reasons for accepting hypotheses]." [2] Braithwaite elaborates: "The solution of these historical problems involves the individual psychology of thinking and the sociology of thought. None of these questions are our business here." [3]

Against this negative chorus, the 'Ayes' have *not* had it. Aristotle (*Prior Analytics* II, 25) and Peirce [4] hinted that in science there may be more problems for the logician than just analyzing the arguments supporting already invented hypotheses. But contemporary philosophers are unreceptive to this. Let us try once again to discuss the distinction F. C. S. Schiller made between the 'Logic of Proof' and the 'Logic of Discovery.' [5] We may be forced, with the majority, to conclude 'Nay.' But only after giving Aristotle and Peirce a sympathetic hearing. Is there *anything* in the idea of a 'logic of discovery' which merits the attention of a tough-minded, analytic logician?

It is unclear what a logic of discovery is a logic of. Schiller intended nothing more than "a logic of inductive inference." Doubtless his colleagues were so busy sectioning syllogisms that they ignored inferences which mattered in science. All the attention philosophers now give to inductive reasoning, probability, and the principles of theory construction would have pleased Schiller. But,

Reprinted by permission of the publisher from *Current Issues in the Philosophy of Science,* ed. by Herbert Feigl and Grover Maxwell, copyright © 1961 by Holt, Rinehart and Winston, Inc.

for Peirce, the work of Popper, Reichenbach, and Braithwaite would read less like a *Logic of Discovery* than like a *Logic of the Finished Research Report.* Contemporary logicians of science have described how one sets out reasons in support of a hypothesis once proposed. They have said nothing about the conceptual context within which such a hypothesis is initially proposed. Both Aristotle and Peirce insisted that the proposal of a hypothesis can be a reasonable affair. One can have good reasons, or bad, for suggesting one kind of hypothesis initially, rather than some other kind. These reasons may differ in type from those which lead one to accept a hypothesis once suggested. This is not to deny that one's reasons for proposing a hypothesis initially may be identical with his reasons for later accepting it.

One thing must be stressed. When Popper, Reichenbach, and Braithwaite urge that there is no logical analysis appropriate to the psychological complex which attends the conceiving of a new idea, they are saying nothing which Aristotle or Pierce would reject. The latter did not think themselves to be writing manuals to help scientists make discoveries. There could be no such manual.[6] Apparently they felt that there is a *conceptual* inquiry, one properly called "a logic of discovery," which is *not* to be confounded with the psychology and sociology appropriate to understanding how some investigator stumbled on to an improbable idea in unusual circumstances. There are factual discussions such as these latter. Historians like Sarton and Clagett have undertaken such circumstantial inquiries. Others—for example, Hadamard and Poincaré—have dealt with the psychology of discovery. But these are not logical discussions. They do not even turn on conceptual distinctions. Aristotle and Peirce thought they were doing something other than psychology, sociology, or history of discovery; they purported to be concerned with a logic of discovery.

This suggests caution for those who reject wholesale any notion of a logic of discovery on the grounds that such an inquiry can *only* be psychology, sociology, or history. That Aristotle and Peirce deny just this has made no impression. Perhaps Aristotle and Peirce were wrong. Perhaps there is no room for logic between the psychological dawning of a discovery and the justification of that discovery *via* successful predictions. But this should come as the conclusion of a discussion, not as its preamble. If Peirce is correct, nothing written by Popper, Reichenbach, or Braithwaite cuts against him. Indeed, these authors do not discuss what Peirce wishes to discuss.

Let us begin this uphill argument by distinguishing

(1) reasons for accepting a hypothesis H, from
(2) reasons for suggesting H in the first place.

This distinction is in the spirit of Peirce's thesis. Despite his arguments, most philosophers deny any *logical* difference between these two. This must be faced. But let us shape the distinction before denting it with criticism.

What would be our reasons for accepting *H*? These will be those we might have for thinking *H* true. But the reasons for suggesting *H* originally, or for formulating *H* in one way rather than another, may not be those one requires before thinking *H* true. They are, rather, those reasons which make *H* a *plausible type of conjecture.* Now, no one will deny *some* differences between what is required to show *H* true, and what is required for deciding *H* constitutes a plausible kind of conjecture. The question is: Are these logical in nature, or should they more properly be called "psychological" or "sociological"?

Or one might urge, as does Professor Feigl, that the difference is just one of refinement, degree, and intensity. Feigl argues that considerations which settle whether *H* constitutes a plausible conjecture are of the *same type* as those which settle whether *H* is true. But since the initial proposal of a hypothesis is a groping affair, involving guesswork amongst sparse data, there *is* a distinction to be drawn; but this, Feigl urges, concerns two ends of a spectrum, ranging all the way from inadequate and badly selected data to that which is abundant, well diversified, and buttressed by a battery of established theories. The issue therefore remains: Is the difference between reasons for accepting *H* and reasons for suggesting it originally one of logical type, or one of degree, or of psychology, or of sociology?

Already a refinement is necessary if our original distinction is to survive. The distinction just drawn must be reset in the following, more guarded, language. Distinguish now

(1′) reasons for accepting a particular, minutely specified hypothesis *H*, from

(2′) reasons for suggesting that, whatever specific claim the successful *H* will make, it will, nonetheless, be a hypothesis of one *kind* rather than another.

Neither Aristotle, nor Peirce, nor (if you will excuse the conjunction) myself in earlier writings,[7] sought this distinction on these grounds. The earlier notion was that it was some particular, minutely specified *H* which was being looked at in two ways: (1) what would count for the acceptance of that *H*, and (2) what would count in favor of suggesting that same *H* initially.

This latter way of putting it is objectionable. The issue is whether, *before* having hit a hypothesis which succeeds in its predictions, one can have good reasons for anticipating that the hypothesis will be one of some particular *kind.* Could Kepler, for example, have had good reasons, *before* his elliptical-orbit hypothesis was established, for supposing that the successful hypothesis concerning Mars' orbit would be of the noncircular kind? [8] He *could* have argued that, whatever path the planet *did* describe, it would be a closed, smoothly curving, plane geometrical figure. Only this *kind* of hypothesis could entail such observation-statements as that Mars' apparent velocities at 90 degrees and at 270 degrees of eccentric anomaly were greater than any circular-type *H* could explain. Other *kinds* of hypotheses were available to Kepler: for example, that Mars' *color* is responsible for its high velocities, or that the dispositions of Jupiter's moons are responsible. But these would not have struck Kepler as capable of explaining such surprising phenomena. Indeed, he would have thought it unreasonable to develop such hypotheses at all, and would have argued thus. [Braithwaite counters: "But exactly which hypothesis was to be rejected was a matter for the 'hunch' of the physicists." [9] However, which *type* of hypothesis Kepler chose to reject was not just a matter of 'hunch.']

I may still be challenged. Some will continue to berate my distinction between reasons for suggesting which type of hypothesis *H* will be, and reasons for accepting *H* ultimately.[10] There may indeed be "psychological" factors, the opposition concedes, which make certain types of hypotheses 'look' as if they might explain phenomena. Ptolemy knew, as well as did Aristarchus before him and Copernicus after him, that a kind of atsronomy which displaced the earth would be theoretically simpler, and easier to manage, than the hypothesis of a geocentric, geostatic universe. *But,* philosophers challenge, for psychological, sociological, or historical reasons, alternatives to geocentricism did not 'look' as if they could explain the absence of stellar parallax. This cannot be a matter of logic, since for Copernicus one such alternative *did* 'look' as if it could explain this. Insofar as scientists have *reasons* for formulating types of hypotheses (as opposed to hunches and intuitions), these are just the kinds of reasons which later show a particular *H* to be true. Thus, if the absence of stellar parallax constitutes more than a psychological reason for Ptolemy's resistance to alternatives to geocentricism, then it is his reason for rejecting such alternatives as *false.* Conversely, his reason for developing a geostatic type of hypothesis (again, absence of parallax) was his reason for taking some such hypothesis as *true.* And Kepler's reasons for rejecting Mars' color or Jupiter's moons as indicating the kinds of hypotheses responsible for Mars' accelerations were reasons which also served later in establishing some hypothesis of the noncircularity type.

So the objection to my distinction is: The only *logical* reason for proposing *H* will be of a certain type is that *data* incline us to think some particular *H* true. What Hanson advocates is psychological, sociological, or historical in nature; it has no logical import for the differences between proposing and establishing hypotheses.

Kepler again illustrates the objection. Every historian of science knows how the idea of uniform circular motion affected astronomers before 1600. Indeed, in 1591 Kepler abandoned a hypothesis because it entailed other-than-uniform circular orbits—something simply inconceivable for him. So psychological pressure against forming alternative types of hypotheses was great. But *logically* Kepler's reasons for entertaining a type of Martian motion other than uniformly circular were his reasons for accepting that as astronomical truth. He first encountered this type of hypothesis on perceiving that no simple adjustment of epicycle, deferent, and eccentric could square Mars' observed distances, velocities, and apsidal positions. These were also reasons which led him to assert that the planet's orbit is not the effect of circular motions, but of an elliptical path. Even after other inductive reasons confirmed the truth of the latter hypothesis, these early reasons were *still* reasons for accepting *H* as true. So they cannot have been reasons merely for proposing which type of hypothesis *H* would be, and nothing more.

This objection has been made strong. If the following cannot weaken it, then we shall have to accept it; we shall have to grant that there is no aspect of discovery which has to do with logical, or conceptual, considerations.

When Kepler published *De Motibus Stellae Martis,* he had established that Mars' orbit was an ellipse, inclined to the ecliptic, and had the sun in one of the foci. Later (in the *Harmonices Mundi*) he generalized this for other planets. Consider the hypothesis *H'*: *Jupiter's* orbit is of the noncircular type.

The reasons which led Kepler to formulate *H'* were many. But they included this: that *H* (the hypothesis that Mars' orbit is elliptical) is true. Since Eudoxos, Mars had been the typical planet. (*We* know why. Mars' retrogradations and its movement around the empty focus—all this we observe with clarity from earth because of earth's spatial relations with Mars.) Now, Mars' dynamical properties are usually found in the other planets. If its orbit is ellipsoidal, then it is reasonable to expect that, whatever the exact shape of the other orbits (for example, Jupiter's), they will all be of the noncircular type.

But such reasons would not *establish* *H'*. Because what makes it reasonable to anticipate that *H'* will be of a certain type is *ana-*

logical in character. (Mars does *x;* Mars is a typical planet; so perhaps all planets do the same kind of thing as *x*.) Analogies cannot establish hypotheses, not even *kinds* of hypotheses. Only observations can do that. In this the hypothetico-deductive account (of Popper, Reichenbach, and Braithwaite) is correct. To establish *H′* requires plotting its successive positions on a smooth curve whose equations can be determined. It may then be possible to assert that Jupiter's orbit is an ellipse, an oviform, an epicycloid, or whatever. But it would not be reasonable to expect this when discussing only what type of hypothesis is likely to describe Jupiter's orbit. Nor is it right to characterize this difference between '*H*-as-illustrative-of-a-type-of hypothesis' and '*H*-as-empirically established' as a difference of psychology only. *Logically,* Kepler's analogical reasons for proposing that *H′* would be of a certain type were good reasons. But, logically, they would not then have been good reasons for asserting the truth of a specific value for *H′*—something which could be done only years later.

What are and are not good reasons for reaching a certain conclusion is a logical matter. No further observations are required to settle such issues, any more than we require experiments to decide, on the basis of one's bank statements, whether one is bankrupt. Similarly, whether or not Kepler's reasons for anticipating that *H′* will be of a certain kind are *good* reasons is a matter for logical inquiry.

Thus, the differences between reasons for expecting that some as yet-undiscovered *H* will be of a certain type and those that establish this *H* are greater than is conveyed by calling them "psychological," "sociological," or "historical."

Kepler reasoned initially by analogy. Other kinds of reasons which make it plausible to propose that an *H*, once discovered, will be of a certain type, might include, for example, the detection of a formal symmetry in sets of equations or arguments. At important junctures Clerk Maxwell and Einstein detected such structural symmetries. This allowed them to argue, before getting their final answers, that those answers would be of a clearly describable type.

In the late 1920's, before anyone had explained the "negative-energy" solutions in Dirac's electron theory, good analogical reasons could have been advanced for the claim that, whatever specific assertion the ultimately successful *H* assumed, it would be of the Lorentz-invariant type. It could have been conjectured that the as yet undiscovered *H* would be compatible with the Dirac explanation of Compton scattering and doublet atoms, and would fail to confirm Schrödinger's hunch that the phase waves within configuration space actually described observable physical phenomena. All this could have been said before Weyl, Oppenheimer, and Dirac formulated the "hole theory of the positive electron." Good analogical reasons for supposing that this *type* of *H* would succeed could have been and, as a matter of fact, were advanced. Indeed, Schrödinger's attempt to rewrite the Dirac theory so that the negative-energy solutions disappeared was *rejected* for failing to preserve Lorentz invariance.

Thus, reasoning from observations of *A*s as *B*s to the proposal "All *A*s are *B*s" is different in type from reasoning analogically from the fact that *C*s are *D*s to the proposal "The hypothesis relating *A*s and *B*s will be of the same type as that relating Cs and Ds." (Here it is the *way* *C*s are *D*s which seems analogous to the way *A*s are *B*s.) And both of these are typically different from reasoning involving the detection of symmetries in equations describing *A*s and *B*s.

Indeed, put this way, what *could* an objection to the foregoing consist of? Establishing a hypothesis and proposing by analogy that a hypothesis is likely to be of a particular type surely follow reasoning which is different in type. Moreover, both procedures have a fundamentally logical or conceptual interest.

An objection: "Analogical arguments, and those based on the recognition of formal symmetries, are used because of inductively established beliefs in the reliability of arguments of that type. So, the cash value of such appeals ultimately collapses into just those accounts given by *H-D* theorists."

Agreed. But we are not discussing the *genesis* of our faith in these types of arguments, only the logic of the arguments themselves. *Given* an analogical premise, or one based on symmetry considerations—or even on enumeration of particulars—one argues *from* these in logically different ways. Consider what further moves are necessary to convince one who doubted such arguments. A challenge to "All *A*s are *B*s" when this is based on induction by enumeration could only be a challenge to justify induction, or at least to show that the particulars are being correctly described. This is inappropriate when the arguments rest on analogies or on the recognition of formal symmetries.

Another objection: "Analogical reasons, and those based on symmetry are still reasons for *H* even after it is (inductively) established. They are reasons *both* for proposing that *H* will be of a certain type and for accepting *H*."

Agreed, again. But, analogical and symmetry arguments could never by *themselves* establish particular *H*s. They can only make it plausible to suggest that *H* (when discovered) will be of a certain type. However, inductive arguments can, by themselves, establish particular hypotheses. So they must differ from arguments of the analogical or symmetrical sort.

H-D philosophers have been most articulate on these matters. So, let us draw out a related issue on which Popper, Reichenbach, and Braithwaite seem to me not to have said the last word.

J. S. Mill was wrong about Kepler (*A System of Logic,* III, 2–3). It is impossible to reconcile the delicate adjustment between theory, hypothesis, and observation recorded in *De Motibus Stellae Martis* with Mill's statement that Kepler's first law is but "a compendius expression for the one set of directly observed facts." Mill did not understand Kepler (as Peirce notes [*Collected Papers,* I, p. 31]). (It is equally questionable whether Reichenbach understood him: "Kepler's laws of the elliptic motion of celestial bodies were inductive generalizations of observed fact . . . [he] observed a series of . . . positions of the planet Mars and found that they may be connected by a mathematical relation . . .")[11] Mill's *Logic* is as misleading about scientific discovery as any account proceeding *via* what Bacon calls *"inductio per enumerationem simplicem ubi non reperitur instantia contradictoria."* (Indeed Reichenbach observes: "It is the great merit of John Stuart Mill to have pointed out that all empirical inferences are reducible to the *inductio per enumerationem simplicem. . . .*")[12] The accounts of *H-D* theorists are equally misleading.

An *H-D* account of Kepler's first law would treat it as a high-level hypothesis in an *H-D* system. (This is Braithwaite's language.) It is regarded as a quasi-axiom, from whose assumption observation-statements follow. If these are true—if, for example, they imply that Uranus' orbit is an ellipse and that its apparent velocity at 90 degrees is greater than at aphelion—then the first law is confirmed. (Thus Braithwaite writes: "A scientific system consists of a set of hypotheses which for a deductive system . . . arranged in such a way that from some of the hypotheses as premises all the other hypotheses logically follow . . . the establishment of a system as a set of true propositions depends upon the establishment of its lowest level hypotheses . . .")[13]

This describes physical theory more adequately than did pre-Baconian accounts in terms of simple enumeration, or even post-Millian accounts in terms of ostensibly not-so-simple enumerations. It tells us about the logic of laws, and what they do in finished arguments and explanations. *H-D* accounts do not, however, tell us anything about the context in which laws are proposed in the first place; nor, perhaps, were they even intended to do so.

The induction-by-enumeration story *did* intend to do this. *It* sought to describe good reasons for initially proposing *H*. The *H-D* account must be silent on this point. Indeed, the two accounts are not strict alternatives. As Braithwaite suggests they are when he remarks of a certain higher-level hypothesis that it "will not have been established by induction by simple enumeration; it will have been obtained by the hypothetico-deductive method. . . .") [14] They are thoroughly compatible. Acceptance of the second is no reason for rejecting the first. A law *might* have been inferred from just an enumeration of particulars (for example, Boyle's law in the seventeenth century, Bode's in the eighteenth, the laws of Ampere and Faraday in the nineteenth, and much of meson theory now). It could *then* be built into an H-D system as a higher order proposition. If there is anything wrong with the older view, *H-D* accounts do not reveal this.

There *is* something wrong. It is false. Scientists do not always discover every feature of a law by enumerating and summarizing observables. (Thus even Braithwaite [15] says: "Sophisticated generalizations (such as that about the proton-electron constitution of the hydrogen atom) . . . [were] certainly not derived by simple enumeration of instances . . .") But *this* does not strengthen the *H-D* account as against the inductive view. There is *no H-D* account of how "sophisticated generalizations" are *derived.* On his own principles, the *H-D* theorist's lips are sealed on this matter. But there are conceptual considerations which help us understand the *reasoning* that is sometimes successful in determining the type of an as-yet-undiscovered hypothesis.

Were the *H-D* account construed as a description of scientific practice, it would be misleading. (Braithwaite's use of "derived" is thus misleading. So is his announcement [p. 11] that he is going to explain "how we come to make use of sophisticated generalizations.") Natural scientists do not "start from" hypotheses. They start from data. And even then not from commonplace data, but from surprising anomalies. (Thus Aristotle remarks [16] that knowledge begins in astonishment. Peirce makes perplexity the trigger of scientific inquiry.[17] And James and Dewey treat intelligence as the result of mastering problem situations.) [18]

By the time a law gets fixed into an *H-D* system, the *original* scientific thinking is over. The pedestrian process of deducing observation-statements begins only after the physicist is convinced that the proposed hypothesis is at least of the right type to explain the initially perplexing data. Kepler's assistant could work out the consequences of *H′* and check its validity by seeing whether Jupiter behaved as *H′* predicts. This was possible because of Kepler's argument that what *H* had done for Mars, *H′* might do for Jupiter. The *H-D* account is helpful here; it analyzes *the argument of a completed research report.* It helps us see how experimentalists elaborate a theoretician's hypotheses. And the *H-D* account illuminates yet another aspect of science, but its proponents have not stressed it. Scientists often dismiss explanations alternative to that which has won their provisional assent along lines that typify the *H-D* method. Examples are in Ptolemy's *Almagest,* when (on observational grounds) he rules out a moving earth, in Copernicus' *De Revolutionibus . . . ,* when he rejects Ptolemy's lunar theory, in Kepler's *De Motibus Stellae Martis,* when he denies that the planes of the planetary orbits intersect in the center of the ecliptic, and in Newton's *Principia,* when he discounts the idea that the gravitational force law might be of an inverse cube nature. These mirror formal parts of Mill's *System of Logic* or Braithwaite's *Scientific Explanation.*

Still, the *H-D* analysis remains silent on reasoning which often conditions the discovery of laws—reasoning that determines which type of hypothesis is likely to be most fruitful to propose.

The induction-by-enumeration story views scientific inference as being from observations to the law, from particulars to the general. There is something true about this which the *H-D* account must ignore. Thus Newton wrote: "The main business of natural philosophy is to argue from phenomena. . . ." [19]

This inductive view, however, ignores what Newton never ignored: the inference is also from *explicanda* to an *explicans.* Why a beveled mirror shows spectra in sunlight is not explained by saying that all beveled mirrors do this. Why Mars moves more rapidly at 270 degrees and 90 degrees than could be expected of circular-uniform motions is not explained by saying that Mars (or even all planets) always move thus. On the induction view, these latter might count as laws. But only when it is explained why beveled mirrors show spectra and why planets apparently accelerate at 90 degrees will we have laws of the type suggested: Newton's laws of refraction and Kepler's first law. And even before such discoveries were made, arguments in favor of those *types* of laws were possible.

So the inductive view rightly suggests that laws are somehow related to inferences *from* data. It wrongly suggests that the resultant law is but a summary of these data, instead of being an explanation of these data. A logic of discovery, then, might consider the structure of arguments in favor of one *type* of possible explanation in a given context as opposed to other *types.*

H-D accounts all agree that laws explain data. (Thus Braithwaite says: "A hypothesis to be regarded as a natural law must be a general proposition which can be thought to explain its instances;

if the reason for believing the general proposition is solely direct knowledge of the truth of its instances, it will be felt to be a poor sort of explanation of these instances . . ." [*op. cit.*, p. 302].) *H-D* theorists, however, obscure the initial connection between thinking about data and thinking about what kind of hypothesis will most likely lead to a law. They suggest that the fundamental inference in science is from higher-order hypotheses to observation-statements. This may characterize the setting out of one's reasons for making a prediction after *H* is formulated and provisionally established. It need not be a way of setting out reasons in favor of proposing originally of what type *H* is likely to be.

Yet the original suggestion of a hypothesis type is often a reasonable affair. It is not as dependent on intuition, hunches, and other imponderables as historians and philosophers suppose when they make it the province of genius but not of logic. If the establishment of *H* through its predictions has a logic, so has the initial suggestion that *H* is likely to be of one kind rather than another. To form the first specific idea of an elliptical planetary orbit, or of constant acceleration, or of universal gravitational attraction does indeed require genius—nothing less than a Kepler, a Galileo, or a Newton. But this does not entail that reflections leading to these ideas are nonrational. Perhaps *only* Kepler, Galileo, and Newton had intellects mighty enough to fashion these notions initially; but to concede this is not to concede that their reasons for first entertaining concepts of such a type surpass rational inquiry.

H-D accounts begin with the hypothesis as given, as cooking recipes begin with the trout. Recipes, however, sometimes suggest, "First catch your trout." The *H-D* account is a recipe physicists often use after catching hypotheses. However, the conceptual boldness which marks the history of physics shows more in the ways in which scientists *caught* their hypotheses than in the ways in which they elaborated these once caught.

To study only the verification of hypotheses leaves a vital part of the story untold—namely, the reasons Kepler, Galileo, and Newton had for thinking their hypotheses would be of one kind rather than another. In a letter to Fabricus, Kepler underlines this:

Prague, July 4, 1603

Dear Fabricius,

> . . . You believe that I start with imagining some pleasant hypothesis and please myself in embellishing it, examining it only later by observations. In this you are very much mistaken. The truth is that after having built up an hypothesis on the ground of observations and given it proper foundations, I feel a peculiar desire to investigate whether I might discover some natural, satisfying combination between the two . .

Had any *H-D* theorist ever sought to give an account of the way in which hypotheses in science *are discovered,* Kepler's words are for him. Doubtless *H-D* philosophers have tried to give just such an account. Thus, Braithwaite [20] writes: "Every science *proceeds* . . . by thinking of general hypotheses . . . from which particular consequences are deduced which can be tested by observation . . . ," and again, "Galileo's deductive system was . . . presented as deducible from . . . Newton's laws of motion and . . . his law of universal gravitation . . ."

How would any *H-D* theorist analyze the law of gravitation?

(1) First, the hypothesis *H*: that between any two particles in the universe exists an attracting force varying inversely as the square of the distance between them ($F = \lambda Mm/r^2$).

(2) Deduce from this (in accordance with the *Principia*)
 (a) *Kepler's* Laws, and
 (b) *Galileo's* Laws.

(3) But particular instances of (a) and (b) square with what is observed.

(4) Therefore *H* is, to this extent, confirmed.

The H-D account says nothing about how *H* was first puzzled out. But now consider why, here, the H-D account is *prima-facie* plausible.

Historians remark that Newton's reflections on this problem began in 1860 when Halley asked: "If between a planet and the sun there exists an attraction varying inversely as the square of their distance, what then would be the path of the planet?" Halley was astonished by the immediate answer: "An ellipse." The astonishment arose not because Newton *knew* the path of a planet, but because he had apparently deduced this from the hypothesis of universal gravitation. Halley begged for the proof; but it was lost in the chaos of Newton's room. Sir Isaac's promise to work it out anew terminated in the writing of the *Principia* itself. Thus the story unfolds as an *H-D* plot: (1) from the suggestion of a hypothesis (whose genesis is a matter of logical indifference—that is, psychology, sociology, or history) to (2) the deduction of observation statements (the laws of Kepler and Galileo), which turn out true, thus (3) establishing the hypothesis.

Indeed, the entire *Principia* unfolds as the plot requires—from propositions of high generality through those of restricted generality terminating in observation-statements. Thus Braithwaite [21] observes: "Newton's *Principia* [was] modelled on the Euclidean analogy and professed to prove [its] later propositions—those which were confirmed by confrontation with experience—by deducing them from original first principles . . ."

Despite this, the orthodox account is suspicious. The answer Newton gave Halley is not unique. He could have said "a circle" or "a parabola," and have been equally correct. The general answer is: "A conic section." The greatest mathematician of his time is not likely to have dealt with so mathematical a question as that concerning the possibility of a formal demonstration with an answer which is but a single value of the correct answer.

Yet the reverse inference, the retrodiction, *is* unique. Given that the planetary orbits are ellipses, and allowing Huygen's law of centripetal force and Kepler's rule (that the square of a planet's period of revolution is proportional to the cube of its distance from the sun), the *type* of the law of gravitation can be inferred. Thus the question, "If the planetary orbits are ellipses, what form will the gravitational force law take?" invites the unique answer, "an inverse square type of law."

Given the datum that Mars moves in an ellipse, one can (by way of Huygen's law and Kepler's third law) explain this uniquely by suggesting how it might follow from a law of the inverse square type, such as the law of universal gravitation was later discovered to be.

The rough idea behind all this is: Given an ellipsoidal eggshell, imagine a tiny pearl moving inside it along the maximum elliptical orbit. What *kind* of force must the eggshell exert on the pearl to keep the latter in this path? Huygen's weights, when whirled on strings, required a force in the string, and in Huygen's arm, of $F_{(k)} \propto r/T^2$ (where r signifies distance, T time, and k is a constant of proportionality). This restraining force kept the weights from flying away like stones from David's sling. And something like this force would be expected in the eggshell. Kepler's third law gives $T^2 \propto r^3$. Hence, $F_{(k)} \propto r/r^3 \propto 1/r^2$. The force the shell exerts on the pearl will be of a kind which varies inversely as the square of the distance of the pearl from that focus of the ellipsoidal eggshell where the force may be supposed to be centered. This is not yet the law of gravitation. But it certainly is an argument which suggests that the law is likely to be of an inverse square type. This follows by what Peirce called 'retrodictive reasoning.' But what *is* this retrodictive reasoning whose superiority over the *H-D* account has been so darkly hinted at?

Schematically, it can be set out thus:

(1) Some surprising, astonishing phenomena $p_1, p_2, p_3 \ldots$ are encountered.[22]

(2) But $p_1, p_2, p_3 \ldots$ would not be surprising were a hypothesis of *H*'s type to obtain. They would follow as a matter of course from something like *H* and would be explained by it.

(3) Therefore there is good reason for elaborating a hypothesis of the type of *H*; for proposing it as a possible hypothesis from whose assumption p_1, p_2, p_3 . . . might be explained.[23]

How, then, would the discovery of universal gravitation fit this account?

(1) The astonishing discovery that all planetary orbits are elliptical was made by Kepler.
(2) But such an orbit would not be surprising if, in addition to other familiar laws, a law of 'gravitation,' of the inverse square type obtained. Kepler's first law would follow as a matter of course; indeed that kind of hypothesis might even explain why (since the sun is in but one of the foci) the orbits are ellipses on which the planets travel with nonuniform velocity.
(3) Therefore there is good reason for further elaborating hypotheses of this kind.

This says something about the rational context within which a hypothesis of *H*'s type might come to be "caught" in the first place. It begins where all physics begins—with problematic phenomena requiring explanation. It suggests what might be done to particular hypotheses once proposed—namely, the *H-D* elaboration. And it points up how much philosophers have yet to learn about the kinds of reasons scientists might have for thinking that one kind of hypothesis may explain initial perplexities; why, for example, an inverse square type of hypothesis may be preferred over others, *if* it throws the initially perplexing data into patterns within which determinate modes of connection can be perceived. At least it appears that the ways in which scientists sometimes reason their way *towards* hypotheses, by eliminating those which are certifiably of the wrong type, may be as legitimate an area for conceptual inquiry as are the ways in which they reason their way *from* hypotheses.

Recently, in the Lord Portsmouth collection in the Cambridge University Library, a document was discovered which bears on our discussion. There, in "Additional manuscripts 3968, No. 41, bundle 2," is the following draft in Newton's own hand:

> And in the same year [1665, twenty years before the *Principia*] I began to think of gravity extending to ye orb of the Moon, and (having found out how to estimate the force with which a globe revolving within a sphere presses the surface of the sphere), from Kepler's rule . . . I deduced that the forces which keep the planets in their Orbs must be reciprocally as the squares of their distances from the centres about which they revolve.

This manuscript corroborates our argument. ("Deduce," in this passage, is used as when Newton speaks of deducing laws from phenomena—which is just what Aristotle and Peirce would call "retroduce.") Newton *knew* how to estimate the force of a small globe on the inner surface of a sphere. (To compare this with Halley's question and our pearl-within-eggshell reconstruction, note that a sphere can be regarded as a degenerate ellipsoid—that is, where the foci superimpose.) From this and from Kepler's rule, $T^2 \propto r^3$, Newton determinded that, whatever the final form of the law of gravitation, it would very probably be of the inverse-square type. These were the reasons which led Newton to think further about the details of universal gravitation. The reasons for accepting one such hypothesis of this type *as a law* are powerfully set out later in the *Principia* itself; and they are much more comprehensive than anything which occurred to him at this early age. But without such preliminary reasoning Newton might have had no more grounds than Hooke or Wren for thinking the gravitation law to be of an inverse-square type.

The morals of all this for our understanding of contemporary science are clear. With such a rich profusion of data and technique as we have, the arguments necessary for *eliminating* hypotheses of the wrong type become a central research inquiry. Such arguments are not always of the *H-D* type; but if for that reason alone we refuse to scrutinize the conceptual content of the reasoning which precedes the actual proposal of definite hypotheses, we will have a poorer understanding of scientific thought in our time. For our own sakes, we must attend as much to how scientific hypotheses are caught, as to how they are cooked.

NOTES

1. Karl Popper, *The Logic of Scientific Discovery*. New York: Basic Book, 1959, pp. 31–32.
2. Hans Reichenbach, *Experience and Prediction*. Chicago: Univ. of Chicago Press, 1938, p. 382.
3. R. B. Braithwaite, *Scientific Explanation*. Cambridge: Cambridge Univ. Press, 1955, pp. 21–22.
4. C. S. Peirce, *Collected Papers*. Cambridge (Mass.): Harvard Univ. Press, 1931, Vol. I, Sec. 188.
5. F. C. S. Schiller, "Scientific Discovery and Logical Proof," Charles Singer, ed., *Studies in the History and the Methods of the Sciences*. Vol. I. Oxford: Clarendon Press, 1917.
6. "There is no science which will enable a man to bethink himself of that which will suit his purpose," J. S. Mill, *A System of Logic*, III, Chap. I.
7. Cf. *Patterns of Discovery*. Cambridge, Mass.: Harvard Univ. Press, 1958, pp. 85–92, "The Logic of Discovery," in *Journal of Philosophy*, LV, 25, 1073–1089, 1958; More on "The Logic of Discovery," *op. cit.*, LVII, 6, 182–188, 1960.
8. Cf. *De Motibus Stellae Martis*. Munich, pp. 250ff.

9. *Op. cit.*, p. 20.
10. Reichenbach writes that philosophy "cannot be concerned with the first, but only with the latter" (*op. cit.*, p. 382).
11. Reichenbach, *op. cit.*, p. 371.
12. *Ibid.*, p. 389.
13. Braithwaite, *op. cit.*, pp. 12–13.
14. *Ibid.*, p. 303.
15. *Ibid.*, p. 11.
16. Aristotle, *Metaphysics* 982b, 11ff.
17. Peirce, *op. cit.*, Vol. II, Book III, Chap. 2, Part III.
18. Cf. John Dewey, *How We Think*. London: Heath & Co., 1909, pp. 12f.
19. Newton, *Principia*, Preface.
20. *Op. cit.*, pp. xv, xi, 18.
21. *Op. cit.*, p. 352.
22. The astonishment may consist in the fact that p is at variance with accepted *theories*—for example, the example, the discovery of discontinuous emission of radiation by hot black bodies, or the photoelectric effect, the Compton effect, and the continuous β-ray spectrum, or the orbital aberrations of Mercury, the refrangibility of white light, and the high velocities of Mars at 90 degrees. What is important here is *that* the phenomena are encountered as anomalous, not *why* they are so regarded.
23. This is a free development of remarks in Aristotle (*Prior Analytics*, II, 25) and Peirce (*op. cit.*). Peirce amplifies: "It must be remembered that retroduction, although it is very little hampered by logical rules, nevertheless, is logical inference, asserting its conclusion only problematically, or conjecturally, it is true, but nevertheless having a perfectly definite logical form" (*op. cit.*, Vol. I, p. 188).

READING 4

Comments on Hanson's "Is There a Logic of Scientific Discovery?"

Paul K. Feyerabend

Hanson's main thesis seems to consist in the following two statements: (1) The invention of a new physical theory need not be a completely irrational process because there may exist arguments to the effect that the theory must be of certain kind; (2) the hypothetico-deductive model (*H-D*-model) of explanation is incomplete because it cannot give an account of these arguments.

My reply is that Hanson is correct when asserting (1), but that he is wrong when asserting (2).

In order to substantiate my reply let me first describe the main features of the *H-D* model. In this model it is assumed that the purpose of science is explanation. Explanation always starts from a situation to be explained, or from an explanandum. The *explanandum* need not refer directly to observables, but it must at least be plausible that it is true. According to the *H-D* model there are three conditions which an *explanans* must satisfy in order to be satisfactory. They are: (1) it must be *relevant*—that is, it must enable us at least to obtain a sentence describing the situation *S* to be explained; (2) it must *not* be *ad hoc*—that is, it must yield more than just *S;* and (3) it must be *testable*—that is, it must be possible to describe an (observational) situation which is such that the hypothesis will be withdrawn as soon as this situation occurs. Let me add here a little comment on condition (2). Professor Hanson has pointed out, and I think with some justification, that it is not sufficient for the hypothesis to give *S* plus something else *in a simple additive matter;* that *all* stones fall to the ground when released would seem to be a very poor

Reprinted by permission of the publisher from *Current Issues in the Philosophy of Science,* ed. by Herbert Feigl and Grover Maxwell, copyright © 1961 by Holt, Rinehart and Winston, Inc.

explanation of the fact that *this* stone falls to the ground when released, although it satisfies (1), (2), (3) above. For quite clearly my question why *this* stone behaves as it does will not be answered by the hint that the behavior of *any* stone will give rise to the same question. This indicates that simple generalization will not do as an explanatory device. Without going into too much detail I think that a necessary condition of a satisfactory explanation will be (2′) that the explaining hypothesis contain terms which do not occur in the explanandum, but which are essential for the derivation of the explanandum from it. It is this necessary condition which forces us to introduce terms that initially do not possess a direct observational significance (although they may become observable as time goes on and people become accustomed to expressing even observational affairs in these new terms) and which thereby forces us to leave the observational domain. Popper has expressed this consequence by saying that explanation is always explanation of what is known in terms of what is not known. Unfortunately most of the highly sophisticated contemporary studies in the logic of explanation and confirmation deal with explanations that do not satisfy (2′).

Conditions (1) through (3), with (2) modified into (2′), are *restrictive* conditions. They do not tell us what hypothesis to choose because they do not uniquely determine the explanans. Indeed, once (2′) has been introduced, it is very difficult to see how a unique determination could be possible. Now if I understand Professor Hanson correctly, he says that in actual scientific practice the choice of proper explanations is further restricted by conditions which are different from all the conditions mentioned so far and which may even make the choice of the explanans *unique*. I suggest that the condition which Hanson has in mind can be stated in the following terms:

Condition (4): the explanans, apart from satisfying (1) through (3), must not be inconsistent with certain more general points of view, theories, or metaphysical principles which are held either by the inventor, or by the (scientific) community in which he lives. I think that in spite of its negative form (which I have chosen quite deliberately) this condition covers what Professor Hanson wants to add to the *H-D* model. Contemporary physicists reject Bohm's general hypothesis about the nature of microscopic particles (the hypothesis developed, for example, in his *Cause and Chance in Modern Physics,* which is very different from the model of 1951), not because they think it violates one of the conditions (1) through (3), but because it is inconsistent with some further assumptions which are thought to be confirmed to such a high degree that it would be foolish to give them up. At a certain stage of its development the hypothesis of the shortest length was criticized, not because it was considered *ad hoc,* or empirically false, but because it was difficult to see how it could be made relativistically invariant. Ptolemaeus rejected the hypothesis of Aristarchus, not because it could not give a satisfactory account of the facts, but because it was inconsistent with the Aristotelian dynamics. It is very important, by the way, to realize that the principles or laws with which a new hypothesis is supposed to be consistent need not always be of the empirical kind. Copernicus' acceptance of Aristarchus' model was prompted by empirical considerations as well as by his desire to make astronomy consistent with Neoplatonism. That esthetic considerations sometimes play an equally strong role may be seen from the reasons which made Galileo reject Kepler's ellipses.

Now all these examples, while showing that rule (4) indeed expresses what Professor Hanson wants to say, seem also to show that physicists, when considering a new theory, do not only pay attention to conditions (1) through (3), but also to a background of partly empirical, partly metaphysical beliefs. Does this fact prove that the *H-D* account is incomplete?

In order to show that it does not, we split condition (4) into two parts, (4a) and (4b). According to (4a) the total set S of hypotheses made about the world, be they now empirical, or metaphysical, must be consistent. According to (4b) a certain subset, S', of S has already been accepted. (4a) and (4b) taken together demand that any newly introduced hypothesis H, apart from satisfying (1), (2), and (3), must also be compatible with S'. I contend that this demand cannot be interpreted as a new methodological rule. For when enquiring about the reasons why we accept S', we shall meet two: the empirical adequacy of some of its elements (which can be accounted for by the *H-D* model), and the (as yet untested) belief in others. Let the set of the latter elements be S''. Quite clearly S'' cannot be justified, apart from its agreement with (1), (2), and (3), by the additional demand raised with respect to H, for this demand presupposes that S' and, therefore, S'' are already given. It follows that the acceptance, *in advance of the carrying out of tests,* of S'' must be regarded as irrational to the extent to which it is not restricted by (1), (2), or (3). But if that is the case, then the acceptance, on the basis of (4), of the hypothesis H is likewise irrational, for the rationality of a procedure is not increased by showing that it is consistent with other irrational procedures. And as is well known, the demand of consistency—namely, (4a)—is not a new rule, but is inherent in the *H-D* model.

To sum up: The new rule which Hanson wants to add to the *H-D* model is dispensable because part of it is already contained in this model and the rest—namely, (4b)—is not a new methodological demand: it is the (sociopsychological) fact that certain beliefs are being held.

Let me finally comment on Hanson's belief that his rule (4) may, under certain circumstances, lead to a unique determination of the hypothesis required for the explanation of a given set of facts. This is not further surprising. After all, if the hypotheses included in the set with which the new hypothesis *H* is to be compatible are rich enough, they will of course allow for the derivation, on the basis of the facts, of assumptions which could not have been derived with the help of the observational statements alone. Now, if the elements of the set are only implicity used, then the impression will arise that a new kind of 'logic,' or new ways or 'reasoning,' have been discovered, whereas all that is going on is a sloppy kind of talking where premises are neither explicitly stated nor consistently used. (In the sense, of course, in which sloppy thinking is a special kind of thinking, it is indeed true that we are here confronted with "new ways of thinking.")

Considerations "by analogy" are a good example of what has just been said. Consider the following argument: "Mars moves in an ellipse: Mars is a planet (I think the word "typical," which Professor Hanson seems to think is important, can be safely omitted); therefore Jupiter (which is a planet) will also move in an ellipse." As it stands, this argument is quite clearly not deductive. Does this show, however, that we are here presented with a new and special type of reasoning, which has not been sufficiently taken into account by formal logicians, and which therefore ought to be an object of special study? It seems to me that such a conclusion, although in agreement with the contemporary fashion to replace criticism according to old and precise standards by the assertion that what one is dealing with requires the recognition of new standards, is utterly unwarranted. After all, the assumption that the planets all obey the same laws of motion is such a basic idea of the planetary astronomy from Eudoxos up to Copernicus that an astronomer could safely omit *stating* this assumption without thereby leaving the domain of deductive reasoning, or running danger of being misunderstood by his colleagues as an innovator of logic. While reading an argument like the one quoted above, they all would know what else was being presupposed, and they would therefore be perfectly able *to read this argument* in the way it was intended—namely, *as an incompletely stated deductive argument.* It was left to Professor Hanson and some other philosophers of science to imply that these astronomers knew in fact much less and were therefore in need of new types of argument. I think that a little more historical research would convince Professor Hanson that none of the examples he discusses can be interpreted in this way and that the *logic of discovery* for which he is looking so eagerly simply does not exist.

How suppression of premises by those who regard them as obvious and well known can cultivate in some of their less well-oriented readers the belief in new ways of reasoning is excellently shown by the case which Professor Hanson himself has chosen to discuss, namely, the alleged 'derivation,' or 'retroduction,' from Kepler's laws of the law of universal gravitation. What are the premises of this argument, and what is its conclusion? The premises are the elliptical motion, in accordance with Kepler's laws, of a single mass point around a center which is *at rest* in one of the foci. The conclusion is that the acceleration, with respect to this center, of the circling mass point is of the $1/r^2$ type. This quite obviously has nothing whatever to do with the planets. For in the case of the planets, the *premise* of the argument is not valid because there is not a *single* mass point circling around the center, but many. Also the center is not at rest (with respect to some inertial frame). And, secondly, the law of gravitation is pronounced quite generally which shows also that the *conclusion* of the above deduction has nothing to do with gravitation. Hence, this deduction by itself can by no means be regarded as a unique determination, on the basis of experience, of the law of universal gravitation. Only if we add to it certain assumptions about the magnitude of disturbances as well as a more general hypothesis about the similarity, with respect to their dynamical properties, of all particles of matter, is it possible to make the conclusion a unique consequence of the thus extended premises. But since the new set of premises already makes use of some very decisive properties of the law of gravitation, this is not any more the great achievement it seemed to be in the beginning. The real achievement lies somewhere else: it lies in the conception of a universal, attractive, central force. This *act* of conception, itself, is not justifiable, either by "retroduction," or by arguments of analogy; for it is an irrational, inventive act, the *result* of which is justified only *after* it has been shown to conform with (1), (2), and (3) above and has had predictive success. A "logic of invention" which helps us to produce such a law simply does not exist.

READING 5

Rejoinder to Feyerabend

Norwood Russell Hanson

In a rejoinder, one rejoins. It is a pleasure to join, or rejoin, issue with my friend Professor Feyerabend—especially when he leaves undefended, indeed indefensible, so many flanks in his attack.

After absorbing Feyerabend's stern moralizing on the virtues of "a little more historical research" and the vices of "sloppy thinking," it gives me some satisfaction to be able to point out that he cannot really have *read* my essay at all. Or, if he did, his scrutiny was sloppy. For he credits me with views I am explicitly at pains to reject.

Thus, on page 70 Feyerabend writes as if my intention had been to propound a 'new methodological rule.' It was the purpose of my sixth footnote to deny the very possibility of such a "rule." He regards as *irrational* the acceptance of untested beliefs *in advance of the carrying out of tests*—and announces this as a hit against me. But this is precisely the point I seek to make on page 56 of my essay, when I write, "But such reasons [involving untested beliefs] would not *establish* [that is, lead to the acceptance of] *H′*." Never once did I suggest that it would be rational to *accept* untested beliefs.

I *did* suggest that it often is rational to *entertain* certain types of untested beliefs, with an eye to exploring them further experimentally. By this I mean that reasons could be given, before actual testing, which would show up the plausibility, or the unplausibility, of exploring further some particular *kind* of hypothesis, rather than others. Indeed, Reichenbach, in his *Philosophic Foundation of Quantum Mechanics,* gives a splendid illustration of this point. After setting out the time-independent Schrödinger equation for waves corresponding to particles in a field of force (p. 71), Reichenbach writes:

Reprinted by permission of the publisher from *Current Issues in the Philosophy of Science,* ed. by Herbert Feigl and Grover Maxwell, copyright © 1961 by Holt, Rinehart and Winston, Inc.

> We certainly do not disparage Schrödinger's work when we do not consider the derivation given, or the more complicated derivations originally given by Schrödinger, as a proof of the validity of the resulting wave equation. Such a derivation—and Schrödinger never meant anything else—can be used to make the wave equation *plausible,* and therefore represents an excellent guide within the context of discovery . . .

Feyerabend then goes on to say that I am opting for a "new methodological demand." I can find in my paper no linguistic basis for this reaction. He then speaks of "Hanson's belief that his rule[1] may, under certain circumstances, lead to a *unique determination of* the *hypothesis* required for the explanation of a given set of facts." (My italics.) Apparently I need not have bothered to set out the analyses of pages 53 and 54–and especially entry (2′). For the "unique determination of the hypothesis" *in advance of experimentation* is precisely what I deny can ever be had. At most, an argument making it plausible to explore one *kind* of hypothesis, rather than others, can be entertained before experiment. This is *all* I argue for, and all I want to argue for.

But it takes two people to argue; one to speak (or write), and one to listen (or read). If Feyerabend will not read what I say, his "Comments" on what I say must be expected to suffer somewhat.

> . . . the assumption that the planets all obey the same laws of motion is such a basic idea of the planetary astronomy from Eudoxos up to Copernicus, that an astronomer could safely omit *stating* this assumption without leaving the domain of deductive reasoning . . . (Feyerabend, p. 71).[2]

Poppycock! No astronomer could ever safely omit stating such an assumption, particularly when orbital noncircularity appeared to be precisely what was going uniquely to *distinguish* Mars from the other planets! Nor, as a matter of historical fact [which Feyerabend chides us for neglecting] was the inference from Mars' elliptical orbit to Jupiter's actually taken as *deductive.* In *De Motibus Stellae Martis* (1609) Kepler settles the issue for Mars. Not until the *Harmonices Mundi,* ten years later, is Jupiter brought into the elliptical orbit generalization, *after observations* on that planet. But why should Kepler have *observed* Jupiter during the interim at all? What reasons had he for examining Jupiter's orbit more closely, rather than just accepting extant sixteenth century studies? *This* it was the function of my essay to explore–a fact Feyerabend seems not to appreciate.

Incidentally, it is as a *result* of "historical research" in seventeenth-century scientific texts that my interest in "plausibility arguments" for as-yet-untested hypotheses was kindled. I am not "looking so eagerly" for a *logic of discovery* as Professor Feyerabend supposes. (Cf., the second paragraph of my paper.) If I misconstrue the texts, this should be easy enough to disclose. Feyerabend may yet find time to do this for me, between his own bouts of "historical research" and "rigorous [that is, nonsloppy] thinking." But, alas, not a word in his "Comments" illuminates the arguments Kepler *actually used.*

The last paragraph of Feyerabend's "Comments" is quixotic in the extreme. Here, I can no longer identify his opponent. Discovering the acceleration of a particle in elliptical motion about a center to be of the $1/r^2$ type—to say that this *"has nothing whatever to do with Newton's law of universal gravitation and with the planets"* is almost irresponsible. On page 65 I set out Newton's own original reflections on this problem. *He* felt the $1/r^2$ type of law involved in estimating "the force with which a globe revolving within a sphere presses the surface of the sphere" *certainly* had something to do with "the forces which keep the planets in their Orbs." *Of course,* this early demonstration does not yield the law of universal gravitation as a conclusion. But it certainly *does* make it plausible to suppose that such a gravitational law, when discovered, would be of a type which describes forces acting on the planets "reciprocally as the squares of their distances from the centres about which they revolve . . ."

Professor Feyerabend's parting shot, namely,

> A "logic of invention" which helps us to produce such a law simply does not exist,

is lofted as if it hit my thesis dead center. But how can this be so, since I agree with this very contention; indeed I stress the point on pages 53 and 56? This fact may stimulate Feyerabend's curiosity. He may even go back and read what I say.

Never mind, someone *could* have said the things he says I say. And then his "Comments" would have been devastating.

NOTES

1. Incidentally, this is Feyerabend's rule, not mine. He hangs it on me and then fires with a rhetorical scatter-gun that certainly hits "the rule." But *my* essay was supposed to have been the target, not Feyerabend's shooting-gallery rewrite of it.
2. My example is drawn straight out of Kepler. What, then, is the relevance of referring to planetary astronomy "from Eudoxos up to Copernicus"?

READING 6

Truth, Rationality, and the Growth of Scientific Knowledge

Karl R. Popper

1. THE GROWTH OF KNOWLEDGE: THEORIES AND PROBLEMS

I

My aim in this lecture is to stress the significance of one particular aspect of science—its need to grow, or, if you like, its need to progress. I do not have in mind here the practical or social significance of this need. What I wish to discuss is rather its intellectual significance. I assert that continued growth is essential to the rational and empirical character of scientific knowledge; that if science ceases to grow it must lose that character. It is the way of its growth which makes science rational and empirical; the way, that is, in which scientists discriminate between available theories and choose the better one or (in the absence of a satisfactory theory) the way they give reasons for rejecting all the available theories, thereby suggesting some of the conditions with which a satisfactory theory should comply.

You will have noticed from this formulation that is not the accumulation of observations which I have in mind when I speak of the growth of scientific knowledge, but the repeated overthrow of scientific theories and their replacement by better or more satisfactory ones. This, incidentally, is a procedure which might be found worthy of attention even by those who see the most important aspect of the growth of scientific knowledge in new experiments and

Reprinted by permission of the author and publisher from Karl R. Popper, *Conjectures and Refutations*, 4th ed., 1972, Routledge and Kegan Paul Ltd., pp. 215–250.

in new observations. For our critical examination of our theories leads us to attempts to test and to overthrow them; and these lead us further to experiments and observations of a kind which nobody would ever have dreamt of without the stimulus and guidance both of our theories and of our criticism of them. For indeed, the most interesting experiments and observations were carefully designed by us in order to *test* our theories, especially our new theories.

In this paper, then, I wish to stress the significance of this aspect of science and to solve some of the problems, old as well as new, which are connected with the notions of scientific progress and of discrimination among competing theories. The new problems I wish to discuss are mainly those connected with the notions of objective truth, and of getting nearer to the truth—notions which seem to me of great help in analysing the growth of knowledge.

Although I shall confine my discussion to the growth of knowledge in science, my remarks are applicable without much change, I believe, to the growth of prescientific knowledge also—that is to say, to the general way in which men, and even animals, acquire new factual knowledge about the world. The method of learning by trial and error—of learning from our mistakes—seems to be fundamentally the same whether it is practised by lower or by higher animals, by chimpanzees or by men of science. My interest is not merely in the theory of scientific knowledge, but rather in the theory of knowledge in general. Yet the study of the growth of scientific knowledge is, I believe, the most fruitful way of studying the growth of knowledge in general. For the growth of scientific knowledge may be said to be the growth of ordinary human knowledge *writ large* (as I have pointed out in the 1958 Preface to my *Logic of Scientific Discovery*).

But is there any danger that our need to progress will go unsatisfied, and that the growth of scientific knowledge will come to an end? In particular, is there any danger that the advance of science will come to an end because science has completed its task? I hardly think so, thanks to the infinity of our ignorance. Among the real dangers to the progress of science is not the likelihood of its being completed, but such things as lack of imagination (sometimes a consequence of lack of real interest); or a misplaced faith in formalization and precision (which will be discussed below in section v); or authoritarianism in one or another of its many forms.

Since I have used the word 'progress' several times, I had better make quite sure, at this point, that I am not mistaken for a believer in a historical law of progress. Indeed I have before now struck various blows against the belief in a law of progress,[1] and I hold that even science is not subject to the operation of anything resembling such a law. The history of science, like the history of all human ideas, is a history of irresponsible dreams, of obstinacy, and of error. But science is one of the very few human activities—perhaps the only one—in which errors are systematically criticized and fairly often, in time, corrected. This is why we can say that, in science, we often learn from our mistakes, and why we can speak clearly and sensibly about making progress there. In most other fields of human endeavour there is change, but rarely progress (unless we adopt a very narrow view of our possible aims in life): for almost every gain is balanced, or more than balanced, by some loss. And in most fields we do not even know how to evaluate change.

Within the field of science we have, however, a *criterion of progress:* even before a theory has ever undergone an empirical test we may be able to say whether, provided it passes certain specified tests, it would be an improvement on other theories with which we are acquainted. This is my first thesis.

To put it a little differently, I assert that we *know* what a good scientific theory should be like, and—even before it has been tested—what kind of theory would be better still, provided it passes certain crucial tests. And it is this (meta-scientific) knowledge which makes it possible to speak of progress in science, and of a rational choice between theories.

II

Thus it is my first thesis that we can know of a theory, even before it has been tested, that *if* it passes certain tests it will be better than some other theory.

My first thesis implies that we have a criterion of relative *potential* satisfactoriness, or of *potential* progressiveness, which can be applied to a theory even before we know whether or not it will turn out, by the passing of some crucial tests, to be satisfactory *in fact*.

This criterion of relative potential satisfactoriness (which I formulated some time ago,[2] and which, incidentally, allows us to grade theories according to their degree of relative satisfactoriness) is extremely simple and intuitive. It characterizes as preferable the theory which tells us more; that is to say, the theory which contains the greater amount of empirical information or *content;* which is logically stronger; which has the greater explanatory and predictive power; and which can therefore be *more severely tested* by comparing predicted facts with observations. In short, we prefer an interesting, daring, and highly informative theory to a trivial one.

All these properties which, it thus appears, we desire in a theory can be shown to amount to one and the same thing: to a higher degree of empirical *content* or of testability.

III

My study of the *content* of a theory (or of any statement whatsoever) was based on the simple and obvious idea that the informative content of the *conjunction, ab,* of any two statements, *a,* and *b,* will always be greater than, or at least equal to, that of any of its components.

Let *a* be the statement 'It will rain on Friday'; *b* the statement 'It will be fine on Saturday'; and *ab* the statement 'It will rain on Friday and it will be fine on Saturday': it is then obvious that the informative content of this last statement, the conjunction *ab,* will exceed that of its component *a* and also that of its component *b.* And it will also be obvious that the probability of *ab* (or, what is the same, the probability that *ab* will be true) will be smaller than that of either of its components.

Writing $Ct(a)$ for 'the content of the statement a', and $Ct(ab)$ for 'the content of the conjunction *a* and *b*', we have

$$(1) \qquad Ct(a) \leqslant Ct(ab) \geqslant Ct(b)$$

This contrasts with the corresponding law of the calculus of probability,

$$(2) \qquad p(a) \geqslant p(ab) \leqslant p(b)$$

where the inequality signs of (1) are inverted. Together these two laws, (1) and (2), state that with increasing content, probability decreases, and *vice versa;* or in other words, that content increases with increasing improbability. (This analysis is of course in full agreement with the general idea of the logical *content* of a statement as the class of *all those statements which are logically entailed* by it. We may also say that a statement *a* is logically stronger than a statement *b* if its content is greater than that of *b*—that is to say, if it entails more than *b*.)

This trivial fact has the following inescapable consequences: if growth of knowledge means that we operate with theories of increasing content, it must also mean that we operate with theories of decreasing probability (in the sense of the calculus of probability). Thus if our aim is the advancement or growth of knowledge, then a high probability (in the sense of the calculus of probability) cannot possibly be our aim as well: *these two aims are incompatible.*

I found this trivial though fundamental result about thirty

years ago, and I have been preaching it ever since. Yet the prejudice that a high probability must be something highly desirable is so deeply ingrained that my trivial result is still held by many to be 'paradoxical'.[3] Despite this simple result the idea that a high degree of probability (in the sense of the calculus of probability) must be something highly desirable seems to be so obvious to most people that they are not prepared to consider it critically. Dr. Bruce Brooke-Wavell has therefore suggested to me that I should stop talking in this context of 'probability' and should base my arguments on a 'calculus of content' and of 'relative content'; or in other words, that I should not speak about science aiming at improbability, but merely say that it aims at maximum content. I have given much thought to this suggestion, but I do not think that it would help: a head-on collision with the widely accepted and deeply ingrained probabilistic prejudice seems unavoidable if the matter is really to be cleared up. Even if, as would be easy enough, I were to base my own theory upon the calculus of content, or of logical strength, it would still be necessary to explain that the probability calculus, in its ('logical') application to propositions or statements, is nothing but a *calculus of the logical weakness or lack of content of these statements* (either of absolute logical weakness or of relative logical weakness). Perhaps a head-on collision would be avoidable if people were not so generally inclined to assume uncritically that a high probability must be an aim of science, and that, therefore, the theory of induction must explain to us how we can attain a high degree of probability for our theories. (And it then becomes necessary to point out that there is something else—a 'truthlikeness' or 'verisimilitude'—with a calculus totally different from the calculus of probability with which it seems to have been confused.)

To avoid these simple results, all kinds of more or less sophisticated theories have been designed. I believe I have shown that none of them is successful. But what is more important, they are quite unnecessary. One merely has to recognize that the property which we cherish in theories and which we may perhaps call 'verisimilitude' or 'truthlikeness' (see section xi below) is *not* a probability *in the sense of the calculus of probability* of which (2) is an inescapable theorem.

It should be noted that the problem before us is not a problem of words. I do not mind what you call 'probability', and I do not mind if you call those degrees for which the so-called 'calculus of probability' holds by any other name. I personally think that it is most convenient to reserve the term 'probability' for whatever may satisfy the well-known rules of this calculus (which Laplace, Keynes, Jeffreys and many others have formulated, and for which I have given various formal axiom systems). If (and only if) we accept this terminology, then there can be no doubt that the absolute probability of a statement *a* is simply the *degree of its logical weakness, or lack of informative content,* and that the relative probability of a statement *a,* given a statement *b,* is simply the degree of the relative weakness, or the relative *lack* of *new* informative content in statement *a,* assuming that we are already in possession of the information *b.*

Thus if we aim, in science, at a high informative content—if the growth of knowledge means that we know more, that we know *a* and *b,* rather than *a* alone, and that the content of our theories thus increases—then we have to admit that we also aim at a low probability, in the sense of the calculus of probability.

And since a low probability of being falsified, it follows that a high degree of falsifiability, or refutability, or testability, is one of the aims of science—in fact, precisely the same aim as a high informative content.

The criterion of potential satisfactoriness is thus testability, or improbability: only a highly testable or improbable theory is worth testing, and is actually (and not merely potentially) satisfactory if it withstands severe tests—especially those tests to which we could point as crucial for the theory before they were ever undertaken.

It is possible in many cases to compare the severity of tests objectively. It is even possible, if we find it worth while, to define a measure of the severity of tests. (See the *Addenda* to this volume.) By the same method we can define the explanatory power and the degree of corroboration of a theory.[4]

IV

The thesis that the criterion here proposed actually dominates the progress of science can easily be illustrated with the help of historical examples. The theories of Kepler and Galileo were unified and superseded by Newton's logically stronger and better testable theory, and similarly Fresnel's and Faraday's by Maxwell's. Newton's theory, and Maxwell's, in their turn, were unified and superseded by Einstein's. In each such case the progress was towards a more informative and therefore logically less probable theory: towards a theory which was more severely testable because it made predictions, which in a purely logical sense, were more easily refutable.

A theory which is not in fact refuted by testing those new and bold and improbable predictions to which it gives rise can be said to be corroborated by these severe tests. I may remind you in this connection of Galle's discovery of Neptune, of Hertz's discovery of electromagnetic waves, of Eddington's eclipse observations, of Elsasser's interpretation of Davisson's maxima as interference fringes of de Broglie waves, and of Powell's observations of the first Yukawa mesons.

All these discoveries represent corroborations by severe tests—by predictions which were highly improbable in the light of our previous knowledge (previous to the theory which was tested and corroborated). Other important discoveries have also been made while testing a theory, though they did not lead to its corroboration but to its refutation. A recent and important case is the refutation of parity. But Lavoisier's classical experiments which show that the volume of air decreases while a candle burns in a closed space, or that the weight of burning iron-filings increases, do not establish the oxygen theory of combustion; yet they tend to refute the phlogiston theory.

Lavoisier's experiments were carefully thought out; but even most so-called 'chance-discoveries' are fundamentally of the same logical structure. For these so-called 'chance-discoveries' are as a rule refutations of theories which were consciously or unconsciously held: they are made when some of our expectations (based upon these theories) are unexpectedly disappointed. Thus the catalytic property of mercury was discovered when it was accidentally found that in its presence a chemical reaction had been speeded up which had not been expected to be influenced by mercury. But neither Oersted's nor Röntgen's nor Becquerel's nor Fleming's discoveries were really accidental, even though they had accidental components: every one of these men was searching for an effect of the kind he found.

We can even say that some discoveries, such as Columbus' discovery of America, corroborate one theory (of the spherical earth) while refuting at the same time another (the theory of the size of the earth, and with it, of the nearest way to India); and that they were chance-discoveries to the extent to which they contradicted all expectations, and were not consciously undertaken as tests of those theories which they refuted.

V

The stress I am laying upon change in scientific knowledge, upon its growth, or its progressiveness, may to some extent be contrasted with the current ideal of science as an axiomatized deductive system. This ideal has been dominant in European epistemology from

Euclid's Platonizing cosmology (for this is, I believe, what Euclid's *Elements* were really intended to be) to that of Newton, and further to the systems of Boscovic, Maxwell, Einstein, Bohr, Schrödinger, and Dirac. It is an epistemology that sees the final task and end of scientific activity in the construction of an axiomatized deductive system.

As opposed to this, I now believe that these most admirable deductive systems should be regarded as stepping stones rather than as ends: [5] as important stages on our way to richer, and better testable, scientific knowledge.

Regarded thus as means or stepping stones, they are certainly quite indispensable, for we are bound to develop our theories in the form of deductive systems. This is made unavoidable by the logical strength, by the great informative content, which we have to demand of our theories if they are to be better and better testable. The wealth of their consequences has to be unfolded deductively; for as a rule, a theory cannot be tested except by testing, one by one, some of its more remote consequences; consequences, that is, which cannot immediately be seen upon inspecting it intuitively.

Yet it is not the marvelous deductive unfolding of the system which makes a theory rational or empirical but the fact that we can examine it critically; that is to say, subject it to attempted refutations, including observational tests; and the fact that, in certain cases, a theory may be able to withstand those criticisms and those tests—among them tests under which its predecessors broke down, and sometimes even further and more severe tests. It is in the rational choice of the new theory that the rationality of science lies, rather than in the deductive development of the theory.

Consequently there is little merit in formalizing and elaborating a deductive non-conventional system beyond the requirements of the task of criticizing and testing it, and of comparing it critically with competitors. This critical comparison, though it has, admittedly, some minor conventional and arbitrary aspects, is largely non-conventional, thanks to the criterion of progress. It is this critical procedure which contains both the rational and the empirical elements of science. It contains those choices, those rejections, and those decisions, which show that we have learnt from our mistakes, and thereby added to our scientific knowledge.

VI

Yet perhaps even this picture of science—as a procedure whose rationality consists in the fact that we learn from our mistakes—is not quite good enough. It may still suggest that science progresses from theory to theory and that it consists of a sequence of better deductive systems. Yet what I really wish to suggest is that science should be visualized as *progressing from problems to problems*—to problems of ever increasing depth.

For a scientific theory—an explanatory theory—is, if anything, an attempt to solve a scientific problem, that is to say, a problem concerned or connected with the discovery of an explanation.[6]

Admittedly, our expectations, and thus our theories, may precede, historically, even our problems. *Yet science starts only with problems.* Problems crop up especially when we are disappointed in our expectations, or when our theories involve us in difficulties, in contradictions; and these may arise either within a theory, or between two different theories, or as the result of a clash between our theories and our observations. Moreover, it is only through a problem that we become conscious of holding a theory. It is the problem which challenges us to learn; to advance our knowledge; to experiment; and to observe.

Thus science starts from problems, and not from observations; though observations may give rise to a problem, especially if they are *unexpected;* that is to say, if they clash with our expectations or theories. The conscious task before the scientist is always the solution of a problem through the construction of a theory which solves the problem; for example, by explaining unexpected and unexplained observations. Yet every worthwhile new theory raises new problems; problems of reconciliation, problems of how to conduct new and previously unthought-of observational tests. And it is mainly through the new problems which it raises that it is fruitful.

Thus we may say that the most lasting contribution to the growth of scientific knowledge that a theory can make are the new problems which it raises, so that we are led back to the view of science and of the growth of knowledge as always starting from, and always ending with, problems—problems of an ever increasing depth, and an ever increasing fertility in suggesting new problems.

2. THE THEORY OF OBJECTIVE TRUTH: CORRESPONDENCE TO THE FACTS

VII

So far I have spoken about science, its progress, and its criterion of progress, without even mentioning *truth*. Perhaps surprisingly, this can be done without falling into pragmatism or instrumentalism: it is perfectly possible to argue in favour of the intuitive satisfactoriness of the criterion of progress in science without ever speaking about the truth of its theories. In fact, before I became acquainted with Tarski's theory of truth,[7] it appeared to me safer to discuss the criterion of progress without getting too deeply involved in the highly controversial problem connected with the use of the word 'true'.

My attitude at the time was this: although I accepted, as almost everybody does, the objective or absolute or correspondence theory of truth—truth as correspondence with the facts—I preferred to avoid the topic. For it appeared to me hopeless to try to understand clearly this strangely elusive idea of a correspondence between a statement and a fact.

In order to recall why the situation appeared so hopeless we only have to remember, as one example among many, Wittgenstein's *Tractatus* with its surprisingly naïve picture theory, or projection theory, of truth. In this book a proposition was conceived as a picture or projection of the fact which it was intended to describe and as having the same structure (or 'form') as that fact; just as a gramophone record is indeed a picture or a projection of a sound, and shares some of its structural properties.[8]

Another of these unavailing attempts to explain this correspondence was due to Schlick, who gave a beautifully clear and truly devastating criticism [9] of various correspondence theories—including the picture or projection theory—but who unfortunately produced in his turn another one which was no better. He interpreted the correspondence in question as a one-to-one correspondence between our designations and the designated objects, although counter examples abound (designations applying to many objects, objects designated by many designations) which refute this interpretation.

All this was changed by Tarski's theory of truth and of the correspondence of a statement with the facts. Tarski's greatest achievement, and the real significance of his theory for the philosophy of the empirical sciences, is that he rehabilitated the correspondence theory of absolute or objective truth which had become suspect. He vindicated the free use of the intuitive idea of truth as correspondence to the facts. (The view that his theory is applicable only to formalized languages is, I think, mistaken. It is applicable to any consistent and even to a 'natural' language, if only we learn from Tarski's analysis how to dodge its inconsistencies; which means, admittedly, the introduction of some 'artificiality'—or caution—into its use. See also *Addendum* 5, below.)

Although I may assume in this assembly some familiarity with

Tarski's theory of truth, I may perhaps explain the way in which it can be regarded, from an intuitive point of view, as a simple elucidation of the idea of *correspondence to the facts.* I shall have to stress this almost trivial point because, in spite of its triviality, it will be crucial for my argument.

The highly intuitive character of Tarski's ideas seems to become more evident (as I have found in teaching) if we first decide explicitly to take 'truth' as a synonym for 'correspondence to the facts', and then (forgetting all about 'truth') *proceed to explain the idea of 'correspondence to the facts'.*

Thus we shall first consider the following two formulations, each of which states very simply (in a metalanguage) under what conditions a certain assertion (of an object language) corresponds to the facts.

(1) The statement, or the assertion, *'Snow is white'* corresponds to to the facts if, and only if, snow is, indeed, white.
(2) The statement, or the assertion, *'Grass is red'* corresponds to the facts if, and only if, grass is, indeed, red.

These formulations (in which the word 'indeed' is only inserted for ease, and may be omitted) sound, of course, quite trivial. But it was left to Tarski to discover that, in spite of their apparent triviality, they contained the solution of the problem of explaining correspondence to the facts.

The decisive point is Tarski's discovery that, in order to speak of correspondence to the facts, as do (1) and (2), we must use a metalanguage in which we can *speak about two things: statements; and the facts to which they refer.* (Tarski calls such a metalanguage 'semantical'; a metalanguage in which we can speak about an object language but not about the facts to which it refers is called 'syntactical'.) Once the need for a (semantical) metalanguage is realized, everything becomes clear. (Note that while (3) ' "John called" *is true'* is essentially a statement belonging to such a metalanguage, (4) 'It is true *that* John called' may belong to the same language as 'John called'. Thus the phrase *'It is true that'*—which, like double negation, is logically redundant—differs widely from the metalinguistic predicate *'is true'*. The latter is needed for general remarks such as, "If the conclusion is not true, the premises cannot all be true' or 'John once made a true statement'.)

I have said that Schlick's theory was mistaken, yet I think that certain comments he made (*loc. cit.*) about his own theory throw some light on Tarski's. For Schlick says that the problem of truth shared the fate of some others whose solutions were not easily seen because they were mistakenly supposed to lie on a very deep level, while actually they were fairly plain and, at first sight, unimpressive. Tarski's solution may well appear unimpressive at first sight. Yet its fertility and its power are impressive indeed.

VIII

Thanks to Tarski's work, the idea of objective or absolute truth—that is truth as correspondence to the facts—appears to be accepted today with confidence by all who understand it. The difficulties in understanding it seem to have two sources: first, the combination of an extremely simple intuitive idea with a certain amount of complexity in the execution of the technical programme to which it gives rise; secondly, the widespread but mistaken dogma that a satisfactory theory of truth should yield a criterion of *true belief*—of well-founded, or rational belief. Indeed, the three rivals of the correspondence theory of truth—the coherence theory which mistakes consistency for truth, the evidence theory which mistakes 'known to be true' for 'true', and the pragmatic or instrumentalist theory which mistakes usefulness for truth—these are all subjective (or 'epistemic') theories of truth, in contradistinction to Tarski's objective (or 'metalogical') theory. They are subjective in the sense that *they all stem from the fundamental subjectivist position which can conceive of knowledge only as a special kind of mental state, or as a disposition, or as a special kind of belief,* characterized, for example, by its history or by its relation to other *beliefs.*

If we start from our subjective experience of believing, and thus look upon knowledge as a special kind of belief, then we may indeed have to look upon truth—that is, true knowledge—as some even more special kind of belief: as one that is well-founded or justified. This would mean that there should be some more or less effective criterion, if only a partial one, of well-foundedness; some symptom by which to differentiate the experience of a well-founded belief from other experiences of belief. It can be shown that all subjective theories of truth aim at such a criterion: they try to define truth in terms of the sources or origins of our beliefs,[10] or in terms of our operations of verification, or of some set of rules of acceptance, or simply in terms of the quality of our subjective convictions. They all say, more or less, that truth is what we are justified in believing or in accepting, in accordance with certain rules or criteria, of origins or sources of our knowledge, or of reliability, or stability, or success, or strength of conviction, or inability to think otherwise.

The theory of objective truth leads to a very different attitude. This may be seen from the fact that it allows us to make assertions such as the following: a theory may be true even though nobody believes it, and even though we have no reason to think that it is true; and another theory may be false even though we have comparatively good reasons for accepting it.

Clearly, these assertions would appear to be self-contradictory from the point of view of any subjective or epistemic theory of truth. But within the objective theory, they are not only consistent, but quite obviously true.

A similar assertion which the objective correspondence theory would make quite natural is this: even if we hit upon a true theory, we shall as a rule be merely guessing, and it may well be impossible for us to know that it *is* true.

An assertion like this was made, apparently for the first time, by Xenophanes [11] who lived 2,500 years ago; which shows that the objective theory of truth is very old indeed—antedating Aristotle, who also held it. But only with Tarski's work has the suspicion been removed that the objective theory of truth as correspondence to the facts may be either self-contradictory (because of the paradox of the liar), or empty (as Ramsey suggested), or barren, or at the very least redundant, in the sense that we can do without it.

In my theory of scientific progress I might perhaps do without it, up to a point. Since Tarski, however, I no longer see any reason why I should try to avoid it. And if we wish to elucidate the difference between pure and applied science, between the search for knowledge and the search for power or for powerful instruments, then we cannot do without it. For the difference is that, in the search for knowledge, we are out to find true theories, or at least theories which are nearer than others to the truth—which correspond better to the facts; whereas in the search for powerful instruments we are, in many cases, quite well served by theories which are known to be false.[12]

So one great advantage of the theory of objective or absolute truth is that it allows us to say—with Xenophanes—that we search for truth, but may not know when we have found it; that we have no criterion of truth, but are nevertheless guided by the idea of truth as a *regulative principle* (as Kant or Peirce might have said); and that, though there are no general criteria by which we can recognize truth—except perhaps tautological truth—there are criteria of progress towards the truth (as I shall explain presently).

The status of truth in the objective sense, as correspondence to the facts, and its role as a regulative principle, may be compared to that of a mountain peak usually wrapped in clouds. A climber may not merely have difficulties in getting there—he may not know when he gets there, because he may be unable to distinguish, in the clouds, between the main summit and a subsidiary peak. Yet this does not affect the objective existence of the summit; and if the climber tells us 'I doubt whether I reached the actual summit', then he does, by implication, recognize the objective existence of the summit. The very idea of error, or of doubt (in its normal straightforward sense) implies the idea of an objective truth which we may fail to reach.

Though it may be impossible for the climber ever to make sure that he has reached the summit, it will often be easy for him to realize that he has not reached it (or not yet reached it); for example, when he is turned back by an overhanging wall. Similarly, there will be cases when we are quite sure that we have not reached the truth. Thus while coherence, or consistency, is no criterion of truth, simply because even demonstrably consistent systems may be false in fact, incoherence or inconsistency do establish falsity; so, if we are lucky, we may discover the falsity of some of our theories.[13]

In 1944, when Tarski published the first English outline of his investigations into the theory of truth (which he had published in Poland in 1944), few philosophers would have dared to make assertions like those of Xenophanes; and it is interesting that the volume in which Tarski's paper was published also contained two subjectivist papers on truth.[14]

Though things have improved since then, subjectivism is still rampant in the philosophy of science, and especially in the field of probability theory. The subjectivist theory of probability, which interprets degrees of probability as degrees of rational belief, stems directly from the subjectivist approach to truth—especially from the coherence theory. Yet it is still embraced by philosophers who have accepted Tarski's theory of truth. At least some of them, I suspect, have turned to probability theory in the hope that it would give them what they had originally expected from a subjectivist or epistemological theory of the attainment of truth *through verification;* that is, a theory of rational and justifiable belief, based upon observed instances.[15]

It is an awkward point in all these subjective theories that they are irrefutable (in the sense that they can too easily evade any criticism). For it is always possible to uphold the view that everything we say about the world, or everything we print about logarithms, should be replaced by a belief-statement. Thus we may replace the statement 'Snow is white' by 'I believe that snow is white' or perhaps even by 'In the light of all the available evidence I believe that it is rational to believe that snow is white'. That we can (in a way) 'replace' assertions about the objective world by one of these subjectivist circumlocutions is trivial, although in the case of the assertions found in logarithm tables—which might well be produced by machines—somewhat unconvincing. (It may be mentioned in passing that the subjective interpretation of logical probability links these subjectivist replacements, exactly as in the case of the coherence theory of truth, with an approach which, on closer analysis, turns out to be essentially 'syntactic' rather than 'semantic'—though it may be presented within a 'semantical system'.)

It may be useful to sum up the relationships between the objective and subjective theories of scientific knowledge with the help of a little table:

Objective or Logical or Ontological Theories	Subjective or Psychological or Epistemological Theories
truth as correspondence with the facts	*truth as property of our state of mind—or knowledge or belief*
objective probability (inherent in the situation, and testable by statistical tests)	*subjective probability (degree of rational belief based upon our total knowledge)*
objective randomness (statistically testable)	*lack of knowledge*
equiprobability (physical or situational symmetry)	*lack of knowledge*

In all these cases I am inclined to say not only that these two approaches should be distinguished, but also that the subjectivist approach should be discarded as a lapse, as based on a mistake—though perhaps a tempting mistake. There is, however, a similar table in which the epistemological (right hand) side is not based on a mistake.

truth	*conjecture*
testability	*empirical test*
explanatory or predictive power	*degree of corroboration (that is, report of the results of tests)*
'verisimilitude'	

3. TRUTH AND CONTENT: VERISIMILITUDE VERSUS PROBABILITY

IX

Like many other philosophers I am at times inclined to classify philosophers as belonging to two main groups—those with whom I disagree, and those who agree with me. I also call them the verificationists or the justificationist philosophers of knowledge (or of belief), and the falsificationists or fallibilists or critical philosophers of knowledge (or of conjectures). I may mention in passing a third group with whom I also disagree. They may be called the disappointed justificationists—the irrationalists and sceptics.

The members of the first group—the verificationists or justificationists—hold, roughly speaking, that whatever cannot be supported by positive reasons is unworthy of being believed, or even of being taken into serious consideration.

On the other hand, the members of the second group—the falsificationists or fallibilists—say, roughly speaking, that what cannot (at present) in principle be overthrown by criticism is (at present) unworthy of being seriously considered; while what can in principle be so overthrown and yet resists all our critical efforts to do so may quite possibly be false, but is at any rate not unworthy of being seriously considered and perhaps even of being believed—though only tentatively.

Verificationists, I admit, are eager to uphold the most important tradition of rationalism—the fight of reason against superstition and arbitrary authority. For they demand that we should accept a belief *only if it can be justified by positive evidence;* that is to say, shown to be true, or at least, to be highly probable. In other words, they demand that we should accept a belief only if it can be *verified,* or probabilistically *confirmed.*

Falsificationists (the group of fallibilists to which I belong) believe—as most irrationalists also believe—that they have discovered logical arguments which show that the programme of the first group cannot be carried out: that we can never give positive reasons which justify the belief that a theory is true. But, unlike irrationalists, we falsificationists believe that we have also discovered a way to realize the old ideal of distinguishing rational science from various forms of superstition, in spite of the breakdown of the original inductivist or justificationist programme. We hold that this ideal can be realized, very simply, by recognizing that the rationality of science lies not in its habit of appealing to empirical evidence in support of its dogmas—astrologers do so too—but solely in the *critical approach:* in an attitude which, of course, involves the critical use, among other arguments, of empirical evidence (especially in refutations). For us, therefore, science has nothing to do with the quest for certainty or probability or reliability. We are not interested in establishing scientific theories as secure, or certain, or probable. Conscious of our fallibility we are only interested in criticizing them and testing them, hoping to find out where we are mistaken; of learning from our mistakes; and, if we are lucky, of proceeding to better theories.

Considering their views about the positive or negative function of argument in science, the first group—the justificationists—may be also nicknamed the 'positivists' and the second—the group to which I belong—the critics or the 'negativists'. These are, of course, mere nicknames. Yet they may perhaps suggest some of the reasons why some people believe that only the positivists or verificationists are seriously interested in truth and in the search for truth, while we, the critics or negativists, are flippant about the search for truth, and addicted to barren and destructive criticism and to the propounding of views which are clearly paradoxical.

This mistaken picture of our views seems to result largely from the adoption of a justification programme, and of the mistaken subjectivist approach to truth which I have described.

For the fact is that we too see science as the search for truth, and that, at least since Tarski, we are no longer afraid to say so. Indeed, it is only with respect to this aim, the discovery of truth, that we can say that though we are fallible, we hope to learn from our mistakes. It is only the idea of truth which allows us to speak sensibly of mistakes and of rational criticism, and which makes rational discussion possible—that is to say, critical discussion in search of mistakes with the serious purpose of eliminating as many of these mistakes as we can, in order to get nearer to the truth. Thus the very idea of error—and of fallibility—involves the idea of an objective truth as the standard of which we may fall short. (It is in this sense that the idea of truth is a *regulative* idea.)

Thus we accept the idea that the task of science is the search for truth, that is, for true theories (even though as Xenophanes pointed out we may never get them, or know them *as true* if we get them). Yet we also stress that *truth is not the only aim of science.* We want more than mere truth: what we look for is *interesting truth*—truth which is hard to come by. And in the natural sciences (as distinct from mathematics) what we look for is truth which has a high degree of explanatory power, in a sense which implies that it is logically improbable truth.

For it is clear, first of all, that we do not merely want truth—we want more truth, and new truth. We are not content with 'twice two equals four', even though it is true: we do not resort to reciting the multiplication table if we are faced with a difficult problem in topology or in physics. Mere truth is not enough; what we look for are *answers to our problems.* The point has been well put by the German humorist and poet Busch, of Max-and-Moritz fame, in a little nursery rhyme—I mean a rhyme for the epistemological nursery: [16]

> Twice two equals four: 'tis true,
> But too empty, and too trite.

What I look for is a clue
To some matters not so light.

Only if it is an answer to a problem—a difficult, a fertile problem, a problem of some depth—does a truth, or a conjecture about the truth, become relevant to science. This is so in pure mathematics, and it is so in the natural sciences. And in the latter, we have something like a logical measure of the depth or significance of the problem in the increase of logical improbability or explanatory power of the proposed new answer, as compared with the best theory or conjecture previously proposed in the field. This logical measure is essentially the same thing which I have described above as the logical criterion of potential satisfactoriness and of progress.

My description of this situation might tempt some people to say that truth does not, after all, play a very big role with us negativists even as a regulative principle. There can be no doubt, they will say, that negativists (like myself) much prefer an attempt to solve an interesting problem by a bold conjecture, *even if it soon turns out to be false,* to any recital of a sequence of true but uninteresting assertions. Thus it does not seem, after all, as if we negativists had much use for the idea of truth. Our ideas of scientific progress and of attempted problem-solving do not seem very closely related to it.

This, I believe, would give quite a mistaken impression of the attitude of our group. Call us negativists, or what you like: but you should realize that we are as much interested in truth as anybody—for example, as the members of a court of justice. When the judge tells a witness that he should speak 'The truth, the *whole truth,* and nothing but the truth', then what he looks for is as much of the *relevant truth* as the witness may be able to offer. A witness who likes to wander off into irrelevancies is unsatisfactory as a witness, even though these irrelevancies may be truisms, and thus part of 'the whole truth'. It is quite obvious that what the judge—or anybody else—wants when he asks for 'the whole truth' is as much *interesting and relevant* true information as can be got; and many perfectly candid witnesses have failed to disclose some important information simply because they were unaware of its relevance to the case.

Thus when we stress, with Busch, that we are not interested in mere truth but in interesting and relevant truth, then, I contend, we only emphasize a point which everybody accepts. And if we are interested in bold conjectures, even if these should soon turn out to be false, then this interest is due to our methodological conviction that only with the help of such bold conjectures can we hope to discover interesting and relevant truth.

There is a point here which, I suggest, it is the particular task of the logician to analyse. 'Interest', or 'relevance', in the sense here intended, can be *objectively* analysed; it is relative to our problems; and it depends on the explanatory power, and thus on the content or improbability, of the information. The measures alluded to earlier (and developed in the *Addenda* to this volume) are precisely such measures as take account of some *relative content* of the information—its content relative to a hypothesis or to a problem.

I can therefore gladly admit that falsificationists like myself much prefer an attempt to solve an interesting problem by a bold conjecture, *even* (*and especially*) *if it soon turns out to be false,* to any recital of a sequence of irrelevant truisms. We prefer this because we believe that this is the way in which we can learn from our mistakes; and that in finding that our conjecture was false, we shall have learnt much about the truth, and shall have got nearer to the truth.

I therefore hold that both ideas—the idea of truth, in the sense of correspondence with the facts, and the idea of content (which may be measured by the same measure as testability)—play about equally important roles in our considerations, and that both can shed much light on the idea of progress in science.

X

Looking at the progress of scientific knowledge, many people have been moved to say that even though we do not know how near or how far we are from the truth, we can, and often do, *approach more and more closely to the truth.* I myself have sometimes said such things in the past, but always with a twinge of bad conscience. Not that I believe in being over-fussy about what we say: as long as we speak as clearly as we can, yet do not pretend that what we are saying is clearer than it is, and as long as we do not try to derive apparently exact consequences from dubious or vague premises, there is no harm whatever in occasional vagueness, or in voicing every now and then our feelings and general intuitive impressions about things. Yet whenever I used to write, or to say, something about science as getting nearer to the truth, or as a kind of approach to truth, I felt that I really ought to be writing 'Truth', with a capital 'T', in order to make quite clear that a vague and highly metaphysical notion was involved here, in contradistinction to Tarski's 'truth' which we can with a clear conscience write in the ordinary way with small letters.[17]

It was only quite recently that I set myself to consider whether the idea of truth involved here was really so dangerously vague and metaphysical after all. Almost at once I found that it was not, and that there was no particular difficulty in applying Tarski's fundamental idea to it.

For there is no reason whatever why we should not say that one theory corresponds better to the facts than another. This simple initial step makes everything clear: there really is no barrier here between what at first sight appeared to be Truth with a capital 'T' and truth in a Tarskian sense.

But can we really speak about *better* correspondence? Are there such things as *degrees* of truth? Is it not dangerously misleading to talk as if Tarskian truth were located somewhere in a kind of metrical or at least topological space so that we can sensibly say of two theories—say an earlier theory t_1 and a later theory t_2, that t_2 has superseded t_1, or progressed beyond t_1, by approaching more closely to the truth than t_1?

I do not think that this kind of talk is at all misleading. On the contrary, I believe that we simply cannot do without something like this idea of a better or worse approximation to truth. For there is no doubt whatever that we can say, and often want to say, of a theory t_2 that it corresponds better to the facts, or that as far as we know it seems to correspond better to the facts, than another theory t_1.

I shall give here a somewhat unsystematic list of six types of case in which we should be inclined to say of a theory t_1 that it is superseded by t_2 in the sense that t_2 seems—as far as we know—to correspond better to the facts than t_1, in some sense or other.

(1) t_2 makes more precise assertions than t_1, and these more precise assertions stand up to more precise tests.
(2) t_2 takes account of, and explains, more facts than t_1 (which will include for example the above case that, other things being equal, t_2's assertions are more precise).
(3) t_2 desribes, or explains, the facts in more detail than t_1.
(4) t_2 has passed tests which t_1 has failed to pass.
(5) t_2 has suggested new experimental tests, not considered before t_2 was designed (and not suggested by t_1, and perhaps not even applicable to t_1); and t_2 has passed these tests.
(6) t_2 has unified or connected various hitherto unrelated problems.

If we reflect upon this list, then we can see that the *contents* of the theories t_1 and t_2 play an important role in it. (It will be remembered that the *logical content* of a statement or a theory a is the class of all statements which follow logically from a, while I have defined the *empirical content* of a as the class of all basic statements which contradict *a*.[18]) For in our list of six cases, the empirical content of theory t_2 exceeds that of theory t_1.

This suggests that we combine here the ideas of truth and of content into one—the idea of a degree of better (or worse) correspondence to truth or of greater (or less) likeness or similarity to truth; or to use a term already mentioned above (in contradistinction to probability) the idea of (degrees of) *verisimilitude*.

It should be noted that the idea that every statement or theory is not only either true or false but has, independently of its truth value, some degree of verisimilitude, does not give rise to any multivalued logic—that is, to a logical system with more than two truth values, true and false; though some of the things the defenders of multi-valued logic are hankering after seem to be realized by the theory of verisimilitude (and related theories alluded to in section 3 of the *Addenda* to this volume).

XI

Once I had seen the problem it did not take me long to get to this point. But strangely enough, it took me a long time to put two and two together, and to proceed from here to a very simple *definition of verisimilitude* in terms of truth and of content. (We can use either logical or empirical content, and thus obtain two closely related ideas of verisimilitude which however merge into one if we consider here only empirical theories, or empirical aspects of theories.)

Let us consider the *content* of a statement *a*; that is, the class of all the logical consequences of *a*. If *a* is true, then this class can consist only of true statements, because truth is always transmitted from a premise to all its conclusions. But if *a* is false, then its content will always consist of both true and false conclusions. (Example: 'It always rains on Sundays' is false, but its conclusion that it rained last Sunday happens to be true.) Thus whether a statement is true or false, *there may be more truth, or less truth, in what it says,* according to whether its content consists of a greater or a lesser number of true statements.

Let us call the class of the true logical consequences of *a* the 'truth-content' of *a* (a German term *'Wahrheitsgehalt'*—reminiscent of the phrase 'there is truth in what you say'—of which 'truth-content' may be said to be a translation, has been intuitively used for a long time); and let us call the class of the false consequences of *a*—but only these—the 'falsity-content' of *a*. (The 'falsity-content' is not, strictly speaking, a 'content', because it does not contain any of the true conclusions of the false statements which form its elements. Yet it is possible—see the Addenda—to define its *measure* with the help of two contents.) These terms are precisely as objective as the terms 'true' or 'false' and 'content' themselves. Now we can say:

Assuming that the truth-content and the falsity-content of two theories t_1 and t_2 are comparable, we can say that t_2 is more closely similar to the truth, or corresponds better to the facts, than t_1, if and only if either

(*a*) *the truth-content but not the falsity-content of t_2 exceeds that of t_1,*
(b) *the falsity-content of t_1, but not its truth-content, exceeds that of t_2,*

If we now work with the (perhaps fictitious) assumption that the content and truth-content of a theory a are in principle *measurable,* then we can go slightly beyond this definition and can define $Vs(a)$, that is to say, a measure of the *verisimilitude* or *truthlikeness* of a. The simplest definition will be

$$Vs(a) = Ct_T(a) - Ct_F(a)$$

where $Ct_T(a)$ is a measure of the truth-content of a, and $Ct_F(a)$ is a measure of the falsity-content of a. A slightly more complicated but in some respects preferable definition will be found in section 3 of the *Addenda* to the present volume.

It is obvious that $Vs(a)$ satisfies our two demands, according to which $Vs(a)$ should increase

(*a*) if $Ct_T(a)$ increases while $Ct_F(a)$ does not, and
(*b*) if $Ct_F(a)$ decreases while $Ct_T(a)$ does not.

Some further considerations of a slightly technical nature and the definitions of $Ct_T(a)$ and especially $Ct_F(a)$ and $Vs(a)$ will be found in the *Addenda.* Here I want only to discuss three nontechnical points.

XII

The first point is this. Our idea of approximation to truth, or of verisimilitude, has the same objective character and the same ideal or regulative character as the idea of objective or absolute *truth.* It is *not an epistemological or an epistemic idea*—no more than truth or content. (In Tarski's terminology, it is obviously a 'semantic' idea, like truth, or like logical consequence, and, therefore, content.) Accordingly, we have here again to distinguish between the question 'What do you intend to say if you say that the theory t_2 has a higher degree of verisimilitude than the theory t_1?', and the question 'How do you know that the theory t_2 has a higher degree of verisimilitude than the theory t_1?'

We have so far answered only the first of these questions. The answer to the second question depends on it, and it is exactly analogous to the following (absolute rather than comparative) question about truth: 'I do *not* know—I only guess. But I can examine my guess critically, and if it withstands severe criticism, then this fact may be taken as a good critical reason in favour of it.'

My second point is this. Verisimilitude is so defined that maximum verisimilitude would be achieved only by a theory which is not only true, but completely comprehensively true: if it corresponds to *all* facts, as it were, and, of course, only to *real* facts. This is of course a much more remote and unattainable ideal than a mere correspondence with *some* facts (as in, say, 'Snow is usually white').

But all this holds only for the maximum degree of verisimilitude, and not for the *comparison of theories with respect to their degree of verisimilitude.* This comparative use of the idea is its main point; and the idea of a higher or lower degree of verisimilitude seems less remote and more applicable and therefore perhaps more important for the analysis of scientific methods than the—in itself much more fundamental—idea of absolute truth itself.

This leads me to my third point. Let me first say that I do not suggest that the explicit introduction of the idea of verisimilitude will lead to any changes in the theory of method. On the contrary, I think that my theory of testability or corroboration by empirical tests is the proper methodological counterpart to this new metalogical idea. The only improvement is one of clarification. Thus I have often said that we prefer the theory t_2 which has passed certain severe tests to the theory t_1 which has failed these tests, because a false theory is certainly worse than one which, for all we know, may be true.

To this we can now add that even after t_2 has been refuted in its turn, we can still say that it is better than t_1, for although both have been shown to be false, the fact that t_2 has withstood tests which t_1 did not pass may be a good indication that the falsity-content of t_1 exceeds that of t_2 while its truth-content does not. Thus we may still give preference to t_2, even after its falsification, because we have reason to think that it agrees better with the facts than did t_1.

All cases where we accept t_2 because of experiments which were crucial between t_2 and t_1 seem to be of this kind, and especially all cases where the experiments were found by trying to think out, with the help of t_2, cases where t_2 leads to other results than did t_1. Thus Newton's theory allowed us to predict some deviations from Kepler's laws. Its success in this field established that it did not fail in cases which refuted Kepler's: at least the now known falsity-content of Kepler's theory was not part of Newton's, while it was pretty clear that the truth-content could not have shrunk, since Kepler's theory followed from Newton's as a 'first approximation.'

Similarly, a theory t_2 which is more precise than t_1 can now be shown to have—always provided its falsity content does not exceed that of t_1—a higher degree of verisimilitude than t_1. The same will hold for t_2 whose numerical assertions, though false, come nearer to the true numerical values than those of t_1.

Ultimately, the idea of verisimilitude is most important in cases where we know that we have to work with theories which are *at best* approximations—that is to say, theories of which we actually know that they cannot be true. (This is often the case in the social sciences.) In these cases we can still speak of better or worse approximations to the truth (and we therefore do not need to interpret these cases in an instrumentalist sense).

XIII

It always remains possible, of course, that we shall make mistakes in our relative appraisal of two theories, and the appraisal will often be a controversial matter. This point can hardly be overemphasized. Yet it is also important that in principle, and as long as there are no revolutionary changes in our background knowledge, the relative appraisal of our two theories, t_1 and t_2, will remain stable. More particularly, our preferences need not change, as we have seen, if we eventually refute the better of the two theories. Newton's dynamics, for example, even though we may regard it as refuted, has of course maintained its superiority over Kepler's and Galileo's theories. The reason is its greater content or explanatory power. Newton's theory continues to explain more facts than did the others; to explain them with greater precision; and to unify the previously unconnected problems of celestial and terrestrial mechanics. The reason for the stability of relative appraisals such as these is quite simple: the logical relation between the theories is of such a character that, first of all, there exist with respect to them those crucial experiments, and these, when carried out, went against Newton's predecessors. And secondly, it is of such a character that the later refutations of Newton's theory could not support the older theories: they either did not affect them, or (as with the perihelion motion of Mercury) they could be claimed to refute the predecessors also.

I hope that I have explained the idea of better agreement with the facts, or of degrees of verisimilitude, sufficiently clearly for the purpose of this brief survey.

XIV

A brief remark on the early history of the confusion between verisimilitude and probability may perhaps be appropriate here.

As we have seen, progress in science means progress towards more interesting, less trivial, and therefore less 'probable' theories (where 'probable' is taken in any sense, such as *lack* of content, or statistical frequency, that satisfies the calculus of probability) and this means, as a rule, progress towards less familiar and less comfortable or plausible theories. Yet the idea of greater verisimilitude, of a better approximation to the truth, is usually confused, intuitively, with the totally different idea of probability (in its various senses of 'more likely than not', 'more often than not', 'seems likely to be true', 'sounds plausible', 'sounds convincing'). The confusion is a very old one. We have only to remember some of the other words for 'probable', such as 'likely' which comes originally from 'like the truth' or 'verisimilar' (*'eoikotōs', 'eikotōs, 'eikos,'* etc., in Greek; *'verisimilis'* in Latin; *'wahrscheinlich'* in German) in order to see some of the traces, and perhaps some of the sources, of this confusion.

Two at least of the earliest of the Presocratic philosophers used *'eoikota'* in the sense of 'like the truth' or 'similar to the truth'. Thus we read in Xenophanes (DK, B 35): 'These things, let us suppose, are like the truth.'

It is fairly clear that verisimilitude or truthlikeness is meant here, rather than probability or degree of incomplete certainty. (Otherwise the words 'let us suppose' or 'let it be conjectured' or 'let it be imagined' would be redundant, and Xenophanes would have written something like, 'These things, let it be *said,* are probable.')

Using the same word (*'eoikota'*), Parmenides wrote (DK, B 8, 60): [19] 'Now of this world thus arranged to seem wholly like truth I shall tell you . . .'

Yet already in the same generation or the next, Epicharmus, in a criticism of Xenophanes, seems to have used the word *'eikotōs'* in the sense of 'plausible', or something like it (DK, 21 A 15); though the possibility cannot be excluded that he may have used it in the sense of 'like the truth,' and that it was Aristotle (our source is *Met.,* 1010a4) who read it in the sense of 'plausible' or 'likely.' Some three generations later, however, *'eikos'* is used quite unambiguously in the sense of 'likely' or 'probable' (or perhaps even of 'more frequently than not') by the sophist Antiphon when he writes (DK, B 60): 'If one begins a thing well it is likely to end well.'

All this suggests that the confusion between verisimilitude and probability goes back almost to the beginning of Western philosophy: and this is understandable if we consider that Xenophanes stressed the fallibility of our knowledge which he described as uncertain guesswork and at best "like the truth." This phrase, it seems, lent itself to misinterpretation as 'uncertain and at best of some fair degree of certainty'—that is, 'probable.'

Xenophanes himself seems to have distinguished clearly between degrees of certainty and degrees of truthlikeness. This emerges from another fragment (quoted above towards the end of the chapter 5, p. 153) which says that even if by chance we were to hit upon, and pronounce, the final truth (that is, we may add, perfect truthlikeness), we should not know it. Thus great uncertainty is compatible with greatest truthlikeness.

I suggest that we return to Xenophanes and re-introduce a clear distinction between *verisimilitude* and *probability* (using this latter term in the sense laid down by the calculus of probability).

The differentiation between these two ideas is the more important as they have become confused; because both are closely related to the idea of truth, and both introduce the idea of an approach to truth by degrees. Logical probability (we do not discuss here physical probability) represents the idea of approaching logical certainty, or tautological truth, through a gradual diminution of informative content. Verisimilitude, on the other hand, represents the idea of approaching comprehensive truth. It thus combines truth and content while probability combines truth with lack of content.[20]

The feeling that it is absurd to deny that science aims at probability stems, I suggest, from a misguided 'intuition'—from the intuitive confusion between the two notions of verisimilitude and of probability which, as it now turns out, are utterly different.

4. BACKGROUND KNOWLEDGE AND SCIENTIFIC GROWTH

XV

People involved in a fruitful critical discussion of a problem often rely, if only unconsciously, upon two things: the acceptance by all parties of the common aim of getting at the truth, or at least nearer to the truth, and a considerable amount of common background knowledge. This does not mean that either of these two things is an indispensible basis of every discussion, or that these two things are themselves *'a priori'*, and cannot be critically discussed in their turn. It only means that criticism starts from nothing, even though every one of its starting points may be challenged, one at a time, in the course of the critical debate.

Yet though every one of our assumptions may be challenged, it is quite impracticable to challenge all of them at the same time. Thus all criticism must be piecemeal (as against the holistic view of Duhem and of Quine); which is only another way of saying that the fundamental maxim of every critical discussion is that we should stick to our problem, and that we should subdivide it, if practicable, and try to solve no more than one problem at a time, although we may, of course, always proceed to a subsidiary problem, or replace our problem by a better one.

While discussing a problem we always accept (if only temporarily) all kinds of things as *unproblematic:* they constitute for the time being, and for the discussion of this particular problem, what I call our *background knowledge.* Few parts of this background knowledge will appear to us in all contexts as absolutely unproblematic, and any particular part of it *may* be challenged at any time, especially if we suspect that its uncritical acceptance may be responsible for some of our difficulties. But almost all of the vast amount of background knowledge which we constantly use in any informal discussion will, for practical reasons, necessarily remain unquestioned; and the misguided attempt to question it all—that is to say, *to start from scratch*—can easily lead to the breakdown of a critical debate. (Were we to start the race where Adam started, I know of no reason why we should get any further than Adam did.)

XVI

The fact that, as a rule, we are at any given moment taking a vast amount of traditional knowledge for granted (for almost all our knowledge is traditional) creates no difficulty for the falsificationist or fallibilist. For he does not *accept* this background knowledge; neither as established nor as fairly certain, nor yet as probable. He knows that even its tentative acceptance is risky, and stresses that every bit of it is open to criticism, even though only in a piecemeal way. We can never be certain that we shall challenge the right bit; but since our quest is not for certainty, this does not matter. It will be noticed that this remark contains my answer to Quine's holistic view of empirical tests; a view which Quine formulates (with reference to Duhem), by asserting that our statements about the external world face the tribunal of sense experience not individually but only as a corporate body.[21] Now it has to be admitted that we can often test only a large chunk of a theoretical system, and sometimes perhaps only the whole system, and that, in these cases, it is sheer guesswork which of its ingredients should be held responsible for any falsification; a point which I have tried to emphasize—also with reference to Duhem—for a long time past.[22] Though this argument may turn a verificationist into a sceptic, it does not affect those who hold that all our theories are guesses anyway.

This shows that the holistic view of tests, even if it were true, would not create a serious difficulty for the fallibilist and falsificationist. On the other hand, it should be said that the holistic argument goes much too far. It is possible in quite a few cases to find which hypothesis is responsible for the refutation; or in other words, which part, or group of hypotheses, was necessary for the derivation of the refuted prediction. The fact that such logical dependencies may be discovered is established by the practice of *independence proofs* of axiomatized systems; proofs which show that certain axioms of an axiomatic system cannot be derived from the rest. The more simple of these proofs consist in the construction, or rather in the discovery, of a *model*—a set of things, relations, operations, or functions—which satisfies all the axioms except the *one* whose independence is to be shown: for this one axiom—and therefore for the theory as a whole—the model constitutes a counter example.

Now let us say that we have an axiomatized theoretical system, for example of physics, which allows us to predict that certain things do not happen, and that we discover a counter example. There is no reason whatever why this counter example may not be found to satisfy most of our axioms or even all our axioms except one whose independence would be thus established. This shows that the holistic dogma of the 'global' character of all tests or counter examples is untenable. And it explains why, even without axiomatizing our physical theory, we may well have an inkling of what has gone wrong with our system.

This, incidentally, speaks in favour of operating, in physics, with highly analysed theoretical systems—that is, with systems which, even though they may fuse all the hypotheses into one, allow us to separate various groups of hypotheses, each of which may become an object of refutation by counter examples. (An excellent recent example is the rejection, in atomic theory, of the law of parity;

another is the rejection of the law of commutation for conjugate variables, prior to their interpretation at matrices, and to the statistical interpretation of these matrices.)

XVII

One fact which is characteristic of the situation in which the scientist finds himself is that we constantly add to our background knowledge. If we discard some parts of it, others, closely related to them, will remain. For example, even though we may regard Newton's theory as refuted—that is, his system of ideas, and the formal deductive system which derives from it—we may still assume, as part of our background knowledge, the approximate truth, within limits, of its quantitative formulae.

The existence of this background knowledge plays an important role in one of the arguments which support (I believe) my thesis that the rational and empirical character of science would vanish if it ceased to progress. I can sketch this argument here only in its barest outline.

A serious empirical test always consists in the attempt to find a refutation, a counter example. In the search for a counter example, we have to use our background knowledge; for we always try to refute first the *most risky* predictions, the '*most unlikely . . .* consequences' (as Peirce already saw [23]); which means that we always look in the *most probable kinds* of places for the *most probable* kinds of counter examples—most probable in the sense that we should expect to find them in the light of our background knowledge. Now if a theory stands up to many such tests, then, owing to the incorporation of the results of our tests into our background knowledge, there may be, after a time, no places left where (in the light of our new background knowledge) counter examples can with a high probability be expected to occur. But this means that the degree of severity of our test declines. This is also the reason why an often repeated test will no longer be considered as significant or as severe: there is something like a law of diminishing returns from repeated tests (as opposed to tests which, in the light of our background knowledge, are of a *new kind,* and which therefore may still be felt to be significant). These are facts which are inherent in the knowledge-situation; and they have often been described—especially by John Maynard Keynes and by Ernest Nagel—as difficult to explain by an inductivist theory of science. But for us it is all very easy. And we can even explain, by a similar analysis of the knowledge-situations, why the empirical character of a very successful theory always grows stale, after a time. We may then feel (as Poincaré did with respect to Newton's theory) that the theory is nothing but a set of implicit definitions or conventions—until we progress again and, by refuting it, incidentally re-establish its lost empirical character. (*De mortuis nil nisi bene:* once a theory is refuted, its empirical character is secure and shines without blemish.)

5. THREE REQUIREMENTS FOR THE GROWTH OF KNOWLEDGE

XVIII

But let us return again to the idea of getting nearer to the truth—to the search for theories which agree better with the facts (as indicated by the list of six comparisons in section X above).

What is the general problem situation in which the scientist finds himself? He has before him a scientific problem: he wants to find a new theory capable of explaining certain experimental facts; facts which the earlier theories successfully explained; others which they could not explain; and some by which they were actually falsified. The new theory should also resolve, if possible, some theoretical difficulties (such as how to dispense with certain *ad hoc* hypotheses, or how to unify two theories). Now if he manages to produce a theory which is a solution to all these problems, his achievement will be a very great.

Yet it is not enough. I have been asked, 'What more do you want?' My answer is that there are many more things which I want; or rather, which I think are required by the logic of the general problem situation in which the scientist finds himself; by the task of getting nearer to the truth. I shall confine myself here to the discussion of three such requirements.

The first requirement is this. The new theory should proceed from some *simple, new, and powerful, unifying idea* about some connection or relation (such as gravitational attraction) between hitherto unconnected things (such as planets and apples) or facts (such as inertial and gravitational mass) or new 'theoretical entities' (such as field and particles). This *requirement of simplicity* is a bit vague, and it seems difficult to formulate it very clearly. It seems to be intimately connected with the idea that our theories should describe the structural properties of the world—an idea which it is hard to think out fully without getting involved in an infinite regress. (This is so because any idea of a particular structure of the world—unless, indeed, we think of a purely *mathematical* structure—already presupposes a universal theory; for example, explaining the laws of chemistry by interpreting molecules as structures of atoms, or of subatomic particles, presupposes the idea of universal laws that regulate the properties and the behaviour of the atoms, or of the particles.) Yet one important ingredient in the idea of simplicity can be logically analysed. It is the idea of testability.[24] This leads us immediately to our second requirement.

For, secondly, we require that the new theory should be *independently testable.*[25] That is to say, apart from explaining all the *explicanda* which the new theory was designed to explain, it must have new and testable consequences (preferably consequences of a *new kind*); it must lead to the prediction of phenomena which have not so far been observed.

This requirement seems to me indispensable since without it our now theory might be *ad hoc;* for it is always possible to produce a theory to fit any given set of explicanda. Thus our two first requirements are needed in order to restrict the range of our choice among the possible solutions (many of them uninteresting) of the problem in hand.

If our second requirement is satified then our new theory will represent a potential step forward, whatever the outcome of the new tests may be. For it will be better testable than the previous theory: the fact that it explains all the explicanda of the previous theory, and that, in addition, it gives rise to new tests, suffices to ensure this.

Moreover, the second requirement also ensures that our new theory will, to some extent, be fruitful as an instrument of exploration. That is to say, it will suggest to us new experiments, and even if these should at once lead to the refutation of the theory, our factual knowledge will have grown through the unexpected results of the new experiments. Moreover, they will confront us with new problems to be solved by new explanatory theories.

Yet I believe that there must be a third requirement for a good theory. It is this. We require that the theory should pass some new, and severe, tests.

XIX

Clearly, this requirement is totally different in character from the previous two. These could be seen to be fulfilled, or not fulfilled, largely by analysing the old and the new theories logically. (They are 'formal requirements'.) The third requirement, on the other hand, can be found to be fulfilled, or not fulfilled, only by testing the new theory empirically. (It is a 'material requirement', a requirement of *empirical success*.)

Moreover, the third requirement clearly cannot be indispensable in the same sense as are the two previous ones. For these two are indispensable for deciding whether the theory in question should be at all accepted as a serious candidate for examination by empirical tests; or in other words, whether it is an interesting and promising theory. Yet on the other hand, some of the most interesting and most admirable theories ever conceived were refuted at the very first test. And why not? The most promising theory may fail if it makes predictions of a new kind. An example is the marvellous theory of Bohr, Kramers and Slater [26] of 1924 which, as an intellectual achievement, might perhaps even rank with Bohr's theory of the hydrogen atom of 1913. Yet unfortunately it was almost at once refuted by the facts—by the coincidence experiments of Bothe and Geiger.[27] This shows that not even the greatest physicist can anticipate the secrets of nature: his inspirations can only be guesses, and it is no fault of his, or of his theory, if it is refuted. Even Newton's theory was in the end refuted; and indeed, we hope that we shall in this way succeed in refuting, and improving upon, every new theory. And if it is refuted in the end, why not in the beginning? One might well say that it is merely a historical accident if a theory is refuted after six months rather than after six years, or six hundred years.

Refutations have often been regarded as establishing the failure of a scientist, or at least of his theory. It should be stressed that this is an inductivist error. Every refutation should be regarded as a great success; not merely a success of the scientist who refuted the theory, but also of the scientist who created the refuted theory and who thus in the first instance suggested, if only indirectly, the refuting experiment.

Even if a new theory (such as the theory of Bohr, Kramers, and Slater) should meet an early death, it should not be forgotten; rather its beauty should be remembered, and history should record our gratitude to it—for bequeathing to us new and perhaps still unexplained experimental facts and, with them, new problems; and for the services it has thus rendered to the progress of science during its successful but short life.

All this indicates clearly that our third requirement is not indispensable: even a theory which fails to meet it can make an important contribution to science. Yet in a different sense, I hold, it is indispensable none the less. (Bohr, Kramers and Slater rightly aimed at more than making an important contribution to science.)

In the first place, I contend that further progress in science would become impossible if we did not reasonably often manage to meet the third requirement; thus if the progress of science is to continue, and its rationality not to decline, we need not only successful refutations, but also positive successes. We must, that is, manage reasonably often to produce theories that entail new predictions, especially predictions of new effects, new testable consequences, suggested by the new theory and never thought of before.[28] Such a new prediction was that planets would under certain circumstances deviate from Kepler's laws; or that light, in spite of its zero mass, would prove to be subject to gravitational attraction (that is, Einstein's eclipse-effect). Another example is Dirac's prediction that there will be an anti-particle for every elementary particle. New predictions of these kinds must not only be produced, but they must also be reasonably often corroborated by experimental evidence, I contend, if scientific progress is to continue.

We do need this kind of success; it is not for nothing that the great theories of science have all meant a new conquest of the unknown, a new success in predicting what had never been thought of before. We need successes such as that of Dirac (whose anti-particles have survived the abandonment of some other parts of his theories), or that of Yukawa's meson theory. We need the success, the empirical corroboration, of some of our theories, if only in order to appreciate the significance of successful and stirring refutations (like that of parity). It seems to me quite clear that it is only through these temporary successes of our theories that we can be reasonably successful in attributing our refutations to definite portions of the theoretical maze. (For we *are* reasonably successful in this—a fact which must remain inexplicable for one who adopts Duhem's and Quine's views on the matter.) An unbroken sequence of refuted theories would soon leave us bewildered and helpless: we should have no clue about the parts of each of these theories—or of our background knowledge—to which we might, tentatively, attribute the failure of that theory.

XX

Earlier I suggested that science would stagnate, and lose its empirical character, if we should fail to obtain refutations. We can now see that for very similar reasons science would stagnate, and lose its empirical character, if we should fail to obtain verifications of new predictions; that is, if we should only manage to produce theories that satisfy our first two requirements but not the third. For suppose we were to produce an unbroken sequence of explanatory theories each of which would explain all the explicanda in its field, including the experiments which refuted its predecessors; each would also be independently testable by predicted new effects; yet each would be at once refuted when these predictions were put to the test. Thus each would satisfy our first two requirements, but all would fail to satisfy the third.

I assert that, in this case, we should feel that we were producing a sequence of theories which, in spite of their increasing degree of testability, were *ad hoc,* and that we were not getting any nearer to the truth. And indeed, this feeling may well be justified: this whole sequence of theories might easily be *ad hoc.* For if it is admitted that a theory may be *ad hoc* if it is not independently testable by experiments of a new kind but merely explains all the explicanda, including the experiments which refuted its predecessors, then it is clear that the mere fact that the theory is also independently testable cannot as such ensure that it is not *ad hoc.* This becomes clear if we consider that it is always possible, by a trivial stratagem, to make an *ad hoc* theory independently testable, *if we do not also require that it should pass the independent tests in question:* we merely have to connect it (conjunctively) in some way or other with any testable but not yet tested fantastic *ad hoc* prediction which may occur to us (or to some science fiction writer).

Thus our third requirement, like the second, is needed in order to eliminate trivial and other *ad hoc* theories.[29] But it is needed also for what seem to me even more serious reasons.

I think that we are quite right to expect, and perhaps even to hope, that even our best theories will be superseded and replaced by better ones (though we may at the same time feel the need for encouragement in our belief that we are making progress). Yet this should certainly not induce in us the attitude of merely producing theories so that they can be superseded.

For our aim as scientists is to discover the truth about our problem; and we must look at our theories as serious attempts to find the truth. If they are not true, they may be, admittedly, important stepping stones towards the truth, instruments for further discoveries. But this does not mean that we can ever be content to look at them as being *nothing but* stepping stones, *nothing but* instruments; for this would involve giving up even the view that they are instruments of theoretical *discoveries;* it would commit us to looking upon them as mere instruments for some observational or pragmatic purpose. And this approach would not, I suspect, be very successful, even from a pragmatic point of view: if we are content to look at our theories as mere stepping stones, then most of them will not even be good stepping stones. Thus we ought not to aim at theories which are mere instruments for the exploration of facts, but we ought to try to find genuine explanatory theories: we should make genuine guesses about the structure of the world. In brief, we should not be satisfied with the first two requirements.

Of course, the fulfillment of our third requirement is not in our hands. No amount of ingenuity can ensure the construction of a successful theory. We also need luck; and we also need a world whose mathematical structure is not so intricate as to make progress impossible. For indeed, if we should cease to progress in the sense of our third requirement—if we should only succeed in refuting our theories but not in obtaining some verifications of predictions of a new kind—we might well decide that our scientific problems have become too difficult for us because the structure (if any) of the world is beyond our powers of comprehension. Even in this case we might proceed, for a time, with theory construction, criticism, and falsification: the *rational* side of the method of science might, for a time, continue to function. Yet I believe that we should feel that, especially for the functioning of its *empirical* side, both kinds of successes are essential: success in refuting our theories, and success on the part of some of our theories in resisting at least some of our most determined attempts to refute them.

XXI

It may be objected that this is merely good psychological advice about the attitude which scientists ought to adopt—a matter which, after all, is their private affair—and that a theory of scientific method worthy of its name should be able to produce logical or methodological arguments in support of our third requirement. Instead of appealing to the attitude or the psychology of the scientist, our theory of science should even be able to explain his attitude, and his psychology, by an analysis of the logic of the situation in which he finds himself. There is a problem here for our theory of method.

I accept this challenge, and I shall produce three reasons: the first from the idea of truth; the second from the idea of getting nearer to the truth (verisimilitude); and the third from our old idea of independent tests and of crucial tests.

(1) The first reason why our third requirement is so important is this. We know that *if we had an independently testable theory which was, moreover, true, then it would provide us with successful predictions* (and *only* with successful ones). Successful predictions—

though they are not, of course, *sufficient* conditions for the truth of a theory—are therefore at least necessary conditions for the truth of an independently testable theory. In this sense—and only in this sense—our third requirement may even be said to be 'necessary', if we seriously accept truth as a regulative idea.

(2) The second reason is this. It is our aim to strengthen the verisimilitude of our theories, or to get nearer to the truth, then we should be anxious not only to reduce the falsity content of our theories but also to strengthen their truth content.

Admittedly this may be done in certain cases simply by constructing the new theory in such a way that the refutations of the old theory are explained ('saving the phenomena', in this case the refutations). But there are other cases of scientific progress—cases whose existence shows that this way of increasing the truth content is not the only possible one.

The cases I have in mind are cases in which there was no refutation. Neither Galileo's nor Kepler's theories were refuted before Newton: what Newton tried to do was to explain them from more general assumptions, and thus to unify two hitherto unrelated fields of inquiry. The same may be said of many other theories: Ptolemy's system was not refuted when Copernicus produced his. And though there was, before Einstein, the puzzling experiment of Michelson and Morley, this had been successfully explained by Lorentz and Fitzgerald.

It is in cases like these that *crucial experiments* become decisively important. We have no reason to regard the new theory as better than the old theory—to believe that it is nearer to the truth—until we have derived from the new theory *new predictions* which were unobtainable from the old theory (the phases of Venus, the perturbations, the mass-energy equation) and until we have found that these new predictions were successful. For it is only this success which shows that the new theory had true consequences (that is, a truth content) where the old theories had false consequences (that is, a falsity content).

Had the new theory been refuted in any of these crucial experiments then we should have had no reason to abandon the old one in its favour—even if the old theory was not wholly satisfactory. (This was the fate of the Bohr-Kramers-Slater theory.)

In all these important cases we need the new theory in order to find out where the old theory was deficient. Admittedly, the situation is different if the deficiency of the old theory is already known before the new theory is invented; but logically the case has enough similarity with the other cases to regard a new theory which leads to *new* crucial experiments (Einstein's mass-energy equation) as superior to one which can only save the known phenomena (Lorentz-Fitzgerald).

(3) The same point—the importance of crucial tests—can be made without appealing to the aim of increasing the verisimilitude of a theory, by using an old argument of mine—the need to make the tests of our explanations independent.[30] This need is a result of the growth of knowledge—of the incorporation of what was new and problematic knowledge into background knowledge, with a consequent loss of explanatory power to our theories.

These are my main arguments.

XXII

Our third requirement may be divided into two parts: first we require of a good theory that it should be successful in some of its new predictions; secondly we require that it is not refuted too soon—that is, before it has been strikingly successful. Both requirements sound strange. The first because the *logical* relationship between a theory and any corroborating evidence cannot, it seems, be affected by the question whether the theory is temporally prior to the evidence. The second because if the theory is doomed to be refuted, its intrinsic value can hardly depend upon delaying the refutation.

Our explanation of this slightly puzzling difficulty is simple enough: the successful new predictions which we require the new theory to produce are identical with the crucial tests which it must pass in order to become sufficiently interesting to be accepted as an advance upon its predecessor, and to be considered worthy of further experimental examination which may eventually lead to its refutation.

Yet the difficulty can hardly be resolved by an inductivist methodology. It is therefore not surprising that inductivists such as John Maynard Keynes have asserted that the value of predictions (in the sense of facts derived from the theory but previously not known) was imaginary; and indeed if the value of a theory would lie merely in its relation to its evidential basis, then it would be logically irrelevant whether the supporting evidence precedes or follows in time the invention of the theory. Similarly the great founders of the hypothetical method used to stress the 'saving of the phenomena', that is to say, the demand that the theory should explain *known* experience. Successful *new* prediction—of new effects—seems to be a late idea, for obvious reasons; I do not know when and with whom it originated; yet the distinction between the prediction of known effects and the prediction of new effects was hardly ever made explicitly. But it seems to me quite indispensable as a part of an epistemology which views science as progressing to better and better explanatory theories; that is, not merely to instruments of exploration, but to genuine explanations.

Keynes' objection (that it is an historical accident whether this support was known before the theory was proposed, or only afterwards so that it could attain the status of a prediction) overlooks the all-important fact that it is through our theories that we learn to observe, that is to say, *to ask questions* which lead to observations and to their interpretations. This is the way our observational knowledge grows. And the questions asked are, as a rule, crucial questions which may lead to answers that decide between competing theories. It is my thesis that it is the *growth* of our knowledge, our way of choosing between theories, in a certain problem situation, which makes science rational. Now both the idea of the growth of knowledge and that of a problem situation are, at least partly, historical ideas. This explains why another *partly historical* idea—that of a genuine prediction of evidence (it may be about past facts) not known when the theory was first proposed—may play an important role here, and why the apparently irrelevant time element may become relevant.[31]

I shall now briefly sum up our results with respect to the epistemologies of the two groups of philosophers I have dealt with, the verificationists and the falsificationists.

While the verificationists or inductivists in vain try to show that scientific beliefs can be justified or, at least, established as probable (and so encourage, by their failure, the retreat into irrationalism), we of the other group have found that we do not even want a highly probable theory. Equating rationality with the critical attitude, we look for theories which, however fallible, progress beyond their predecessors; which means that they can be more severely tested, and stand up to some of the new tests. And while the verificationists laboured in vain to discover valid positive arguments in support of their beliefs, we for our part are satisfied that the rationality of a theory lies in the fact that we choose it because it is better than its predecessors; because it can be put to more severe tests; because it may even have passed them, if we are fortunate; and because it may, therefore, approach nearer to the truth.

APPENDIX: A PRESUMABLY FALSE YET FORMALLY HIGHLY PROBABLE NON-EMPIRICAL STATEMENT

In the text of this chapter I have drawn attention to the criterion of progress and of rationality based on the comparison of *degrees of testability or degrees of the empirical content or explanatory power of theories.* I did so because these degrees have been little discussed so far.

I always thought that the comparison of these degrees leads to a criterion which is more important and more realistic than the simpler *criterion of falsifiability* which I proposed at the same time, and which has been widely discussed. But this simpler criterion is also needed. In order to show the need for the falsifiability or testability criterion as a criterion of the empirical character of scientific theories, I will discuss, as an example, a simple, purely existential statement which is formulated in purely empirical terms. I hope this example will also provide a reply to the often repeated criticism that it is perverse to exclude purely existential statements from empirical science and to classify them as metaphysical.

My example consists of the following purely existential theory:

'There exists a finite sequence of Latin elegiac couplets such that, if it is pronounced in an appropriate manner at a certain time and place, this is immediately followed by the appearance of the Devil—that is to say, of a man-like creature with two small horns and one cloven hoof.'

Clearly, this untestable theory is, in principle, verifiable. Though according to my criterion of demarcation it is excluded as non-empirical and nonscientific or, if you like, metaphysical, it is not so excluded by those positivists who consider all well-formed statements and especially all verifiable ones as empirical and scientific.

Some of my positivist friends have indeed assured me that they consider my existential statement about the Devil to be empirical. It is empirical though false, they said. And they indicated that I was mistaking false empirical statements for non-empirical ones.

However, I think that the confusion, if any, is not mine. I too believe that the existential statement is false: but I believe that it is a false *metaphysical* statement. And why, I ask, should anybody who takes it for *empirical* think that it is *false?* Empirically, it is irrefutable. No observation in the world can establish its falsity. There can be no empirical grounds for its falsity.

Moreover, it can be easily shown to be highly probable: like all existential statements, it is in an infinite (or sufficiently large) universe *almost logically true,* to use an expression of Carnap's. Thus, if we take it to be empirical, we have no reason to reject it, and every reason to accept it and to believe in it—especially upon a subjective theory of probable belief.

Probability theory tells us even more: it can be easily proved not only that empirical evidence can *never refute* an almost logically true existential statement, but that it can *never reduce its probability.*[32] (Its probability could be reduced only by some information which is at least 'almost logically false,' and therefore not by an observational evidence statement.) So the empirical probability or degree of empirical confirmation (in Carnap's sense) of our statement about the devil-summoning spell must forever remain equal to unity, whatever the facts may be.

It would of course be easy enough for me to amend my criterion of demarcation so as to include such purely existential statements among the empirical statements. I merely should have to admit not only testable or falsifiable statements among the empirical ones, but also statements which may, in principle, be empirically 'verified.'

But I believe that it is better not to amend my original falsifiability criterion. For our example shows that, if we do not wish to accept my existential statement about the spell that summons the devil, we must deny its empirical character (notwithstanding the fact that it can easily be formalized in any model language sufficient for the expression of even the most primitive scientific assertions). By denying the empirical character of my existential statement, I make it possible to reject it on grounds other than observational evidence. (See chapter 8, section 2, for a discussion of such grounds, and chapter 11, especially pp. 275–277, for a formalization of a similar argument.)

This shows that it is preferable, as I have been trying to make clear for some considerable time, not to assume uncritically that the terms 'empirical' and 'well-formed' (or 'meaningful') must coincide —and that the situation is hardly improved if we assume, uncritically, that probability or probabilistic 'confirmability' may be used as a criterion of the empirical character of statements or theories. For a non-empirical and presumably false statement may have a high degree of probability, as has been shown here.

NOTES

1. See especially my *Poverty of Historicism* (2nd edn., 1960), and ch. 16 of the present volume.

2. See the discussion of degrees of testability, empirical content, corroborability, and corroboration in my L.Sc.D., especially sections 31 to 46; 82 to 85; new

appendix *ix; also the discussion of degrees of explanatory power in this appendix, and especially the comparison of Einstein's and Newton's theories (in note 7 on p. 401). In what follows, I shall sometimes refer to testability, etc., as the 'criterion of progress', without going into the more detailed distinctions discussed in my book.
3. See for example J. C. Harsanyi, 'Popper's Improbability Criterion for the Choice of Scientific Hypotheses', *Philosophy,* 35, 1960, pp. 332 ff. Incidentally, I do not propose any 'criterion' for the choice of scientific hypotheses: every choice remains a risky guess. Moreover, the theoretician's choice is the hypothesis most worthy of *further critical discussion* (rather than of *acceptance).*
4. See especially appendix *ix to my *L.Sc.D.*
5. I have been influenced in adopting this view by Dr J. Agassi who, in a discussion in 1956, convinced me that the attitude of looking upon the finished deductive systems as an end is a relic of the long domination of Newtonian ideas (and thus, I may add, of the Platonic, and Euclidean, tradition). For an even more radical view of Dr Agassi's see the last footnote to this chapter.
6. Compare this and the following two paragraphs with my *Poverty of Historicism,* section 28, pp. 121 ff., and chs. 1 and 16 of this volume.
7. See my *L.Sc.D,* especially section 84, and my *Open Society,* especially pp. 369–374.
8. Cp. Wittgenstein's *Tratatus,* especially 4.0141; also 2.161; 2.17; 2.223; 3.11.
9. See especially pp. 56–7 of his remarkable *Erkenntnislehre,* 2nd edn., 1925.
10. See my Introduction to this volume, 'On the Sources of Knowledge and of Ignorance'.
11. See my Introduction, p. 26, and ch. 5, p. 152 f., above.
12. See the discussion of the 'second view' (called 'instrumentalism') in ch. 3, above.
13. See Alfred Tarski's paper, 'The Semantic Conception of Truth', in *Philosophy and Phenom. Research,* 4, 1943–4, pp. 341 ff. (Cp. especially section 21.)
14. See the volume referred to in the preceding note, especially pp. 279 and 336.
15. Cp. Carnap, *Logical Foundations of Probability,* 1950, p. 177, and my *L.Sc.D.,* especially section 84.
16. From W. Busch, *Schein und Sein* (first published posthumously in 1909; p. 28 of the *Insel* edition, 1952). My attention has been drawn to this rhyme by an essay on Busch as a philosopher which my late friend Julius Kraft contributed to the volume *Erziehung und Politik* (Essays for Minna Specht); see p. 262. My translation makes it perhaps more like a nursery rhyme than Busch intended.
17. Similar misgivings are expressed by Quine when he criticizes Peirce for operating with the idea of approaching to truth. See W. V. Quine, *Word and Object,* New York, 1960, p. 23.
18. This definition is logically justified by the theorem that, so far as the 'empirical part' of the logical content is concerned, comparison of empirical contents and of logical contents always yield the same results; and it is intuitively justified by the consideration that a statement a tells the more about our world of experience the more possible experiences it excludes (or forbids). About basic statements see also the *Addenda* to this volume.
19. In this fragment *'eoikota'* has been most frequently translated as 'probable' or 'plausible'. For example W. Kranz, in Diels-Kranz, *Fragmente der Vorsokratiker,* 6th edn., translates it *'wahrscheinlich-einleuchtend'* that is, 'probable and plausible'; he reads the passage thus: 'This world-arrangement (or world-order) I shall expound to you in all its parts as something probable and plausible.' In translating '(wholly) like truth' or '(wholly) like the truth', I am somewhat influenced by the line (DN,, B 35) quoted above from Xenophanes (and also by K. Reinhardt's *Parmenides,* p. 5 f., where Wilamowitz is referred to). See also section vii of the Introduction to the present volume; the quotation from Osiander in section i of ch. 3; section xii of ch. 5. above; and *Addendum 6,* below.
20. This, incidentally, holds for both absolute probability, $p(a)$, and relative probability, $p(a,b)$; and there are corresponding absolute and relative concepts of verisimilitude.

21. See W. V. Quine, *From a Logical Point of View,* 1953, p. 41.
22. See my *L.Sc.D.,* especially sections 19 to 22; and this volume, ch. 3, text to note 28.
23. See the *Collected Papers of C. S. Peirce,* vol. VII, 7.182 and 7.206. I owe this reference to W. B. Gallie (cp. *Philosophy,* 35, 1960, p. 67), and a similar one to David Rynin.
24. See sections 31–46 of my L.ScD. More recently I have stressed (in lectures) the need to *relativize* comparisons of simplicity to those hypotheses which compete *qua solutions of a certain problem, or set of problems.* The idea of simplicity, though intuitively connected with the idea of a unified theory that springs from *one* intuitive picture of the facts, cannot be analysed in terms of numerical paucity of hypotheses. For every (finitely axiomatizable) theory can be formulated in one statement; and it seems that, for every theory and every *n,* there is a set of *n* independent axioms (though not necessarily 'organic' axioms in the Warsaw sense).
25. For the idea of an *independent test* see my paper 'The Aim of Science', Ratio, 1, 1957.
26. *Phil. Mag.,* 47, 1924, pp. 785 ff.
27. *Zeitschr. f. Phys.,* 32, 1925, pp. 63 ff.
28. I have drawn attention to 'new' predictions of this kind and to their philosophical significance in ch. 3. See especially pp. 117 f.
29. Dr Jerzy Giedymin (in a paper 'A Generalization of the Refutability Postulate', *Studia Logica,* 10, 1960, see especially pp. 103 ff.) has formulated a general methodological principle of empiricism which says that our various rules of scientific method must not permit what he calls a 'dictatorial strategy'; that is they must exclude the possibility that we shall always win the game played in accordance with these rules: Nature must be able to defeat us at least sometimes. If we drop our third requirement, then we can always win, and need not consider Nature at all, as far as the construction of 'good' theories is concerned: speculations about answers which Nature may give to our questions will play no role in our problem situation which will always be fully determined by our past failures alone.
30. See especially my paper 'The Aim of Science,' *Ratio,* 1, 1957.
31. Verificationists may think that the preceding discussion of what I have called here the third requirement quite unnecessarily elaborates what nobody contests. Falsificationists may think otherwise; and personally I feel greatly indebted to Dr Agassi for drawing my attention to the fact that I have previously never explained clearly the distinction between what are called here the second and third requirements. He thus induced me to state it here in some detail. I should mention, however, that he disagrees with me about the third requirement which, as he explained to me, he cannot accept because he can regard it only as a residue of verificationist modes of thought. (See also his paper in the *Australasian Journal of Philosophy,* 39, 1961, where he expresses his disagreement on p. 90.) I admit that there may be a whiff of verificationism here; but this seems to me a case where we have to put up with it, if we do not want a whiff of some form of instrumentalism that takes theories to be mere instruments of exploration.
32. This is a consequence of the 'principle of stability' of the probability calculus; see theorem (26), section V, of my paper 'Creative and Non-Creative Definitions in the Calculus of Probability,' *Synthese,* 15, 1963, No. 2, pp. 167 ff.

SUGGESTIONS FOR FURTHER READING

Blackwell, R.J. *Discovery in the Physical Sciences.* Notre Dame, Ind.: University of Norte Dame Press, 1969.

Bunge, M. (ed.) *The Critical Approach to Science and Philosophy.* New York: The Free Press, 1964.

Campbell, N. *What Is Science?* New York: Dover Publications, Inc., 1952, pp. 58–76.

Caws, P. "The Structure of Discovery." *Science,* 166 (1969), pp. 1375–1380.

Colodny, R.G. (ed.) *Mind and Cosmos.* Pittsburgh: University of Pittsburgh Press, 1966, pp. 22–40.

Diesing, P. *Patterns of Discovery in the Social Sciences.* Chicago: Aldine-Atherton, 1971.

Lakatos, I. and Musgrave, A. (eds.) *Criticism and the Growth of Knowledge.* Cambridge: Cambridge University Press, 1970.

Michalos, A.C. *The Popper-Carnap Controversy.* The Hague: Martinus Nijhoff, 1971, pp. 3–15, 70–82.

Popper, K. R. *The Logic of Scientific Discovery.* New York: Science Editions, Inc., 1961.

Toulmin, S. *The Philosophy of Science.* London: Hutchinson University Library, 1953, pp. 17–56.

SECTION III

Types of Scientific Explanation

The primary aim of the first article in this section is, as Hempel states, "to propose, and to elaborate to some extent, a distinction of two basic modes of explanation—and similarly of prediction and retrodiction—which will be called the deductive and the inductive mode." No one doubts that one of the aims of the scientific enterprise is to explain various features of the world in which we live. For example, the stick in water looks bent, but we know it does not bend and straighten itself out as we lower and raise it in the water. So we would like to have a scientific explanation of why it looks bent. Again, a psychiatrist might have a patient who fears horses, as Freud's "Little Hans" feared horses, for no apparent reason; and the psychiatrist might claim, as Freud did, that he can explain the patient's fear scientifically by analyzing the patient's relation to his father.

When one examines more or less typical explanations of why sticks look bent and why some people have an extraordinary fear of horses, one finds a "logical neatness" about the first sort of explanation which seems to be absent in the second sort. Confronted by such apparent differences, one can take several positions. One could say, "How about that!" and forget it. (That one just failed the course!) Or, one could say, "Well, we obviously have two kinds of scientific explanation, both of which are respectable, but are not equally strong, definitive, conclusive (or some such thing)." That is roughly Hempel's position. Or, one could say, "There is only one kind of scientific explanation, but one or more kinds of pseudo explanations. Bent sticks in water aren't the only kinds of objects that are not what they seem to be." That is roughly May Brodbeck's position. (References may be found at the end of the section.)

In order to appreciate Hempel's article you must understand that he is presenting not only two modes of explanation, but a massive defense against severe attacks (including his own) of his proposals. Beginning students have a tendency to find contradictions behind every bush, as it were, because they lose track of Hempel's line of reasoning. Typically he proposes a view, considers objections to it and then either explains why he does not believe the objection requires a change in his view or he changes the view. Thus, you should try to get his view straight in the first place, try to see exactly what feature of his model is being challenged by any objection and whether or not his response to the objection is satisfactory.

The objective of the Weingartner article is "to show that what appears to be a dispute about the precise nature of historical explanation is in fact the product of a disagreement about the nature of philosophic method, with all the far-reaching implications that such a disagreement implies." Here is an excellent case study of the interaction between the *reportive* and the *stipulative* functions of philosophic analyses, with Hempelians leaning toward the stipulative side and anti-Hempelians leaning toward the reportive side. Anti-Hempelians point out, for example, that "Hempel's scheme does not reflect what historians actually do when they explain," but it would seem that a satisfactory model of historical explanation ought to capture *something* of what historians do. Models of anything ought to look something like whatever it is they are modeling. Otherwise, they are caricatures, cartoons or, maybe, just fantasies. Hempelians, on the other hand, claim that their scheme does "reflect" some facets of actual historical explanations well enough to be regarded as a model. It is at least a model in the sense that many, but by no means all, historians would be happy to pattern their explanations after it *if they could.* Unfortunately, as Hempel states, they generally cannot come very close to the model because they simply do not have the laws and other data required to do the job.

Weingartner's article can serve as a touchstone for your own view of the proper methodology for studying philosophic problems of science and technology. It is a fact that professional philosophers may be distinguished on the basis of their leaning toward reporting or stipulating. If, following Weingartner, we take Hempel's work as illustrative of one who leans toward the stipulative aspect of philosophy, we may take Bunge's work (especially in articles like that on simplicity—Reading 18) as illustrative of one who leans toward the reportive aspect. I do not mean to suggest that one *ought* to lean one way or the other, or that you *ought* to make a choice and encourage one approach while discouraging the other. Rather, I want to emphasize that these different approaches or tendencies exist and that by being aware of them, some kinds of arguments can be averted. Furthermore, I am persuaded that we are better off with *both* sorts of inclinations than we would be with only one. We need reports to remind us of what the world is really like, but we need stipulations too, to remind us of what the world might be or ought to be like.

READING 7

Deductive-Nomological vs. Statistical Explanation

Carl G. Hempel

1. OBJECTIVES OF THIS ESSAY

This essay is concerned with the form and function of explanation in the sense in which it is sought, and often achieved, by empirical science. It does not propose to examine all aspects of scientific explanation; in particular, a closer study of historical explanation falls outside the purview of the present investigation. My main object is to propose, and to elaborate to some extent, a distinction of two basic modes of explanation—and similarly of prediction and retrodiction—which will be called the deductive and the inductive mode.[1]

The structure of deductive explanation and prediction conforms to what is now often called the covering-law model: it consists in the deduction of whatever is being explained or predicted from general laws in conjunction with information about particular facts. The logic of this procedure was examined in some earlier articles of mine, and especially in a study carried out in collaboration with P. Oppenheim.[2]

Since then, various critical comments and constructive suggestions concerning those earlier efforts have appeared in print, and these as well as discussions with interested friends and with my students have led me to reconsider the basic issues concerning the deductive model of scientific explanation and prediction. In the first of the two principal parts of this essay, I propose to give a brief survey of those issues, to modify in certain respects the ideas

Reprinted by permission of the publisher from Hempel, Carl G. "Deductive-Nomological vs. Statistical Explanation", *Minnesota Studies in the Philosophy of Science, Volume III,* edited by H. Feigl and G. Maxwell, University of Minnesota Press, Minneapolis. Copyright © 1962, University of Minnesota.

set forth in the earlier articles, and to examine some new questions concerning deductive explanation, deductive prediction, and related procedures.

The second major part of the present study is an attempt to point out, and to shed some light on, certain fundamental problems in the logic of inductive explanation and prediction.

PART I. DEDUCTIVE-NOMOLOGICAL SYSTEMATIZATION

2. The Covering-Law Model of Explanation

The deductive conception of explanation is suggested by cases such as the following: The metal screwtop on a glass jar is tightly stuck; after being placed in warm water for a short while, it can be readily removed. The familiar explanation of this phenomenon is, briefly, to the effect that the metal has a higher coefficient of thermal expansion than glass, so that a given rise in temperature will produce a larger expansion of the lid than of the neck of the glass jar; and that, in addition, though the metal is a good conductor of heat, the temperature of the lid will temporarily be higher than that of the glass—a fact which further increases the difference between the two perimeters. Thus, the loosening of the lid is here explained by showing that it came about, by virtue of certain antecedent circumstances, in accordance with certain physical laws. The explanation may be construed as an argument in which the occurrence of the event in question is inferred from information expressed by statements of two kinds: (a) general laws, such as those concerning the thermal conductivity of metal and the coefficients of expansion for metal and for glass, as well as the law that heat will be transferred from one body to another of lower temperature with which it is in contact; (b) statements describing particular circumstances, such as that the jar is made of glass, the lid of metal; that initially, at room temperature, the lid fitted very tightly on the top of the jar; and that then the top with the lid on it was immersed in hot water. To show that the loosening of the lid occurred "by virtue of" the circumstances in question, and "in accordance with" those laws, is then to show that the statement describing the result can be validly inferred from the specified set of premises.

Thus construed, the explanation at hand is a deductive argument of this form:

$$(2.1) \quad \frac{\begin{array}{c} L_1, L_2, \ldots, L_r \\ C_1, C_2, \ldots, C_k \end{array}}{E}$$

Here, $L_1, L_2, \ldots, L_r$ are general laws and $C_1, C_2, \ldots, C_k$ are statements of particular occurrences, facts, or events; jointly, these premises form the explanans. The conclusion E is the explanandum statement; it describes the phenomenon (or event, etc.) to be explained, which will also be called the explanandum phenomenon (or event, etc.); thus, the word 'explanandum' will be used to refer ambiguously either to the explanandum statement or to the explanandum phenomenon. Inasmuch as the sentence E is assumed to be a logical consequence of the premises, an explanatory argument of form (2.1) deductively subsumes the explanandum under *"covering laws."* [3] I will say, therefore, that (2.1) represents the *covering-law model of explanation.* More specifically, I will refer to explanatory arguments of the form (2.1) as *deductive-nomological,* or briefly as deductive, explanations: as will be shown later, there are other explanations invoking general laws that will have to be construed as inductive rather than as deductive arguments.

In my illustration, the explanandum is a particular event, the loosening of a certain lid, which occurs at a definite place and time. But deductive subsumption under general laws can serve also to explain general uniformities, such as those asserted by laws of nature. For example, the uniformity expressed by Galileo's law for free fall can be explained by deduction from the general laws of mechanics and Newton's law of gravitation, in conjunction with statements specifying the mass and radius of the earth. Similarly, the uniformities expressed by the law of geometrical optics can be explained by deductive subsumption under the principles of the wave theory of light.[4]

3. Truth and Confirmation of Deductive Explanations

In *SLE* (Section 3) two basic requirements are imposed upon a scientific explanation of the deductive-nomological variety.[5] (i) It must be a deductively valid argument of the form (2.1), whose premises include at least one general law essentially, i.e., in such a way that if the law were deleted, the argument would no longer be valid. Intuitively, this means that reliance on general laws is essential to this type of explanation; a given phenomenon is here explained, or accounted for, by showing that it conforms to a general nomic pattern. (ii) The sentences constituting the explanans must be true, and hence so must the explanandum sentence. This second requirement was defended by the following consideration: suppose we required instead that the explanans be highly confirmed by all the relevant evidence available, though it need not necessarily be true. Now it might happen that the explanans of a given argument of the form (2.1) was well confirmed at a certain earlier stage of scientific research, but strongly disconfirmed by the more comprehensive evidence available at a later time, say, the present. In this event, we would have to say that the explanandum was correctly explained by the given argument at the earlier stage, but not at the later one. And this seemed counterintuitive, for common usage appeared to construe the correctness of a given explanation as no more time dependent than, say, the truth of a given statement. But this justification, with its reliance on a notion of correctness that does not appear in the proposed definition of explanation, is surely of questionable merit. For in reference to explanations as well as in reference to statements, the vague idea of correctness can be construed in two different ways, both of which are of interest and importance for the logical analysis of science: namely, as truth in the semantical sense, which is independent of any reference to time or to evidence; or as confirmation by the available relevant evidence —a concept which is clearly time dependent. We will therefore distinguish between *true explanations,* which meet the requirement of truth for their explanans, and *explanations that are more or less well confirmed* by a given body of evidence (e.g., by the total evidence available). These two concepts can be introduced as follows:

First, we define a *potential explanation* (of deductive-nomological form) [6] as an argument of the form (2.1) which meets all the requirements indicated earlier, except that the statements forming its explanans and explanandum need not be true. But the explanans must still contain a set of sentences, L_1, L_2, . . ., L_r, which are lawlike, i.e., which are like laws except for possibly being false.[7] Sentences of this kind will also be called *nomic,* or *nomological, statements.* It is this notion of potential explanation which is involved, for example, when we ask whether a tentatively proposed but as yet untried theory would be able to explain certain puzzling empirical findings.

Next, we say that a given potential explanation is more or less highly confirmed by a given body of evidence according as its explanans is more or less highly confirmed by the evidence in question. If the explanation is formulated in a formalized language for which an adequate quantitative concept of degree of confirmation or of inductive probability is available, we might identify the probability of the explanation relative to e with the probability of the explanans relative to e.

Finally, by a *true explanation* we understand a potential explanation with true explanans—and hence also with true explanandum.

4. Causal Explanation and the Covering-Law Model

One of the various modes of explanation to which the covering-law model is relevant is the familiar procedure of accounting for an event by pointing out its "cause." In our first illustration, for example, the expansion of the lid might be said to have been caused by its immersion in hot water. Causal attributions of this sort presuppose appropriate laws, such as that whenever metal is heated under constant pressure, it expands. It is by reason of this implicit presupposition of laws that the covering-law model is relevant to the analysis explanation. Let us consider this point more closely.

We will first examine general statements of causal connections, i.e., statements to the effect that an event of a given kind A—for example, motion of a magnet near a closed wire loop—will cause an event of some specified kind B—for example, flow of a current in the wire. Thereafter, we will consider statements concerning causal relations among individual events.

In the simplest case, a general statement asserting a causal connection between two kinds of events, A and B, is tantamount to the statement of the general law that whenever and wherever an instance of A occurs, it is accompanied by an instance of B. This analysis fits, for example, the statement that motion of a magnet causes a current in a neighboring wire loop. Many general statements of causal connection call for a more complex analysis, however. Thus, the statement that in a mammal, stoppage of the heart will cause death presupposes that certain "normal" conditions prevail, which are not explicitly stated, but which are surely meant to preclude, for example, the use of a heart-lung machine. "To say that X causes Y is to say that under proper conditions, an X will be followed by a Y," as Scriven[8] puts it. But unless the "proper conditions" can be specified, at least to some extent, this analysis tells us nothing about the meaning of 'X causes Y.' Now, when this kind of causal locution is used in a given context, there usually is at least some general understanding of the kind of background conditions that have to be assumed; but still, to the extent that those conditions remain indeterminate, a general statement of causal connection falls short of making a definite assertion and has at best the character of a promissory note to the effect that there are further background factors whose proper recognition would yield a truly general connection between the "cause" and "effect" under consideration.

Sentences concerning causal connections among individual events show similar characteristics. For example, the statement that the death of a certain person was caused by an overdose of phenobarbital surely presupposes a generalization, namely, a statement of a general causal connection between one kind of event, a person's taking an overdose of phenobarbital, and another, the death of that person.

Here again, the range of application for the general causal statement is not precisely stated, but a sharper specification can be given by indicating what constitutes an overdose of phenobarbital for a person—this will depend, among other things, on his weight and on his habituation to the drug—and by adding the proviso that death will result from taking such an overdose if the organism is left to itself, which implies, in particular, that no countermeasures are taken. To explain the death in question as having been caused by the antecedent taking of phenobarbital is therefore to claim that the explanandum event followed according to law upon certain antecedent circumstances. And this argument, when stated explicitly, conforms to the covering-law model.

Generally, the assertion of a causal connection between individual events seems to me unintelligible unless it is taken to make, at least implicitly, a nomological claim to the effect that there are laws which provide the basis for the causal connection asserted. When an individual event, say b, is said to have been caused by a certain antecedent event, or configuration of events, a, then surely the claim is intended that whenever "the same cause" is realized, "the same effect" will recur. This claim cannot be taken to mean that whenever a recurs then so does b; for a and b are individual events at particular spatio-temporal locations and thus occur only once. Rather, a and b are, in this context, viewed as particular events of certain *kinds*—e.g., the expansion of a piece of metal or the death of a person—of which there may be many further instances. And the law tacitly implied by the assertion that b, as an event of kind B, was caused by a, as an event of kind A, is a general statement of causal connection to the effect that, under suitable circumstances, an instance of A is invariably accompanied by an instance of B. In most causal explanations offered in other than advanced scientific contexts, the requisite circumstances are not fully stated; for these cases, the import of the claim that b, as an instance of B, was caused by a may be suggested by the following approximate formulation: event b was in fact preceded by an event a of kind A, and by certain further circumstances which, though not fully specified or specifiable, were of such a kind that an occurrence of an event of kind A under such circumstances is universally followed by an event of kind B. For example, the statement that the burning (event of kind B) of a particular haystack was caused by a lighted cigarette carelessly dropped into the hay (particular event of kind A) asserts, first of all, that the latter event did take place; but a burning cigarette will set a haystack on fire only if certain further conditions are satisfied, which cannot at present be fully stated; and thus, the causal attribution at hand implies, second, that further conditions of a not fully specifiable kind were realized, under which an event of kind A will invariably be followed by an event of kind B.

To the extent that a statement of individual causation leaves the relevant antecedent conditions—and thus also the requisite explanatory laws—indefinite, it is like a note saying that there is a treasure hidden somewhere. Its significance and utility will increase as the location of the treasure is narrowed down, as the relevant conditions and the corresponding covering laws are made increasingly explicit. In some cases, such as that of the barbiturate poisoning, this can be done quite satisfactorily; the covering-law structure then emerges, and the statement of individual causal connection becomes amenable to test. When, on the other hand, the relevant conditions or laws remain largely indefinite, a statement of causal connection is rather in the nature of a program, or of a sketch, for an explanation in terms of causal laws; it might also be viewed as a "working hypothesis" which may prove its worth by giving new, and fruitful, direction to further research.

I would like to add here a brief comment on Scriven's observation that "when one asserts that X causes Y one is certainly committed to the generalization that an identical cause would produce an identical effect, but this in no way commits one to any necessity for producing laws not involving the term 'identical,' which justify this claim. Producing laws is one way, not necessarily more conclusive, and usually less easy than other ways, of supporting the causal statement." [9] I think we have to distinguish here two questions, namely (i) what is being claimed by the statement that X causes Y, and in particular, whether asserting it commits one to a generalization, and (ii) what kind of evidence would support the causal statement, and in particular, whether such support can be provided only by producing generalizations in the form of laws.

As for the first question, I think the causal statement does imply the claim that an appropriate law or set of laws hold by virtue of which X causes Y; but, for reasons suggested above, the law or laws in question cannot be expressed by saying that an identical cause would produce an identical effect. Rather, the general claim implied by the causal statement is to the effect that there are certain "relevant" conditions of such a kind that whenever they occur in conjunction with an event of kind X, they are invariably followed by an event of kind Y.

In certain cases, some of the laws that are claimed to connect X and Y may be explicitly statable—as, for example, in our first illustration, the law that metals expand upon heating; and then, it will be possible to provide evidential support (or else disconfirmation) for them by the examination of particular instances; thus, while laws are implicitly claimed to underlie the causal connection in question, the claim can be supported by producing appropriate empirical evidence consisting of particular cases rather than of general laws. When, on the other hand, a nomological claim made by a causal statement has merely the character of an existential statement to the effect that there are relevant factors and suitable laws connecting X and Y, then it may be possible to lend some credibility to this claim by showing that under certain conditions an event of kind X is at least very frequently accompanied by an event of kind Y. This might justify the working hypothesis that the background conditions could be further narrowed down in a way that would eventually yield a strictly causal connection. It is this kind of statistical evidence, for example, that is adduced in support of such claims as that cigarette smoking is "a cause of" or "a causative factor in" cancer of the lung. In this case, the supposed causal laws cannot at present be explicitly stated. Thus, the nomological claim implied by this causal conjecture is of the existential type; it has the character of a working hypothesis that gives direction to further research. The statistical evidence adduced lends support to the hypothesis and justifies the program, which clearly is the aim of further research, of determining more precisely the conditions under which smoking will lead to cancer of the lung.

The most perfect examples of explanations conforming to the covering-law model are those provided by physical theories of deterministic character. A theory of this kind deals with certain specified kinds of physical systems, and limits itself to certain aspects of these, which it represents by means of suitable parameters; the values of these parameters at a given time specify the state of the system at that time; and a deterministic theory provides a system of laws which, given the state of an isolated system at one time, determine its state at any other time. In the classical mechanics of systems of mass points, for example, the state of a system at a given time is specified by the positions and momenta of the component particles at that time; and the principles of the theory—essentially the Newtonian laws of motion and of gravitation—determine the state of an isolated system of mass points at any time provided that its state at some one moment is given; in particular, the state at a specified moment may be fully explained, with the help of the theoretical principles in question, by reference to its state at some earlier time. In this theoretical scheme, the notion of a cause as a more or less narrowly circumscribed antecedent event has been replaced by that of some antecedent state of the total system, which provides the "initial conditions" for the computation, by means of the theory, of the later state that is to be explained; if the system is not isolated, i.e., if relevant outside influences act upon the system during the period of time from the initial state invoked to the state to be explained, then the particular circumstances that must be stated in the explanans include also those "outside influences"; and it is these "boundary conditions" in conjunction with the "initial" conditions which replace the everyday notion of cause, and which have to be thought of as being specified by the statements C_1, C_2, . . ., C_k in the schematic representation (2.1) of the covering-law model.

Causal explanation in its various degrees of explicitness and precision is not the only type of explanation, however, to which the covering-law model is relevant. For example, as was noted earlier, certain empirical regularities, such as that represented by Galileo's law, can be explained by deductive subsumption under more comprehensive laws or theoretical principles; frequently, as in the case of the explanation of Kepler's laws by means of the law of gravitation and the laws of mechanics, the deduction yields a conclusion of which the generalization to be explained is only an approximation. Then the explanatory principles not only show why the presumptive general law holds, at least in approximation, but also provide an explanation for the deviations.

Another noncausal species of explanation by covering laws is illustrated by the explanation of the period of swing of a given pendulum by reference to its length and to the law that the period of a mathematical pendulum is proportional to the square root of its length. This law expresses a mathematical relation between the length and the period (a dispositional characteristic) of a pendulum *at the same time;* laws of this kind are sometimes referred to as *laws of coexistence,* in contradistinction to *laws of succession,* which concern the changes that certain systems undergo in the course of time. Boyle's, Charles's, and Van der Waal's laws for gases, which concern concurrent values of pressure, volume, and temperature of a gas; Ohm's law; and the law of Wiedemann and Franz (according to which, in metals, electric conductivity is proportional to thermal conductivity) are examples of laws of coexistence. Causal explanation in terms of antecedent events clearly calls for laws of succession in the explanans; in the case of the pendulum, where only a law of coexistence is invoked, we would not say that the pendulum's having such and such a length at a given time *caused* it to have such and such a period.[10]

It is of interest to note that in the example at hand, a statement of the length of a given pendulum in conjunction with the law just referred to will much more readily be accepted as explaining the pendulum's period than a statement of the period in conjunction with the same law would be considered as explaining the length of the pendulum; and this is true even though the second argument has the same logical structure as the first: both are cases of deductive subsumption, in accordance with the schema (2.1), under a law of coexistence. The distinction made here seems to me to result from the consideration that we might change the length of the pendulum at will and thus control its period as a "dependent variable," whereas the reverse procedure does not seem possible. This idea is open to serious objections, however; for clearly, we can also change the period of a given pendulum at will, namely, by changing its length; and in doing so, we will change its period. It is not possible to retort that in the first case we have a change of length independently of a change of the period; for if the location of the pendulum, and thus the gravitational force acting on the pendulum bob, remains unchanged, then the length cannot be changed without also changing the period. In cases such as this, the common-sense conception of explanation appears to provide no clear and reasonably defensible grounds on which to decide whether a given argument that deductively subsumes an occurrence under laws is to qualify as an explanation.

The point that an argument of the form (2.1), even if its premises are assumed to be true, would not always be considered as constituting an explanation is illustrated even more clearly by the following example, which I owe to my colleague, Mr. S. Bromberger. Suppose that a flagpole stands vertically on level ground and subtends an angle of 45 degrees when viewed from the ground level at a distance of 80 feet. This information, in conjunction with some elementary theorems of geometry, implies deductively that the pole is 80 feet high. The theorems in question must here be understood as belonging to physical geometry and thus as having the status of general laws, or, better, general theoretical principles, of physics. Hence, the deductive argument is of the type (2.1). And yet, we would not say that its premises *explained* the fact that the pole is 80 feet high, in the sense of showing why it is that the pole has a height of 80 feet. Depending on the context in which it is raised, the request for an explanation might call here for some kind of causal account of how it came about that the pole was given this height, or perhaps for a statement of the purpose for which this height was chosen. An account of the latter kind would again be a special case of causal explanation, invoking among the antecedent conditions certain dispositions (roughly speaking, intentions, preferences, and beliefs) on the part of the agents involved in erecting the flagpole.

The geometrical argument under consideration is not of a causal kind; in fact, it might be held that if the particular facts and the geometrical laws here invoked can be put into an explanatory connection at all, then at best we might say that the height of the pole—in conjunction with the other particulars and the laws—explains the size of the subtended angle, rather than vice versa. The consideration underlying this view would be similar to that mentioned in the case of the pendulum: It might be said that by changing the height of the pole, a change in the angle can be effected, but not vice versa. But here as in the previous case, this contention is highly questionable. Suppose that the other factors involved, especially the distance from which the pole is viewed, are kept constant; then the angle can be changed, namely by changing the length of the pole; and thus, if the angle is made to change, then, trivially, the length of the pole changes. The notion that somehow we can "independently" control the length and thus make the angle a dependent variable, but not conversely, does not seem to stand up under closer scrutiny.

In sum then, we have seen that among those arguments of the form (2.1) which are not causal in character there are some which would not ordinarily be considered as even potential explanations; but ordinary usage appears to provide no clear general criterion for those arguments which are to be qualified as explanatory. This is not surprising, for our everyday conception of explanation is strongly influenced by preanalytic causal and teleological ideas; and these can hardly be expected to provide unequivocal guidance for a more general and precise analysis of scientific explanation and prediction.

5. Covering Laws: Premises or Rules?

Even if it be granted that causal explanations presuppose general laws, it might still be argued that many explanations of particular occurrences as formulated in everyday contexts or even in scientific discourse limit themselves to adducing certain particular facts as the presumptive causes of the explanandum event, and that therefore a formal model should construe these explanations as accounting for the explanandum by means of suitable statements of particular fact, $C_1, C_2, \ldots, C_k$, alone. Laws would have to be cited, not in the context by *giving* such an explanation, but in the context of *justifying* it; they would serve to show that the antecedent circumstances specified in the explanans are indeed connected by causal laws with the explanandum event. Explanation would thus be comparable to proof by logical deduction, where explicit reference to the rules or laws of logic is called for, not in stating the successive steps of the proof, but only in justifying them, i.e., in showing that they conform to the principles of deductive inference. This conception would construe general laws and theoretical principles, not as scientific statements, but rather as extralogical rules of scientific inference. These rules, in conjunction with those of formal logic, would govern inferences—explanatory, predictive, retrodictive, etc.—that lead from given statements of particular fact to other statements of particular fact.

The conception of scientific laws and theories as rules of inference has been advocated by various writers in the philosophy of science.[11] In particular, it may be preferred by those who hesitate, on philosophic grounds, to accord the status of bona fide statements, which are either true or false, to sentences which purport to express either laws covering an infinity of potential instances or theoretical principles about unobservable "hypothetical" entities and processes.[12]

On the other hand, it is well known that in rigorous scientific studies in which laws or theories are employed to explain or predict empirical phenomena, the formulas expressing laws and theoretical principles are used, not as rules of inference, but as statements—especially as premises—quite on a par with those sentences which presumably describe particular empirical facts or events. Similarly, the formulas expressing laws also occur as conclusions in deductive arguments; for example, when the laws governing the motion of the components of a double star about their common center of gravity are derived from broader laws of mechanics and of gravitation.

It might also be noted here that a certain arbitrariness is involved in any method of drawing a line between those formulations of empirical science which are to count as statements of particular fact and those which purport to express general laws, and which accordingly are to be construed as rules of inference. For any term representing an empirical characteristic can be construed as dispositional, in which case a sentence containing it acquires the status of a generalization. Take, for example, sentences which state the boiling point of helium at atmospheric pressure, or the electric conductivity of copper: are these to be construed as empirical statements or rather as rules? The latter status could be urged on the grounds that (i) terms such as 'helium' and 'copper' are dispositional, so that their application even to one particular object involves a universal assertion. and that (ii) each of the two statements attributes a specific disposition to *any* body of helium or of copper at *any* spatiotemporal location, which again gives them the character of general statements.

The two conceptions of laws and theories—as statements or as rules of inference—correspond to two different formal reconstructions, or models, of the language of empirical science; and a model incorporating laws and theoretical principles as rules can always be replaced by one which includes them instead as scientific statements.[13] And what matters for our present purposes is simply that in either mode of representation, explanations of the kind here considered "presuppose" general theoretical principles essentially: either as indispensable premises or as indispensable rules of inference.

Of the two alternative construals of laws and theories, the one which gives them the status of statements seems to me simpler and more perspicuous for the analysis of the issues under investigation here; I will therefore continue to construe deductive-nomological explanations as having the form (2.1).

6. Explanation, Prediction, Retrodiction, and Deductive Systematization—a Puzzle about 'About'

In a deductive-nomological explanation of a particular past event, the explanans logically implies the occurrence of the explanandum event; hence we may say of the explanatory argument that it could also have served as a predictive one in the sense that it could have been used to predict the explanandum event if the laws and particular circumstances adduced in its explanans had been taken into account at a suitable earlier time.[14] Predictive arguments of the form (2.1) will be called *deductive-nomological predictions,* and will be said to conform to the covering-law model of prediction. There are other important types of scientific prediction; among these, statistical prediction, along with statistical explanation, will be considered later.

Deductive-nomological explanation in its relation to prediction is instructively illustrated in the fourth part of the *Dialogues Concerning Two New Sciences.* Here, Galileo develops his laws for the motion of projectiles and deduces from them the corollary that if projectiles are fired from the same point with equal initial velocity, but different elevations, the maximum range will be attained when the elevation is 45°. Then, Galileo has Sagredo remark: "From accounts given by gunners, I was already aware of the fact that in the use of cannon and mortars, the maximum range . . . is obtained when the elevation is 45° . . . but to understand why this happens far outweighs the mere information obtained by the testimony of others or even by repeated experiment." [15] The reasoning that affords such understanding can readily be put into the form (2.1); it amounts to a deduction, by logical and mathematical means, of the corollary from a set of premises which contains (i) the fundamental laws of Galileo's theory for the motion of projectiles and (ii) particular statements specifying that all the missiles considered are fired from the same place with the same initial velocity. Clearly then, the phenomenon previously noted by the gunners is here *explained,* and thus *understood,* by showing that its occurrence was to be expected, under the specified circumstances, in view of certain general laws set forth in Galileo's theory. And Galileo himself points with obvious pride to the predictions that may in like fashion be obtained by deduction from his laws; for the latter imply "what has perhaps never been observed in experience, namely, that of other shots those which exceed or fall short of 45° by equal amounts have equal ranges." Thus, the explanation afforded by Galileo's theory "prepares the mind to understand and ascertain other facts without need of recourse to experiment," [16] namely, by deductive subsumption under the laws on which the explanation is based.

We noted above that if a deductive argument of the form (2.1) explains a past event, then it could have served to predict it if the information provided by the explanans had been available earlier. This remark makes a purely logical point; it does not depend on any empirical assumptions. Yet it has been argued, by Rescher, that the thesis in question "rests upon a tacit but unwarranted assumption as to the nature of the phsyical universe." [17]

The basic reason adduced for this contention is that "the explanation of events is oriented (in the main) towards the past, while prediction is oriented towards the future," [18] and that, therefore, before we can decide whether (deductive-nomological) explanation and prediction have the same logical structure, we have to ascertain whether the natural laws of our world do in fact permit inferences from the present to the future as well as from the present to the past. Rescher stresses that a given system might well be governed by laws which permit deductive inferences concerning the future, but not concerning the past, or conversely; and on this point he is quite right. As a schematic illustration, consider a model "world" which consists simply of a sequence of colors, namely, Blue (B), Green (G), Red (R), and Yellow (Y), which appear on a screen during successive one-second intervals i_1, i_2, i_3, . . . Let the succession of colors be governed by three laws:

(L_1) B is always followed by G.
(L_2) G and R are always followed by Y.
(L_3) Y is always followed by R.

Then, given the color of the screen for a certain interval, say i_3, these laws unequivocally determine the "state of the world," i.e., the screen color, for all later intervals, but not for all earlier ones. For example, given the information that during i_3 the screen is Y, the laws predict the colors for the subsequent intervals uniquely as RYRYRY . . .; but for the preceding states i_1 and i_2, they yield no unique information, since they allow here two possibilities: BG and YR.

Thus, it is possible that a set of laws governing a given system should permit unique deductive *predictions* of later states from a given one, and yet not yield unique deductive *retrodictions* concerning earlier states; conversely, a set of laws may permit unique retrodiction, but no unique prediction. But—and here lies the flaw in Rescher's argument—this is by no means the same thing as to say that such laws, while permitting deductive prediction of later states from a given one, do not permit explanation; or, in the converse case, that while permitting explanation, they do not permit prediction. To illustrate by reference to our simple model world: Suppose that during i_3 we find the screen to be Y, and that we seek to explain this fact. This can be done if we can ascertain, for example, that the color for i_1 had been B; for from the statement of this particular antecedent fact we can infer, by means of L_1, that the color for i_2 must have been G and hence, by L_2, that the color for i_3 had to be Y. Evidently, the same argument, used before i_3 could serve to predict uniquely the color for i_3 on the basis of that for i_1. Indeed, quite generally, any predictive argument made possible by the laws for our model world can also be used for explanatory purposes and vice versa. And this is so although those laws, while permitting unique predictions, do not always permit unique retrodictions. Thus, the objection under consideration misses its point because it tacitly confounds explanation with retrodiction.[19]

The notion of scientific retrodiction, however, is of interest in its own right; and, as in the case of explanation and prediction, one important variety of it is the deductive-nomological one. It has the form (2.1), but with the statements C_1, C_2, . . ., C_k referring to circumstances which occur later than the event specified in the conclusion E. In astronomy, an inference leading, by means of the laws of celestial mechanics, from data concerning the present positions and movements of the sun, the earth and Mars to a statement of the distance between earth and Mars a year later or a year earlier illustrates deductive-nomological prediction and retrodiction respectively; in this case, the same laws can be used for both purposes because the processes involved are reversible.

It is of interest to observe here that in their predictive and retrodictive as well as in their explanatory use, the laws of classical mechanics, or other sets of deterministic laws for physical systems, require among the premises not only a specification of the state of the system for some time, t_0, earlier or later than the time, say t_1, for which the state of the system is to be inferred, but also a statement of the boundary conditions prevailing between t_0 and t_1; these specify the external influences acting upon the system during the time in question. For certain purposes in astronomy, for example, the disturbing influence of celestial objects other than those explicitly considered may be neglected as insignificant, and the system under consideration may then be treated as "isolated"; but this should not lead us to overlook the fact that even those laws and theories of the physical sciences which provide the exemplars of deductively nomological prediction do not enable us to forecast certain future events strictly on the basis of information about the present: the predictive argument also requires certain premises concerning the future—e.g., absence of disturbing influences, such as a collision of Mars with an unexpected comet—and the temporal scope of these boundary conditions must extend up to the very time at which the predicted event is to occur. The assertion therefore that laws and theories of deterministic form enable us to predict certain aspects of the future from information about the present has to be taken with a considerable grain of salt. Analogous remarks apply to deductive-nomological retrodiction and explanation.

I will use the term 'deductive-nomological systematization' to refer to any argument of the type (2.1), irrespective of the temporal relations between the particular facts specified by $C_1, C_2, \ldots, C_k$ and the particular events, if any, described by E. And, in obvious extension of the concepts introduced in Section 3 above, I will speak of *potential* (deductive-nomological) *systematizations,* of *true systematizations,* and of *systematizations* whose joint premises are more or less well confirmed by a given body of evidence.

To return now to the characterization of an explanation as a potential prediction: Scriven [20] bases one of his objections to this view on the observation that in the causal explanation of a given event (e.g., the collapse of a bridge) by reference to certain antecedent circumstances (e.g., excessive metal fatigue in one of the beams) it may well happen that the only good reasons we have for assuming that the specified circumstances were actually present lie in our knowledge that the explanandum event did take place. In this situation, we surely could not have used the explanans predictively since it was not available to us before the occurrence of the event to be predicted. This is an interesting and important point in its own right; but in regard to our conditional thesis that an explanation could have served as a prediction *if* its explanans had been taken account of in time, the argument shows only that the thesis is sometimes counterfactual (i.e., has a false antecedent), but not that it is false.

In a recent article, Scheffler [21] has subjected the idea of the structural equality of explanation and prediction to a critical scrutiny; and I would like to comment here briefly on at least some of his illuminating observations.

Scheffler points out that a prediction is usually understood to be an assertion rather than an argument. This is certainly the case; and we might add that, similarly, an explanation is often formulated, not as an argument, but as a statement, which will typically take the form 'q because p.' But predictive statements in empirical science are normally established by inferential procedures (which may be deductive or inductive in character) on the basis of available evidence; thus, there arises the question as to the logic of predictive arguments in analogy to the problem of the logic of explanatory arguments; and the idea of structural equality should be understood as pertaining to explanatory, predictive, retrodictive, and related argements in science.

Scheffler also notes that a scientific prediction statement may be false, whereas, under the requirement of truth for explanations as laid down in Section 3 of *SLE,* no explanation can be false. This remark is quite correct; however, I consider it to indicate, not that there is a basic discrepancy between explanation and prediction, but that the requirement of truth for scientific explanations is unduly restrictive. The restriction is avoided by the approach that was proposed above in Section 3, and again in the present section in connection with the general characterization of scientific systematization; this approach enables us to speak of explanations no less than of predictions as being possibly false, and as being more or less well confirmed by the empirical evidence at hand.

Another critical observation Scheffler puts forth concerns the view, presented in *SLE,* that the difference between an explanatory and a predictive argument does not lie in its logical structure, but is "of a pragmatic character. If . . . we know that the phenomenon described by E has occurred, and a suitable set of statements C_1, C_2, . . ., C_k, L_1, L_2, . . ., L_r is provided afterwards, we speak of an explanation of the phenomenon in question. If the latter statements are given and E is derived prior to the occurrence of the phenomenon it describes, we speak of a prediction." [22] This characterization would make explanation and prediction mutually exclusive procedures, and Scheffler rightly suggests that they may sometimes coincide, since, for example, one may reasonably be said to be both predicting and explaining the sun's rising when, in reply to the question 'Why will the sun rise tomorrow?' one offers the appropriate astronomical information.[23]

I would be inclined to say, therefore, that in an explanation of the deductive-nomological variety, the explanandum event—which may be past, present, or future—is taken to be "given," and a set of laws and particular statements is then adduced which provides premises in an appropriate argument of type (2.1); whereas in the case of prediction, it is the premises which are taken to be "given," and the argument then yields a conclusion about an event to occur after the presentation of the predictive inference. Retrodiction may be construed analogously. The argument referred to by Scheffler about tomorrow's sunrise may thus be regarded, first of all, as predicting the event on the basis of suitable laws and presently available information about antecedent circumstances; then, taking the predicted event as "given," the premises of the same argument constitute an explanans for it.

Thus far, I have dealt with the view that an explanatory argument is also a (potentially) predictive one. Can it be held equally that a predictive argument always offers a potential explanation? In the case of deductive-nomological predictions, an affirmative answer might be defended, though as was illustrated at the end of Section 4, there are some deductive systematizations which one would readily accept as predictions while one would find it at least awkward to qualify them as explanations. Construing the question at hand more broadly, Scheffler, and similarly Scriven,[24] have rightly pointed out, in effect, that certain sound predictive arguments of the nondeductive type cannot be regarded as affording potential explanations. For example, from suitable statistical data on past occurrences, it may be possible to "infer" quite soundly certain predictions concerning the number of male births, marriages, or traffic deaths in the United States during the next month; but none of these arguments would be regarded as affording even a low-level explanation of the occurrences they serve to predict. Now, the inferences here involved are inductive rather than deductive in character; they lead from information about observed finite samples to predictions concerning as yet unobserved samples of a given population. However, what bars them from the role of potential explanations is not their inductive character (later I will deal with certain explanatory arguments of inductive form) but the fact that they do not invoke any general laws either of strictly universal or of statistical form: it appears to be characteristic of an explanation, though not necessarily of a prediction, that it present the inferred phenomena as occurring in conformity with general laws.

In concluding this section, I would like briefly to call attention to a puzzle concerning a concept that was taken for granted in the preceding discussion, for example, in distinguishing between prediction and retrodiction. In drawing that distinction, I referred to whether a particular given statement, the conclusion of an argument of form (2.1), was "about" occurrences at a time earlier or later than some specified time, such as the time of presentation of that argument. The meaning of this latter criterion appears at first to be reasonably clear and unproblematic. If pressed for further elucidation, one might be inclined to say, by way of a partial analysis, that if a sentence explicitly mentions a certain moment or period of time then the sentence is about something occurring at that time. It seems reasonable, therefore, to say that the sentence 'The sun rises on July 17, 1958,' says something about July 17, 1958, and that, therefore, an utterance of this sentence on July 16, 1958, constitutes a prediction.

Now the puzzle in question, which might be called the puzzle of 'about,' shows that this criterion does not even offer a partially satisfactory explication of the idea of what time a given statement is about. For example, the statement just considered can be equivalently restated in such a way that, by the proposed criterion, it is about July 15 and thus, if uttered on July 16, is about the past rather than about the future. The following rephrasing will do: 'The sun plus-two-rises on July 15,' where plus-two-rising on a given date is understood to be the same thing as rising two days after that date. By means of linguistic devices of this sort, statements about the future could be reformulated as statements about the past, or conversely: we could even replace all statements with temporal reference by statements which are, all of them, ostensibly "about" one and the same time.

The puzzle is not limited to temporal reference, but arises for spatial references as well. For example, a statement giving the mean temperature at the North Pole can readily be restated in a form in which it speaks ostensibly about the South Pole; one way of doing this is to attribute to the South Pole the property of having, in such and such a spatial relation to it, a place where the mean temperature is such and such; another device would be to use a functor, say 'm,' which, for the South Pole, takes as its value the mean temperature at the North Pole. Even more generally there is a method which, given any particular object *o*, will reformulate any statement in such a way that it is ostensibly about *o*. If, for example, the given statement is 'The moon is spherical,' we introduce a property term, 'moon-spherical,' with the understanding that it is to apply to *o* just in case the moon is spherical; the given statement then is equivalent to '*o* is moon-spherical.'

The puzzle is mentioned here in order to call attention to the difficulties that face an attempt to explicate the idea of what a statement is "about," and in particular, what time it refers to; and that idea seems essential for the characterization of prediction, retrodiction, and similar concepts.[25]

PART II. STATISTICAL SYSTEMATIZATION

7. Law of Strictly General and Statistical Form

The nomological statements adduced in the explanans of a deductive-nomological explanation are all of a strictly general form: they purport to express strictly unexceptionable laws or theoretical principles interconnecting certain characteristics (i.e., qualitative or quantitative properties or relations) of things or events. One of the simplest forms a statement of this kind can take is that of a universal conditional: 'All (instances of) F are (instances) of G.' When the attributes in question are quantities, their interconnections are usually expressed in terms of mathematical functions, as is illustrated by many of the laws and theoretical principles of the physical sciences and of mathematical economics.

On the other hand, there are important scientific hypotheses and theoretical principles which assert that certain characters are associated, not unexceptionally or universally, but with a specified long-range frequency; we will call them statistical generalizations, or laws (or theoretical principles) of statistical form, or (statistical) probability statements. The laws of radioactive decay, the fundamental principles of quantum mechanics, and the basic laws of genetics are examples of such probability statements. These statistical generalizations, too, are used in science for the systemization of various empirical phenomena. This is illustrated, for example, by the explanatory and predictive applications of quantum theory and of the basic laws of genetics as well as by the postdictive use of the laws of radioactive decay in dating archeological relics by means of the radio-carbon method.

The rest of this essay deals with some basic problems in the logic of statistical systematizations, i.e., of explanatory, predictive, or similar arguments which make essential use of statistical generalizations.

Just as in the case of deductive-nomological systematization, arguments of this kind may be used to account not only for particular facts or events, but also for general regularities, which, in this case, will be of a statistical character. For example, from statistical generalizations stating that the six different results obtainable by rolling a given die are equiprobable and statistically independent of each other, it is possible to deduce the statistical generalization that the probability of rolling two aces in succession is 1/36; thus the latter statistical regularity is accounted for by subsumption (in this case purely deductive) under broader statistical hypotheses.

But the peculiar logical problems concerning statistical systematization concern the role of probability statements in the explanation, prediction, and postdiction of individual events or infinite sets of such events. In preparation for a study of these problems, I shall now consider briefly the form and function of statistical generalizations.

Statistical probability hypotheses, or statistical generalizations, as understood here, bear an important resemblance to nomic statements of strictly general form; they make a universal claim, as is suggested by the term 'statistical law,' or 'law of statistical form.' Snell's law of refraction, which is of strictly general form, is not simply a descriptive report to the effect that a certain quantitative relationship has so far been found to hold, in all cases of optical refraction, between the angle of incidence and that of refraction: it asserts that that functional relationship obtains universally, in all cases of refraction, no matter when and where they occur.[26] Analogously, the statistical generalizations of genetic theory or the probability statements specifying the half lives of various radioactive substances are not just reports on the frequencies with which certain phenomena have been found to occur in some set of past instances; rather, they serve to assert certain peculiar but universal modes of connection between certain attributes of things or events.

A statistical generalization of the simplest kind asserts that the probability for an instance of F to be an instance of G is r, or briefly that $p(G,F) = r$; this is intended to express, roughly speaking, that the proportion of those instances of F which are also instances of G is r. This idea requires clarification, however, for the notion of the proportion of the (instances of) G among the (instances of) F has no clear meaning when the instances of F do not form a finite class. And it is characteristic of probability hypotheses with their universal character, as distinguished from statements of relative frequencies in some finite set, that the reference class—F in this case—is not assumed to be finite; in fact, we might underscore their peculiar character by saying that the probability r does not refer to the class of all actual instances of F but, so to speak, to the class of all its potential instances.

Suppose, for example, that we are given a homogeneous regular tetrahedron whose faces are marked 'I,' 'II,' 'III,' 'IV.' We might then be willing to assert that the probability of obtaining a III, i.e., of the tetrahedron's coming to rest on that face, upon tossing it out of a dice box is $\frac{1}{4}$; but while this assertion would be meant to say something about the frequency with which a III is obtained as a result of rolling the tetrahedron, it could not be construed as simply specifying that frequency for the class of all tosses which are in fact ever performed with the tetrahedron. For we might well maintain our probability hypothesis even if the given tetrahedron were tossed only a few times throughout its existence, and in this case, our probability statement would certainly not be meant to imply that exactly or even nearly, one fourth of those tosses yielded the result III. In fact, we might clearly maintain the probability statement even if the tetrahedron happened to be destroyed without ever having been tossed at all. We might say, then, that the probability hypothesis ascribes to the tetrahedron a certain disposition, namely, that of yielding a III in about one out of four cases in the long run. That disposition may also be described by a subjunctive or counterfactual statement: If the tetrahedron were to be tossed (or had been tossed) a large number of times, it would yield (would have yielded) the result III in about one fourth of the cases.[27]

Let us recall here in passing that nomological statements of strictly general form, too, are closely related to corresponding subjunctive and counterfactual statements. For example, the lawlike statement 'All pieces of copper expand when heated' implies the subjunctive conditional 'If this copper key were heated it would expand' and the counterfactual statement, referring to a copper key that was kept at constant temperature during the past hour, 'If this copper key had been heated half an hour ago, it would have expanded.'[28]

To obtain a more precise account of the form and function of probability statements, I will examine briefly the elaboration of the concept of statistical probability in contemporary mathematical theory. This examination will lead to the conclusion that the logic of statistical systematization differs fundamentally from that of deductive-nomological systematization. One striking symptom of the difference is what will be called here *the ambiguity of statistical systematization.*

In Section 8, I will describe and illustrate this ambiguity in a general manner that presupposes no special theory of probability; then in Section 9, I will show how it reappears in the explanatory and predictive use of probability hypotheses as characterized by the mathematical theory of statistical probability.

8. The Ambiguity of Statistical Systematization

Consider the following argument which represents, in a nutshell, an attempt at a statistical explanation of a particular event: "John Jones was almost certain to recover quickly from his streptococcus infection, for he was given penicillin, and almost all cases of streptococcus infection clear up quickly upon administration of penicillin." The second statement in the explanans is evidently a statistical generalization, and while the probability value is not specified numerically, the words 'almost all cases' indicate that it is very high.

At first glance, this argument appears to bear a close resemblance to deductive-nomological explanations of the simplest form, such as the following: This crystal of rock salt, when put into a Bunsen flame, turns the flame yellow, for it is a sodium salt, and all sodium salts impart a yellow color to a Bunsen flame. This argument is basically of the form:

(8.1)

All F are G.

x is F.

x is G.

The form of the statistical explanation, on the other hand, appears to be expressible as follows:

(8.2)

Almost all F are G.

x is F.

x is almost certain to be G.

Despite this appearance of similarity, however, there is a fundamental difference between these two kinds of argument: A nomological explanation of the type (8.1) accounts for the fact that x is G by stating that x has another character, F, which is uniformly accompanied by G, in virtue of a general law. If in a given case these explanatory assumptions are in fact true, then it follows logically that x must be G; hence x cannot possibly possess a character, say H, in whose presence G is uniformly absent; for otherwise, x would have to be both G and non-G. In the argument (8.2), on the other hand, x is said to be almost certain to have G because it has a character, F, which is accompanied by G in almost all instances. But even if in a given case the explanatory statements are both true, x may possess, in addition to F, some other attribute, say H, which is almost always accompanied by non-G. But by the very logic underlying (8.2), this attribute would make it almost certain that x is not G.

Suppose, for example, that almost all, but not quite all, penicillin-treated, streptococcal infections result in quick recovery, or briefly, that almost all P are R; and suppose also that the particular case of illness of patient John Jones which is under discussion—let us call it j—is an instance of P. Our original statistical explanation may then be expressed in the following manner, which exhibits the form (8.2):

(8.3a)

Almost all P are R.

j is P.

---

j is almost certain to be R.

Next, let us say that an event has the property P* if it is either the event j itself or one of those infrequent cases of penicillin-treated streptococcal infection which do not result in quick recovery. Then clearly j is P*, whether or not j is one of the cases resulting in recovery, i.e., whether or not j is R. Furthermore, almost every instance of P* is an instance of non-R (the only possible exception being j itself). Hence, the argument (8.3a) in which, on our assumption, the premises are true can be matched with another one whose premises are equally true, but which by the very logic underlying (8.3a), leads to a conclusion that appears to contradict that of (8.3a):

(8.3b)

Almost all P* are non-R.

j is P*.

---

j is almost certain to be non-R.

If it should be objected that the property P* is a highly artificial property and that, in particular, an explanatory statistical law should not involve essential reference to particular individuals (such as j in our case), then another illustration can be given which leads to the same result and meets the contemplated requirement. For this purpose, consider a number of characteristics of John Jones at the onset of his illness, such as his age, height, weight, blood pressure, temperature, basal metabolic rate, and IQ. These can be specified in terms of numbers; let n_1, n_2, n_3, . . . be the specific numerical values in question. We will say that an event has the property S if it is a case of streptococcal infection in a patient who at the onset of his illness has the height n_1, age n_2, weight n_3, blood pressure n_4, and so forth. Clearly, this definition of S in terms of numerical characteristics no longer makes reference to j. Finally, let us say that an event has the property P** if it is either an instance of S or one of those infrequent cases of streptococcal infection treated with penicillin which do not result in quick recovery. Then evidently j is P** because j is S; and furthermore, since S is a very rare characteristic, almost every instance of P** is an instance of non-R. Hence, (8.3a) can be matched with the following argument, in which the explanatory probability hypothesis involves no essential reference to particular cases:

(8.3c)

Almost all P** are non-R.
j is P**.
―――――――――――――――
j is almost certain to be non-R.

The premises of this argument are true if those of (8.3a) are, and the conclusion again appears to be incompatible with that of (8.3a).

The peculiar phenomenon here illustrated will be called the *ambiguity of statistical explanation.* Briefly, it consists in the fact that if the explanatory use of a statistical generalization is construed in the manner of (8.2), then a statistical explanation of a particular event can, in general, be matched by another one, equally of the form (8.2), with equally true premises, which statistically explains the nonoccurrence of the same event. The same difficulty arises, of course, when statistical arguments of the type (8.2) are used for predictive purposes. Thus, in the case of our illustration, we might use either of the two arguments (8.3a) and (8.3c) in an attempt to predict the effect of penicillin treatment in a fresh case, j, of streptococcal infection; and even though both followed the same logical pattern—that exhibited in (8.2)—and both had true premises, one argument would yield a favorable, the other an unfavorable forecast. We will, therefore, also speak of the *ambiguity of statistical prediction* and, more inclusively, of the *ambiguity of statistical systematization.*

This difficulty is entirely absent in nomological systematization, as we noted above; and it evidently throws into doubt the explanatory and predictive relevance of statistical generalizations for particular occurrences. Yet there can be no question that statistical generalizations are widely invoked for explanatory and predictive purposes in such diverse fields as physics, genetics, and sociology. It will be necessary, therefore, to examine more carefully the logic of the arguments involved and, in particular, to reconsider the adequacy of the analysis suggested in (8.2). And while for a general characterization of the ambiguity of statistical explanation it was sufficient to use an illustration of statistical generalization of the vague form 'Almost all F are G,' we must now consider the explanatory and predictive use of statistical generalizations in the precise form of quantitative probability statements: 'The probability for an F to be a G is r.' This brings us to the question of the theoretical status of the statistical concept of probability.

9. The Theoretical Concept of Statistical Probability and the Problem of Ambiguity

The mathematical theory of statistical probability [29] seeks to give a theoretical systematization of the statistical aspects of random experiments. Roughly speaking, a random experiment is a repeatable process which yields in each case a particular finite or infinite set of "results," in such a way that while the results vary from repetition to repetition in an irregular and practically unpredictable manner, the relative frequencies with which the different results occur tend to become more or less constant for large numbers of repetitions. The theory of probability is intended to provide a "mathematical model," in the form of a deductive system, for the properties and interrelations of such long-run frequencies, the latter being represented in the model by probabilities.

In the mathematical theory of probability, each of the different outcomes of a random experiment which have probabilities assigned to them is represented by a set of what might be called elementary possibilities. For example, if the experiment is that of rolling a die, then getting an ace, a deuce, and so forth, would normally be chosen as elementary possibilities; let us refer to them briefly as I, II, . . ., VI, and let F be the set of these six elements. Then any of those results of rolling a die to which probabilities are usually assigned can be represented by a subset of F: getting an even number, by the set (II, IV, VI); getting a prime number, by the set (II, III, V); rolling an ace, by the unit set (I); and so forth. Generally, a random experiment is represented in the theory by a set F and a certain set, F*, of its subsets, which represent the possible outcomes that have definite probabilities assigned to them. F* will sometimes, but not always, contain all the subsets of F. The mathematical theory also requires F* to contain, for each of its member sets, its complement in F; and also for any two of its member sets, say G_1 and G_2, their sum, $G_1 \vee G_2$, and their products $G_1 \cdot G_2$. As a consequence, F* contains F as a member set.[30] The probabilities associated with the different outcomes of a random experiment then are represented by a real-valued function $p_F(G)$ which ranges over the sets in F*.

The postulates of the theory specify that p_F is a nonnegative additive set function such that $p_F(F) = 1$; i.e., for all G in F*, $p(G) \geqslant 0$; if G_1 and G_2 are mutually exclusive sets in F* then $p_F(G_1 \vee G_2) = p_F(G_1) + p_F(G_2)$. These stipulations permit the proof of the theorems of elementary probability theory; to deal with experiments that permit infinitely many different outcomes, the requirement of additivity is suitably extended to infinite sequences of mutually exclusive member sets of F*.

The abstract theory is made applicable to empirical subject matter by means of an interpretation which connects probability statements with sentences about long-run relative frequencies associated with random experiments. I will state the interpretation in a form which is essentially that given by Cramér,[31] whose book *Mathematical Methods of Statistics* includes a detailed discussion of the foundations of mathematical probability theory and its applications. For convenience, the notation '$p_F(G)$' for the probability of G relative to F will now be replaced by '$p(G, F)$.'

(9.1) *Frequency interpretation of statistical probability:* Let F be a given kind of random experiment and G a possible result of it; then the statement that $p(G, F) = r$ means that in a long series of repetitions of F, it is practically certain that the relative frequency of the result G will be approximately equal to r.

Evidently, this interpretation does not offer a precise definition of probability in statistical terms: the vague phrases 'a long series,' 'practically certain,' and 'approximately equal' preclude that. But those phrases are chosen deliberately to enable formulas stating precisely fixed numerical probability values to function as theoretical representations of near-constant relative frequencies of certain results in extended repetitions of a random experiment.

Cramér also formulates two corollaries of the above rule of interpretation; they refer to those cases where r differs very little from 0 or from 1. These corollaries will be of special interest for an examination of the question of ambiguity in the explanatory and predictive use of probability statements, and I will therefore note them here (in a form very similar to that chosen by Cramér):

(9.2a) If $0 \leqslant p(G, F) < \epsilon$, where ϵ is some very small number, then, if a random experiment of kind F is performed one single time, it can be considered as practically certain that the result G will not occur.[32]

(9.2b) If $1 - \epsilon < p(G, F) \leqslant 1$, where ϵ is some very small number, then if a random experiment of kind F is performed one single time, it can be considered as practically certain that the result G will occur.[33]

I now turn to the explanatory use of probability statements. Consider the experiment, D, of drawing, with subsequent replacement and thorough mixing, a ball from an urn containing one white ball and 99 black ones of the same size and material. Let us suppose that the probability, p(W, D), of obtaining a white ball as a result of a performance of D is .99. According to the statistical interpretation, this is an empirical hypothesis susceptible of test by reference to finite statistical samples, but for the moment, we need not enter into the question how the given hypothesis might be established. Now rule (9.2b) would seem to indicate that this hypothesis might be used in statistically explaining or predicting the results of certain individual drawings from the urn. Suppose for example, that a particular drawing, d, produces a white ball. Since p(W, D) differs from 1 by less than, say, 0.15, which is a rather small number, (9.2b) yields the following argument, which we might be inclined to consider as a statistical explanation of the fact that d is W:

(9.3)

$$\frac{1 - .015 < p(W, D) \leqslant 1;\ \text{and } .015 \text{ is a very small number.} \qquad d \text{ is an instance of } D.}{\text{It is practically certain that } d \text{ is } W.}$$

This type of reasoning is closely reminiscent of our earlier argument (8.3a), and it leads into a similar difficulty, as will now be shown. Suppose that besides the urn just referred to, which we will assume to be marked '1,' there are 999 additional urns of the same kind, each containing 100 balls, all of which are black. Let these urns be marked '2,' '3' . . . '1000.' Consider now the experiment E which consists in first drawing a ticket from a bag containing 1000 tickets of equal size, shape, etc., bearing the numerals '1,' '2' . . . '1000,' and then drawing a ball from the urn marked with the same numeral as the ticket drawn. In accordance with standard theoretical considerations, we will assume that p(W, E)

$= .00099$. (This hypothesis again is capable of confirmation by statistical test in view of the interpretation (9.1).) Now, let e be a particular performance of E in which the first step happens to yield the ticket numbered 1. Then, since e is an instance of E, the interpretative rule (9.2a) permits the following argument:

$$\text{(9.4a)} \quad \frac{\begin{array}{c} 0 \leqslant p(W, E) < .001; \text{ and } .001 \text{ is a very small number.} \\ \text{e is an instance of E.} \end{array}}{\text{It is practically certain that e is not W.}}$$

But on our assumption, the event e also happens to be an instance of the experiment D of drawing a ball from the first urn; we may therefore apply to it the following argument:

$$\text{(9.4b)} \quad \frac{\begin{array}{c} 1 - .015 < p(W, D) \leqslant 1; \text{ and } .015 \text{ is a very small number.} \\ \text{e is an instance of D.} \end{array}}{\text{It is practically certain that e is W.}}$$

Thus, in certain cases the interpretative rules (9.2a) and (9.2b) yield arguments which again exhibit what was called above the ambiguity of statistical systematization.

This ambiguity clearly springs from the fact that (a) the probability of obtaining an occurrence of some specified kind G depends on the random experiment whose result G is being considered, and that (b) a particular instance of G can normally be construed as an outcome of different kinds of random experiment, with different probabilities for the outcome in question; as a result, under the frequency interpretation given in (9.2a) and (9.2b), an occurrence of G in a particular given case may be shown to be both practically certain and practically impossible. This ambiguity does not represent a flaw in the formal theory of probability: it arises only when the empirical interpretation of that theory is brought into play.

It might be suspected that the trouble arises only when an attempt is made to apply probability statements to individual events, such as one particular drawing in our illustration: statistical probabilities, it might be held, have significance only for reasonably large samples. But surely this is unconvincing since there is only a difference in degree between a sample consisting of just one case and a sample consisting of many cases. And indeed, the problem of ambiguity recurs when probability statements are used to account for the frequency with which a specified kind of G of result occurs in finite samples, no matter how large.

For example, let the probability of obtaining recovery (R) as the result of the "random experiment" P of treating cases of streptococcus infection with penicillin be $p(R, P) = .75$. Then, assuming statistical independence of the individual cases, the frequency interpretation yields the following consequence, which refers to more or less extensive samples: For any positive deviation d, however small, there exists a specifiable sample size n_d such that it is practically certain that in one single series of n_d repetitions of the experiment P, the proportion of cases of R will deviate from .75 by less than d.[34] It would seem therefore that a recovery rate of close to 75 per cent in a sufficiently large number of instances of P could be statistically explained or predicted by means of the probability statement that $p(R, P) = .75$. But any such series of instances can also be construed as a set of cases of another random experiment for which it is practically certain that almost all the cases in the sample recover; alternatively, the given cases can be construed as a set of instances of yet another random experiment for which it is practically certain that none of the cases in a sample of the given size will recover. The arguments leading to this conclusion are basically similar to those presented in connection with the preceding illustrations of ambiguity; the details will therefore be omitted.

In its essentials, the ambiguity of statistical systematization can be characterized as follows: If a given object or set of objects has an attribute A which with high statistical probability is associated with another attribute C, then the same object or set of objects will, in general, also have an attribute B which, with high statistical probability, is associated with non-C. Hence, if the occurrence of A in the particular given case, together with the probability statement which links A with C, is regarded as constituting adequate grounds for the predictive or explanatory conclusion that C will almost certainly occur in the given case, then there exists, apart from trivial exceptions, always a competing argument which in the same manner, from equally true premises, leads to the predictive or explanatory conclusion that C will not occur in that same case. This peculiarity has no counterpart in nomological explanation: If an object or set of objects has a character A which is invariably accompanied by C then it cannot have a character B which is invariably accompanied by non-C.[35]

The ambiguity of statistical explanation should not, of course, be taken to indicate that statistical probability hypotheses have no explanatory or predictive significance, but rather that the above analysis of the logic of statistical systematization is inadequate. That analysis was suggested by a seemingly plausible analogy between the systematizing use of statistical generalizations and that of nomic ones—an analogy which seems to receive strong support from the interpretation of statistical generalizations which is offered in current statistical theory. Nevertheless, that analogy is deceptive, as will now be shown.

10. The Inductive Character of Statistical Systematization and the Requirement of Total Evidence

It is typical of the statistical systematizations considered in this study that their "conclusion" begins with the phrase 'It is almost certain that,' which never occurs in the conclusion of a nomological explanation or prediction. The two schemata (8.1) and (8.2) above exhibit this difference in its simplest form. A nomological systematization of the form (8.1) is a deductively valid argument: if its premises are true then so is its conclusion. For arguments of the form (8.2), this is evidently not the case. Could the two types of argument be assimilated more closely to each other by giving the conclusion of (8.1) the form 'It is certain that x is G'? This suggestion involves a misconception which is one of the roots of the puzzle presented by the ambiguity of statistical systematization. For what the statement 'It is certain that x is G' expresses here can be restated by saying that the conclusion of an argument of form (8.1) cannot be false if the premises are true, i.e., that the conclusion is a logical consequence of the premises. Hence, the certainty here in question represents not a property of the conclusion that x is G, but rather a relation which that conclusion bears to the premises of (8.1). Generally, a sentence is certain, in this sense, relative to some class of sentences just in case it is a logical consequence of the latter. The contemplated reformulation of the conclusion of (8.1) would therefore be an elliptic way of saying that

(10.1) 'x is G' is certain relative to, i.e., is a logical consequence of, the two sentences 'All F are G' and 'x is F.'[36]

But clearly this is not equivalent to the original conclusion of (8.1); rather, it is another way of stating that the entire schema (8.1) is a deductively valid form of inference.

Now, the basic error in the formulation of (8.2) is clear: near certainty, like certainty, must be understood here not as a property but as a relation; thus, the "conclusion" of (8.2) is not a complete statement but an elliptical formulation of what might be more adequately expressed as follows:

(10.2) 'x is G' is almost certain relative to the two sentences 'Almost all F are G' and 'x is F.'

The near certainty here invoked is sometimes referred to as (high) probability; the conclusion of arguments like (8.2) is then expressed by such phrases as '(very) probably, x is G,' or 'it is (highly) probable that x is G'; a nonelliptic restatement would then be given by saying that the sentences 'Almost all F are G' and 'x is F' taken jointly lend strong support to, or confer a high probability or a high degree of rational credibility upon, 'x is G.' The probabilities referred to here are logical or inductive probabilities, in contradistinction to the statistical probabilities mentioned in the premises of the statistical systematization under examination. The notion of logical probability will be discussed more fully a little later in the present section.

As soon as it is realized that the ostensible "conclusions" of arguments such as (8.2) and their quantitative counterparts, such as (9.3), are elliptic formulations of relational statements, one puzzling aspect of the ambiguity of statistical systematization vanishes: the apparently conflicting claims of matched argument pairs such as (8.3a) and (8.3b) or (9.4a) and (9.4b) do not conflict at all. For what the matched arguments in a pair claim is only that each of two contradictory sentences such as 'j is R' and 'j is not R' in the pair (8.3), is strongly supported by certain other statements, which, however, are quite different for the first and for the second sentence in question. Thus far then, no more of a "conflict" is established by a pair of matched statistical systematizations than, say, by the following pair of deductive arguments, which show that each of two contradictory sentences is even conclusively supported, or made certain, by other suitable statements which, however, are quite different for the first and for the second sentence in question:

(10.3a)

All F are G.
a is F.
―――――
a is G.

(10.3b)

No H is G.
a is H.
―――――
a is not G.

The misconception thus dispelled arises from a misguided attempt to construe arguments containing probability statements among their premises in analogy to deductive arguments such as (8.1)—an attempt which prompts the construal of formulations such as 'j is almost certain to be R' or 'probably, j is R' as self-contained complete statements rather than as elliptically formulated statements of a relational character.[37]

The idea, repeatedly invoked in the preceding discussion, of a statement or set of statements e (the evidence) providing strong grounds for asserting a certain statement h (the hypothesis), or of e lending strong support to h, or making h nearly certain is, of course, the central concept of the theory of inductive inference. It might be conceived in purely qualitative fashion as a relation S which h bears to e if e lends strong support to h; or it may be construed in quantitative terms, as a relation capable of gradations which represents the extent to which h is supported by e. Some recent theories of inductive inference have aimed at developing rigorous quantitative conceptions of inductive support: this is true especially of the systems of inductive logic constructed by Keynes and others and recently, in a particularly impressive form by Carnap.[38] If—as in Carnap's system—the concept is construed so as to possess the formal characteristics of a probability, it will be referred to as the logical (or inductive) probability, or as the degree of confirmation, c(h, e), of h relative to e. (This inductive probability, which is a function of statements, must be sharply distinguished from statistical probability, which is a function of classes of events.) At a general phrase referring to a quantitative notion of inductive support, but not tied to any one particular theory of inductive support or confirmation, let us use the expression '(degree of) inductive support of h relative to e.' [39]

An explanation, prediction, or retrodiction of a particular event or set of events by means of principles, which include statistical generalizations has then to be conceived as an inductive argument. I will accordingly speak of *inductive systematization* (in contradistinction to *deductive systematization,* where whatever is explained, predicted, or retrodicted is a deductive consequence of the premises adduced in the argument).

When it is understood that a statistical systematization is an inductive argument, and that the high probability or near certainty mentioned in the conclusions of such arguments as (8.3a) and (8.3b) is relative to the premises, then, as shown, one puzzle raised by the ambiguity of statistical explanation is resolved, namely, the impression of a conflict, indeed a near incompatibility, of the claims of two equally sound inductive systematizations.

But the same ambiguity raises another, more serious, problem, which now calls for consideration. It is very well to point out that in (8.3a) and (8.3b) the contradictory statements 'j is R' and 'j is not R' are shown to be almost certain by referring to different sets of "premises": it still remains the case that both of these sets are true. Here, the analogy to (10.3a) and (10.3b) breaks down: in these deductive arguments with contradictory conclusions the two sets of premises cannot both be true. Thus, it would seem that by statistical systematizations based on suitably chosen bodies of true information, we may lend equally strong support to two assertions which are incompatible with each other. But then—and this is the new problem—which of such alternative bodies of evidence is to be relied on for the purposes of statistical explanation or prediction?

An answer is suggested by a principle which Carnap calls *the requirement of total evidence.* It lays down a general maxim for all applications of inductive reasoning as follows: "in the application of inductive logic to a given knowledge situation, the total evidence available must be taken as basis for determining the degree of confirmation." [40] Instead of the total evidence, a smaller body, e_1, of evidence may be used on condition that the remaining part, e_2, of the total evidence is inductively irrelevant to the hypothesis h whose confirmation is to be determined. If, as in Carnap's system, the degree of confirmation is construed as an inductive probability, the irrelevance of e_2 for h relative to e_1 can be expressed by the condition that $c(h, e_1 \cdot e_2) = c(h, e_1)$.[41]

The general consideration underlying the requirement of total evidence is obviously this: If an investigator wishes to decide what credence to give to an empirical hypothesis or to what extent to rely on it in planning his actions, then rationality demands that he take into account all the relevant evidence available to him; if he were to consider only part of that evidence, he might arrive at a much more favorable, or a much less favorable, appraisal, but it would surely not be rational for him to base his decision on evidence he knew to be selectively biased. In terms of the concept of degree of confirmation, the point might be stated by saying that the degree of confirmation assigned to a hypothesis by the principles of inductive logic will represent the rational credibility of the hypothesis for a given investigator only if the argument takes into account all the relevant evidence available to the investigator.

The requirement of total evidence is not a principle of inductive logic, which is concerned with relations of potential evidential support among statements, i.e., with whether, or to what degree, a given set of statements supports a given hypothesis. Rather, the requirement is a maxim for the *application* of inductive logic; it might be said to state a necessary condition of rationality in forming beliefs and making decisions on the basis of available evidence. The requirement is not limited to arguments of the particular form of statistical systematizations, where the evidence represented by the "premises," includes statistical generalizations: it is a necessary condition of rationality in the application of any mode of inductive reasoning, including, for example, those cases in which the evidence contains no generalizations, statistical or universal, but only data on particular occurrences.

Let me note here that in the case of deductive systematization, the requirement is automatically satisfied and thus presents no special problem.[42] For in a deductively valid argument whose premises constitute only part of the total evidence available at the time, that part provides conclusive grounds for asserting the conclusion; and the balance of the total evidence is irrelevant to the conclusion in the strict sense that if it were added to the premises, the resulting premises would still constitute conclusive grounds for the conclusion. To state this in the language of inductive logic: the logical probability of the conclusion relative to the premises of a deductive systematization is 1, and it remains 1 no matter what other parts of the total evidence may be added to the premises.

The residual problem raised by the ambiguity of probabilistic explanation can now be resolved by requiring that if a statistical systematization is to qualify as a rationally acceptable explanation or prediction (and not just as a formally sound *potential* explanation or prediction), it must satisfy the requirement of total evidence. For under this requirement, the "premises" of an acceptable statistical systematization whose "conclusion" is a hypothesis h must consist either of the total evidence e or of some subset of it which confers on h the same inductive probability as e; and the same condition applies to an acceptable systematization which has the negation of h as its "conclusion." But one and the same evidence, e, cannot—if it is logically self-consistent—confer a high probability on h as well as on its negation, since the sum of the two probabilities is unity. Hence, of two statistical systematizations whose premises confer high probabilities on h and on the negation of h, respectively, at least one violates the requirement of total evidence and is thus ruled out as unacceptable.

The preceding considerations suggest that a statistical systematization may be construed generally as an inductive argument showing that a certain statement or finite set of statements, e, which includes at least one statistical law, gives strong but not logically conclusive support to a statement h, which expresses whatever is being explained, predicted, retrodicted, etc. And if an argument of this kind is to be acceptable in science as an empirically sound explanation, prediction, or the like—rather than only a formally adequate, or potential one—then it will also have to meet the requirement of total evidence.

But an attempt to apply the requirement of total evidence to statistical systematizations of the simple kind considered so far encounters a serious obstacle. This was noted, among others, by S. Barker with special reference to "statistical syllogisms," which are inductive arguments with two premises, very similar in character to the arguments (9.4a) and (9.4b) above. Barker points out, in effect, that the statistical syllogism is subject to what has been called here the ambiguity of statistical systematization, and he goes on to argue that the principle of total evidence will be of no avail as a way to circumvent this shortcoming because generally our total evidence will consist of far more than just two statements, which would moreover have to be of the particular form required for the premises of a statistical syllogism.[43] This observation would not raise a serious difficulty, at least theoretically speaking, if an appropriate general system of inductive logic were available: the rules of this system might enable us to show that that part of our total evidence which goes beyond the premises of our simple statistical argument is inductively irrelevant to the conclusion in the sense specified earlier in this section. Since no inductive logic of the requisite scope is presently at hand, however, it is a question of great interest whether some more manageable substitute for the requirement of total evidence might not be formulated which would not presuppose a full system of inductive logic and would be applicable to simple statistical systematizations. This question will be examined in the next section on the basis of a closer analysis of simple statistical systematizations offered by empirical science. . . .

NOTES

1. This distinction was developed briefly in Hempel [25], Sec. 2.
2. See Hempel [24], especially Secs. 1–4; Hempel [25]; and Hempel and Oppenheim [26]. This latter article will henceforth be referred to as *SLE*. The point of these discussions was to give a more precise and explicit statement of the deductive model of scientific explanation and to exhibit and analyze some of the logical and methodological problems to which it gives rise: the general conception of explanation as deductive subsumption under more general principles had been set forth much earlier by a variety of authors, some of whom are listed in *SLE*, fn. 4. In fact, in 1934 that conception was explicitly presented in the following passage of an introductory textbook: "Scientific explanation consists in subsuming under some rule or law which expresses an invariant character of a group of events, the particular event it is said to explain. Laws themselves may be explained, and in the same manner, by showing that they are consequences of more comprehensive theories." (Cohen and Nagel [10], p. 397.) The conception of the explanation of laws by deduction from theories was developed in great detail by N. R. Campbell; for an elementary account see his book [4], which was first published in 1921. K. R. Popper, too, has set forth this deductive conception of explanation in several of his publications (cf. fn. 4 in *SLE*); his earliest statement appears in Sec. 12 of his book [38], which has at long last been published in a considerably expanded English version [40].
3. The suggestive terms 'covering law' and 'covering-law model' are borrowed from Dray, who, in his book [13], presents a lucid and stimulating critical discussion of the question whether, or to what extent, historical explanation conforms to the deductive pattern here considered. To counter a misunderstanding

that might be suggested by some passages in Ch. II, Sec. 1 of Dray's book, I would like to emphasize that the covering-law model must be understood as permitting reference to *any number of laws* in the explanation of a given phenomenon: there should be no restriction to just *one* "covering law" in each case.

4. More accurately the explanation of a general law by means of a theory will usually show (1) that the law holds only within a certain range of application, which may not have been made explicit in its standard formulation; (2) that even within that range, the law holds only in close approximation, but not strictly. This point is well illustrated by Duhem's emphatic reminder that Newton's law of gravitation, far from being an inductive generalization of Kepler's laws, is actually incompatible with them, and that the credentials of Newton's theory lie rather in its enabling us to compute the perturbations of the planets, and thus their deviations from the orbits assigned to them by Kepler. (See Duhem [14], pp. 312ff, and especially p. 317. The passages referred to here are included in the excerpts from P. P. Wiener's translation of Duhem's work that are reprinted in Feigl and Brodbeck [15], under the title "Physical Theory and Experiment.")

Analogously, Newtonian theory implies that the acceleration of a body falling freely in a vacuum toward the earth will increase steadily, though over short distances it will be very nearly constant. Thus, strictly speaking, the theory contradicts Galileo's law, but shows the latter to hold true in very close approximation within a certain range of application. A similar relation obtains between the principles of wave optics and those of geometrical optics.

5. No claim was made that this is the only kind of scientific explanation; on the contrary, at the end of Sec. 3, it was emphasized that "Certain cases of scientific explantion involve 'subsumption' of the explanandum under a set of laws of which at least some are statistical in character. Analysis of the peculiar logical structure of that type of subsumption involves difficult special problems. The present essay will be restricted to an examination of the causal type of explanation . . ." A similar explicit statement is included in the final paragraph of Sec. 7 and in Sec. 5.3 of the earlier article, Hempel [24]. These passages seem to have been overlooked by some critics of the covering-law model.

6. This was done already in *SLE,* Sec. 7.

7. The term 'lawlike sentence' and the general characterization given here of its intended meaning are from Goodman [20]. The difficult problem of giving an adequate general charaterization of those sentences which if true would constitute laws will not be dealt with in the present essay. For a discussion of the issues involved, see, for example, *SLE,* Secs. 6–7; Braithwaite [3], Ch. IX, where the central question is described as concerning "the nature of the difference, if any, between 'nomic laws' and 'mere generalizations' "; and the new inquiry into the subject by Goodman [20, 21]. All the sentences occurring in a potential explanation are assumed, of course, to be empirical in the broad sense of belonging to some language adequate to the purposes of empirical science. On the problem of characterizing such systems more explicity, see especially Scheffler's stimulating essay [46].

8. [49], p. 185.

9. *Ibid.,* p. 194.

10. Note, however, that from a law of coexistence connecting certain parameters it is possible to derive laws of succession concerning the rates of change of those parameters. For example, the law expressing the period of a mathematical pendulum as a function of its length permits the derivation, by means of the calculus, of a further law to the effect that if the length of the pendulum changes in the course of time, then the rate of change of its period at any moment is proportional to the rate of change of its length, divided by the square root of its length, at that moment.

11. Among these is Schlick [48], who gives credit to Wittgenstein for the idea that a law of nature does not have the character of a statement, but rather that of an instruction for the formation of statements. Schlick's position in this

article is prompted largely by the view that a genuine statement must be definitively verifiable–a condition obviously not met by general laws. But this severe verifiability condition cannot be considered as an acceptable standard for scientific statements.

More recently, Ryle–see, for example, [44], pp. 121–123–has described law statements as statements which are true or false, but one of whose jobs is to serve as inference tickets: they license their possessors to move from the assertion of some factual statements to the assertion of others.

Toulmin [53], has taken the view, more closely akin to Schlick's, that laws of nature and physical theories do not function as premises in inferences leading to observational statements, but serve as modes of representation and as rules of inference according to which statements of empirical fact may be inferred from other such statements. An illuminating discussion of this view will be found in E. Nagel's review of Toulmin's book, in *Mind,* 63:403–412 (1954); it is reprinted in Nagel [35], pp. 303–315.

Carnap [5], par. 51, makes explicit provision for the construction of languages with extralogical rules of inferences. He calls the latter physical rules, or P-rules, and emphasizes that whether, or to what extent, P-rules are to be countenanced in constructing a language is a question of expedience. For example, adoption of P-rules may oblige us to alter the rules—and thus the entire formal structure—of the language of science in order to account for some new empirical findings which, in a language without P-rules, would prompt only modification or rejection of certain statements previously accepted in scientific theory.

The admission of material rules of inference has been advocated by W. Sellars in connection with his analysis of subjunctive conditionals; see [51, 52]. A lucid general account and critical appraisal of various reasons that have been adduced in support of construing general laws as inferences rules will be found in Alexander [1].

12. For detailed discussions of these issues, see Barker [2], especially Ch. 7; Scheffler [46], especially Secs. 13–18; Hempel [25], especially Sec. 10.

13. On this point, see the review by Nagel mentioned in fn. 11.

14. This remark does not hold, however, when all the laws invoked in the explanans are laws of coexistence (see Sec. 4) and all the particular statements adduced in the explanans pertain to events that are simultaneous with the explanandum event. I am indebted to Mr. S. Bromberger for having pointed out to me this oversight in my formulation.

15. [18], p. 265.

16. *Ibid.*

17. [42], p. 282.

18. *Ibid.,* p 286.

19. In Sec. 3 of *SLE,* to which Rescher refers in his critique, an explanation of a past event is explicitly construed as a deductive argument inferring the occurrence of the event from "antecedent conditions" and laws; so that the temporal direction of the inference underlying explanation is the same as that of a predictive nomological argument, namely, from statements concerning certain initial (and boundary) conditions to a statement concerning the *subsequent* occurrence of the explanandum event.

I should add, however, that although all this is said unequivocally in *SLE* there is a footnote in *SLE,* Sec. 3, which is certainly confusing, and which, though not referred to by Rescher, might have encouraged him in his misunderstanding. The footnote, numbered 2a, reads: "The logical similarity of explanation and prediction, and the fact that one is directed towards past occurrences, the other towards future ones, is well expressed in the terms 'postdictability' and 'predictability' used by Reichenbach [in *Philosophic Foundations of Quantum Mechanics,* p. 13]." To reemphasize the point at issue: postdiction, or retrodiction, is not the same thing as explanation.

20. [50].

21. [47].

22. *SLE,* Sec. 3.

23. Scheffler [47], p. 300.
24. See *ibid.*, p. 296; Scriven [49].
25. Professor Nelson Goodman, to whom I had mentioned my difficulties with the notion of a statement being "about" a certain subject, showed me a draft of an article entitled "About," which has now appeared in *Mind*, 70:1–24 (1961); in it, he proposes an analysis of the notion of aboutness which will no doubt prove helpful in dealing with the puzzle outlined here, and which may even entirely resolve it.
26. It is sometimes argued that a statement asserting such a universal connection rests, after all, only on a finite, and necessarily incomplete, body of evidence; that, therefore, it may well have exceptions which have so far gone undiscovered, and that, consequently, it should be qualified as probabilistic, too. But this argument fails to distinguish between the claim made by a given statement and the strength of its evidential support. On the latter score, all empirical statements have to count as only more or less well supported by the available evidence; but the distinction between laws of strictly universal form and those of statistical form refers to the claim made by the statements in question: roughly speaking, the former attribute a certain character to all members of a specified class; the latter, to a fixed proportion of its members.
27. The characterization given here of the concept of statistical probability seems to me to be in agreement with the general tenor of the "propensity interpretation" advocated by Popper in recent years. This interpretation "differs from the purely statistical or frequency interpretation only in this–that it considers the probability as a characteristic property of the experimental arrangement rather than as a property of a sequence"; the property in question is explicitly construed as *dispositional*. (Popper [39], pp. 67–68. See also the discussion of this paper at the Ninth Symposium of the Colston Research Society, in Körner [30], pp. 78–89 passim.) However, the currently available statements of the propensity interpretation are all rather brief (for further references, see Popper [40]; a fuller presentation is to be given in a forthcoming book by Popper.
28. In fact, Goodman [20], has argued very plausibly that one symptomatic difference between lawlike and nonlawlike generalizations is precisely that the former are able to lend support to corresponding subjunctive or counterfactual conditionals; thus the statement 'If this copper key were to be heated it would expand' can be supported by the law mentioned above. By contrast, the statement 'All objects ever placed on this table weigh less than one pound' is nonlawlike, i.e., even if true, it does not count as a law. And indeed, even if we knew it to be true, we would not adduce it in support of corresponding counterfactuals; we would not say, for example, that if a volume of Merriam-Webster's Unabridged Dictionary had been place on the table, it would have weighed less than a pound. Similarly, it might be added, general statements of this latter kind possess no explanatory power: this is why the sentences $L_1, L_2, \ldots, L_n$ in the explanans of any deductive-nomological explanation are required to be lawlike. The preceding considerations suggest the question whether there is a category of statistical probability statements whose status is comparable to that of accidental generalizations. It would seem clear, however, that insofar as statistical probability statements are construed as dispositional in the sense suggested above, they have to be considered as being analogous to lawlike statements.
29. The mathematical theory of statistical probability has been developed in two major forms. One of these is based on an explicit definition of probabilities as limits of relative frequencies in certain infinite reference sequences. The use of this limit definition is an ingenious attempt to permit the development of a simple and elegant theory of probability by means of the apparatus of mathematical analysis, and to reflect at the same time the intended statistical application of the abstract theory. The second approach, which offers certain theoretical advantages and is now almost generally adopted, develops the formal theory of probability as an abstract theory of certain set-functions and then specifies rules for its application to empirical subject matter. The brief characterization of the theory of statistical probability given in this section follows the second

approach. However, the problem posed by the ambiguity of statistical systematization arises as well when the limit definition of probability is adopted.
30. See, for example, Kolmogoroff [31], Sec. 2.
31. [11], pp. 148–149. Similar formulations have been given by other representatives of this measure-theoretical conception of statistical probability, for example, by Kolmogoroff [31], p. 4.
32. Cf. Cramér [11], p. 149; see also the very similar formulation in Kolmogoroff [31], p. 4.
33. Cf. Cramér [11], p. 150.
34. *Ibid.*, pp. 197–198.
35. My manuscript here originally contained the phrase 'is invariably (or even in some cases) accompanied by non-C.' By reading the critique of this passage as given in the manuscript of Professor Scriven's contribution to the present volume, I became aware that the claim made in parentheses is indeed incorrect. Since the point is entirely inessential to my argument, I deleted the parenthetical remark after having secured Professor Scriven's concurrence. However, Professor Scriven informed me that he would not have time to remove whatever references to this lapse his manuscript might contain: I therefore add this note for clarification.
36. A sentence of the form 'It is certain that x is G' ostensibly attributes the modality of certainty to the proposition expressed by the conclusion in relation to the propositions expressed by the premises. For the purposes of the present study, involvement with propositions can be avoided by construing the given modal sentence as expressing a logical relation that the conclusion, taken as a sentence, bears to the premise sentences. Concepts such as near certainty and probability can, and will here, equally be treated as applying to pairs of sentences rather than to pairs of propositions.
37. These remarks seem to me to be relevant, for example, to C. I. Lewis's notion of categorical, as contradistinguished from hypothetical, probability statements. For in [32], p. 319, Lewis argues as follows: "Just as 'If D then (certainly) P, and D is the fact,' leads to the categorical consequence, 'Therefore (certainly) P'; so too, 'If D then probably P, and D is the fact,' leads to a categorical consequence expressed by 'It is probable that P.' And this conclusion is not merely the statement over again of the probability relation between 'P' and 'D'; any more than 'Therefore (certainly) P' is the statement over again of 'If D then (certainly) P.' 'If the barometer is high, tomorrow will probably be fair; and the barometer *is* high,' categorically assures something expressed by 'Tomorrow will probably be fair.' This probability is still relative to the grounds of judgment; but if these grounds are actual, and contain all the available evidence which is pertinent, then it is not only categorical but may fairly be called *the* probability of the event in question."

This position seems to me to be open to just those objections which have been suggested in the main text. If 'P' is a statement, then the expressions 'certainly P' and 'probably P' as envisaged in the quoted passage are not statements: if we ask how one would go about trying to ascertain whether they were true, we realize that we are entirely at a loss unless and until a reference set of statements or assumptions is specified relative to which P may then be found to be certain, or to be highly probable, or neither. The expressions in question, then, are essentially incomplete; they are elliptic formulations of relational statements; neither of them can be the conclusion of an inference. However plausible Lewis's suggestion may seem, there is no analogue in inductive logic to *modus ponens*, or the "rule of detachment" of deductive logic, which, given the information that 'D,' and also 'if D then P,' are true statements, authorizes us to detach the consequent 'P' in the conditional premise and to assert it as a self-contained statement which must then be true as well.

At the end of the quoted passage, Lewis suggests the important idea that 'probably P' might be taken to mean that the total relevant evidence available at the time confers high probability upon P; but even this statement is relational in that it tacitly refers to some unspecified time; and besides, his general notion of a categorical probability statement as a conclusion of an argument is not made dependent on the assumption that the premises of the argument include all the relevant evidence available.

It must be stressed, however, that elsewhere in his discussion, Lewis emphasizes the relativity of (logical) probability, and thus the very characteristic which rules out the conception of categorical probability statements.

38. See especially [7, 8], and, for a very useful survey, [6].

39. In a recent study, Kemeny and Oppenheim [29], have proposed, and theoretically developed, an interesting concept of "degree of factual support" (of a hypothesis by given evidence), which differs from Carnap's concept of degree of confirmation, or inductive probability, in important respects; for example, it does not have the formal character of a probability function. For a suggestive distinction and comparison of different concepts of evidence, see Rescher [43].

40. Carnap [7], p. 211. In his comments, pp. 211–213, Carnap points out that in less explicit form, the requirement of total evidence has been recognized by various authors at least since Bernoulli. The idea also is suggested in the passage from Lewis [32], quoted in fn. 36. Similarly, Williams, whose book *The Ground of Induction* centers about various arguments that have the character of statistical systematizations, speaks of "the most fundamental of all rules of probability logic, that 'the' probability of any proposition is its probability in relation to the known premises and them only." (Williams [55], p. 72.)

I wish to acknowledge here my indebtedness to Professor Carnap, to whom I turned in 1945, when I first noticed the ambiguity of statistical explanation, and who promptly pointly pointed out to me in a letter that this was but one of several apparent paradoxes of inductive logic which result from violations of the requirement of total evidence.

In his recent book, Barker [2], pp. 70–78, concisely and lucidly presents the gist of the puzzle under consideration here and examines the relevance to it of the principle of total evidence.

41. Cf. Carnap [7], pp. 211, 494.

42. Carnap [7], p. 211, says "There is no analogue to this requirement [of total evidence] in deductive logic"; but it seems more accurate to say that the requirement is automatically met here.

43. See Barker [2], pp. 76–78. The point is made in a more general form by Carnap [7], p. 404.

REFERENCES

1. Alexander, H. Gavin. "General Statements as Rules of Inference," in *Minnesota Studies in the Philosophy of Science,* Vol. II, H. Feigl, M. Scriven, and G. Maxwell, eds. Minneapolis: University of Minnesota Press, 1958. Pp. 309–329.
2. Barker, S. F. *Induction and Hypothesis.* Ithaca: Cornell University Press, 1957.
3. Braithwaite, R. B. *Scientific Explanation.* Cambridge: Cambridge University Press, 1953.
4. Campbell, Norman. *What Is Science?* New York: Dover Press, 1952.
5. Carnap, R. *The Logical Syntax of Language.* New York: Harcourt, Brace, and Co., 1937.
6. Carnap, R. "On Inductive Logic," *Philosophy of Science,* 12:72–97 (1945).
7. Carnap, R. *Logical Foundations of Probability.* Chicago: University of Chicago Press, 1950.
8. Carnap, R. *The Continuum of Inductive Methods.* Chicago: University of Chicago Press, 1952.

9. Carnap, R., and Y. Bar-Hillel. *An Outline of a Theory of Semantic Information.* Massachusetts Institute of Technology, Research Laboratory of Electronics. Technical Report No. 247. 1952.
10. Cohen, M. R., and E. Nagel. *An Introduction to Logic and Scientific Method.* New York: Harcourt, Brace, and Co., 1934.
11. Cramér, H. *Mathematical Methods of Statistics.* Princeton: Princeton University Press, 1946.
12. De Finetti, Bruno. "Recent Suggestions for the Reconciliations of Theories of Probability," in *Proceedings of the Second Berkeley Symposium on Mathematical Statistics and Probability,* J. Neyman, ed. Berkeley: University of California Press, 1951. Pp. 217–226.
13. Dray, W. *Laws and Explanation in History.* London: Oxford University Press, 1957.
14. Duhem, Pierre. *La Théorie physique, son objet et sa structure.* Paris: Chevalier et Rivière, 1906.
15. Feigl, H., and May Brodbeck, eds. *Readings in the Philosophy of Science.* New York: Appleton-Century-Crofts, 1953.
16. Feigl, H., and W. Sellars, eds. *Readings in Philosophical Analysis.* New York: Appleton-Century-Crofts, 1949.
17. Feigl, H., M. Scriven, and G. Maxwell, eds. *Minnesota Studies in the Philosophy of Science,* Vol. II. Minneapolis: University of Minnesota Press, 1958.
18. Galilei, Galileo. *Dialogues Concerning Two New Sciences.* Transl. by H. Crew and A. de Salvio. Evanston, Ill.: Northwestern University, 1946.
19. Gardiner, Patrick, ed. *Theories of History.* Glencoe, Ill.: Free Press, 1959.
20. Goodman, Nelson. "The Problem of Counterfactual Conditionals," *Journal of Philosophy,* 44:113–128 (1947). Reprinted, with minor changes, as the first chapter of Goodman [20].
21. Goodman, Nelson. *Fact, Fiction, and Forecast.* Cambridge, Mass.: Harvard University Press, 1955.
22. Goodman, Nelson. "Recent Developments in the Theory of Simplicity," *Philosophy and Phenomenological Research,* 19:429–446 (1959).
23. Hanson, N. R. "On the Symmetry between Explanation and Prediction," *Philosophical Review,* 68:349–358 (1959).
24. Hempel, C. G. "The Function of General Laws in History," *Journal of Philosophy,* 39:35–48 (1942). Reprinted in Feigl and Sellars [16], and in Jarrett and McMurrin [27].
25. Hempel, C. G. "The Theoretician's Dilemma," in *Minnesota Studies in the Philosophy of Science,* Vol. II, H. Fiegl, M. Scriven, and G. Maxwell, eds. Minneapolis: University of Minnesota Press, 1958. Pp. 37–98.
26. Hempel, C. G., and P. Oppenheim. "Studies in the Logic of Explanation," *Philosophy of Science,* 15:135–175 (1948). Secs 1-7 of this article are reprinted in Feigl and Brodbeck [15].
27. Jarrett, J. L., and S. M. McMurrin, eds. *Contemporary Philosophy.* New York: Henry Holt, 1954.
28. Jeffrey, R. C. "Valuation and Acceptance of Scientific Hypotheses," *Philosophy of Science,* 23:237–246 (1956).
29. Kemeny, J. G., and P. Oppenheim. "Degree of Factual Support," *Philosophy of Science,* 19:307–324 (1952).

30. Körner, S., ed. *Observation and Interpretation.* Proceedings of the Ninth Symposium of the Colston Research Society. New York: Academic Press Inc., 1957. London: Butterworth, 1957.
31. Kolmogoroff, A. *Grundbegriffe der Wahrscheinlichkeitrechnung.* Berlin: Springer, 1933.
32. Lewis, C. I. *An Analysis of Knowledge and Valuation.* La Salle, Ill.: Open Court Publishing Co., 1946.
33. Luce, R. Duncan, and Howard Raiffa. *Games and Decisions.* New York: Wiley, 1957.
34. Mises, Richard von. *Positivism. A Study in Human Understanding.* Cambridge, Mass.: Harvard University Press, 1951.
35. Nagel, E. *Logic without Metaphysics.* Glencoe, Ill.: The Free Press, 1956.
36. Neumann, John von, and Oskar Morgenstern. *Theory of Games and Economic Behavior.* Princeton: Princeton University Press, 2d ed., 1947.
37. Neyman, J. *First Course in Probability and Statistics.* New York: Henry Holt, 1950.
38. Popper, K. R. *Logik der Forschung.* Vienna: Springer, 1935.
39. Popper, K. R. "The Propensity Interpretation of the Calculus of Probability, and the Quantum Theory," in *Observation and Interpretation,* S. Körner, ed. Proceedings of the Ninth Symposium of the Colston Research Society. New York: Academic Press Inc., 1957. London: Butterworth, 1957. Pp 65–70.
40. Popper, K. R. *The Logic of Scientific Discovery.* London: Hutchinson, 1959.
41. Reichenbach, H. *The Theory of Probability.* Berkeley and Los Angeles: University of California Press, 1949.
42. Rescher, N. "On Prediction and Explanation," *British Journal for the Philosophy of Science,* 8:281–290 (1958).
43. Rescher, N. "A Theory of Evidence," *Philosophy of Science,* 25:83–94 (1958).
44. Ryle, G. *The Concept of Mind.* London: Hutchinson, 1949.
45. Savage, L. J. *The Foundations of Statistics.* New York: Wiley, 1954.
46. Scheffler, I. "Prospects of a Modest Empiricism," *Review of Metaphysics,* 10:383–400, 602–625 (1957).
47. Scheffler, I. "Explanation, Prediction, and Abstraction," *British Journal for the Philosophy of Science,* 7:293–309 (1957).
48. Schlick, M. "Die Kausalität in der gegenwärtigen Physik," *Die Natur-Wissenschaften,* 19:145–162 (1931).
49. Scriven, M. "Definitions, Explanations, and Theories," in *Minnesota Studies in the Philosophy of Science,* Vol. II, H. Feigl, M. Scriven, and G. Maxwell, eds. Minneapolis: University of Minnesota Press, 1958. Pp. 99–195.
50. Scriven, M. "Explanations, Predictions, and Laws," in this volume of *Minnesota Studies in the Philosophy of Science,* pp. 170–230.
51. Sellars, W. "Inference and Meaning," *Mind,* 62:313–338 (1953).
52. Sellars, W. "Conterfactuals, Dispositions, and the Causal Modalities," in *Minnesota Studies in the Philosophy of Science,* Vol. II, H. Feigl, M. Scriven, and G. Maxwell, eds. Minneapolis: University of Minnesota Press, 1958. Pp. 225–308.
53. Toulmin, S. *The Philosophy of Science.* London: Hutchinson, 1953.
54. Wald, A. *Statistical Decision Functions.* New York: Wiley, 1950.
55. Williams, D. C. *The Ground of Induction.* Cambridge, Mass.: Harvard University Press, 1947.

READING 8

The Quarrel About Historical Explanation

Rudolph H. Weingartner

I

An historically minded reader of the current literature on historical explanation might well assess that literature in something like the following way: "We have here a discussion of a philosophic question in its early and lively phase. The temporal and substantive starting point is a strong proposal made by Carl G. Hempel in an article called 'The Function of General Laws in History.'[1] The literature which follows it includes every possible kind of reaction: Hempel has found his adherents; his analysis has provoked criticism of every variety, and it has stimulated a wide range of alternative proposals—some of them closely resembling that of Hempel, others differing radically. In a few years' time interest in the issue will die down, for the various explications of historical explanation will converge and disagreement will be confined to more and more minute details."

Some of the things noted by this imaginary observer of the passing philosophical scene are quite correct. The discussion, now so lively, did for all practical purposes begin with Hempel's article [2]; moreover, almost every paper written on this question makes Hempel's analysis of historical explanation its own starting point. Nevertheless, our observer's assessment and prognosis of the discussion is in error. It fails, I think, to go beneath the surface and to see the one fundamental issue which divides the disputants, an issue which will not be settled in the course of more precise and detailed analyses of historical explanation. It will be the burden of this paper to show that what appears to be a dispute about the precise

Reprinted by permission of the author and the *Journal of Philosophy*, 58, No. 2, 1961, pp. 29–45.

nature of historical explanation is in fact the product of a disagreement about the nature of philosophic method, with all the far-reaching implications that such a disagreement implies.

To begin with, the discussion of historical explanation is not so free a free-for-all as it might at first seem to be. There are clearly two sides: Hempel and Hempelians on one and anti-Hempelians on the other. What accounts for the appearance of a war of all against all is the lack of a general on the anti-Hempelian side. The Hempelians are fewer in number, but their ranks are ordered—or, rather, as ordered as philosophical ranks can be expected to be.[3] While the opponents of Hempelianism seldom fight among each other, they make very little effort at concerted action. Indeed, the anti-Hempelians rarely even mention each other. In order, now, to be in a position to see what the quarrel is about, it will be necessary to summarize the positions of the two sides, starting with the Hempel proposal itself.

II

Hempel begins with an outline of the structure of explanation in the natural sciences.[4] According to this scheme, an event (as distinct from a law) is explained if and only if the statement asserting its occurrence (E) can be logically deduced from premises consisting of (1) a set of well confirmed[5] statements expressing instantial or determining conditions (C_1, C_2, . . . C_n) and (2) a set of well confirmed[5] universal hypotheses, i.e., general laws (L_1, L_2, . . . L_n). E is the explanandum and the two sets of statements, C_1, C_2, . . . C_n and L_1, L_2, . . . L_n, are the explanans, where E is deducible from C_1, C_2, . . . C_n and L_1, L_2, . . . L_n.

But is this also the structure of an historical explanation? Certainly not in the sense that one will actually find in history books statements which conform to the pattern just indicated. Aside from the niceties of rigor and precision, explanations found in works of history rarely state instantial conditions with the requisite completeness and accuracy. Even less frequently do historians offer explanations which mention the general law or laws in virtue of which the explanandum is deducible.

These observations about actual explanations are all Hempel's. Indeed, Hempel is at pains to point out further how serious the gap is between his model and the explanations historians actually provide. Although no law may in fact be mentioned, it would occasionally be no trouble to state it—merely a bore. The Normandy sank to the bottom of New York Harbor because a bomb was set off inside it. The law, "Whenever a bomb is set off in a ship (and certain circumstances prevail), the ship will sink," is familiar and not worthy of mention. Often, however, the required hypotheses are so complex, one would be hard put to it even to formulate them. Think of the stuff histories are made of: outbreaks of wars and revolutions, spreads of religions, failures of foreign policies. No simple laws will do as generalizations belonging to the *explanantia* of such events. And finally, there is the problem of truth or confirmation. Even when a law is stated, how likely is it that it can be well confirmed? The historian is in no position to conduct experiments; the number of instances of any generalization he might have wanted to use may forever be insufficient to establish it.

Not only, then, does Hempel recognize that historians do not provide explanations which conform in so many words to the pattern outlined, but he is careful to note that, even if he wanted to, the historian would seldom be able to do so. All that the historian does (and, in most cases, can) provide is an *explanation sketch* which "consists of a more or less vague indication of the laws and initial conditions considered as relevant and [which] needs 'filling out' in order to turn it into a full-fledged explanation."[6]

But if it is true that in the vast majority of cases historians merely offer explanation *sketches,* it follows that the paradigm first set down *is* the proper analysis of an historical *explanation.* To explain why an event occurred is to deduce the statement expressing what has occurred from instantial condition statements and general laws. To the extent to which and in any way in which the historian falls short of these requirements, he fails to explain, but merely provides a sketch of an explanation. Stated in somewhat exaggerated terms, Hempel's article must be construed not as an argument in support of the thesis that historical explanation has a certain pattern, but rather as a claim that the explanations offered in history are in certain ways and in varying degrees inadequate. Full-fledged explanations always have the same structure, whether they occur in science or history.[7]

III

All anti-Hempelians agree in making less of a demand upon historical explanation than do Hempel and Hempelians. They emphatically do not agree on the extent to which the demands upon the historian as explainer should be reduced. It is convenient for our purposes to divide the alternative analyses[8] of historical explanation into three groups, with the first closest to and the third furthest from the Hempel proposal.

(1) There are those who agree with Hempel that the explanans of an historical explanation must contain not only a set of instantial conditions, but general statements of some kind as well. According to these writers, however, an historical explanation may be complete and emphatically not a mere sketch that requires further "filling out," even when its generalization or generalizations fall short in various degrees and ways from being universal hypotheses.

And when the requirement of a law is relaxed sufficiently, it no longer makes sense to speak of deducibility. According to Patrick Gardiner, for example, historians make use of generalizations which make imprecise correlations, to which exceptions may be granted, the terms of which may be "open." Such statements—Gardiner prefers to call them "assessments" or "judgments"—fashion some sort of links between the "details" of the instantial conditions and the event to be explained; they serve as "guiding threads." In fulfilling this function *they do all that is needed for a complete historical explanation.* Statements of this kind are not "made, or accepted, in default of something 'better'; we should rather insist that their formulation represents the *end* of historical inquiry, not that they are stages on the journey towards that end." [9]

(2) A proposal by T. A. Goudge [10] will serve to exemplify a second type of alternative to Hempel's proposal. According to him, historical explanations [11] can be expected to achieve in practice no more than the following pattern: the explanation indicates conditions, temporally prior to or simultaneous with E, which are jointly sufficient but not independently necessary for the occurrence of E. An historical explanation would be complete if the conditions specified were independently necessary for the occurrence of E.

In this, the sufficient condition model of historical explanation, laws or law-like statements are explicitly said not to play a role. The event to be explained *is* explained in virtue of the fact that it "falls into place" as the terminal phase of a sequence which, in its entirety, constitutes the sufficient condition for the occurrence of the event. The logical relations among the components of an explanation are said to be not implicative, but conjunctive. The deductive model is "the wrong model to have in mind." Instead, coherent narrative is taken to be the model of explanation.

(3) The coherence of a narrative does not, of course, depend upon the *sufficiency* of the conditions said to bring about an event. Accordingly, writers can be found who depart still further from the Hempelian model by claiming that an event may be adequately

explained by citing *necessary* conditions for its occurrence. W. B. Gallie, for example, requires of what he calls a "characteristically genetic [or historical] explanation" of events that it refer "us to one or a number of their temporally prior necessary conditions. . . . In such cases explanation commences from our recognition of the event to be explained as being of such a kind that *some one* of a disjunction of describable conditions is necessary to its occurrence; and the explanation consists in elucidating *which one* of this disjunctive set is applicable, in the sense of being necessary, to the event in question."[12] If some particular historical explanation (following this pattern) is taken to be inadequate, Gallie would consider it irrelevant and of no help at all if general laws were mentioned in an attempt to improve the explanation. If any "filling out" should be needed, it is to be done *within* the necessary condition model, by specifying different or more such conditions.

Essentially the same position is held by both Arthur C. Danto[13] and William Dray.[14] For both of them the detailing of necessary conditions explains the occurrence of an event, though both of them go further than Gallie in regarding coherent narrative as a model of explanation. Following the model of "the continuous series," Dray considers an event explained when he *"can trace the course of events by which it came about."*[15] The explanation of Caesar's behavior, writes Danto, is "merely . . . the narrative we might construct describing Caesar's career, . . . describing the causes and conditions of the occurrences."[16] No general statements are required. Explanation by means of such a narrative is the end (or at least one of the ends) of historical inquiry, and not an initial step on the road to the deductive model.[17]

IV

If it is correct that the dispute about historical explanation is one in which Hempelians oppose anti-Hempelians, it is also true that the arguing on both sides has been curiously ineffectual. The various attacks of the anti-Hempelians have merely been countered by a reiteration of the original Hempelian position. Neither side has made any admissions. To explain this strange lack of progress, we must turn away from the actual proposals made about historical explanation and consider two underlying conceptions of what philosophy is and does.

There is some common ground shared by both parties to the quarrel about historical explanation. They are both genuinely doing philosophy *of* history. They philosophize, that is, about a subjective matter or activity as it is given; neither side is content to deduce statements about history from some antecedently established *a priori* metaphysical system. This common ground, however, is not enough to guarantee that the arguments by the disputants about historical explanation will meet. Given this very broad agreement, it may still be true that the "distances" at which philosophers stand from the object of their philosophic reflection vary considerably.[18] More specifically, the "distance" from which Hempelians look upon historical explanation is much greater than that from which their critics regard it.

Not even the starting point is precisely the same on both sides of the dispute. Hempel and Hempelians make no attempt to survey explanations as they are actually proffered—whether in history or in science. Induction is no part of their argument: no argument is made (and none can be) that their reconstruction says, more clearly and more elegantly perhaps, what historians (or scientists) say and do.[19] They proceed along different lines. A sample explanation is kept in mind or cited.[20] It is, presumably, a good sample, a paragon of an explanation, one which satisfies. The logical character of this explanation is sketched out quickly, and from this point to the finished product (a reconstruction of historical explanation) it is not reference to further examples of explanation which determine what is said, but principles which are, so to speak, purely philosophical. Put less cryptically, the starting point of philosophic reflection is an *insight* into what an explanation is; all that follows constitutes a reconstruction and elaboration of that insight in terms of a philosophic position that does not directly depend upon an understanding of the particular thing (historical explanation) being examined, but is grounded in philosophic considerations of a much broader sort.

All this can be made a great deal clearer by an attempt that makes explicit the train of thought that leads to the Hempelian reconstruction of historical explanation. Explanations are offered in response to questions such as "Why did this happen?" They are designed to reduce curiosity to understanding; they resolve a kind of tension. When the explanation is (psychologically) successful, the hearer gratefully exclaims, "Aha!," or, in Friedrich Waismann's image, he doffs his hat.

It seems clear that, if the much advertised Aha! experience is to occur, the explanans and the explanandum must in some way "hang together." No questioner, presumably, will be satisfied at the end of an exchange such as this: Q. "Why was Napoleon exiled to Elba?" A. "Because the area of Elba is eighty-six square miles." It is true, however, that for some people, in some contexts, an explanans and explanandum "hang together," while for others, or the same people under different circumstances, the same components fail to cohere and do not lead to the desired Aha! "Because she wanted to sleep" will satisfy a child's question, "Why did the doll close her eyes?," but it does not evoke an Aha! from an adult. Moreover, if such a response *did* satisfy an adult (and explanations no less superstitious have satisfied adults), we should emphatically come back with "it *ought* not to satisfy anyone!"

At this point the Hempelian may be imagined to take leave of the psychological discussion of explanations and the myriad of different contexts in which explanations may be offered. We know that explanans and explanandum must hang together; what must now be investigated is that in virtue of which these components hang together—not for the child or for a ninth-century serf, but for any rational being. A *philosophic* demand is being put upon explanations—a demand that the cohesion of its components should not depend upon the background of the person who asks for the explanation, nor upon the context in which it is given, but upon relations which are, so to speak, intrinsic to the explanation.

We need not look far for such relationships; there are not many. The explanandum must be *deducible* from the explanans. When that is the case, the explanation's coherence is independent of the speaker, the hearer, and the context. Only when that is the case is the philosopher empowered to say, "You *ought* to be satisfied with this explanation." More important still, only if deducibility is understood to be a necessary feature of an adequate explanation, is the philosopher in a position to say to the child or the superstitious person, "You *ought not* to be satisfied with this explanation," when an Aha! experience is reported as evoked by an explanans and explanandum which do *not* have the logical coherence of deducibility.[21]

Another general principle is operative in the requirement of the deducibility of the explanandum. The question, "Why did E occur?," must be understood as a demand for the cause or causes of E. Moreover, under numerous circumstances, the reply actually consists of a reference to some condition or conditions and a "that's why!" For the Hempelian this is not enough, no matter how frequently hearers of such "explanations" indicate their satisfaction with answers of this kind.[22] As a philosopher who has a *theory* of causality, he wants to know that the conditions mentioned *are* causes. And this theory—in general terms a Humean one—can find such a guarantee only in the statement (well established) of a *regularity;* he requires that the statements of the alleged cause and effect be mentioned by the statement of a law. Only then does he have the assurance that the alleged cause is, in fact, a cause.

This point may be put in another way. When the occurrence of E is explained by references to causes C_1, C_2, . . . C_n, a cognitive claim is made to the effect that C_1, C_2, . . . C_n did bring about E. Whatever the reaction of the plain man may be, the philosopher of the Hempelian persuasion is not yet satisfied. Even if it is supposed that the statements asserting that E and that C_1, C_2, . . . C_n are well established, nothing about these statements gives us any assurance that C_1, C_2, . . . C_n are the causes of E. To obtain this kind of knowledge (which is crucial to the explanation) logical procedures have long since been worked out, quite independently of historical explanation. To know that C_1, C_2, . . . C_n caused E, we must know—that is, have verified—a statement to the effect that "Whenever C_1, C_2, . . . C_n, then E." General principles, in other words, which rest upon a position which is applicable to a much broader area than, and is genetically independent of, the problems raised by historical explanation operate in the construction of the model of historical explanation.

A final philosophic principle which seems operative in the Hempelian analysis of historical explanation is of a much more general sort and might well be said to be assumed in the foregoing. To explain—whether in science, daily affairs, or history—is to provide knowledge. Knowledge, as distinct from mere belief, opinion, guessing, surmising, feeling sure, and so forth, must be objective. Ever since Plato, a dominant tradition in philosophy has insisted that knowledge must be logically independent of the particular state of mind of him who possesses it, the society of which its possessor is a part, the state of mind of the audience to which he imparts it, and of any other factor besides the relationship between the proposition said to be known and what is the case. When Hempel and Oppenheim, in their paper on explanation, [23] demand by an historian fully meets the demands the philosopher places that the constituent statements of an explanation be true, they are expressing their adherence to this tradition. For a perfectly adequate explanation, it is not enough to say that its constituents are well confirmed. For this is compatible with their falsity and hence with the inadequacy of the explanation. Only when its constituents are true does the explanation hold, regardless of time, place, speaker, or hearer.

And knowledge is one.[24] When historians explain, they claim to provide knowledge. Explanation in history, therefore, must fulfill the same requirements as explanations in other areas, above all in science. More generally, historical explanation must meet the criteria which any claim to knowledge must meet, regardless of its context. It may be that no actual explanation ever offered upon claims to knowledge. In that case, the philosopher's model serves as a measure of the historian's success and indicates the direction of possible improvement. The model, however, is established on a firmer and wider basis than an examination of historical explanations: if such explanations do not live up to the model, the philosopher cannot give up the model short of giving up the philosophic position in which it is grounded. Accordingly, if the historian insists that he cannot meet the requirements laid down by the philosopher for explanation in history, the philosopher has no choice but to reply, "to that extent and in that respect, so much the worse for history."[25]

V

A self-conscious anti-Hempelian might summarize his objections to Hempel's proposal in the following way. First and worst of all, Hempel's scheme does not reflect what historians actually do when they explain. No matter how much and how carefully one reads works of history, one is not likely to turn up Hempelian explanations. Indeed, there is more than one reason to suppose that historians could not, even if they so chose, always provide explanations of the Hempelian sort; [26] and some reasons may be given in support of the thesis that they never could. Secondly, even if an historian were to provide a set of statements which fulfilled all the requirements laid down by Hempel, it would by no means follow that this set would actually *explain*. Such a set would have the form of an explanation (according to Hempel) and, paradoxically enough, perhaps not satisfy a single reader. Thirdly (and this is what gives teeth to the first two objections), the anti-Hempelian would note that historians do, as a matter of fact, successfully explain many events, using methods quite different from those recommended by Hempel.

Given objections of this sort (and given the actual analyses made by anti-Hempelians on the basis of which these objections were formulated in the first place), we may single out three related principles or, better, considerations, which underlie and function in most of the analyses made by anti-Hempelians. By discussing each of them in turn, it may be possible to determine the "distance" from which these philosophers look upon historical explanation.

(1) In an important sense, the philosopher must not go beyond his data. The philosopher's job is to ascertain as clearly as

he knows how the "logic" of historical explanations as they are actually offered. Just as it is the task of the topographer to reproduce the exact outline of a mountain—and not to recommend that the mountain's silhouette be converted into an isosceles triangle with an apex angle of exactly 45°—so the philosopher must remain faithful to the "outline" of his data. In less metaphorical terms, there must be a sense (and not a trivial one) in which the analysandum and the analysans are equivalent. If the philosopher seeks clarity, what he wants to get clear is what historians actually do and not what they might do, no matter how good a thing it would be if historians followed such a recommendation. The primary role of philosophy is *de*scriptive and not *pre*scriptive.

(2) If analyses are to meet such requirements, the philosopher cannot approach the analysandum with theories of his own. His job (in this case) is to find out what *historians* do when they explain. To explain may very well be to provide knowledge; this fact, however, does not constitute a justification for coming to this act of historians with a *general theory* as to the nature of knowledge. It may well turn out that giving an historical explanation involves the stating of causes of an event.[27] But it is the philosopher's business to find out what the *historian* means by "cause" and not to impose some philosophic doctrine—Humean or Aristotelian—upon the historian's work. He must recognize that "cause" may well have a different meaning when used in the context of history from that which it has in science. It is up to him to *reveal* this meaning and not to change it.[28]

In the most extreme version of this view all of philosophy consists of analyses of this kind. According to it, the philosopher cannot rightfully come into possession of any theories of his *own*. To find out what an historical explanation is involves only a determination of how the term "explanation" is used in the "language game" played by historians. The introduction of theories derived from elsewhere—including, presumably, other language games—would be worse than inadvisable; it would patently contradict the philosopher's basic conviction about the nature of philosophy.

(3) So far restrictions have been emphasized: we have focused upon what the anti-Hempelians consider not to be a proper part of philosophic analysis. But important things must be said on the positive side as well. If philosophic analysis is analysis of a particular *language, all* of the facets of the language must be considered. The language of history is a language used by someone to speak to someone. In the specific case of an historical explanation, the historian explains the occurrence of an event to a certain audience. In doing so, he manifests certain intentions; in the

audience to which he speaks he finds or arouses certain expectations. The enterprise as a whole is carried on to satisfy certain human interests. The anti-Hempelian emphatically rejects Hempelian austerity; he rules in, as part of the data to be clarified, the entire context in which historical explanations are offered. "Explanations are practical, context bound affairs. . . ."[29]

This can be put more formally. The Hempelian is interested only in the syntactics and the semantics of historical explanation. For him, the "logic" of a term has but these two dimensions. The anti-Hempelian's rejoinder is (in part) that such an analysis is most incomplete and hence in error. "Explanation is not a syntactical but a pragmatic notion."[30] Explanations are offered to provide understanding; and the understanding, as an event taking place in the hearer's mind, is a part of what must be considered by the analyst.

We can thus easily see that in several different ways the anti-Hempelian conducts his investigations from a position much closer to the data than does the Hempelian. His stance, one might say, is more empirical: he seeks to discover how "explanation" is actually used in the related set of contexts which make up the field of history. The immediately obvious manifestation of this "empiricism" is the procedure that is most frequently used: the citing of example after example of explanations, both "genuine" (that is, having actually occurred) and madeup for the occasion. Faithfulness to the data in all of their shifting complexity and sensitivity to the nuances of different uses are the prime desiderata of such an analysis.

Earlier we saw that the Hempelian's analysis is a product of an insight into usage *and* of general philosophic principles. Since there is considerable agreement among Hempelians about these principles, it comes as no surprise, then, that in this group there is also a large measure of agreement about a single model of historical explanation. Now, considering the many different contexts in which historical explanations are given, it becomes understandable why those who take into consideration the interest and knowledge of the hearer as well as the intention of the speaker should find that there are many different types of historical explanation.[31]

VI

If two people looked at the world wearing spectacles of differently colored glass, we should not be astonished at their not agreeing about the colors of the things they saw. Moreover, if they never removed their glasses, we should have no expectation that they would ever come to an agreement, no matter how strenuously they peered at the world's furniture. Hempelians and anti-Hempelians use different methods of analysis in their scrutiny of historical explanation.[32] Regardless of how diligently they continue their work, if each camp persists in its methods, we cannot expect them to arrive at even approximately similar conclusions. But clearly, the methods of analysis used are themselves subject to controversy. While so far we have given "persuasive expositions" of both methods, it will be well to conclude with a few partisan remarks.

But before giving full rein to partisanship, it must, I think, be granted that both modes of philosophic analysis have led to important results. The actual work done in close to half a century provides sufficient evidence to refute a claim on either side that *the* philosophic method has at last been found. Yet from this it by no means follows that either method is equally appropriate for the solution of *any* kind of philosophic problem.

The method employed by the anti-Hempelians was developed as part of and in response to a conception of philosophy which sees as its function the resolution of certain kinds of puzzles. Language is infinitely complex, a fact which not only makes it a flexible instrument for a great variety of purposes, but which also traps its users into strange corners from which escape comes hard. Not everyone is trapped, but most philosophers, at one time or another, have been led astray. For them this method proposes to point a way out: linguistic therapy. The point of this kind of analysis seems to be to find the linguistic roots of a puzzle: to determine where an analogy might have been misleading, where grammatical form might have hidden a different logical form, to spell out, in other words, how the trap was laid. When all this has been made obvious, the trap ceases to be a trap; the problem ceases to be.

Unquestionably there is evidence that a method sensitive to all the nuances of usage, a method which can note and measure every drift of language, is capable of achieving therapeutic results. But it is notorious that "the professional philosopher deliberately and methodically [causes] the headaches which he is subsequently going to cure."[33] Only when prodded and no longer plain, will the plain man come to doubt the existence of the external world; only when coaxed will he begin to wonder about the unity and continuity of the self. And, while such prodding and coaxing is often successful, it is by no means the case that every linguistic situation can be converted into a trap; not every question can be twisted into a puzzle.

Now, the question as to the nature of historical explanation does not seem to be a puzzle at all. To be sure, there are questions about history which have the quality of the kind of muddle for which the anti-Hempelian method seems suitable: Can history be objective? How can we know what is no more? and the like. The question about historical explanation, however, does not seem to involve this (to some) puzzling aspect of history; certainly the writers on this subject have not understood the problem to be so tainted. Moreover, in any form similar to the present one, the problem of historical explanation is not very ancient. It goes back no further than the latter part of the 19th century, when methodology (as distinct from epistemology) was isolated as a subject matter. In spite of the fact that Hempel tacitly addresses himself to this earlier discussion, neither he nor subsequent writers find, there or elsewhere, headaches to cure.

An anti-Hempelian, while perhaps admitting that his method first was and primarily is used to resolve puzzles, may still reply that this fact does not rule out the method's serviceability in problems of another kind. But this seems not to be the case; for when there is no puzzle to be resolved, it becomes difficult to see what philosophical task the method could accomplish. Historians, scientists, teachers, parents, and all the rest of mankind have used many different words in explaining many different things. No doubt, the number of explanatory patterns is finite and it might well be possible to discern different genera, species, and varieties of explanations. The resulting taxonomy might be of some interest to psychologists, sociologists, or what are known as communication specialists; some of these might even wish to go on to correlate different branches of the classification with the cultural background, education, sex, and annual income of the explainer. But why this should be of interest to the philosopher—any more than *any* report a scientist might give about the workings of men and nature—is indeed hard to see. Still less plausible is it that a philosopher should do such work himself. In the end, he lacks the training, as a philosopher, to do an adequate job.

And the continued use of this method on the problem of historical explanation inevitably leads to such a taxonomy. The reason is simple: there is nothing to stop it. On the other hand, there is no puzzle to be resolved. The need to do so would give direction to an analysis; the alleviation of the need would signal its conclusion. On the other hand, the method precludes philosophy's playing a systematic role. The analysis cannot be conducted with the aid of theories derived from other analyses, nor can it be a part of the task of the analysis to relate the analysandum to other concepts and principles, with a view to arriving at a systematic reconstruction of an entire domain. This leaves only the function of "revealing" by means of an indefinite number of examples. When order is made among the examples, we are well on the way to a taxonomy.

There is no doubt that the method of the Hempelians involves dangers. When one comes to a problem *with* theories, the possibility always exists that the analysandum is lost sight of in a network of concepts and principles. Reconstruction, if one does not take heed, may become construction. There is no magic formula for gauging the "distance" the philosopher must stand from the problem of his concern. When that interval shrinks to the vanishing point, philosophy, we have seen, becomes a mere reporting. When, through the interposition of too high a stack of theories, the "distance" becomes too great, philosophy relapses into the idle *a priorism* of ages hopefully gone by. Neither alternative is acceptable. The anti-Hempelians, however, come close to embracing the former, while the Hempelians are still trying to maintain distance without losing sight of their object.

NOTES

*A shorter version of this paper was read at the 1960 International Congress for Logic, Methodology, and Philosophy of Science, held at Stanford University.

1. *The Journal of Philosophy,* Vol. 39 (1942). Reprinted in Herbert Feigl and Wilfrid Sellars, eds., *Readings in Philosophical Analysis* (New York: Appleton-Century-Crofts, 1949), pp. 459–471. Page references are to the latter printing.

2. Careful historians of philosophy will rightly complain that (1) Hempel's proposal of 1942 was by no means new and that (2) the keen interest in the problem of historical explanation, to which philosophers writing in the decades on either side of the turn of the century gave expression, had never completely died down. In support of the first contention Karl Popper must above all be cited. He himself correctly claims priority for the view Hempel sets forth (see n. 7 to ch. 25 of *The Open Society and Its Enemies,* Princeton University Press, 1950, pp. 720–723). Moreover, depending on how broadly the Hempelian position is conceived, ancestors even of Popper may be found. Alan Donagan points out that Morris Cohen, in *Reason and Nature,* 1st ed. (1931), bk. I, ch I, sec. 2, maintains a similar view. (See Alan Donagan, "Explanation in History," *Mind,* Vol. 66 (1957), reprinted in Patrick Gardiner, ed., *Theories of History,* The Free Press, 1959, n. 2, p. 428). But, as is so frequently the case in the history of philosophy, what matters is not merely the inventing and holding of a view, but the power and precision of its formulation. In Hempel's article the position was for the first time brought into such sharp focus that it had to be accepted, adopted, modified, or rejected; it could no longer be ignored. With respect to the second point, we must keep in mind the great change that has taken place in the method and style of philosophy since the work of Dilthey, Simmel, Weber, Rickert, and others. There are links between these continental philosophers of history and those currently concerned with this topic (Colling-

wood and Mandelbaum, for example), but Hempel's article is the earliest written wholly in the analytic style of philosophy.
3. There is some disagreement even among Hempelians. In this paper, however, intra-Hempelian disputes will be ignored.
4. This outline is both expanded and refined in another *locus classicus,* Carl G. Hempel and Paul Oppenheim, "The Logic of Explanation," *Philosophy of Science,* Vol. 15 (1948), reprinted in Herbert Feigl and May Brodbeck, eds *Readings in the Philosophy of Science* (New York: Appleton-Century-Crofts, 1953), pp. 319–352. Page references are to the latter printing.
5. In the later paper *(ibid.,* p. 322) "well confirmed" is changed to "true."
6. *Op. cit.,* p. 465.
7. Among those who have, with no or few reservations, accepted Hempel's analysis of historical explanation are: Ernest Nagel (see, especially, "Determinism in History," *Philosophy and Phenomenological Research,* Vol. 20 (1960) and Morton White (see "Historical Explanation," *Mind,* Vol. 52 (1943), reprinted in Gardiner, *op. cit.,* pp. 357–373).
8. Some writers grant that Hempel's proposal is an analysis of *some* historical explanations; they then make their suggestions as *additions* to that of Hempel. Most anti-Hempelians offer their proposals *in place* of Hempel's. The proposals to be discussed here are all analyses of *causal* explanations. As such they are at least in some ways comparable to Hempel's scheme. Some anti-Hempelians, however, also recognize historical explanations which are, in their view, not causal at all. See Sec. V, below.
9. Patrick Gardiner, *The Nature of Historical Explanation* (Oxford: Oxford University Press, 1952), pp. 95–96. Also, see Alan Donagan, *loc. cit.,* as well as Michael Scriven, "Truisms as the Grounds for Historical Explanations," in Gardiner's *Theories of History,* pp. 443–475, and "The Logic of Criteria," *The Journal of Philosophy,* Vol. 56 (1959).
10. "Causal Explanation in Natural History," *The British Journal for the Philosophy of Science,* Vol. 9 (1958).
11. Goudge's actual subject matter is natural history. There seems to be no reason, however, why his proposal should not be extended so as to apply to history.
12. W. B. Gallie, "Explanations in History and the Genetic Sciences," *Mind,* Vol. 64 (1955), reprinted in Gardiner, *op. cit.,* p. 387; italics in original.
13. "On Explanations in History," *Philosophy of Science,* Vol. 23 (1956).
14. *Laws and Explanation in History* (Oxford: Oxford University Press, 1957).
15. *Ibid.,* p. 68, italics in original.
16. Danto, *loc. cit.,* p. 29.
17. Scriven may also be accociated with this view. See the papers cited above, as well as "Explanation and Prediction in Evolutionary Theory," *Science,* Vol. 130 (1959), pp. 477–482.
18. This useful metaphor stems from Georg Simmel. See his "On the Nature of Philosophy," in Kurt H. Wolff, ed., *Georg Simmel, 1858–1918* (Columbus: Ohio State University Press, 1959), pp. 282–309.
19. The closest that Hempelians come to "verifying" their proposal by reference to explanations actually offered by historians is to restate the historian's explanation and give it the form of a Hempelian explanation sketch (see Nagel, *loc. cit.).* This, however, amounts to the *imposition* of a scheme upon the words of historians. Anti-Hempelians will be quick to point out that from the fact that the imposition is possible (that the historian's words do not resist being ordered into a Hempelian framework) it does not follow that there may not be an indefinite number of other schemata which are still more adequate philosophic reconstructions of historical explanation.
20. Hempel and Oppenheim, *loc. cit.,* pp. 320–321.
21. Only the question of coherence is here under discussion. Nothing is being said about when one ought or ought not to be satisfied as far as other features of explanations are concerned.

22. See a discussion of this point in Jack Pitt, "Generalizations in Historical Explanation," *The Journal of Philosophy,* Vol. 56 (1959).
23. *Op. cit.*
24. Without explicitly saying so, Hempel addresses himself to the view current around the turn of the century, that there were essentially two kinds of knowledge, that given by the *Naturwissenchaften* and that found in the *Geisteswissenschaften.* Hempel's effort to show that there was no essential difference (as far as knowledge is concerned) between these two sets of disciplines is an expression of the unity of science principle held by the members of the Vienna Circle.
25. It might be remarked parenthetically that this is not as horrible as it seems. (1) There are lots of things men do that they might do more adequately and yet will never do with complete adequacy. (2) Historians do, among other things, explain—some more than others. It does not, however, seem to be the main business of historians to explain, but rather, to describe. Explanation looms nowhere near as large in history as does historical explanation in philosophy.
26. Anti-Hempelians often forget that Hempel admits all this. Nevertheless, this first point becomes an objection when it is taken in conjunction with the second and (particularly) third objections and with the conception of analysis (to be discussed below) which underlies all of the objections of the anti-Hempelian.
27. Though, according to some anti-Hempelians, not all historical explanations are causal explanations by any means. See n. 31 below.
28. See, for example, Gardiner, *The Nature of Historical Explanation,* Parts I and III.
29. Scriven, "Truisms as the Grounds for Historical Explanation," *loc. cit.*, p. 450.
30. *Ibid.*, p. 452.
31. Not only do anti-Hempelians as a group suggest numerous "models" of historical explanations, but particular members of that group, notably Dray and Scriven, also see several types of historical explanation. See, for example, William Dray, " 'Explaining What' in History," in Gardiner, *Theories of History,* pp. 403–408, and Scriven, "Truisms as the Grounds for Historical Explanations," *loc. cit.*
32. These two methods correspond approximately to what P. F. Strawson calls the methods of the American School and the English School respectively. See his "Construction and Analysis," in *The Revolution in Philosophy* (London: Macmillan; New York: St. Martin's Press, 1957), pp. 97–110.
33. H. H. Price, "Clarity Is Not Enough," *Proceedings of the Aristotelian Society,* Supplementary Volume 19 (1945), p. 5.

SUGGESTIONS FOR FURTHER READING

Borger, R. and Cioffi, F. (ed.) *Explanation in the Behavioural Sciences.* Cambridge: Cambridge University Press, 1970.

Brodbeck, M. (ed.) *Readings in the Philosophy of the Social Sciences.* New York: The Macmillan Co., 1968, pp. 363–397.

Brody, B.A. (ed.) *Readings in the Philosophy of Science.* Englewood Cliffs, N.J.: Prentice-Hall, Inc., 1970, pp. 8–179.

Brown, R. *Explanation in Social Science.* Chicago: Aldine Publishing Co., 1963.

Campbell, N. *What Is Science?* New York: Dover Publications, Inc., 1952, pp. 77–108.

Kahl, R. (ed.) *Studies in Explanation.* Englewood Cliffs, N.J.: Prentice-Hall, Inc., 1963.

Kaplan, A. *The Conduct of Inquiry.* San Francisco: Chandler Publishing Co., 1964, pp. 327–369.

Martin, M. *Concepts of Science Explanation.* Glenview, Ill.: Scott, Foresman and Co., 1972, pp. 44–74.

Nagel, E. *The Structure of Science.* New York: Harcourt, Brace and World, Inc., 1961, pp. 15–46.

Rescher, N. *Scientific Explanation.* New York: The Free Press, 1970.

Smart, J.J.C. *Between Science and Philosophy.* New York: Random House, Inc., 1968, pp. 53–120.

Wartofsky, M.W. *Conceptual Foundations of Scientific Thought.* New York: The Macmillan Co., 1968, pp. 240–291.

SECTION IV

The Nature and Function of Scientific Laws

The objective of the first article herein is to introduce several different types of scientific laws and then to propose a definition that captures all and only those features that should be captured in a definition of "scientific laws." All laws, like airlines, aspirins and refrigerators, are not alike, and the easiest way to go astray in attempting to provide an adequate account of laws is to focus on a single example. Indeed, if asked to cite a good reason for *not* introducing science to students by putting them through the introductory course in some particular science such as, for example, physics, chemistry or biology, one could hardly find a better reason than the *diversity* that exists in the various sciences. And nowhere does this diversity reveal itself with more vividness than in the nature of laws. As you consider, with Mehlberg, Kepler's laws of planetary motion, Comte's law of the three stages of the development of man, Maxwell's laws of electromagnetism, Newton's laws of mechanics and gravitation, laws of economics and psychology, statistical and causal laws, and so on, you will begin to appreciate this diversity. Moreover, you will probably be very suspicious when you read Mehlberg's conclusion that "Scientific laws should be defined as essentially quantified statements." If that proposal were accepted, then "All saints are in heaven and devils are in hell" would be a scientific law! Hence, other alternatives are presented in the next two articles in this section.

The primary aim of the Jobe article is to elucidate five proposed explications of "scientific law," reveal their inadequacies and defend an allegedly satisfactory proposal offered by Reichenbach several years ago. When you review each of these accounts, you should pay special attention to the suggested "paradigm cases" of laws. For example, Braithwaite claims that a hypothesis becomes lawlike if it "occurs in an established scientific deductive system as a higher-level hypothesis" or "as a deduction from higher-level hypotheses." But Jobe replies that such a requirement is too strong *because* "Water expands upon freezing" and "Sugar is (water)-soluble" are "paradigm cases" that do not meet the requirement. The question is: Are they paradigm laws or merely true generalizations about the world? They are obviously not very technical or scientifically sophisticated propositions. But should we suggest that some degree of sophistication is necessary in order for a statement to be a law? How could one ever specify the appropriate degree—or measure sophistication at all, for that matter? The selection of paradigm cases is absolutely crucial here. Is Parkinson's law that work expands to fit the time allotted for it *really* a law? Is the iron law of oligarchy a law, i.e., that in any group of people, leaders and followers are bound to emerge? We want our explication to capture all and only genuine laws, but how can we identify them? (Will you feel better if I assure you that Plato wrestled with precisely this kind of question over 2,000 years ago over the nature of piety, and that the United States Supreme Court has wrestled with it repeatedly over the nature of pornography?)

In the last article of this section Alexander criticizes four attempts to treat scientific laws as merely rules of inference; as, if you like, instruments or tools permitting logical transformations or deductions rather than as statements that can be true or false. The difference between these alternatives is easy to illustrate. Consider, for example, the alleged law "Water expands upon freezing." If we think of this alleged law as a proposition, claim, assertion or a declarative sentence about the world, then we are committed to the view that it has a truth-value. It must, in this view, be either true or false. Alternatively, however, we might say that insofar as the quoted expression is treated as a law, it does not have propositional status. Rather, it should be understood as a shorthand way of writing the rule "*Given* some freezing water, *infer* that the water is expanding (or has expanded or will expand)." This rule, then, is not true or false. It can't be because it does not say that anything is or is not the case any more than "How's your sister?" says so. Furthermore, establishing laws-as-rules, then, cannot be the same as proving that they are true. And if theories are sets of laws, then theories can't be true either. In short, lots of interesting conclusions follow from the decision to explicate "scientific law" as some sort of a rule, and Alexander's aim is to show that there are many more unwanted than wanted conclusions.

READING 9

Types of Scientific Laws

Henryk Mehlberg

The idea that the universe is governed by comprehensive and exceptionless laws has impressed the human mind more than any other article of the scientific credo. The supply of universal laws which science has succeeded in discovering constitutes its most significant and least debatable acquisition in the eyes both of educated laymen and of most professional students of science. Yet the procedures the scientist uses to discover and establish scientific laws are more complicated and less understood than those he applies to discover scientific facts. This difference can be explained by several circumstances. Basic fact-finding methods of science are taken over directly from pre-scientific usage with refinements which are negligible in comparison with the features common to scientific and pre-scientific ways of handling factual information. Laws, on the other hand, though present in embryo in pre-scientific knowledge, are, in the main, an acquisition of science. The methods that science had to develop in order to acquire its present extensive knowledge of laws supposed to "govern the universe" were also taken over to some degree from pre-scientific procedures of generalization. But the refinements, the departures, and the extensions have been much greater than in the case of scientific fact-finding methods. The methodology of scientific laws raises, therefore, a host of problems which simply do not arise in connection with the scientific continuation of the pre-scientific quest for facts.

Scientific laws are, further, more complicated logical entities than scientific facts, even than indirectly ascertainable facts. It is one thing to produce an oral or written picture of a fact witnessed or remembered by the observer; it is quite another thing to condense into a single concise statement (i.e., a law) a theo-

Reprinted by permission of University of Toronto Press from H. Mehlberg, *The Reach of Science,* pp. 156–165, 180–193. © Canada 1958 by University of Toronto Press.

retically infinite number of facts, a negligible fraction of which may have been actually observed in the most favourable cases. We must therefore expect, in surveying critically the procedures applied in discovering and establishing scientific laws, greater difficulties than we encountered while studying fact-finding methods. However, this examination may bring us closer to an understanding of what are generally admitted to be the most characteristic aspects of the scientist's entire procedure.

The first point worthy of consideration is that to ascertain whether a given method for establishing a particular law or group of laws is reliable, we must examine both the *nature* and the *function* of this law or group of laws. Similarly, to determine whether a given set of validating methods is sufficient to establish the sum-total of scientific laws, we need to know just what constitutes such a law. We must therefore postpone our discussion of the scientific methods actually employed to establish laws, until we analyse the main types of laws (§ 19) and their main functions (§ 20), and frame a definition of scientific law (§ 21).

The term "scientific law" will be construed, in this study, as referring to a law established (or capable of being established) in any science whatsoever, either empirical or demonstrative, either natural or humanistic. The phrase "law of nature," on the other hand, will be applied to what J. S. Mill [1] calls "ultimate laws" of the natural sciences, i.e., a selection of laws of these sciences which suffice for deriving all their remaining laws and meet some additional requirements. Laws of natural sciences derivable from these ultimate laws are called "derivative"; the ultimate and the derivative laws of natural sciences together add up to the class of "natural laws." Thus, every law of nature is a natural law, but not conversely. All scientific laws are not natural; economic and sociological laws in the empirical field, and laws of pure mathematics in the demonstrative field, are not so classifiable. For instance, Newton's axioms of mechanics exemplify laws of nature. The laws of planetary motions, deducible from these axioms (and a few additional assumptions), are (derivative) natural laws. The sociological law of cultural lag is a scientific law, without being either a law of nature or a natural law.

There are several important types of scientific laws:

(1) Conceptually Universal Laws

We have distinguished fact-finding statements from well-established facts. The former attribute, wrongly or rightly, a particular property to a particular individual, or assert a particular relation among a few particular individuals. The latter are selected from the sum total of fact-like statements in accordance with a criterion based on the existence of an adequate body of evidence. We shall, accordingly, distinguish law-like statements from laws, by recourse to the same criterion; a law-like statement is a well-established law if its truth is adequately supported by available evidence.[2] In other words, a well-established law is a law-like statement known to be true. That this piece of iron expands after having been heated, is a particular or local fact; that every piece of iron would do so under similar circumstances is a law.

Yet, in spite of the basic similarity between the two pairs of concepts, there is an additional difficulty in the case of law-like statements. It is by no means easy to delimit all law-like statements in a satisfactory way, without either distorting the meaning usually attached to the phrase or indulging in a merely lexicographic exercise. The real problem is to provide a definition of the concept of law-like statements which will remain reasonably close to actual usage and shed some light on the considerable role allotted to such statements in the scientific framework since the birth of modern science during the seventeenth cenury.

One point can be taken for granted: in contradistinction to particular and local facts, laws are "general" or "universal." This much seems to be accepted in all attempts at clarifying the concept of scientific law. The trouble is, however, that this requirement of the universality of scientific laws can be construed in at least three distinct and equally important ways, which refer respectively to the universality of concepts, of quantification, and of spatio-temporal scope. In this section, we shall survey the types of laws which meet these three requirements; the bearing of universality on the definition of scientific law will be discussed on pages 196–208.

Thus, in the first case, a statement is said to be law-like if all its constituent terms are universal, but fact-like if even one of these terms is local or individual. By local terms, I mean proper names of objects and events, such as "the earth," "Napoleon," and "the Second World War," whereas universal terms will be construed as those which, by their very nature, are applicable always and everywhere, such as "green," "atom," and "weight." "The earth is a planet" is a particular fact in this sense. On the other hand "Some stars have no satellites" is a law, in so far as the requirement of conceptual universality is concerned.

Yet a proposal to confine the concept of scientific law to statements which are universal in this sense would certainly run counter to many a well-established way of speaking. Thus Kepler's laws of planetary motions refer explicitly to the individual called "sun" and to some of its satellites, but it is rather awkward not to call his discoveries "scientific laws." Similarly, A. Comte's sociological "law of the three stages," which allegedly governs the succession of the religious, the metaphysical, and the scientific (or "positive") stages in the historical development of man, though rather questionable in view of its vagueness and the inadequacy of the support which Comte has derived for it from historical facts, can hardly be denied the status of a law-like statement.

The tendency which is indicated by the requirement of conceptual universality of scientific laws is nevertheless of great importance for the present outlook of science. The fact is that in the most basic sciences, especially in physics, there is an unmistakable trend to get rid of local terms and constants in the formulation of basic principles. This does not mean that all the implications which Eddington, for example, has read into this trend are justified, but its existence is obvious and philosophically significant. Thus, to quote a simple example, the fundamental standards of the ramified system of physical quantities were based until recently upon terrestrial characteristics and were consequently of a local nature; for example, a second has been, by definition, a definite fraction of the time the earth takes in carrying out a complete revolution around its axis. Since similar considerations were applicable to other standards of physical quantities, the laws of nature involving such standards (inclusive of Newton's laws of mechanics and gravitation and of Maxwell's laws of electro-magnetism) did not meet the requirements of conceptual universality. These theories would have to be classified as systems of fact-like, rather than of law-like statements should conceptual universality be construed as a prerequisite of law-like status. The "universalist" tendency just referred to has, however, changed this situation. The basic assumptions of the aforementioned fundamental physical theories can now be expressed in purely universal terms, since the basic standards of quantities which they involve are so expressible. This is why it seems to be rather in keeping with modern scientific tendencies to stress the importance of conceptual universality, by basing on it the distinction between fundamental and derivative laws. The fundamental laws would, by definition, have to meet the requirement of conceptual universality. Law-like statements which are not fundamental but can be deduced from fundamental laws, may be termed "derivative." It remains to be seen, however, whether this dichotomy accounts for all scientific laws.

(2) Formally Universal Laws

Since a law is a general statement and every general statement has a number of "instances" (which are typically, though not necessarily, fact-like), one may wonder whether every law-like statement should claim to be true of *all* its instances. Such a requirement has often been formulated as essential to laws, and the most conspicuous laws are universal in this sense. For example, Einstein's mass-energy convertibility law asserts that the energy content of *every* material system is equal to the product of its mass and the square of the speed of light for *every* possible value of mass and of energy. The law is applicable to every physical system, regardless of its mass and energy content. Hence, all the instances of the law which correspond to particular values of energy and mass and to a particular choice of the system are included. Yet one may feel perhaps that the speed of light is a constant number, not a variable, and that, consequently, only *one* speed is actually involved in the law. The answer to this objection is plain and, in a sense, already alluded to in what has just been said about the universalistic tendency of contemporary science. We have to distinguish between *local constants,* which are simply the local values of spatio-temporal co-ordinates of a single individual (and correspond, in the mathematical language of science, to proper names of individuals in the vernacular) and *universal constants,* such as the speed of light, the magnitude of the quantum of action, or the smallest electrical charge in existence. Universal constants are expressed by a constant, single number, just as local constants are. But their physical significance is completely different, since they refer to any and every space-time region. For example, the speed of light is always and everywhere expressible by the same number of km/sec., provided it is measured from an appropriate "inertial" frame of reference. Thus, universality of quantification implies merely, if properly interpreted, that all *individual variables* are within the scope of a universal quantifier. Non-individual expressions, for instance numbers specifying the speed of light or the quantum of action, may be represented either by a constant value (which would be inadmissible in the case of individual expressions, because of the requirement of conceptual universality) or by a variable operated on by any quantifier, either universal or existential.

As a matter of fact, the requirement of quantificational universality, though ostensibly met by the fundamental laws of nature, determines the appearance rather than the actual meaning of such laws. This requirement is violated by statistical laws, some of which have come to play a decisive role in fundamental theories, for example, in Quantum Theory. A statistical law of the simplest kind specifies the probability p that any object belonging to a class A will also belong to the class B (e.g., the probability that any hydrogen atom that is "excited" at the time t will radiate within the time $t + t_0$). On the now predominant frequency-interpretation of the concept of probability, the meaning of the statement that whatever is A is also B with the probability p can be rendered explicit by the following paraphrase: "If A_n is an appropriately chosen infinite and increasing sequence of finite subsets of A, and p_n is the relative frequency of objects in A_n that also belong to B, then the sequence of fractions p_n tends towards the limit p as the number n increases indefinitely." Moreover, the convergence of the fractions p_n towards the limit p means that for every positive number ϵ however small there *exists* an integer N such that whenever the integer q exceeds N the fraction p_q differs from p by less than ϵ. Hence, existential quantifiers are indispensable to define the concept of limit, and, in particular, the concept of probability as a limiting frequency.

There is little doubt that existential quantifiers occur also in non-statistical laws of nature although the presence of such quantifiers is concealed by the equational form given to most scientific laws for reasons of convenience already referred to (§ 16). The point is that the fundamental laws of nature are mostly ordinary or partial differential equations; the concept of derivative which is involved in such laws is also a limiting concept whose definition includes a mixed quantificational pattern.[3] Consequently, whenever a law of nature has the form of a differential equation, the requirement of quantificational universality does not apply to it.

(3) Laws Universal in Respect to Spatio-Temporal Scope

Finally, a law is often said to be "universal" because it applies to all relevant objects or events in the cosmic space-time, regardless of when and where they exist or occur. In contrast, particular or local facts are concerned with a single space-time region, which happens to be occupied by the object or event referred to in the fact-like statement in question. Yet, as a matter of fact, two sciences, physics and chemistry—or perhaps we should say, "the science of physics," in view of the unmistakable tendency of chemistry to be absorbed by physics?—have the monopoly of establishing laws of nature which possess such universality of scope.[4] The laws of astronomy, of geology, of biology, and of human and animal psychology are valid within certain space-time regions, but not in the whole of space-time. I have already referred to Kepler's laws of planetary motions, which apply only to the solar system. Other laws of contemporary astronomy have a spatio-temporal scope vastly exceeding that of Kepler's laws, but finite nevertheless; they certainly fail to cover the whole of space-time. Similar remarks apply to biology. Thus, Darwin's theory is an attempt at reconstructing and explaining a local story which took place on the surface of the earth (or close to it) during less than a billion years. The laws of economics claim validity with regard to human societies, i.e., for a region whose spatial dimensions are surpassed by those of Darwin's theory, and whose temporal dimensions have an order of magnitude of a few thousand years.

The fact that a scientific law applies only within a specifiable region may be due to two circumstances of a completely different nature: in the first place, it may simply happen that entities which by their very nature should be governed by the law under consideration do not occur beyond a specifiable region, so that the law may be said to apply "vacuously" beyond this region. On the other hand, the law may not have a universal scope because entities eligible to be governed by this law which do occur beyond the specifiable region are there precisely at variance with the requirements of the law. Thus, should the statistical interpretation of the law of cosmic increase of entropy (Second Principle of Thermodynamics) be correct and should the entropy keep increasing only within the present period of cosmic history and start decreasing afterwards, then the law would be local in the second sense of the word. For all that is presently known, biological laws may be vacuously universal with regard to their scope. The bulk of physical laws are taken to be non-vacuously universal.

We have thus to distinguish, with respect to spatio-temporal scope, genuinely universal laws from those which are only vacuously universal. Laws which fail to meet the requirement of (genuine or vacuous) universality of scope, and apply only within a limited spatio-temporal region, may be subdivided in turn into regional and individual laws, depending upon whether they refer to all the individuals within such a region, or to single individuals. Darwin's theory is a set of regional laws; those of Kepler are individual.

(4) Finitist versus Infinitist Laws

The number of instances of a scientific law may be finite or infinite; if infinite, it may be "denumerably" or "non-denumerably" so. Thus, the biological law estimating at about one billion the number of variations needed to lead from the proto-organisms supposedly responsible for the start of life on earth to the present multiplicity of species is a finitist one. Every law concerning human psychology is finitist, since the number of humans is finite. On the other hand, the hypothesis of an infinite space amounts to assuming the geometrical law-like statement to the effect that, for every natural number N, there is an object whose distance from the earth exceeds N units of length. The relevant law would be denumerably infinitist. Those who assume the literal applicability of Euclidean geometry to the cosmic space (even on the assumption that this geometry is reinterpreted so as to deal only with finite regions, for example, in accordance with Whitehead's Method of Extensive Abstraction)[5] and assert accordingly that every finite region includes a non-denumerable class of spatial points are thereby committed to an infinitist law of the second kind.

(5) Statistical versus Causal Laws

I have already mentioned that the simplest statistical laws specify the probability that any element of a class A will also belong to the class B. In general, statistical laws deal with functional relations among the statistical distributions of a set of quantities in a given population. Thus, if the statistical distribution of a given quantity Q in a given population P happens to be uniquely determined by the distributions of a set of other quantities Q_1, $Q_2, \ldots, Q_n$ in the same population, a statistical law of the aforementioned type will state that the distribution of Q in P is a given function of all the distributions of the quantities Q_j $(i = 1,2, \ldots n)$. If we disregard the unobjectifiability of quantum-theoretical concepts (§ 21) then all the basic laws of this theory are of the statistical kind. A typical law of Quantum Theory can be formulated, under the above simplifying assumption, as follows: "If both the distribution of positions and the rate of change of this distribution are given in a population of similar physical entities, then the distribution of any other quantity within the same population can be effectively determined."[6]

Laws which do not deal with populations and the interdependence of statistical distributions in the populations concerned, but rather with single individuals, are often referred to as "causal." Yet this interpretation of "causal laws," though frequent, is hardly felicitous, since it has little in common with the problems of causality and determinism. For example, any statement of the form "Every S is P" would be causal in this sense. We propose, therefore, simply to distinguish statistical from non-statistical laws in the cases just referred to.

Another distinction between statistical and causal laws, more closely connected with the problems of determinism, is involved when a statistical law is said to be causal if it associates a sharp[7] distribution of the determined quantity Q with sharp distributions of the determining quantities Q_i. In this case, the value of Q for any member of the population is uniquely determined by all the corresponding values Q_i, and the latter may be considered to be the "cause" of the former.

(6) Separately versus Contextually Justifiable Laws

One important distinction is connected with the possibility of validating the law under consideration either separately, i.e., without utilizing any other law which is also not separately validatable, or only in conjunction with other laws. For example, Newton's law of equality of action and reaction has been shown by Poincaré to be only contextually capable of validation. The full sense of this distinction will be clarified only after the methods of validating laws and theories have been discussed in some detail. But it may be pointed out at this juncture that so-called "empirical laws" are just separately justifiable laws. In particular, a law is empirical if all its instances are facts which can be validated by applying the fact-finding methods discussed in §§ 15–18.

Not every quantified statement can be admitted as law-like, if the concept of scientific law is to be of any interest in epistemology and the philosophy of science. "Every electron includes an invisible goblin" is a quantified, but hardly a law-like, statement. The source of our hesitation to consider it as possibly eligible for the status of a scientific law is the circumstance that its instances do not represent discoverable facts. Yet the same holds true of statements which we do admit as laws, although they are only contextually justifiable: their instances do not represent discoverable facts etiher. Thus the distinction between empirical laws, contextually justifiable laws, and quantified statements which are altogether inaccessible to scientific validation is an important aspect of the concept of law, to be discussed in connection with the Principle of Verifiability. . . .

ON DEFINING SCIENTIFIC LAWS

What exactly do we mean by a scientific law? We have surveyed, in the two preceding sections, the main types and functions of those statements which are usually granted the status of scientific laws, to prepare for our major objective in the present chapter: a critical survey of scientific methods of discovering and establishing laws. As further preparation, we must consider, on the basis of the preceding discussion, the definition of the concept of scientific law. It must apply to all types of scientific laws listed in § 19 and take into account the main uses to which laws are being put within science, as discussed in § 20. Since the most significant attempts at defining the concept of scientific law have aimed at isolating the types of laws which meet the requirement of *universality,* we shall start with an examination of the definitional value of this requirement.

The main difficulty in any attempt at differentiating law-like from fact-like statements in terms of universality is that the concept of universality is ambiguous. The phrase "universality of statements" was seen to be interpretable in at least three ways, so that there are several definitions of a law-like statement, no two of which are equivalent to each other. (1) A statement can be defined as law-like if it is conceptually universal, that is, all its constituent terms are universal, not local. (2) A statement may be said to be law-like because its spatio-temporal scope is universal; in other words, if the statement applies to any relevant object everywhere and always. (3) A statement may be classified as law-like if it has several instances and is asserted to hold true of all these instances; in other words, if all its quantifiers are universal. Thus, universality of concepts, universality of scope, and universality of quantification are three distinct criteria of law-like status. Moreover, any two, or even all three, of these requirements of universality can be combined in order to define law-like statements. Most of these definitional possibilities have actually been suggested by various authors as a means of circumscribing the meaning of scientific laws. We shall see, however, that these versions of the requirement of universality are all violated by well-known laws of nature.

(1) Universality of Quantification

The requirement of universal quantification, to begin with, is disregarded in all those numerous laws whose quantificational pattern is at least partly existential. "There are liquid crystals," "there exists an indivisible unit of electrical charge" are laws of nature which clearly involve existential quantifiers. Laws which clearly combine universal and existential quantification are legion, since, as pointed out in § 19, all statistical laws and all law-like statements expressible by differential equations belong in this category. There are also basic laws of nature which are neither statistical nor differential and nonetheless involve a mixed quantificational pattern. For instance, Dalton's Law of Multiple Proportions in chemistry, and Hauy's Law of Rational Indices in crystallography come under this heading. So does the law asserting the existence of atoms of electricity, since it amounts to a statement that the electrical charge of *every* material object is *some* multiple of a specifiable minimum charge.

As a matter of fact, a purely existential quantification would hardly be of value in a scientific law, since such a law would follow from any single fact which happens to instantiate it. The knowledge of this particular fact would make its law-like competitor superfluous. Laws are introduced if they are able to say more, not less, than facts, in approximately the same number of words; if they were less informative than facts, they would hardly be resorted to in science. Actually, purely existential laws are employed primarily in axiomatic presentations of already available theories, since, under such circumstances, the weakest sufficient assumptions are the most desirable. Thus, Veblen, in his classical axiomatic foundation of Euclidean geometry, uses an existential postulate to the effect that "there is a least one point."[8] But an empirical theory, even when exclusively devoted to the search for laws and relegating particular facts to the role of crude, evidential material, is unlikely to care about such existential postulates whose content is more meagre than that of particular facts.

Let us conclude, therefore, that the requirements of universal quantification is certainly incompatible with the actual supply of scientific laws and that even purely existential laws, though scarce, nevertheless exist. Needless to say, while rejecting the requirement of quantificational universality with regard to law-like statements, we do not deny thereby the existence of some quantificational pattern in every law-like statement. Such statements are always quantified, although no formal restriction on their quantificational pattern seems warranted by examining their logical structure.

(2) Universality of Scope

With respect to the spatio-temporal scope of validity or applicability, three groups of law-like statements can be distinguished (cf. § 19): (*a*) *Universal laws* which are applicable to all relevant objects always and everywhere; (*b*) *Regional laws,* which apply to all relevant objects within, but not without, a limited spatio-temporal region; (*c*) *Individual laws,* applicable to a single individual or a group of a few individuals. It is obvious that if the requirement of universal scope is accepted, only universal laws fully deserve their law-like status, and neither regional nor individual laws are genuine. Newton's Law of Gravitation is universal in this sense; but biological evolution is governed by regional laws; Kepler's Laws of Planetary Motions have individual scope.

Yet, as already pointed out, it would be an arbitrary fiat to deny law-like status to well-established regional and individual laws. That is why several attempts have been made to reconcile the requirement of universal scope with these two types of laws. One of these attempts consists in partitioning the body of scientific laws in such a way that the dominating role of universal laws is taken into account, while providing some accommodation for laws of a more restricted scope. This is to be achieved by stipulating that a law of limited scope (regional or individual) must always be deducible from universal laws; the latter would then deserve the status of ultimate laws and the former would be only derivative.

Such a defence of universality of scope is hardly satisfactory, however, because, as a matter of fact, regional law-like statements are not always deducible from universal premisses. Thus, according to the statistical interpretation of the Second Principle of Thermodynamics, the law asserting an over-all increase of cosmic entropy is valid during certain periods of cosmic history and invalid during the remaining periods. In our own period the law is apparently valid and thus the delusion as to its universality arises. But the regional law of increasing entropy, valid for that space-time region consisting of the whole of space and a considerable time-interval about the present moment, is not deducible from a universal law, since its universal law-like counterpart is false, and hence is not a law. This example shows that if the distinction between ultimate and derivative laws is defined in terms of scope, then the dichotomy "ultimate-derivative" is not exhaustive. A regional law may be neither universal nor derivative. The classification is therefore unable to assure a privileged rank to universal laws. The intention of confining law-like status to ultimate universal laws and those deducible therefrom cannot be carried out.

Another attempt at vindicating the requirement of universality of scope could be made by utilizing the distinction between full-fledged and vacuous universality. In § 19 I have referred to vacuously universal laws of astronomy, biology, psychology, and human history, which are useful for limited regions where the relevant objects can be found, but useless outside these regions because of the absence of such objects. Thus, we may grant vacuous universality to Darwin's theory on the assumption that the entities and events governed by the laws of this theory are not to be found beyond a rather restricted spatio-temporal region.

Strictly speaking, two kinds of vacuously universal laws should be distinguished. A law may be prevented from applying beyond a specified region because of the contingent fact that no relevant objects are available outside this region although there is good reason to assume that if such objects were available, the law would apply to them; we shall then say that the law is vacuously universal in the *intrinsic sense.* Thus laws governing the motion of man-made clocks not only hold true of clocks on the earth, but would certainly also apply to non-terrestrial clocks if there were such. On the other hand, a vacuously universal law may be such that there are no grounds for claiming its potential applicability to relevant, though actually non-existent entities beyond a specified region; we shall then say that the law is vacuously universal in an *extrinsic sense.* Thus, the law that the size of any man-made clock never exceeds a specific limit is vacuously universal in the extrinsic sense, because its extra-terrestrial validity is simply due to the lack of relevant objects. There are some reasons for assuming that if the laws of biological evolution are vacuously universal at all, that is, if there is no life outside the earth, then their vacuous universality is at least intrinsic: life everywhere would take the evolutionary course it did take on earth. It goes without saying that vacuous universality in the extrinsic sense is of no interest in connection with the concept of scientific laws. On the other hand, if every scientific law could be shown to be at least vacuously universal in the intrinsic sense as far as its scope is concerned, the requirement of universality could be considered to be fulfilled.

I do not think that this attempt stands up under closer scrutiny, interesting as the concept of vacuous universality may prove to be from other points of view. As just pointed out and illustrated by the statistical interpretation of the Second Principle of Thermodynamics, there are laws which are neither universal nor deducible from universal laws for the simple reason that they are invalid outside a particular region. Hence, such laws are not even vacuously universal, because though relevant objects are available outside their region of validity, these laws just do not apply to them. Hence regional laws, which are not even vacuously universal, are available, and consequently, this line of defence proves insufficient.

Yet even if we are prepared to grant law-like status to both universal and regional laws, we may feel some qualms about individual laws. Strictly speaking, the statement "Napoleon never seriously contemplated invading Great Britain" is not a particular fact, because of the universal quantifier "never" which stretches over a time-interval and prevents the whole statement from being replaceable by an unquantified, molecular, fact-like statement, under the usual infinitist assumptions concerning the time-variable.

The status of essentially quantified statements about individuals is, of course, arbitrary to some extent. A refusal to grant them law-like status is open to the objection that no essential logical difference is apparent between such statements and those which have regional universality. On the other hand, if we grant them law-like status, we run the risk of blurring the boundary between particular facts and general laws. Actually, none of these difficulties is insuperable because a significant logical difference can be pointed out between non-controversial individual laws and other quantified statements about individuals; such differences may warrant investing some quantified statements about individuals with the majesty of law.

Why do we not hesitate to refer Kepler's conclusions from Tycho Brahe's observations as genuine laws in spite of their applicability to but a small number of individuals, while we are rather inclined to consider the quantified statement about Napoleon as not being law-like? The point is that Kepler's laws, though not deducible from the universal laws of mechanics and gravitation, can be derived from these laws provided they are supplemented by a finite number of fact-like premisses describing the basic structure of the planetary system, that is, the masses and mutual distances of the sun and its satellites. In other words, Kepler's individual laws are an individual instance of universally valid laws which apply to any and every planetary system; the instantiation rests on the individual data concerning our particular planetary system and is provided by the fact-like premisses to be added to the universal laws of mechanics and of gravitation. The quantified statement about Napoleon is not known to be derivable from universal laws supplemented by a few relevant facts about Napoleon, and that is why we are reluctant to grant it law-like status. At bottom, the vague intuition of being law-like which underlies the subdivision of quantified statements about individuals into those which are and those which are not law-like, seems to be a feeling that whenever a quantified statement about individuals instantiates a universal law, it reflects so faithfully and remains so close to this law, that it naturally acquires law-like status by contact, as it were.

To comply with this vague intuition of the law-likeness of some quantified statements about individuals and with our intuitive reluctance to grant law-like status to all quantified statements about individuals, we have tentatively reinstated the distinction between derivative laws governing individuals and ultimate laws of universal scope from which individual laws would have to be derivable. Previously, however, we found that a similar treatment of regional laws is impossible because some of them are not derivable from universal laws. This circumstance suggests that it may be preferable to apply the same procedure to statements about individuals and to grant to these statements law-like status whenever they are essentially quantified, that is, not equivalent to any finite combination of facts and calling for inductive methods of validation instead of the fact-finding methods discussed in chapter Two. The following arguments favour the extension of law-like status to all essentially quantified statements about individuals:

(*a*) A regional law to the effect that all the individuals in the region *R* have the property *P* can always be re-formulated as meaning that the region *R* has an associated property *P′* (on the understanding that to have the property *P′* means to be populated by individuals endowed with the property *P*). This new formulation is actually required in order to ensure the applicability of inductive logic to the regional law, since, according to the usual assumptions, inductive inferences are valid for statements whose subjects involve local constants but may fail to apply to statements with locally defined predicates.[9] This requirement of inductive logic is met by the changed formulation of regional laws but not by the initial one. The new formulation, however, is hardly distinguishable from a statement about individuals. As a matter of fact, we have found reasons for re-formulating statements about individuals in terms of the spatio-temporal regions they occupy. (cf. § 18).

(*b*) Our intuitive reluctance to grant law-like status to quantified statements about individuals whenever these statements are not known to be derivable from universal laws seems to be due, partly at least, to the fact that individuals have come to be so strongly associated with fact-like statements that we are unwilling to consider any statement about individuals as law-like unless its logical proximity to a universal law compels us to do so. This explanation suggests that the whole dichotomy of statements about individuals into those that do and those that do not qualify as law-like is based upon fluctuating linguistic usage and vague psychological associations rather than upon sound logical grounds.

(*c*) Another source of our reluctance to consider all quantified statements about individuals as law-like is traceable to the circumstance that such statements usually seem to lack a certain flavour of necessity which we have come to associate with other more conspicuous examples of scientific laws. However, we may also feel reluctant to grant law-like status to statements which meet all the three requirements of conceptual, spatio-temporal, and quantificational universality. Consider the statement: "There are no gamma-ray microscopes in the universe." It has certainly universality of concepts and of quantification, and, in all probability, universality of scope as well. The less dignified example "No human male has ever had a hat of more than 10 yards in diameter" also meets the three requirements in a similar fashion, the first two with certainty, the third in all probability (in view of the claim about the dimensions of human headgear being made only in regard to the male population). Yet, in spite of all the three requirements being fulfilled, we feel hesitant about considering the two statements as laws, because they lack the quality of *necessity,* often associated with scientific laws. Similarly, the three-dimensional nature of space is frequently felt to be "accidental" although the corresponding statement about space meets all the three requirements of universality and can hardly be denied law-like status. Actually, this "necessity" of scientific laws, analysed almost incessantly since the time of Hume, has the unpleasant habit of vanishing without notice, at the slightest attempt at logical analysis.[10]

(3) Requirement of Conceptual Universality

This requirement is ostensibly incompatible with the proposal to grant, under suitable circumstances, the status of law to quantified statements about individuals. Such statements must necessarily include local constants or individual proper names if the scientific theory is expressed in the vernacular. Similarly, the formulation of a regional law must involve local constants since, in view of the homogeneity of the spatio-temporal continuum, the particular region where the law under consideration is valid can only be specified by using local constants. In other words, the requirement of conceptual universality would automatically entail the universality of scope of all ultimate laws and provide at most for derivative laws of limited scope. We have argued, however, that not all laws of limited scope are derivative. The same argument can serve to refute the requirement of conceptual universality.

Yet, in view of the controversial nature of quantified statements of limited scope, it seems worthwhile to point out that serious reasons in support of laws which violate the requirement of conceptual universality can be adduced independently of the attitude one takes up towards such quantified statements.

The requirement of conceptual universality is tantamount to a ban on local constants in scientific laws. What justification can be given for such a ban? We have seen that well-established laws of nature can easily be quoted which contravene it. As in the previous case, two different lines of defence can be devised to protect conceptual universality. These two defensive strategies will be referred to briefly as the distinction between a categorematic and a syncategorematic use of local constants or proper names in scientific laws, and the division of all scientific laws into ultimate and derivative ones, with the intention of reducing laws involving local constants to the latter category.

Let us examine the first line of defence. I have already pointed out that all the quantitative laws of classical science involved local constants or proper names, because the standards of basic quantities, such as length, duration, mass, were defined in local terms. I grant that this is not any longer so, and that there is an unmistakable, successful tendency to get rid of local terms in formulating the basic laws of nature. Yet, one can hardly refuse to apply the term "law" to the classical regularities only because they are formulated in a way involving local terms, either explicitly, or at least implicitly. The circumstance that such local terms have proved dispensable later, when the classical laws of nature have been reformulated by means of universal terms only, is no real justification for denying the title of laws to the classical regularities in their original formulation. One has to notice that the very possibility of this contemporary "universalistic" re-formulation depends upon an empirical combination of circumstances, not upon the logical nature of the terms involved. If one realizes, moreover, that the basic contentions of, say, Newton and Maxwell would be affected immediately by a ban on local constants in laws of nature and re-classified as fact-like statements, one cannot help concluding that the mere presence of local terms in empirical propositions could not deprive them automatically of law-like status, unless some additional conditions were fulfilled. Some of these conditions refer to inductive validating methods and have already been touched upon.

For the time being, it may suffice to distinguish between statements in which local terms occur as *syncategorematic* components of other constituent expressions and those whose local terms function in a *categorematic* capacity. Thus, the fact that the basic units of physical quantities used to be defined in local terms does not by any means imply that the laws which refer to these quantities are *about* any single individual, because the role these terms play is but syncategorematic. A centimetre may be defined in relation to the individual "earth," but the statement to the effect that the size of an atom is of the order of 10^{-8} cm. is about atoms, not about the earth. On the other hand, Kepler's Laws of Planetary Motions refer explicitly to certain individuals, viz., the sun and its planets; they involve a categorematic use of individual constants.

A merely syncategorematic use of local constants may seem to be more easily reconcilable with the universalist tendency, even if these constants are not replaceable by universal terms. For statements involving only such use of local constants can have a quantificational pattern which is universal throughout, and a universal scope of validity as well. On the other hand, a categorematic use of local terms obviously implies that the statement is about single individuals; its specific referential import is a direct consequence of the categorematic role the local signs play in its structure.

Now, the first line of defence of conceptual universality of scientific laws might consist in the claim that local terms, if occurring at all in scientific laws, are able to do so at most in a syncategorematic capacity. Yet, strangely enough, it is precisely this syncategorematic use of local constants that must be submitted to essentially restrictive rules, on grounds of inductive logic (cf. § 22). On the other hand, the categorematic use of such constants is both inductively unobjectionable and essential in regional and individual laws, say, in the geological laws that govern the succession of eras and periods in the history of the earth. There is no possibility of accounting for the presence of local constants in any law of this type by referring to their allegedly syncategorematic function.

Another attempt can be made to save the requirements of conceptual universality by subdividing, once more, all scientific laws into ultimate and derivative laws. This time the ultimate laws would abide by the requirement of conceptual universality, and the derivative laws are those which follow from ultimate laws. Thus a statement would acquire law-like status if it were either conceptually universal or deducible from other conceptually universal statements. For instance, Kepler's laws would owe their law-like status to their deducibility from mechanical and gravitational laws which do not contain a single individual name.

The trouble with this line of defence is exactly the one which arose in the previous case. Of course, regional and individual laws would now have to be derivative in a new sense, that is, deducible from conceptually universal laws. However, since conceptual universality is a more stringent requirement than universality of scope,[11] any quantified statement which can be deduced from conceptually universal premisses is also derivable from premisses whose scope is universal and would therefore owe its eligibility for law-like rank to its derivability from laws of universal scope. The same consideration would apply to regional laws. We have seen, however, that not all individual and regional laws are so derivable. Accordingly, the requirement of conceptual universality cannot be saved, even when confined to ultimate laws, since the latter do not provide a sufficient basis for deriving all scientific laws.

To sum up: neither conceptual nor quantificational nor spatio-temporal universality is relevant when establishing the law-like status of a statement; nor can such status be confined to statements which are deducible from premisses whose universality is either conceptual or quantificational or spatio-temporal. Scientific laws should be defined as essentially quantified statements. According to this definition, the main difference between laws and facts consists in the methods of their validation. Particular facts are established by applying the non-inductive fact-finding methods described in chapter Two. The validation of laws is necessarily inductive.

(4) Some Alternative Definitions of Scientific Laws

The concept of scientific law plays such a prominent role in the whole structure of science that it is no wonder several attempts have been made to formulate a precise definition of it, which would be true to its actual use and also provide for some reliable conclusions to be drawn therefrom about the meaning, nature, and functions of scientific laws. There seems to be rather general agreement among analysts of science that a law-like statement is essentially quantified or general, in contradistinction to a fact, which is a particular unquantified statement about a single individual or a few individuals. The artificial terminology which the older positivists have tried to introduce by designating scientific laws as "general facts," in contrast to genuine "particular facts," did not strike any roots in language. Since we are anxious not to blur the important differences in methods of validation applicable to particular facts and general laws respectively, we have drawn a rather sharp boundary between the two kinds of propositions.

Yet, apart from the consensus about the role played by quantification in the logical structure of law-like statements, there is little agreement as to the exact shape of the quantificational patterns characteristic of law-like statements. Thus some writers require that all the quantifiers be universal in such statements (e.g., Hempel and Oppenheim,[12] and Popper[13]) and propose to designate as "theories" those statements, some or all of whose quantifiers are either partly or all of them existential. Burks[14] suggests that a statement be called law-like if all its quantifiers are of the same kind—either all existential or all universal—on the understanding that statements with a different pattern of quantifiers should be called "theories." According to Bergmann,[15] any quantifiers are admissible in a law-like statement, provided that all its terms be "empirical constructs," that is, linked up by a chain of full-fledged or conditional definitions with terms of the ostensive level. This variegated sample of views concerning quantificational patterns admissible in law-like statements could easily be enlarged.

The reasons for the various definitional proposals are rarely explicitly stated. It seems that the main motive is the desire either to come as close as possible to the actual meaning of a "scientific law" or to ensure the possibility of validating scientific laws on observational grounds. For example, if it seems desirable that a law be either finitely provable or finitely disprovable by observational findings then the requirement of quantificational homogeneity will be used to define laws. Bergmann establishes an inferential relation between laws and observational facts by his stipulation that the constituent terms of a law be empirical constructs, and this is why he can afford to be very liberal in regard to the admissible quantificational patterns of scientific laws. On the other hand, the desire to be true to the actual use of the term "law" may be responsible for the requirement of quantificational universality suggested by some writers: the most conspicuous examples of natural laws are universal in regard to quantification, although those which do not meet this requirement are far from negligible. The same motive is likely to be responsible for the ban against local constants and individual names made by several authors. The objections to which all these proposals are open have been discussed in the three previous sections.

An interesting feature of some of these proposals is that scientific laws, by definition, turn out to be a special case of scientific theories: every law is a theory, though the converse does not hold. We shall see later that there are serious reasons for separating laws from theories by a wider gap: theories will prove to be rather deductive systems consisting of laws than single statements which differ from law-like statements in regard to their quantificational pattern. According to Pap, only separately provable or disprovable statements can be laws, whereas other statements of similar logical structure should be considered as "theories." This amounts to classifying as theories the class of contextually justifiable laws and provides, once more, for too narrow a gap between laws and theories.

The difference between law-like statements and genuine laws has been discussed by Hempel. The question is whether laws should be defined as true law-like statements, or as law-like statements known to be only probably true, that is, supported by adequate evidence. The first alternative is in closer agreement with ordinary usage, but raises several objections: for example, since we can never be absolutely sure whether a law-like statement is actually true, neither can we be absolutely sure whether any preassigned statement is a law, if laws are defined in terms of truth. The same difficulty obviously applies to the difference between fact-like statements and actual facts. Since we did explain the latter difference in terms of truth, it seems reasonable to take the same course with respect to the former. The difficulty that no statement can be known definitively to be either a fact or a law must not be overrated, however, for two reasons. In the first place, it does not follow that logical and epistemological questions concerning the nature, meaning, and validation of laws and of facts depend upon the empirical question of whether a given statement or class of statements fall under the category of facts or of laws. What is involved in such epistemological and logical investigations relates to whether a given statement is law-like or fact-like, and this can be settled without presupposing any empirical findings. In the second place, any misgivings as to the emergence, in this context, of the concept of truth, which is unfortunately charged with a long tradition of speculative vagueness and haziness, are unwarranted, since modern logic has paved the way towards a precise analysis of this concept, in accordance with its actual use. The meaning of truth is of decisive importance for our whole investigation and will be discussed in detail in Part III.

The fact that several investigators have altogether denied the applicability of the concept of truth to law-like statements is easily accounted for. Thus Ramsey and Schlick have assumed that only sentences meeting a very stringent requirement of empirical verifiability express genuine statements and can be labelled "true" or "false."[16] Since infinitist laws of nature do no meet this requirement they would have to be interpreted as recipes for predicting facts rather than as true statements. We shall try to show that this view of scientific laws, which would deny them any genuine cognitive function, is incompatible with an appropriately re-formulated Principle of Verifiability (cf. § 34).

Toulmin[17] seems to have been led to deny the applicability of the concept of truth to laws by his assumption that laws are mere formulae, that is, incomplete statements which do not specify the conditions of their validity. For example, Boyle's law concerning the relation between the volume V and pressure P of a gas at a constant temperature would be identified with the formula $V \cdot P = C$ without any indication being added as to the class of gases to which this formula applies. This is correct as far as it goes since nobody would ascribe a truth-value to an incomplete statement. However, there is no reason for denying truth-value to laws if they are interpreted as complete statements (as laws usually are) with all the conditions of validity either explicitly stated or implied by the meanings of the relevant terms.

NOTES

1. Mill's classification of laws raises several difficulties to be outlined in the sequel. The distinction between "fundamental" and "derivative" laws, defined by C. G. Hempel and P. Oppenheim ("The Logic of Explanation," *Philosophy of Science*, vol. 15, 1948), renders the same services as Mill's classification, without giving rise to these objections. Cf. J. S. Mill, *A System of Logic* (1874).
2. A law, whether or not well established, can be defined simply as a true law-like statement.
3. The logical structure of limiting concepts, such as the concept of derivative, is explained, for example, in R. Carnap, *Einführung in die Symbolische Logik* (1954).
4. In accordance with his metaphysical views of "cosmic epochs," A. N. Whitehead has denied universal scope even to physico-chemical laws and admitted only regional laws of nature. Cf. his *Science and the Modern World* (1926).
5. Cf. A. N. Whitehead, *The Concept of Nature* (1920); A. Grünbaum, "Whitehead's Method of Extensive Abstraction," *British Journal for the Philosophy of Science*, vol. 4 (1954).
6. The mathematically trained reader may be interested in a more precise formulation which involves several technicalities. It runs: "If the distribution $D(p)$ of the spatial position p in a population of similar physical systems (i.e., of systems associated with the same Hamiltonian function H) and the rate of

change of this distribution $d\,D\,(p)\,/\,d\,t$ are given—$D(p)$ indicating the fraction of the population which occupies the position p—then the average A of any quantity q (such as velocity, momentum, energy, etc.) in this population is determined by the expression $A\,(q) = \int q^{op}\Psi.\bar{\Psi}dp$ where q^{op} is the operator corresponding to the quantity q and the state-function Ψ is uniquely determined by the distribution $D(p)$ and its rate of change through the Schroedinger equation of the physical systems: $H^{op}\ \Psi = \dot{\Psi}$." Cf. H. Reichenbach, *Philosophic Foundations of Quantum Mechanics* (1946); H. Margenau, *The Nature of Physical Reality* (1950).
7. The distribution of a quantity in a given population is said to be sharp if this quantity takes on the same value within the whole population.
8. Cf. O. Veblen, "A System of Axioms for Geometry," *Trans. Amer. Math. Soc.*, vol. 5 (1904).
9. Cf. R. Carnap, "On the Application of Inductive Logic," *Philosophy and Phenomenological Research,* vol. 8 (1947).
10. During the last two decades, a considerable effort has been made by several authors to rationalize our intuitive reluctance to grant law-like status to some quantified statements of individual, regional, and universal scope. The objective was to explain the feeling of necessity often associated with the concept of law and to lay down a precise criterion capable of separating genuine law-like generalizations from merely accidental regularities even if the latter should be on a cosmic scale. These investigations have shed new light on related problems (e.g., the validity of inductive inference and the adequacy of conditional definitions for the analysis of dispositional concepts) but seem to have failed to bring the basic objective any closer. Cf. N. Goodman, *Fact, Fiction and Forecast* (1955), R. B. Braithwaite, *Scientific Explanation* (1955), W. Kneale, *Probability and Induction* (1949).
11. If a true statement includes no local constants (and is thus a conceptually universal law), then it does not involve any reference to a single spatio-temporal region either, and is consequently of universal scope; the converse does not necessarily hold, as shown by our previous discussion of local standards of measurement implicit in the laws of nature.
12. C. G. Hempel and P. Oppenheim, "The Logic of Explanation," *Philosophy of Science,* vol. 15 (1948).
13. K. Popper, *Logik der Forschung* (1935).
14. A. Burks, "Justification in Science," *Proc. Amer. Phil. Assn.*, Eastern Div., vol. 2 (1953).
15. G. Bergmann, "Outline of an Empiricist Philosophy of Physics," *Amer. Jour. of Physics* (1943).
16. F. P. Ramsey, *Foundations of Mathematics* (1931); M. Schlick, "Die Kausalität in der gegenwärtigen Physik," *Die Naturwissenschaften* (1931).
17. S. Toulmin, *The Philosophy of Science* (1953).

READING 10

Some Recent Work on the Problem of Law

Evan K. Jobe

It is widely agreed that 'scientific law' (or 'law of nature') is one of the key scientific terms which any adequate philosophy of science must attempt to clarify or define. The importance of the concept 'law' is made evident by the fact that the distinctive functions of science—explanation and prediction—are usually analyzed with reference to laws. Thus events are *explained* by showing that descriptions of them are deducible from laws (conjoined with suitable statements of "initial conditions"), and laws are utilized in deducing descriptions of unknown future events, thereby permitting their *prediction*. Moreover, it has recently become clear that the concept 'law' is relevant to the analysis of counterfactual conditionals, disposition terms, and the physical modalities. An adequate explication of 'law,' then, will further the analysis of these important concepts.

The explication of 'scientific law' proposed by Hans Reichenbach is the result of what is probably the most systematic and comprehensive attack that has yet been made on this problem. It seems to me that only Reichenbach has both attacked the problem in something like its full complexity and also offered a precise solution. The earlier version of this work was presented in 1947 in his *Elements of Symbolic Logic*. His later, considerably refined, analysis was published in 1954 as *Nomological Statements and Admissible Operations*. This last monograph is rather poorly written and abounds in typographical errors. It has been widely ignored. Nevertheless, it may well contain the most satisfactory solution thus far proposed to the problem of explicating 'scientific law'.

Reprinted by permission of the author and *Philosophy of Science, 34,* 1967, pp. 363–381.

Reichenbach's solution is technically quite complex. Other recent work on the problem of law has taken the form of proposals of a simpler and prima facie perhaps more appealing nature. Representative of these are the approaches of R. B. Braithwaite (1953), S. Körner (1954), A. J. Ayer (1957), W. Sellars (1958), and M. Bunge (1961).

The purpose of this paper is twofold. First, to present and criticize the five approaches just cited. Second, to evaluate recent work that has been explicitly or implicitly critical of Reichenbach's solution. I hope both to point out inadequacies in the non-Reichenbachian approaches and to show that none of the points prominently urged against Reichenbach's solution is well taken. We may now turn to the first major task of this paper—the presentation and criticism of recent representative approaches to the problem of law.

The first view to be considered is that of R. B. Braithwaite. Braithwaite wishes to uphold the Humean view that laws are essentially expressions of constant conjunction between properties. All the various types of generalizations which fall under the category of scientific law, Braithwaite says, "may be brought under concomitance generalizations—that everything which is *A* is *B*—by allowing A and B to be sufficiently complex properties" ([2], pp. 9–10). He notes that it is tempting to suppose that laws must assert something more than mere constant conjunctions, but such a supposition is rejected as unnecessary. What is distinctive about laws lies not in what they assert but rather in the role they play within a scientific system. He says:

> The combination of the constant-conjunction view that scientific laws are only generalizations with a doctrine of the function of such generalizations within scientific systems put the constant-conjunction view in a new light. It answers the complaint that the constant-conjunction view underestimates the place of reason in science; by stressing the importance of the logical relationships between generalizations at different levels within the system, it goes some way to satisfy those who hanker after logically necessary scientific laws ([2], p. 11).

Braithwaite's proposed criterion of lawlikeness is as follows:

> The condition for an established hypothesis *h* being *lawlike* (i.e., being, if true, a natural law) will then be that the hypothesis either occurs in an established scientific deductive system as a higher-level hypothesis containing theoretical concepts or that it occurs in an established scientific deductive system as a deduction from higher-level hypotheses which are supported by empirical evidence which is not direct evidence for *h* itself ([2], p. 301f.).

An "established hypothesis," for Braithwaite, is one which is well confirmed, i.e., "The evidence makes it reasonable to accept the hypothesis" ([2], p. 14). Braithwaite continues:

> This account of natural law makes the application of the notion dependent upon the way in which the hypothesis is regarded by a particular person at a particular time as having been established: "lawlike" may be thought of as a honorific epithet which is employed as a mark of origin. If the hypothesis that all men are mortal is regarded as also being supported by being deduced from the higher-level hypothesis that all animals are mortal, the evidence for this being also that horses have died, dogs have died, etc., then it will be accorded the honorific title "law of nature" which will then indicate that there are other reasons for believing it than evidence of its instances alone ([2], p. 302).

Pap subjects Braithwaite's view to the following criticism:

> ...on this analysis there are no accidental universals at all. Consider, for instance, (1) "all men now in this room are bald," as compared with (2) "all tall men now in this room are bald," where actually some of the men now in this room are tall and some are not. Clearly (2) is deducible from (1), and there is more instantial evidence for (1) than for (2). This is exactly analogogous to Braithwaite's argument... that "all men are mortal" is regarded as a law of nature because it is deducible from "all animals are mortal," for which generalization there is more instantial evidence ([8], p. 204).

Pap's objection, however, is invalidated by Braithwaite's prior clarification of how he is using the term 'scientific hypothesis'. Braithwaite says:

> Scientific hypotheses, which, if true, are scientific laws, will then, for the purpose of my exposition, be taken as equivalent to generalizations of unrestricted range in space and time . . .([2], p. 12).

And a little later, in a footnote, he says:

> Generalizations with only a limited number of instances, which can be proved from a knowledge of these instances... present no logical problem; since they are of little interest in science, they will not be considered further. All scientific hypotheses will be taken to be generalizations with an unlimited number of instances ([2], p. 14).

Rather than being too weak, as Pap appears to think, Braithwaite's criterion for laws is an exceedingly strong one. In the first place, a statement is not, according to Braithwaite, a law, unless it is actually incorporated in some established scientific deductive system. This seems to imply that, whatever other characteristics a statement has, it is not a law unless it is actually regarded or treated as one. On this explication, the notion of an "undiscovered law" would appear to be nonsense. This suggests that the notion which Braithwaite is explicating is not 'law of nature' but rather 'law of nature recognized as such'. In the second place, however, his requirement is too strong even for this latter notion. There seem to be almost paradigm cases of natural laws that have been justifiably treated as such even though they contained no theoretical terms and were not derivable from higher-level hypotheses within a known scientific system. Consider, for example, the generalization 'Water expands upon freezing'. This has not (at least until quite recently) been derivable from higher-level hypotheses. Yet, this seems to represent an almost paradigm case of a law of nature. Other examples of this type may be 'Iron is harder than lead' and 'Sugar is (water)-soluble'. There would seem to be a large class of non-theoretical generalizations performing the distinctive functions of laws, which are excluded by the deducibility requirement.

It may well be that the task of a science is not complete until it can be cast into the form of a deductive system in which all of its laws are directly or indirectly derivable from a set of highest-level laws containing theoretical terms. Braithwaite seems to be working with such a model in mind, and his explication appears to be most nearly adequate for those branches of science that approach this ideal. He might be said to have given an explication of 'law occurring within an ideal scientific system'. The task of such an explication, however, represents a considerable simplification of the task of defining law of 'nature'.

The second view to be considered is that of S. Körner. In order to furnish a satisfactory definition of 'law' Körner believes it is necessary to introduce what he calls "inexact" logical relations. The neglect of such relations in the past has been the result, he says, of the preoccupation of modern logicians with mathematical concepts to which only "exact" logical relations are applicable ([5], p. 227). To follow Körner's example, suppose that the word 'swan' is defined notion of an inexact logical relation is best illustrated by means of concepts defined simply by pointing to particular swans, explaining that 'swan' is applicable to any examples like those pointed to, that 'white' is similarly defined by ostensive reference to white things, and that 'black' is defined by reference to black things. The question may then be raised as to what the relation is between, say, the concept of swan and the concept of white, so defined (assuming all the swans observed were white and that white things other than swans were observed). It is clear that the relation is not that of exclusion. But purely on the basis of the ostensive definitions it cannot be said that the relation between the concepts is that of inclusion, since the definitions do not rule out the possibility that there are in fact swans that are not white. The relation between the concepts so defined must therefore be considered to be the inexact one of "inclusion-or-overlap." Körner points out that we must not confuse the relation of "inclusion-or-overlap" with the relation of "inclusion or of overlap" ([5], p. 224). If two concepts stand in the relation of overlap, then, of course, they stand in the relation of inclusion or of overlap, but two concepts stand in the relation of inclusion-or-overlap only if their definitions determine that they do *not* stand in the relation of exclusion but do *not* determine that they stand in the relation of overlap and also do *not* determine that they stand in the relation of inclusion. To take another example, the ostensive definitions referred to determine that the concepts 'swan' and 'black' stand in the relation of exclusion-or-overlap.

Körner takes the problem of defining 'law of nature' to be essentially that of clarifying the nature of *hypotheses.* He says:

> The question of the logical status of laws of nature is not a question of their truth or falsehood. I shall, therefore, discuss hypotheses of laws of nature or, briefly, hypotheses rather than laws of nature. To state a law of nature is to state a true or established hypothesis, while to state a hypothesis is not necessarily to state a law of nature ([5], p. 216f).

Körner proceeds to lay down five requirements which he believes any satisfactory account of hypotheses must insure for such statements. First, a hypothesis must state a relation between two concepts or classes. Second, the protasis and apodosis of a hypothesis are "either ostensive concepts, or else classes of particulars to which ostensive concepts are applicable" ([5], p. 217). He clarifies his use of 'ostensive concept' by saying, "A concept is ostensive if among the rules governing its application there is an ostensive rule" ([5], p. 217). Third, a hypothesis is a general hypothetical proposition. Körner makes it clear, however, that 'general hypothetical proposition' does not, for him, stand for what is usually called a 'general material implication'. He says:

> We often use grammatically conditional sentences for propositions which are not hypothetical in the sense in which entailment-propositions and hypotheses are hypothetical. We may, for instance, use the sentence "If anything is a man then it is a biped" for the conjunction of propositions that either Socrates is not a man or he is a biped, and that either Plato is not a man or he is a biped *and so on.* To see that this general proposition is not hypothetical, we consider the proposition that if Hercules is a man then he is a biped. This proposition, which is an application of the preceding general proposition, is logically equivalent to the alternation that either Hercules is not a man or he is a biped. Since, as a matter of fact,

> Hercules is not a man and since the first member of the alternation is true, the whole alternation must be true. The general proposition is not hypothetical, because any of its unfulfilled applications are necessarily true ([5], p. 218f).

Fourth, a hypothesis must be an empirical statement. This means that "every hypothesis is incompatible with some true or false proposition which describes possible experience" ([5], p. 220). Fifth, a hypothesis, if true, "entitles us to draw inferences from unobserved instances of its protasis to unobserved instances of its apodosis" ([5], p. 221).

Körner rejects both the view that hypotheses may be logical entailments and the view that hypotheses are mere factual implications. His own view is that a hypothesis is a combination of a statement of logical relation between two concepts together with a statement which amounts to a material implication. He says:

> To use a grammatically conditional sentence such as "If anything is a P then it is a Q" or "If anything is a P then it is not a Q", for a hypothesis is to state a conjunction of two propositions. To state the *first* of these is to state the logical proposition that the ostensive concepts for which "P" and "Q" are used stand in the relation of inclusion-or-overlap (or exclusion-or-overlap). To state the second of these is to state the proposition that anything is in fact either not a P or a Q (either not a P or not a Q). This second proposition is a factual implication and a factual implication of any kind is empirical ([5], p. 225f).

To use one of Körner's examples, in asserting the statement 'If anything is a magnetized piece of iron then it attracts iron filings, what we are stating, on this view, is

> that the ostensive concepts for which "attract iron filings" and "magnetized piece of iron" are being used stand in the logical relation of inclusion-or-overlap and that as a matter of fact anything is either not an instance of the latter concept or an instance of the former ([5], p. 226).

Körner claims that his account satisfies his five requirements cited above. Examination of this claim is perhaps dispensable, however, in view of other grave difficulties which the proposed definition must face. First, it is extremely puzzling to meet with the assumption that the concepts involved in scientific hypotheses are exclusively of the ostensive type. This seems to ignore the vast array of theoretical predicates such as 'is an electron'. Even more serious, however, is the fact that the proposed definition appears to offer no protection against the acceptance as hypotheses of statements which, if true, could only be considered as "accidentally" true. For example, if it happens to be true that all gold lumps are smaller than one cubic mile, then the ostensive definitions of 'gold lump' and 'smaller than one cubic mile' will insure that these concepts stand in the relation of inclusion-or-overlap. Also it will in fact be the case that anything is either not a gold lump or is smaller than a cubic mile. The statement 'If anything is a gold lump then it is smaller than one cubic mile' is therefore not only a "hypothesis" in Körner's special sense but it is also a true one, since both the logical and the factual statements to whose conjunction the hypothesis is equivalent are true. But a true hypothesis, on Körner's account, qualifies as a law of nature, and so the cited statement is, if true, a law of nature.

To take another example, consider the statement 'If anyone is inside house H then he is less than seven feet tall'. Unless someone at least seven feet tall happened to walk into house H the ostensive definitions of 'less than seven feet tall' and 'person inside house H' will insure that the former concept stands in the relation of inclusion-or-overlap to the latter concept. It will also be true that anyone is either not a person in house H or is less than seven feet tall. The conditions for the cited statement's being a true hypothesis are fulfilled, and this statement, too, qualifies as a law of nature. This second example illustrates the case of a hypothesis that might not only be true but could even be established as true (by painstaking observation). Körner says that a law of nature is a hypothesis that is "true or established" ([5], p. 216f)—he does not seem to care which is the case—but it seems well to offer a counterexample that could be both true *and* established, which is the point of this example.

It appears then that the proposed definition, even on the strongest possible interpretation, admits a large class of statements that are obviously not suitable candidates for the title 'law of nature'.

The next approach to be considered is that of A. J. Ayer. Ayer notes that many philosophers are content to explain the "necessity" of natural laws by saying that they hold for all possible, as well as actual, instances, and that the hallmark of generalizations of law as opposed to generalizations of fact is their support of subjunctive conditionals ([1], p. 162).

> . . . while this is correct so far as it goes, I doubt if it goes far enough. Neither the notion of possible, as opposed to actual instances nor that of the subjunctive conditional is so pellucid that these references to them can be regarded as bringing all our difficulties to an end. It will be well to try to take our analysis a little further if we can ([1], p. 162).

He goes on to suggest that the difference between generalizations of law and those of fact "lies not so much on the side of the facts which make them true or false, as in the attitude of those who put them forward" ([1], p. 162). His theory, then, will itself contain dispositional terms, but the dispositional talk will be confined to dispositions concerning people's attitudes. He says:

> In short I propose to explain the distinction between generalizations of law and generalizations of fact, and thereby to give some account of what a law of nature is, by the indirect method of analysing the distinction between treating a generalization as a statement of law and treating it as a statement of fact ([1], p. 162).

Ayer offers his theory, then, not as an actual definition of 'natural law' but rather as a step, admittedly sketchy, toward such a definition. More precisely, he proposes a set of *sufficient* conditions for a statement's being *treated* as a law of nature.

Ayer's proposal is the following:

> ...I suggest that for someone to treat a statement of the form "if anything has Φ it has Ψ" as expressing a law of nature it is sufficient (i) that subject to a willingness to explain away exceptions he believes that in a non-trivial sense everything which in fact has Φ has Ψ (ii) that his belief that something which has Φ has Ψ is not liable to be weakened by the discovery that the object in question also has some other property X, provided (a) that X does not logically entail not-Ψ (*b*) that X is not a manifestation of not-Ψ (c) that the discovery that something has X would not in itself seriously weaken his belief that it has Φ (d) that he does not regard the statement "if anything has Φ and not-X it has Ψ" as a more exact statement of the generalization that he was intending to express ([1], p. 165).

The requirement that the conditional be believed in a "non-trivial" sense is meant to exclude acceptance of the conditional on the basis of its believed vacuity—or so it appears from his earlier use of the term 'trivial' ([1], p. 157).

While Ayer proposes these conditions not as necessary but only as sufficient for a statement's being treated as a law, it can be shown that they are not even sufficient. To construct a simple counterexample, consider that anyone who knows of three buildings that are taller than fifty stories is entitled to belief in the statement 'The three tallest buildings are taller than fifty stories'. (He need not, of course, believe that any of the buildings he knows about are actually in the class of the three tallest buildings.) This statement is an excellent example of a non-lawlike generalization, and it seems clear that no one would be tempted to treat the statement as a law of nature. This statement does, nevertheless, fulfill Ayer's set of supposedly sufficient conditions for being treated as a law. To see this it is only necessary to take 'Φ x' as 'x is one of the three tallest buildings' and 'ψ x' as 'x is more than fifty stories tall'. The conditional in question then becomes

(1) For all x, if x is one of the three tallest buildings then x is more than fifty stories tall.

It is clear that knowing of any three buildings that they are in fact higher than fifty stories forms a firm basis for believing (1). It it also clear that this belief in (1) is not liable to be weakened by the attribution of additional properties to buildings in the class of the three tallest buildings, so long as those additional properties are compatible with the property 'Φ', i.e., compatible with being a member of the class of the three tallest buildings.

It appears that the task of defining even a set of *sufficient* conditions for a statement's being *treated* as a law is not an especially simple one. It seems clear that Ayer's proposal for the solution of this problem cannot be considered successful.

The fourth view to be considered is that of W. Sellars ([12]). Sellars does not actually offer a definition of law, but he does advocate a particular way of regarding laws, which might form the basis for further clarification. He proposes that we take the physical modalities such as physical necessity, possibility, and entailment at their face value ([12], p. 280). He compares the use of modal terms in empirical contexts with the use of terms such as 'ought' in everyday moral contexts. Both have perfectly legitimate functions and form a part of rational discourse, although it is true that a pure description of the world would be lacking in such terms. He goes on to say:

> It is sometimes thought that modal statements do not describe states of affairs in the world, because they are *really* metalinguistic. This won't do at all if it is meant that instead of describing states of affairs in the world, they describe linguistic habits. It is more plausible if it is meant that statements involving modal terms have the force of *prescriptive* statements about the use of certain expressions in the object language ([12], p. 283).

It is not Sellars' aim in the essay under consideration to trace the connections between statements involving physical modalities and purely descriptive statements about the world. He accepts statements of law as statements of the form 'If anything *were A*, it *would be B*', and he regards such a statement as an *inference ticket* ([12], p. 286). Another way of stating a law is simply to say that being *A* physically entails being *B*, or more briefly, that *A* P–entails *B* ([12], p. 281). This is opposed to the more customary view in which a law is assumed to be the form '(x) $(Ax \supset Bx)$'. On the latter view the step from a particular '*Ak*' to '*Bk*' is, with the aid of the generalization, a purely logical one. But on Sellars' view, induction is thought of as "establishing principles *in accord-*

ance with which we reason, rather than major premises from which we reason" ([12], p. 286). He says:

> My point, of course, is not that we couldn't use '(x)(Ax ⊃ Bx)' to represent lawlike statements. It is, rather, that given that we use this form to represent general statements which do not have a lawlike force, to use it *also* to represent lawlike statements is to imply that lawlike statements are a special case of non-lawlike statements ([12], p. 297).

The essential difference, for Sellars, between a statement of constant conjunction and a statement such as '*A* P-entails *B*' is that the latter statement implies that "*simply from the fact that something is A* one is entitled to infer that it is B" ([12], p. 285), whereas the former type statement does not imply just that. He also says, however, that this does not mean that "one can simply *equate* '*A* P-entails *B*' with 'One is entitled to infer that something is *B*, given that it is *A*' ([12], p. 282). It is not even clear that, on Sellars' view, statements of the form '*A* P-entails *B*' can be said to have a truth value. He says that ". . . the question as to the applicability of the terms 'true' or 'false' to the conclusions of inductive inference is . . . complex" ([12], p. 292). But he goes on to say that certain considerations give ". . . aid and comfort to the idea that the conclusions of carefully and correctly drawn inductive inferences are logically true . . ." ([12], p. 293).

Sellars attempts to justify the use of the term 'entailment' in these empirical contexts by reminding us that in the process of actual inquiry the meanings of words do become enriched, they acquire modifications of use and associations while yet retaining the substratum of original meaning. The point seems to be that in the process of acquiring grounds for espousing the "inference ticket '*A* P-entails *B*' the meanings of '*A*' and '*B*' are enriched in such a way that it might be correct to say that knowing that *A* P-entails *B* does have some connection with understanding the meanings of '*A*' and '*B*', i.e., understanding their new or enriched meanings ([12], p. 287f). He points out that such new meanings must not be thought of as involving a change in explicit definition, but that

> . . . the new role played by '*B*' and'*A*' does warrant the statement that the 'meaning' of '*A*' involves the 'meaning' of '*B*'; for they are now 'internally related' in a way in which they were not before ([12], p. 288).

He goes on to point out that the relation between the new and old meanings is a logical one, being based on a scientific decision to espouse the inference ticket, and that

> ... scientific terms have, as part of their logic, a 'line of retreat' as well as a 'plan of advance'—a fact which makes meaningful the claim that in an important sense A and B are the 'same' properties they were 'before' ([12], p. 288).

Sellars attempts to find a parallel for his notion of physical entailment in the field of statistical induction. He finally suggests that the induction of non-statistical laws is a special case of statistical induction where all of the observed sample of cases has the property in question. He represents such induction by the schema:

> Observed *As* have been found, without exception, to be *B*. So, (in all probability) that *K* is an *unexamined, restricted* class of *As* P-entails that the proportion of *Bs* in *K* is 1 ([12], p. 295).

Sellars seems to consider his way of looking at laws to be preferable in that it permits us to make sense of the distinction between uniformities that are exceptionless but physically contingent and uniformities that are physically necessary. He says:

> Surely, to be in a position to say
>
> ...
>
> So, (in all probability) *A-C* P-entails *B;* whereas *A-C′* P-entails *-B* or, more simply,
>
> ...
>
> So, (in all probability) *A* is P-compatible with both *B* and *-B* is to be in a position to say that, *although in the nature of the case we have reason to think that they have not done so,* things might have so worked out that *A* was constantly conjoined with *B* (or, for that matter, with *-B*). And where our reason for thinking that *A* is compatible with both *B* and *-B* is *not* that we have observed *As* which are *B* and *As* which are *-B,* but consists, rather, in considerations of extrapolation and analogy, we are in a position to say that, although in the nature of the case we may have reason to think they have not done so, things *may actually be* such that *A* is constantly conjoined with *B* (or *-B*) ([12], p. 301f).

He seems to believe that a constant-conjunction view of natural law cannot make sense of a distinction between the notions of an exceptionless conjunction and a physically necessary conjunction ([12], p. 294f). He also appears to think that on such a view one could not have reason to suppose that *it is physically possible that* (∃x) (Ax·-Bx) without having reason to suppose that *it is the case* that (∃x) (Ax·-Bx) ([12], p. 293).

Since Sellars does not claim to be furnishing a definition of law, it is not easy to evaluate his contribution to this problem. He might, in fact, be construed as rejecting the problem as is usually envisaged, since he decides to take the physical modalities at their face value. Rather than offering any kind of explication or explanation of the modalities, he presents instead a kind of descriptive commentary—almost in the spirit of the later Wittgenstein—on the way these terms function in actual usage. From the viewpoint of those concerned with giving an *explication* of 'law' Sellars might be said to have furnished an exceptionally sensitive description of what it is that needs explicating.

A few points in Sellar's discussion perhaps call for special consideration. He appears to make much of the notion that the meanings of terms change as they are incorporated in additional laws or "inference tickets." He even says, "The motto of the age of science might well be: *Natural philosophers have hitherto sought to understand* 'meanings'; *the task is to change them*" ([12], p. 288). While the fact of change in meaning is supposed to do something toward making reasonable the use of the term 'entailment' in the expression of laws, it is not clear that there is anything in this notion that could not be accepted on *any* view of law. Given the vagueness of the term 'meaning' it would seem rash to deny that the meanings of terms may change as they are incorporated in new laws. But, as Sellars himself points out, the lines of advance and retreat of these meanings are governed by scientific decisions concerning the espousal of the inference tickets involved. But if this is admitted, it then seems justifiable to regard these changes as essentially epiphenomenal with respect to these decisions and the inductive results on which they are ultimately based. This is not to deny that familiar, well-established laws may occasionally pass over into definitions—their empirical burden being shifted elsewhere. The point is rather that expressing physical laws as entailments is not a particularly clarifying or helpful move.

Sellars appears to offer the schema

> Observed *As* have been found, without exception, to be *B*. So, (in all probability) that K is an *unexamined, restricted* class of *As* P-entails that the proportion of *Bs* in *K* is 1 ([12], p. 295).

as a model for the induction of non-statistical laws. He appears to recognize, however, that such a schema "would have to be tidied up and qualified to meet the demands of a theory of primary statistical induction" ([12], p. 290). He does not mention that such tidying up and qualifying may involve such controversial matters as Goodman's theory of projectibility or Reichenbach's theory of concatenated inductions. In the absence of such tidying up, it would appear to be sound for anyone to reason as follows: All of the millions of observed diurnal rotations of the earth have been rotations during which I did not die. So, in all probability, that the class of diurnal rotations during the next hundred years is an unexamined, restricted class of diurnal rotations of the earth P-entails that the proportion of rotations during which I shall not die within the class of rotations is 1.[1]

Sellars seems to underestimate the capability of a well-developed constant-conjunction theory to handle the distinction between modal and non-modal statements. In Reichenbach's theory, for example, the distinction is made metalinguistically by stating whether or not certain correlated statements are "nomological" (a technical term in Reichenbach's theory) ([11], p. 6). One does *not* have to know that $(\exists x)\ (Ax \cdot -Bx)$ in order to know that it is *physically possible* that $(\exists x)\ (Ax \cdot -Bx)$. The latter is true just in case the statement '$(x)\ (Ax \supset Bx)$' is *not* a nomological statement. Also, on such a theory one may entertain or even be inclined to believe such purely universal statements of constant conjunction as 'All gold lumps are smaller than one cubic mile' without being committed to recognize the statements as laws. It is true that one could not on Reichenbach's view maintain that this statement is *verifiably true* and yet not nomological, but the distinction between the notion of a true, universal statement of constant conjunction and a law of nature is one which is in fact stressed in a sophisticated constant-conjunction theory. The advantage which Sellars seems to claim for his view appears to be illusory.

The last contribution to the problem of the definition of scientific law to be considered is that of M. Bunge ([3]). Bunge begins his discussion of laws by developing an elaborate system of classification of *lawlike statements.* By 'lawlike statements' he appears to mean

> . . . statements overtly or tacitly general in some respect and to some extent, occuring in factual science and referring in some way or other to the regular behavior of parts of the world (including ourselves) or even to the universe as a whole ([3], p. 261).

Under the heading of lawlike statements he apparently wishes to include all statements that anyone might possibly wish to call "lawful" in any sense. For example, he cites the statement 'All the books in the Dean's library are serious' as a lawlike statement that will not support a counterfactual ([3], p. 269). Some other examples of lawlike statements which he cites are 'Inscrutable predicates are not to be employed in science' ([3], p. 262), 'There is practically no atmosphere beyond 1.000 Km high' ([3], p. 266), and 'Workers spend more on food than do clerks with the same income' ([3], p. 270). Taking the term 'lawlike statement' in this broad sense, Bunge finds that lawlike statements can be classified in more than seven dozen (not mutually exclusive) ways. The details of this taxonomy will not concern us here.

Having sketched the broad class of lawlike statements, Bunge turns to the problem of setting up criteria or requisites of lawfulness that will determine which of the numerous class of lawlike statements actually deserve the title of 'laws'. The setting up of these criteria, he says, must be done "without danger of mutilating the corpus of science or of inhibiting research in the younger sciences" ([3], p. 279). He proposes the following as a *liberal criterion of lawfulness*:

> A proposition is a law statement if and only if it is a posteriori (i.e., not logically true), general in some respect (i.e., does not refer to unique objects), has been satisfactorily corroborated for the time being in some domain, and belongs to some theory (whether adult or embryonic) ([3], p. 280).

Bunge's criterion does not include a requirement that law statements be true. He does not, in fact, make it clear whether by 'law statement' he intends to cover statements which, if true, would be laws, or whether he considers that a statement may be a law even though it is false. This point is perhaps not of great importance, however, since the requirement of truth could always be added if the criterion is otherwise satisfactory. The criterion, however, includes the somewhat puzzling requirement that the proposition belong to some *theory*. This requirement is an odd one, first of all, because one normally expects that an explication of 'law' will help make clear why laws are the sort of statements that would be worth including in theories and what sort of function they are fit to perform when they are so included. Secondly, the requirement that the proposition belong to some theory adds little or nothing to the other requirements so far as determining the character of the proposition itself. For example, one might devise a theory to explain why over 40,000 Americans are killed in automobile accidents each year. One of the statements in an embryonic theory which one might devise is the statement 'Some people do not know the facts about braking distances and reaction times.' This statement is clearly a posteriori. It is general in that it does not refer to unique objects. It is easily corroborated and in fact is almost continually being corroborated through operator licensing tests. It would therefore appear to be qualified as a law statement by Bunge's criterion.

Bunge meant his criterion to be a liberal one. Whether he meant to furnish such results as the example just cited cannot be determined from his essay. If he did not, then the criterion needs revision in some manner or other. If results of this kind were intended, then it is clear that the problem of the definition of law is not here conceived as the pinpointing of characteristics that explain or clarify why certain statements are fit to perform certain distinctive functions. The explicans in this case appears to be no more precise than the explicandum.

We may now turn to the second major task of this discussion—the defense of Reichenbach's approach to the problem of law. First of all, it should be stated that Reichenbach's analysis resists compendious summary. The reader must therefore be referred directly to Reichenbach's exposition. By way of barest outline it may be said, however, that Reichenbach's answer to the problem of law is explicitly presented in the form of an explication. The vague term 'law' is to be replaced by the more precise term 'nomological statement'. The scope of this explicans is in turn defined by means of a set of requirements which state the necessary and sufficient conditions which a statement must meet in order to qualify as a nomological statement. Reichenbach first lays down requirements defining a class of *original nomological statements.* The class of nomological statements in general comprises all those statements which are logically derivable from the class of *original* nomological statements. The remainder of this paper will be devoted to an evaluation of some recent criticisms that have been directed against Reichenbach's explication, and also to an evaluation of recent objections which, while not directed explicitly at Reichenbach's work, appear to count against the sort of solution he proposes.

Criticism of Reichenbach's system as set forth in his later monograph appears to be confined to a review of *Nomological Statements and Admissible Operations,* by Carl Hempel ([4], pp. 50–54). This review, which is highly critical, deserves careful consideration.

Hempel objects to requirement 1.1 for original nomological statements, because ". . . it replaces the customary concept of law as a true statement by an historic-pragmatic one . . ." ([4], p. 52) and would ". . . permit us to say that some laws of nature may, in fact, be false" ([4], p. 52). This criticism is based on Hempel's interpretation of Reichenbach's definition of the expression 'verifiably true'.

Reichenbach's definition is as follows:

> A statement is *verifiably true* if it is verified as practically true at some time during the past, present, or future history of mankind ([11], p. 3f).

It may be admitted that this definition is not perfectly clear. Two interpretations seem possible. According to one interpretation, Reichenbach is here defining a *species* of true statement. He is stipulating the conditions under which a true statement is to be called "verifiably" true. Such a clarification of his use of 'verifiably' is needed, because verifiably true statements are not, as one might think, statements that are true and *verifiable,* but rather (for Reichenbach) they are statements which are true and *verified* (at some time or other). It would perhaps have been better if Reichenbach had italicized only the word 'verifiably' in this definition. By italicizing both words he abets a second interpretation, the one which Hempel assumes—that 'verifiably true' is being defined as a unit, and that we are not to assume that verifiably true statements need be true at all, since truth is not mentioned in the definiens. This interpretation seems prima facie rather implausible, but it must be admitted that the matter cannot be decided merely by examining the definition itself. There is, however, a considerable amount of external evidence pointing to the first interpretation and none at all pointing to the second. Reasons for favoring the first interpretation will now be presented.

Reichenbach's formulation, as noted earlier, is a revision of his earlier explication set forth in his *Elements of Symbolic Logic.* In that book he said:

> We now turn to the definition of original nomological statements. The first requirement for these statements is that they be *demonstrable as true.* By the phrase 'demonstrable as true' we mean that there is a method, of the deductive or the inductive kind, proving the statements to be true.... In requiring that the statements be not only true but also demonstrable as true, we exclude statements that are true but that cannot be proved to be true ([10], p. 368).

It is abundantly clear, therefore, that in the earlier version of the theory nomological statements were required to be both true and demonstrable as true. The definition of 'original nomological statement' in this earlier version reads: "An original nomological statement is an all-statement that is demonstrably true, fully exhaustive, and universal" ([10], p. 369). The phrase 'demonstrably true' is clearly, therefore, intended in the sense of 'true and demonstrable as true'.

In his later book Reichenbach notes a number of differences, some rather inconsequential, between his later and his earlier explications. Nowhere, however, does he hint that in the later system the important requirement of *truth* is being dropped. Instead he explicitly makes statements to the contrary. For example, he says:

> Being laws of nature, nomological statements, of course, must be true; they must even be verifiably true, which is a stronger requirement than truth alone ([11], p. 11).

A footnote on the same page reads:

> In ESL, p. 369, I used the term 'demonstrably true'. Since 'demonstrable' usually refers only to deductive proof I will not use the above term. The term 'verifiable' alone would not suffice because it is now generally used in the neutral meaning 'verifiable as true or false'.

He then goes on to explain that since 'verifiable' includes a reference to possibility he will have to confine his usage to the sense of actual verification at some time or other. In all this careful explanation of the change from 'demonstrable' to 'verifiable' and of why the latter word is to be taken in a special sense, there is nothing to indicate that the original requirement of *truth* is to be abandoned.

The above would seem to establish fairly conclusively that by 'verifiably true' Reichenbach means 'true and verified (highly confirmed at some time or other)' rather than merely 'highly confirmed at some time or other'—which would, of course, be compatible with falsity. But there is more evidence. An essential step in the construction of his system is the proof that true, closed existential units are excluded by the joint operation of requirements 1.1, 1.4, 1.5 and 1.6. The role played by requirement 1.1 is based on the fact that any statement satisfying it is true and therefore entails only true statements. Speaking of the existential statement q which is possibly derivable from a statement p satisfying the requirements for original nomological statements, Reichenbach says: "If $-q$ is derivable, we know that q is false, and thus q is no longer dangerous" ([11], p. 43). If requirement 1.1 did not include a requirement of *truth* the proof that true, closed existential units are excluded would collapse.

Still further evidence can be cited. After defining the three orders of truth Reichenbach makes the following comment concerning the phrase 'true of the order k': "It is not necessary here to write 'verifiably true', because the qualifier 'verifiably' is already included in the definition of 'true of the order k' " ([11], p. 66). Here we have a clear indication that the term 'verifiably' in the phrase 'verifiably true' is a *qualifier*. It would hardly be correct to speak of 'verifiably' as a qualifier of 'true' unless it represents a qualification on the condition of being true—unless the condition of being verifiably true is a species of the condition of being true.

Finally, everything else in Reichenbach's system is compatible with the interpretation of 'verifiably true' as involving truth. Any other interpretation undermines both the plausibility and the internal structure of the system. If Reichenbach is to be criticized at all on this point, it should be for the somewhat careless wording of the definition of 'verifiably true', not for the oddities that result when the definition is misinterpreted.

Hempel concludes his discussion of requirement 1.1 with the following: "The various difficulties thus engendered can be avoided by requiring of a statement expressing a law that it be true rather than that it be verified" ([4], p. 52). This suggestion, however, is not workable. We cannot substitute the requirement that the statement be merely true rather than verifiably true. To do so would be to admit such statements as 'All gold lumps are smaller than a cubic mile' as a law of nature (if it is in fact true).

In his criticism of Reichenbach's requirement 1.2 Hempel says that we can, under Reichenbach's definition, take any term, say 'extended', to be an individual-term. He says:

> . . . it suffices to introduce, into the language at hand, two new primitives, 'P_1' and 'P_2' which are synonymous with, but not explicitly defined by, 'Extended•Sublunar' and 'Extended•—Sublunar' respectively; these primitives are then individual-terms, and 'Extended' can be defined by '$P_1 \vee P_2$' ([4], p. 52f).

Hempel does not explain how one goes about introducing a new *primitive* term into a language in such a way that it is synonymous with a given term already in the language. At any rate, in the sort of rational reconstruction of language presupposed by Reichenbach's explication there is no place for a notion of "synonymy" of less precise nature than that conferred (directly or indirectly) by explicit definition. In such a reconstruction, then, a new primitive term will clearly not be "synonymous with" any term already in the language. Hempel's proposal does not, therefore, vitiate the distinction between universal and non-universal statements within the framework of such a reconstruction.

Hempel rejects as too strong that aspect of requirement 1.4 which excludes predicates whose extensions are verifiably confined to a limited portion of space-time ([4], p. 53). He argues that this would result in excluding nearly all predicates, ". . . because according to evolutionary theories of the universe, most characterstics are limited in their occurrence to sufficiently 'late' parts of space-time" ([4], p. 53). Against this objection it might be pointed out that no theory concerning the evolution of the universe has yet reached the status of being verifiably true and therefore capable of forming a basis for verifiably true statements concerning the spatio-temporal distribution of characteristics. Further, even if it became highly confirmed that the present state of the universe has evolved from an earlier simpler state in which nearly all present-day characteristics were absent, it seems impossible that any statements concerning states *prior* to that state could become verified, since such a state might well represent the obliteration of all evidence of prior complex states.

Hempel has no criticism of requirement 1.5 (or of requirement 1.3), but he gives a mistaken presentation of it. According to Hempel, requirement 1.5 stipulates that an original nomological statement ". . . must not be capable of being written as a conjunction one of whose terms is not universal in the sense of R2 . . ." ([4], p. 52). This is not the case. Universality in the sense of R2 (requirement 1.2) is irrelevant to the conditions required in requirement 1.5. The latter has to do with generality, not universality. Hempel was probably led astray by the fact that the term 'universal' is often used (outside Reichenbach) to mean the condition of being an all-statement.

Hempel objects to requirements 1.6 through 1.9 on the grounds that their imposition results in the circumstance that a statement logically equivalent to an original nomological statement may not be an original nomological statement. He says:

> Since in all scientific uses of a nomological statement it is its deductive consequences rather than its form that matters, this aspect of Reichenbach's analysis of nomological statements seems problematic, and all the more so because the restrictions in question, especially R7-R9, give the impression of a complex *ad hoc* device intended to secure some consequences of, at most, minor significance ([4], p. 35).

This objection loses its force when one notes that any statement logically equivalent to a nomological statement will, in fact, be a nomological statement. The fact that the property of being an *original* nomological statement is not an invariant property is irrelevant. since no particular scientific significance attaches to being an *original* nomological statement. Requirements 1.7 through 1.9, to which Hempel takes particular exception, are to help insure that original nomological statements fall within the class of admissible statements. Since the definition of this class is an essential step in the attempt to explicate the notion of "reasonable operation"—one of the major aims of Reichenbach's study—it is not clear why Hempel should regard these requirements as having consequences "of, at most, minor significance."

Hempel considers Reichenbach's treatment of counterfactuals of noninterference as untenable because it does not require that such a counterfactual entail the negation of a regular counterfactual—it does not make the statement 'if *a* had been the case, *b* would still be the case' entail the negation of 'If *a* had been the case, *b* would not be the case' ([4], p. 53). It can readily be shown, however, that except in the special case in which the probability of '*b*' is strictly zero, the disputed entailment does in fact hold. Since a counterfactual of noninterference is used only when '*b*' is known to be true, it is questionable whether this type statement is ever used in circumstances which furnish a strictly zero probability for '*b*'. It will now be shown that for all non-zero probabilities of '*b*' the entailment does hold.

What is required is to show that the statement

(1) If *a* had been the case, *b* would still be the case does, on Reichenbach's interpretation, entail the statement

(2) It is not the case that: if *a* had been the case, *b* would not be the case.

Now, a necessary and sufficient condition for the entailment of (2) by the (1) is that the negation of (2) entail the negation of (1). Since, on Reichenbach's interpretation, the truth condition of (1) is given (in Reichenbach's probability notation) by

(3) $$P\,(a, b) \geqslant P\,(b)$$

it is required to show that the negation of (2), which is

(4) If *a* had been the case, *b* would not be the case entails the negation of (3), which is given by

(5) $$P\,(a, b) < P\,(b)$$

Now, (4), being a regular counterfactual, must be based on an exceptionless regularity of conjunction between the presence of situations of the kind described by '*a*' and the absence of situations of the kind described by '*b*'. In other words, (4) entails that the probability of '*b*', given '*a*', is zero. That is, from (4) we may conclude that

(6) $$P\,(a, b) = 0$$

But (6) in turn entails (5), except in the special case where $P\,(b)$ is itself zero. It appears then that on Reichenbach's interpretation a counterfactual of noninterference *does*, except in a limiting case of doubtful applicability, entail the negation of a regular counterfactual.

It may be concluded that Hempel's critique, though prima facie devastating, does not actually prove Reichenbach's system to be inadequate in any respect.

An example of a thinker who came to despair of the project of furnishing an extensional treatment of laws is Arthur Pap. One of the difficulties with which he is concerned is that of distinguishing universal conditionals which are to be considered laws from those which may be regarded as "accidentally" true. He says:

> It may be the case that all the people who ever inhabited a certain house H before H was torn down died before the age of 65. The statement "for any x, if x is an inhabitant of H, then x dies before 65" would then be true, yet nobody would want to say that it expresses a law ([8], p. 204).

Pap is, of course, aware of the extensionalist reply that a conditional of this type is excluded because it involves the individual constant 'H' and further is not derivable from any set of well-confirmed universal statements that are free from such individual reference. In reply to this objection, Pap says:

> However, a serious criticism must be raised against this approach. Just suppose that H were uniquely characterized by a property P which is purely general in the sense that it might be possessed by an unlimited number of objects. P might be the property of having a green roof; that is, it might happen that H and only H has P. In that case the accidental universal could be expressed in terms of purely general predicates: for any x, if x is an inhabitant of a house that has a green roof, then x dies before 65 ([8], p. 205).

Pap speaks here of *the* accidental universal, as though he had given the same statement in two different versions. This is highly misleading. The statement

(1) For any x, if x is an inhabitant of H, then x dies before 65

and the statement

(2) For any x, if x is an inhabitant of a house that has a green roof, then x dies before 65

are not logically equivalent. In order to derive one from the other, we need the additional premiss

(3) H is the only house that has ever had, has now, or ever will have a green roof

which is a good example of a statement that we could not possibly know to be true. From Reichenbach's point of view, (1) is verifiable but not universal, whereas (2) is universal but not verifiable. Hence both are rejected as laws.

The other major line of reasoning that leads Pap to question the possibility of any extensional treatment of law is as follows ([8], pp. 204–215). A statement of the form '$(x)(Fx \supset Gx)$' cannot be an adequate expression of a law, since such a statement is entailed by '$-(\exists x)Fx$', and '$-\exists x)Fx$' entails both '$(x)(Fx \supset Gx)$' and '$(x)(Fx \supset -Gx)$', and so no vacuously true conditional can support a given counterfactual as opposed to the contrary counterfactual. But many laws of nature appear to be vacuously true, and yet the support of counterfactuals is an essential function of laws. These facts cannot be reconciled on an extensionalist analysis.

The kernel of truth contained in this line of reasoning consists (on Reichenbach's view) in the fact that if it is true of order k that a given conditional true of order k is vacuous, then that conditional will not support a counterfactual. It may, nevertheless, be a law of nature, since, as Reichenbach shows, being a nomological statement is a necessary but not a sufficient condition for support of counterfactuals. Pap seems to confuse the condition of being *vacuous* with the condition of being *verifiably vacuous.* Moreover, he seems to be mistaken in thinking that certain laws of nature which do support counterfactuals are known to be vacuous. For example, he mentions laws which seem to refer to idealized entities or conditons ([8], p. 204).

While some natural laws are most picturesquely expressible in terms of "ideal entities" it seems a mistake to regard such laws as being "about" such non-existent entities. For example, while the law of the conservation of energy is most simply expressed by stating that the total energy of an isolated system remains constant, the mention of such a fictitious system is not essential, and the function of the law is the prediction of the behavior of actual systems. Likewise, the gas law 'PV=RT' has as its purpose the prediction of the behavior of actual quantities of gas. The fact that it only approximates the actual behavior of the gases should not lead us to suppose that it is a law of nature that is really "about" ideal gases rather than actual ones—even though this manner of speaking may be convenient for some purpose. Again, Pap thinks that Newton's first law must be vacuously true, since the existence of bodies acted on by no external forces ". . . is incompatible with the universal presence of gravitation" ([8], p. 204). In a note he continues:

> It might be objected that the law of inertia can be formulated in such a way that it is not contrary to fact: If no *unbalanced* forces act on a body, then it is at rest or in uniform motion relative to the fixed stars. But when the law is used for the derivation of the orbit of a body moving under the influence of a central force, it is used in the contrary-to-fact formulation since the tangential velocities are computed by making a thought experiment: how would the body move at this moment if the central force ceased to act on it and it moved solely under the influence of its inertia? ([8], p. 222).

The suggested re-statement of the law of inertia may be phrased:

(1) Bodies acted on by no unbalanced forces are either at rest or in uniform motion relative to the fixed stars.

In supposing that the counterfactual of his "thought experiment" is not supported by the non-vacuous statement (1), Pap is committing an elementary logic mistake. The counterfactual in question would be of the form:

(2) If body *B* were acted on by no forces it would be at rest or in uniform motion relative to the fixed stars.

But 'acted on by no forces' is merely a special case of 'acted on by no unbalanced forces'. In other words, the class of *bodies acted on by no forces* is included in the class of *bodies acted on by no unbalanced forces.* Perhaps because 'acted on by unbalanced forces' is subordinate to 'acted on by forces', Pap was misled into thinking that their negations were similarly related. They are not. Therefore (2) requires merely (1) for its support, and the law of inertia is not, after all, required in a form in which it is verifiably vacuous. In fact, Pap has failed to show that any law which supports counterfactuals *need* be stated in a form in which it is *known* to be vacuously true, i.e., in which it is *verifiably* vacuous. In the light of Reichenbach's analysis, none of Pap's points against the feasibility of an extensionalist treatment of law appear to be well taken.

The very project of attempting to define 'scientific law' has been criticized by Ernest Nagel, who in his own discussion restricts himself to indicating ". . . some of the more prominent grounds upon which a numerous class of statements is commonly assigned a special status" ([7], p. 50). He says:

> The term 'law of nature' is undoubtedly vague. In consequence, any explication of its meaning which proposes a sharp demarcation between lawlike and non-lawlike statements is bound to be arbitrary . . . . not only is the term 'law' vague in its current usage, but its historical meaning has undergone many changes. We are certainly free to designate as a law of nature any statement we please ([7], p. 49).

This attitude is an eminently reasonable one, but it perhaps fails to do justice to the notion of explication as practiced by Reichenbach. It is true that an explication may be arbitrary in certain respects, but it is normally not intended to reflect all aspects of current usage, let alone attempt a faithfulness to historical meanings. What is required is greater precision and the performance of certain characteristic functions—everything else comes under the head of what Quine has called 'don't cares' ([9], p. 259).

Nagel's discussion includes an interesting criticism of the notion that laws of nature must be either free of individual references or else derivable from other laws which are devoid of such reference. He cites the case of Kepler's laws of planetary motion. While these laws make reference to a particular individual, i.e., the sun, Nagel points out that they are not derivable from Newtonian mechanics and gravitational theory alone, but require for such derivation additional premisses "which state the relative masses and the relative velocities of the planets and the sun" ([7], p. 58). If, on the other hand, we relax the requirement to permit the addition of non-lawlike premisses, Nagel points out that we are faced with the task of explaining why a statement such as 'All the screws in Smith's car are rusty' is not to be called a law. Nagel says:

> ... this statement follows from the presumably fundamental law that all iron screws exposed to oxygen rust, conjoined with the additional premises that all the screws in Smith's current car are iron and have been exposed to oxygen ([7], p. 58).

According to Reichenbach's explication, Kepler's laws are not, strictly speaking, laws at all, being only *relative* nomological statements ([11], p. 94). In this conflict with hallowed usage, logical consistency appears to be on the side of the explication. In calling a statement a "law" one normally implies that it is the very opposite of a statement that can be falsified by a mere accident. And yet a mere accident in the form of, say, a near approach or collision with an interstellar body could impart to one or more of the planets velocities such that, contrary to Kepler's first law, they would pursue hyperbolic rather than elliptical orbits with respect to the sun.

But if Kepler's laws are, strictly speaking, only relative nomological statements, it is nevertheless understandable that the application of the title 'law' to these statements generally goes unchallenged. They are, after all, nomological relative to a set of conditions of exceedingly great stability and permanence—viewed anthropocentrically, at any rate. Barring cosmic catastrophes, they may for all practical purposes be regarded as laws of nature.

On the other hand, the statement 'All the screws in Smith's car are rusty' is nomological relative to the quite ephemeral circumstance that those particular screws are made of iron and received a certain exposure to air and water. The manner in which relative nomological statements are regarded is thus understandably a function of the nature of the circumstances relative to which the statements are nomological. It is therefore only to be expected that an explication which, like Reichenbach's, is concerned primarily with the logical rather than the practical functioning of statements, should at various points deviate from common usage or historical precedent. It does not appear therefore that counterexamples of the type represented by Kepler's laws should be regarded as presenting difficulties for Reichenbach's explication.

Some of the criticisms considered have been directed specifically at Reichenbach's analysis. The other criticisms, while not directed at this analysis in particular, appear to be such that they need to be taken account of in the process of any evaluation of Reichenbach's explication. None of the criticisms appears to have decisively shown the inadequacy of any aspect of this attempt toward a definition of scientific law.

NOTE

1. This example is a variation of one Nagel uses in [6], p. 692.

SUGGESTIONS FOR FURTHER READING

1. Ayer, Alfred J. "What Is a Law of Nature? " *Revue Internationale de Philosophie,* X (Fasicule 2, 1956), 144–165.
2. Braithwaite, Richard B. *Scientific Explanation.* Cambridge: Cambridge University Press, 1953.
3. Bunge, Mario. "Kinds and Criteria of Scientific Laws," *Philosophy of Science,* XXVIII (July, 1961), 260–281.
4. Hempel, Carl G. Review of Reichenbach's *Nomological Statements and Admissible Operations.* In: *Journal of Symbolic Logic,* XX (April 6, 1955), 50–54.
5. Körner, S. "On Laws of Nature," *Mind,* LXII (April, 1953), 216–229.
6. Nagel, Ernest. Review of D. Williams' *The Ground of Induction.* In *The Journal of Philosophy,* XLIV (December 4, 1947), 685–693.
7. Nagel, Ernest. *The Structure of Science.* New York: Harcourt, Brace and World, 1961.
8. Pap, Arthur. "Disposition Concepts and Extensional Logic," *Minnesota Studies in the Philosophy of Science,* ed. H. Feigl, M. Scriven, and G. Maxwell. Minneapolis: University of Minnesota, 1958. II, 196–224.
9. Quine, Willard V. O. *Word and Object.* Cambridge, Mass.: MIT Press, 1960.
10. Reichenbach, Hans. *Elements of Symbolic Logic.* New York: MacMillan, 1947.
11. Reichenbach, Hans. *Nomological Statements and Admissible Operations.* Amsterdam: North Holland Publishing Company, 1954.
12. Sellars, Wilfrid. "Counterfactuals, Dispositions, and the Causal Modalities," *Minnesota Studies in the Philosophy of Science, ed.* H. Feigl, M. Scriven, and G. Maxwell. Minneapolis: University of Minnesota Press, 1958. II, 225–308.

READING 11

General Statements as Rules of Inference?

H. Gavin Alexander

Several philosophers recently have suggested that it is often profitable to regard universal general statements as rules of inference, as material rules rather than formal rules, as P-rules of our language rather than L-rules (to use Carnap's terms). The view is held in different forms. At one extreme there is the comparatively innocuous claim that in some contexts general statements perform a rule analogous to that of deductive rules of inference; at the other, there is the contention that implicit in both everyday language and scientific discourse are P-rules of inference which are in some sense more fundamental than the general statements which can be derived from them.

The considerations which have led philosophers to take up these various connected positions can be listed under four heads. First, there is the fact—noted already by Descartes and Locke—that in everyday discourse we seldom argue in explicit syllogisms. In the same field of ordinary discourse there are the familiar difficulties in interpreting universal statements extensionally. Second, there is the general problem of induction. The old view according to which we need a supreme major premise for all inductive inferences is unattractive, and this is said to be avoided if one treats general laws as principles rather than premises. The third line of argument is the one which has been strongly advocated by Toulmin—that in scientific practice, deductions are not made from laws but in accordance with them. Fourth—although obviously closely related to the first—is the puzzle of counterfactual

Reprinted by permission of the publisher from Alexander, H. Gavin. "General Statements as Rules of Inference," *Minnesota Studies in the Philosophy of Science, Volume II,* edited by H. Feigl, M. Scriven and G. Maxwell, University of Minnesota Press, Minneapolis. © 1958, University of Minnesota.

conditionals: if general statements did not have the force of rules of inference, it would be impossible to derive counterfactual conditions from them.

It may be interesting to examine the validity of these four lines of thought. As we shall see, there are considerable difficulties that have not been so clearly stated.

I

"This egg has been boiling for six minutes now, and so it will be hard."
"President Eisenhower is a man; therefore he is mortal."

Traditional logic books treat these as enthymemes of the first order, that is as syllogisms with their major premises suppressed. Critics of traditional logic claim that it is pointless and even misleading to try to squeeze all inference into the strait jacket of the syllogistic forms. Among such critics Mill held that in arriving at the conclusion that Eisenhower is mortal we do not need to presuppose that all men are mortal; all we need to do is argue by analogy with cases such as Socrates and the Duke of Wellington, both of whom have died. Contemporary philosophers such as Ryle and Black, on the other hand, say that we infer directly from the single premise to the conclusion, using a non-logical rule of inference.[1]

Mill's view clearly will not do by itself. Similarity is an incomplete predicate and therefore in order to know that Eisenhower is mortal, we have to know in what important respects he is like Socrates and the Duke of Wellington. It is not in being more than five feet long or in weighing more than 140 pounds; it is in being a man, or perhaps an animal. So in a way we do presuppose a major premise. The contemporary view is much more plausible. Certainly one of the main functions of general laws is to enable us to infer from one singular statement to another; and this function is so important that it is tempting to follow Ryle and call general laws *inference tickets.*

But if we thus abolish the traditional category of enthymemes of the first order, why should we not do the same with enthymemes of the second order? "All cows are ruminants; so Flossie is a ruminant." "All cows are mammals; therefore, Flossie is a mammal." Following Ryle and Black we now say that it is wrong to regard these as syllogisms with a suppressed minor premise, "Flossie is a cow." No, these are direct inferences which have a rule of inference corresponding to "Flossie is a cow." It might be thought that this is too particular to be a rule of inference. But this is not the case, for we have already seen that there are two different inferences in which it can be used as a governing rule of inference, and there are obviously infinitely many more. The rule can in fact be put in logical form as (Φ) $[\{(x) \cdot (x \text{ is a cow}) \supset \Phi x\} \supset \Phi x\} \text{Flossie}]$. A possible objection is that though it is quite common to suppress minor premises, and thus to use a special rule of inference, this is not so frequent as the use of a general statement as a principle of inference. This may be true, but the difference is not very considerable. In fact, both general statements and singular statements are sometimes used as conclusions of arguments (inductive or deductive) to 'state facts,' sometimes as explicit or implicit steps in an argument. If we wish to say that in the latter use the general statements are principles of inference whenever they are not explicitly stated, then there is no reason why we should not say the same about the singular statements. If, as I would suggest, this seems an excessive multiplication of rules of inference, then the fault lies at the beginning. The traditional account in terms of enthymemes seems preferable to an account that considers only the explicit form of the argument.

This is not the only difficulty of the Ryle-Black approach. No one has shown more clearly than Ryle himself that the vocabulary suitable for talking about rules is different from that suitable for statements. Statements can be true or false, probable or improbable; rules can be correct or incorrect, useful or useless.[2] Normally we say that some general statements are true, others false, or that some are probably true. If then we wish to say that general statements are rules of inference, it has to be shown that by saying that they are correct rules we are doing much the same as when we said that the corresponding general statements were true. What then does it mean to say that some rules are correct, others incorrect? It cannot mean, as it does with rules of grammar, that correct rules are the ones which are accepted and conformed to by the majority (or by the French Academy). The correctness or incorrectness of these P-rules must somehow depend on empirical facts. The only possible meaning is that a rule is correct if the world is such that by following the rule we arrive at true conclusions.

Thus instead of saying that a general statement is universally true, we say that the corresponding rule of inference always enables us to reach true conclusions from true premises; instead of saying that most Φ are ψ's, we say that the rule of inference (x) $(\Phi x \supset \psi x)$ usually enables us to reach a true conclusion when we are given a true premise of the form Φx. But if this is accepted, it becomes obvious that one of the advantages we had hoped for has been lost. Certainly some philosophers, like Ramsey, have been puzzled by the logic of general statements which are so obviously not truth-functional and have hoped that the puzzle might be bypassed by regarding general statements as rules of inference. But we have only been able to do this at the cost of bringing in other general statements; for we now have the general statement that in all cases the use of the rule of inference enables us to arrive at true conclusions. So we are no better off.[3]

Of course, this difficulty might be avoided by holding that, when we say a rule is correct, all that we mean is that it is a rule observed by all users of a certain language (in the philosophical sense of language), and that when we say a rule is probably correct, what we mean is that it might be useful to adopt this as a rule of our language. This would be to make such rules meaning rules or what are normally called analytic truths. This line of thought is the one put forward by Wilfrid Sellars. It is closely connected with the problem of counterfactual conditionals and will therefore be discussed in the last part of this paper. At the moment it is sufficient to notice that it goes considerably beyond the arguments of writers like Ryle and Black and therefore cannot be used by them as a way out of this difficulty unless they are willing to change much of their account. Unless they are prepared to deny that there are any purely empirical general statements or rules—that is statements (or rules) which are not true (or correct) *ex vi terminorum*—then it must follow that there are some general statements, either in the object language or the metalanguage, which cannot be interpreted as rules.

II

I have thus been claiming that it is unhelpful and, in fact, misleading to speak of material rules of inference rather than of implied major premises. But a strong argument against this position can be drawn from a discussion of induction. Many classical logicians regarded inductive inferences as concealed deductive arguments with a suppressed major premise about the uniformity of nature. In this case, however, such an analysis appears artificial. The only apparent alternative is to regard inductive inferences as arguments governed by a non-deductive rule of inference—and if we allow a non-deductive rule of inference here, why not accept them elsewhere? Another advantage claimed by Braithwaite and Black[4] for this approach is that it may enable us to show that induction can be justified inductively without arguing in a circle. Unfortunately, when one examines this last claim, one realizes that it raises considerable difficulties which make the principle-of-inference interpretation of induction as unpalatable as the suppressed-major-premise view.

In his exposition, which is rather more easily summarized than Braithwaite's, Black does not actually subscribe to the view that there is a single supreme inductive rule or principle. He does however consider two forms of a rule which is a very strong candidate for such a position. Rule 1 (R_1) is to argue from *All examined A's have been B* to *All A's are B*. R_2 is to argue from *Most A's examined in a wide variety of conditions have been found to be B* to (probably) *The next A will be B.* Now it is possible to construct the following self-supporting argument:

(a_1) All examined instances of the use of R_1 in arguments with true premises have been instances in which R_1 has been successful.

Hence, all instances of the use of R_1 in arguments with true premises are instances in which R_1 is successful.

This argument uses R_1 to reach the conclusion that R_1 is always successful; a similar argument can clearly be constructed for R_2. The obvious criticism is that these arguments are circular. But, Black points out, this is not quite so obvious as it seems at first. A circular argument, in the usual sense, is either one in which one of the premises is identical with the conclusion or else one in which it is impossible to get to know the truth of one of the premises without first getting to know the truth of the conclusion. But here there is only one premise, which is clearly independent of the conclusion, so that the argument is not circular in the usual way. At this point Braithwaite cites the difference in a deductive system between a rule of inference and a formula, and remarks that the rule of detachment may be used in order to derive the formula $p \cdot (p \supset q) \supset q$.

But in spite of this analogy, one is reluctant to accept this self-supporting 'justification' of induction. One reason is that if one accepts this, then it seems one would also have to accept arguments like the following:

> Rule 4: to argue from *The Pope speaking ex cathedra says p* to *p*.
>
> (a_4) The Pope speaking *ex cathedra* has stated that whenever the Pope speaks *ex cathedra* what he says is true. *Hence* whatever the Pope says *ex cathedra* is true.
>
> Or Rule 5: to argue from *The Bible states that p* to *p*.

(a_5) In 2 Tim. 3:16 it is stated that all scripture is given by inspiration of God and is profitable for doctrine, *hence* etc.[5]

These are clearly circular and yet they seem of exactly the same form as Black's argument.

Wesley Salmon has given an even more convincing reason for rejecting Black's argument as a possible justification of induction.[6] This is that by citing the same evidence as Black would give in support of R_2 one could produce a self-supporting argument for R_3: to argue from *Most examined A's have been B* to (probably) *The next A will* NOT *be B*. Salmon thus holds that Black's arguments are in fact circular although not in the usual way. He writes:

> For an argument to establish its conclusion, either inductively or deductively, two conditions must be fulfilled. First, the premises must be true, and second, the rules of inference used by the argument must be correct.... Unless we are justified in accepting the premises as true and in accepting the rules of inference as correct, the argument is inconclusive.... To regard the facts in the premises as evidence for the conclusion is to assume that the rule of inference used in the argument is a correct one. And [in the case of the self-supporting arguments] this is precisely what is to be proved.

But this view has difficulties of its own. One is that it involves us in something very similar to the principle of the uniformity of nature. Before, writers like Mill wanted such a principle as a major premise; now, writers like Salmon want it as a metastatement telling us that the principle of inductive inference is always (or usually) successful. The difference here is minimal.

A second difficulty in Salmon's account lies in its application to deduction. For according to him, in order to reach conclusions deductively, we must have a justification of the rules of inference. Now it is not easy to see that deductive rules can be justified. It is true that Reichenbach purports to give a justification of the rule of detachment in his *Elements of Symbolic Logic;*[7] but all that he succeeds in showing is that it is inconsistent to admit that a statement 'If . . . then . . .' can be represented by $p \subset q$ (where '$\supset$' is defined truth-functionally) and at the same time to assert both the hypothetical statement and its antecedent but deny its consequent. Actually we only can tell that a given statement in words can be represented by $p \supset q$, if we first see that it would be falsified if the antecedent were true and the consequent false. But even if we were to grant the validity of Reichenbach's 'justification,' we should still be left with the paradoxical conclusion that no deductive arguments were conclusive before Reichenbach's time. God has not been so sparing to men as to make them barely two-legged creatures and left it to Reichenbach to enable them to reach conclusions.

We thus seem to have reached an impasse. For the three theories of inductive inference that we have considered are all untenable. Mill's view that all inductive arguments are enthymemes, with the suppressed major premise that nature is uniform, seems completely artificial as an analysis of the inference from 'Usually when the wind goes round to the S.W. with a falling glass, rain follows' to 'Probably when the wind next goes round to the S.W., etc.' Black avoids this artificiality by regarding this and similar arguments as valid because of material rules of inference. But this leads to the conclusion that there is a large number of non-circular self-supporting arguments such as our two theological examples—which is absurd. Salmon escapes this absurdity by demanding that the rules of inference be justified. But if this demand is extended to ALL rules of inference, it makes nonsense of deduction. If it is applied only to non-deductive rules, it completely spoils the analogy that is being drawn between these so-called material rules of inference and the genuine (deductive) rules of inference. The one positive conclusion that one can draw from the statement of this impasse is that here also the introduction of the idea of material rules of inference raises more difficulties than it solves.

However it may be possible to be slightly more constructive and to indicate a way of escape from this puzzling situation. Here it is perhaps illuminating to consider a special use of the concept of necessity. 'All men are mortal, President Eisenhower is a man, so President Eisenhower is necessarily mortal' or 'he must be mortal.' Here the word 'must' does nothing more than signal the fact that this is the conclusion of a deductive argument. Often we use the same phrasing without ever stating the premises—'He must be ill' or 'It must be going to rain.' Here the function of 'must' is again to indicate that these statements could be exhibited as the conclusions of deductive arguments. If asked why, we would outline the argument by quoting either the major or the minor premise or both: 'He's pale and shivering and can't concentrate,' 'Always when the wind goes to the S.W. and the glass is falling, rain follows.' Thus to say in this sense of 'must' A must be B' is not to say that A and B are connected by some special sort of close tie; it is only to say that A is B, and that a deductive argument which has as a conclusion 'A is B' can be constructed with acceptable premises.

I would suggest that 'probably' plays a somewhat similar role. When we say 'It will probably rain' what we mean is that it is possible to construct an argument of the form 'Usually when conditions x, y, z hold, rain follows within 24 hours; conditions x, y, z hold now; therefore it will probably rain.' When we say 'All observed A's are B; therefore it is probable that all A's are B' we mean that it is possible to construct an argument of the form 'Usually generalizations made on the basis of evidence of such and such a kind have proved true; our observations of A's which are B constitute evidence of this kind; therefore it is probable that all A's are B.' This is the basic use of 'probable.' The frequency concept of probability is a derived use which makes the concept more precise, but at the price of making it inapplicable in those situations where it is impossible to gather statistics.

On this view there is a rule of inference which enables one to move from 'Most A's have been B' to 'Probably this A is B.' But it is a rule of inference which is deductive rather than inductive. There is now no problem of justifying the move, any more than there is a problem of justifying the move from 'Most A's have been B' to 'It is reasonable to expect this A to be B.' The moves are valid because of the meanings of the phrases 'probable' and 'reasonable expectation.'

One can also see that on this analysis Black's self-supporting arguments become less perplexing. Now it is necessary to distinguish the two different cases separately—first, the inference to '(Probably) all uses of R_1 will be successful' where R_1 is the rule of arguing from all observed cases to all cases; and second, the inference to 'Probably the next use of R_2 will be successful' where R_2 is the rule of arguing from most observed cases to the next case. The second of these arguments is now non-circular but unexciting—just as unexciting as any other case of immediate inference. The first argument is, however, invalid if the conclusion contains the word 'probably' and illegitimate if it does not contain it. For if it contains the word 'probably' then, as we have seen above, the implication is that we could construct an argument with a higher order generalization which stated that most other general hypotheses posited on the basis of an amount of evidence similar to that which we possess here have not been disproved later. But clearly this particular case is unique so that it is quite impossible to formulate a generalization about similar cases. So the claim implied by the word 'probably' cannot be fulfilled and the argument is therefore invalid.

If, on the other hand, the word 'probably' is omitted, then the argument conforms to no pattern of inference at all—except the pattern of inference of arguments contained in the induction chapters of logic textbooks. Except when we are discussing induction, we never argue from 'All observed A's are B' to 'All A's are B' but rather to 'Probably all A's are B' or 'It is reasonable to assume that . . .' or 'We can expect . . .' It is true that if the generalizations continue to be confirmed we will drop the qualifying phrase, but this is either because we cease to question the truth of the generalization and instead use it as a premise for other arguments or else because we see that the generalization can be seen as a particular case of a higher order generalization. The words 'probably' and 'necessarily' (in the sense I have discussed above) are used to indicate the line of argument we have followed; when we go on toward some different conclusion starting from this first conclusion as a premise, then we no longer need these indicator words.

III

It is impossible within the space of a few pages to discuss at all adequately the way in which laws function in natural science. The most that can be done is to indicate some of the lines of thought that have led writers such as Toulmin to conclude that natural laws are more like rules than like statements.

One possible argument, though not one that is now often explicitly held, is that science is interested in the prediction and control of nature. Now all prediction and all control must involve singular statements. Thus general laws only function as steppingstones or rules of inference, which enable us to move from the set of singular observation statements to prediction statements. This argument is clearly inconclusive. For it is not clear that the intermediate general truths are rules rather than statements. And further, the account only applies to some scientists, especially applied scientists. Many pure scientists are just interested in discovering laws of nature or general truths—and truths or statements, not just patterns of inference.

Another argument that I think has influenced Toulmin starts from the fact that many scientific laws are such that no one conceivable state of affairs would count as a conclusive refutation. I have suggested elsewhere[8] that this is one of the main considerations that has led Kneale to call natural laws 'principles of natural necessitation' and Quine to posit a continuum from particular empirical statements at one end to the laws of logic at the other. But although it may be impossible to refute laws of nature—they are abandoned rather than refuted—it is possible to give evidence in their support. But it is clearly impossible (logically) to give evidence for rules—one can give evidence for the statement that certain rules are observed; one can give empirical grounds in support of the claim that certain rules should be accepted; but one cannot give evidence for the rules themselves. And it does not make sense to say rules are true—whether true empirically or true *ex vi terminorum*. The fact that scientific laws are not refutable in the same way as empirical generalizations must be explained by discussing the way in which the statements of science hang together (cf. Duhem and Quine) and cannot be verified or falsified individually, rather than by evading the issue by calling them rules.

It is also possible that some writers who take this view of scientific laws have been influenced by the same sort of consideration that we considered above—that in drawing an inference the law is usually not stated explicitly. Thus in Britain one might look at an electric light bulb, see that it was marked 60 watts, and then say, 'It must have a resistance of about 880 ohms.' This is clearly drawn by means of the formula, derived from Ohm's Law, 'Power $= E^2/R$,' but this formula is not stated. Thus it could be said that this formula served as a principle of inference rather than as a suppressed premise. But equally the argument depends on the fact that the standard British power supply is 230 volts. If this is a suppressed premise rather than a rule of inference, the same can equally well be said of the law that the power dissipated (in watts) is equal to the square of the voltage divided by the resistance in ohms.

The principal argument used by Toulmin [9] is more interesting than any of these. It is that "laws of nature tell us nothing about phenomena, if taken by themselves, but rather express the form of a regularity whose scope is stated elsewhere" (*Introduction to the Philosophy of Science.* p. 86). As it stands this is unexceptionable; for, at most, it represents a slight narrowing of normal scientific usage in confining the name *law* to the mathematical formula which is, as it were, the core of the statement. But Toulmin goes on to draw a parallel between these laws—in his narrow sense—and rules or principles. "In this respect, laws of nature resemble other kinds of laws, rules, and regulations. These are not themselves true or false, though statements about their range of application can be" (p. 79). On p. 93 he cites with approval Ryle's metaphorical phrase 'inference-ticket', and on p. 102 he also claims that there is a similarity between laws of nature and principles of deductive inference.

Now as a matter of scientific usage, it is just false to say that the statement of a law never mentions its scope.[10] If one thumbs through a textbook of optics, one finds the following statements:

"Kirchhoff's law of radiation states that the ratio of the radiant emittance to the absorptance is the same for all bodies at a given temperature."

"Bouguer's law of absorption states that in a semi-transparent medium layers of equal thickness absorb equal fractions of the intensity of the light incident upon them, whatever this intensity may be."

"According to Stokes' law, the wavelength of the fluorescent light is always longer than that of the absorbed light."

"Fermat's principle should read: the path taken by a light ray in going from one point to another through any set of media is such as to render its optical path equal, in the first approximation, to other paths closely adjacent to the actual one."

In all these the scope is included in the statement of the law. But it is true that Toulmin's account does apply in some cases. Thus, as he points out, a textbook will say, "In calcite, Snell's law of refraction holds for one ray but not for the other"; it does not say that when calcite was discovered, Snell's law was found to be not strictly true. Does it follow from this that Snell's law is a rule or an inference license?

The first thing to be noticed here is that any scientific law—unlike simple generalizations of natural history—is multiply general; that is, if we were to represent it by a logical formula, we would need several quantifiers. At the first stage Snell's law gives us a formula which sums up the relations between the angles of refraction and the angles of incidence for *all* angles of incidence. Second, Snell's law—if we use the name now to include not only the formula but the statements of its scope—tells us that for *all* different specimens of the same two media, the formula holds and the constant of proportionality between the sines of the angles is the same. Third, Snell's law tells us that the formula holds for *all*—or almost *all*—transparent media but that the constant of proportionality is different for different pairs of media. And perhaps fourth one might say this was true at *all* times. Thus *theoretically* the law could be falsified in four different ways: (1) by finding, say, that the refractive index of substances varied systematically with time; (2) by finding a transparent medium for which the law did not hold; (3) by finding that there were certain pieces of flint glass which, though identical in all other respects, had different refractive indexes; (3a) by finding that there were certain pieces of flint glass identical in all other respects with normal specimens for which the relation was $\sin^2 i/\sin^2 r = \text{constant}$; (4) by finding that all previous observations had been wildly mistaken, and that the law which held for all transparent media was

$\sin^2 i/\sin^2 r = \text{constant}$. Now because of this multiple generality of law statements, it is understandable that a shortcoming in one respect should not lead us to abandon it altogether.

In practice, the fourth kind of 'falsification' will always be such that the first formula will be a valid first approximation to the more accurate formula. There will therefore be cases in which the law in its first inaccurate form will still be used. The third kind of falsification never, I believe, occurs in the physical sciences. Two specimens will never differ in *only one* respect, and therefore if two pieces of glass have different refractive indexes we know they will exhibit other differences so that they cannot be called instances of the same kind of glass. 'Falsification' in the second way is not infrequent; as in the case of abnormal refraction, we do not abandon the law completely but note that its scope is restricted. The first kind of disproof has never yet been observed, but if Whitehead is right it may sometime be found.

But this feature of multiple generality is not confined to the physical sciences. As a British visitor to the United States, I am expected to report some general facts about American university life. These reports could be doubly general, that is, true about *all* (or most) students or faculty on *all* (or most) campuses. Basing my reports on my observations on the Minnesota campus, I might reach conclusions such as "Almost all men undergraduate students dress informally, with no ties and no jackets (coats)" or "Most full professors and university administrators wear dark suits." An American friend will then tell me that the latter generalization is true of all American campuses, but that the former only holds in the Midwest. So what I might do is make generalizations like those in the accompanying list.

Generalization	*Scope*
All professors dress formally	Everywhere
Undergraduates dress very informally	Midwest
Undergraduates take outside work in term time	Not in liberal arts colleges

Now if I constructed such a list, Toulmin would presumably say that the sentences on the left were rules or inference licenses. But this is ridiculous. If anything, they are analogous to propositional functions. That is, they are incomplete statements which can be completed by either supplying a particular context. e.g., 'on the Harvard campus' or else by quantifying, 'on all campuses', 'on some campuses', or 'on most campuses'.

This, I think, is sufficient to show that Toulmin's argument is invalid. But there is still a problem as to the way we apply laws to particular situations. Nagel, for instance, points out that we can infer the motion of a projectile from the laws of mechanics. This, presumably, might be done by treating the projectile as a homogeneous body of mass m, assuming that the air resistance is a simple function of the velocity and can be represented by a simple force acting in the line of motion, and neglecting any effects due to the spin of the projectile. We then write down the equations, substitute in the initial conditions, and deduce that the projectile will land one thousand yards away in the direction in which it was fired.

This deduction therefore takes two steps. The first is to the intermediate conclusion that certain laws, and only these laws, are applicable to this instance; the second is the deduction made by substituting initial conditions into these laws. Often the first step proves to be unjustified as in the case when one applies the simple laws of mechanics to the flight of a cricket ball or baseball and reaches the conclusion that such balls never 'swerve in the air.' Sometimes the deduction may be wrong because the formula used in the deduction is only true to a first approximation and in this particular case the second order differences are important.

But an exactly similar account would apply to my deduction that Franklin Smith usually wears a T shirt. First, I would notice that he was an undergraduate at a Midwestern state university; then I would see which laws held for Midwestern state university campuses and draw my deduction. Of course, in this example, both steps of the inference are of the same kind and both are very obvious. In the case of the projectile, the first step is by means of a verbal argument, the second by mathematical reasoning. It is therefore understandable that the attention of scientists should have been focused on the second stage and that as a consequence philosophers of science should have been misled into thinking that the mathematical formula by itself was the law. In fact, the formula by itself is only a formula; when the scope is supplied it becomes part of a statement. But in no sense is it like a rule.

IV

A. The fourth argument in support of material rules of inference is one put forward by Wilfrid Sellars and expressed most clearly in his paper "Inference and Meaning."[11] Briefly, it is that subjunctive conditionals are expressions in the material mode of material rules of inference in the formal mode. He writes thus:

" 'If anything were red and square, it would be red' cannot plausibly be claimed to assert the same as '(In point of fact) all red and square things are red'; this subjunctive conditional conveys the same information as the logical rule permitting the inference of *x is red* from *x is red and x is square*" (p. 323).

> 'If there were to be a flash of lightning, there would be thunder' gives expression to some such rule as '*There is thunder at time t + n* may be inferred from *there is lightning at time t*' (p. 323).
>
> Whenever we assert a subjunctive conditional of the latter form ('if x were A, x would be B'), we would deny that it was merely in point of fact that all A's are B. (p. 324).
>
> Unless some way can be found of interpreting such subjunctive conditionals in terms of logical principles of inference, we have established not only that they are the expression of material rules of inference, but that the authority of these rules is not derivative from formal rules. (p. 325).
>
> Material transformation rules determine the descriptive meaning of the expressions of a language within the framework established by its logical transformation rules. In other words where 'Ψa' is P-derivable from 'Φa', it is as correct to say that 'Φa ⊃ Ψa' is true by virtue of the meanings of 'Φ' and 'Ψ', as it is to say this where 'Ψa' is L-derivable from 'Φa' (p. 336).

Thus Sellars' view, as expressed in his 1953 paper, is that a necessary and sufficient condition for a generalization to provide grounds for asserting a subjunctive conditional is that the generalization shall be the expression in the object language of a material rule of inference. That is, the generalization has either to be one that contains the word *necessarily* explicitly or else has to be in such a context that *necessarily* is implied. Generalizations of this kind can then be said to be true *ex vi terminorum,* if we take the widest possible concept of meaning; for such generalizations can all be ranged in a continuum which ends at one extreme in 'truths' such as 'All bachelors are unmarried.'

It appears then that we can place all possible generalizations into four classes. At what might be termed the lower extremity, we have those generalizations which can only be established by complete enumeration—of these the standard example is 'All the coins in my pocket were minted after 1940.' From this first class one can only derive what Goodman calls *counter-identical conditionals* such as 'If this coin had been identical with one of the coins in my pocket, it would have been minted after 1940.'

The second class contains those empirical generalizations which are arrived at by genuine induction but which are not 'necessarily true,' not expressions of material rules of inference. These are the truths which Sellars regards as "merely true in point of fact." They also are below the salt, for from them, Sellars holds, one can again derive nothing stronger than counter-identical conditionals. A special subclass of such generalizations would appear to be general statements which mention a particular individual.

The third class—which, as we have seen, merges into the tautologies of which the fourth class is made up—comprises the laws of nature, necessary truths which hold by virtue of the meanings of the words. From such statements (and from the tautologies) one can derive genuine counterfactuals.

(It is however important to realize that one cannot always derive the most usual type of counterfactual. Thus, if one assumes that solar disturbances are a necessary and sufficient condition for the occurrence some 24 hours later of auroral displays, we can assert that 'If solar disturbances had occurred on Tuesday, auroral displays would have taken place on Wednesday.' On the other hand, it would sound odd to say 'If auroral displays had taken place on Wednesday, solar disturbances would have occurred on Tuesday.' Rather we would say 'If auroras had occurred, *this would have meant that* solar disturbances had occurred.' Similarly we might on the basis of our knowledge of segregation in the United States in general and Jonesville in particular, say "If Smith had been colored, he would have lived in northeast Jonesville." But we should not say, "If Smith had lived in northeast Jonesville, he would have been colored"; we could only say, "If . . . it would have meant that . . .")

It is however easy to show that this theory, as I have interpreted it, is untenable. For consider the case of a husband who, after twenty years of a happy marriage, remarks to his wife 'You would have been furious if you had been there. The youths were torturing that cat out of pure sadistic pleasure.' Here we have an example of a counterfactual being asserted by the husband—let us call him Mr. Smith—because he knows by experience that his wife becomes angry whenever she sees wanton cruelty to animals. I would suggest that this is clearly a generalization of the second class. It mentions an individual; it does not explicitly or implicitly include the adverb 'necessarily'; and it can by no stretch of the imagination be said to be true *ex vi terminorum* because of the meaning of "Mrs. Smith." Yet the counterfactual is clearly a genuine one.[12]

It might be suggested in reply that although this generalization is empirical, the counterfactual really rests on a more fundamental generalization, some psychological law about the reaction of certain types of people in certain circumstances. But this is unplausible. Are we going to say that Mr. Smith is able to make such counterfactual statements about his wife because he has majored in psychology at college, but that poor Mr. Brown, married equally long and equally happily, is unable to do so because he never has opened a psychology textbook or even attended a university? Or are we going to say Mr. Brown *implicitly* knows and relies on the psychological law even though he explicitly refuses to assent to the law when it is quoted to him?

It is true that one might say that both Mr. Smith and Mr. Brown are making the assumption that their respective wives have acted in consistent ways and will continue to do so. But this would only be to refer to the assumption of the uniformity of nature, the assumption which may be said to be implicit in every generalization reached by genuine ampliative induction and thus implicit in every generalization of the second group. In fact, this line of defense would make the second class empty, for it would ascribe a necessary nomological connection to each and every generalization in virtue of the fact that it was reached by ampliative induction.

There is an even simpler argument which enables us to reach this same conclusion. If one considers any particular counterfactual such as 'If this stone had been released, it would have fallen' or 'If Grey had clearly and publicly explained Britain's foreign policy, the first war would never have occurred,' one sees that the way in which one would justify such assertions is exactly similar to the way in which one would justify a prediction. Mr. Dulles, knowing that President Eisenhower is about to announce the Eisenhower doctrine, tries to predict what world reaction will be; Mr. Smith, seeing his small son about to drop a stone out of the window, predicts what will happen; and the way in which they do so is the same as the way in which counterfactuals are established. The only difference is that predictions can usually be conclusively verified (or falsified) later; counterfactual conditionals never can be. Now it is obvious that in predicting the future we use empirical generalizations (or the second class); it is clear to me that we also use them in order to arrive at counterfactuals.

We have now seen that having the characteristics of the third class is not a necessary condition for a generalization to give rise to subjunctive conditionals. I want to go on to make the stronger claim that the third class, as defined by Sellars, is empty, that there are no laws of necessary connection. My claim is that when a generalization includes the word *necessarily* or when we say a general statement is necessary, one of a variety of different things is meant. One use is that mentioned above. We often say 'All A's are necessarily B' when we are claiming not only that 'All A's are B' but also that this fact can be justified by a deductive argument from a more general premise which is accepted by the speaker and usually by the listeners also. This is exactly the same as particular statements such as 'It must be going to rain.' Closely allied to this is the case when the generalization is justified by reference to considerations of a different kind as when we say, 'All voters are necessarily over 21' because of the law that no one under 21 may vote. Also similar is the case when the generalization is justified by reference to the meaning of the constituent words—the case of tautologies. What is common to all these is that in inserting the word *necessarily* we are indicating that our assertion is not based on examination of individual A's which are B, and that it is therefore useless to search for A's which are not B.

This, I believe, covers all the cases in which the word *necessarily* is actually included in the sentence itself. The same claim could conceivably be made by saying that the general statement 'All A's are B' was necessary—but this would sound rather unnatural. There are, however, other situations in which philosophers might call a general statement necessary, although they would hesitate to insert the modal adverb *necessarily* in the object-language sentence. These are the cases, mentioned above, when a scientific generalization becomes so well established that no one conceivable state of affairs would be taken as constituting a refutation of it. This is what is sometimes called the functional or the pragmatic a priori. But this sort of necessity consists in the way in which scientists regard certain laws, the way in which they use them; it is not in any way an 'intrinsic property' of the statements. And since, if questioned, scientists would cite empirical evidence in support of these laws, it would, as we have seen, be misleading to say that they were true *ex vi terminorum*.

B. There is, however, another way in which some philosophers have attempted to distinguish a special class of 'necessary' or 'basic' laws. This distinction is most easily explained by considering a universe which consists of a perfect billiard table and two perfect billiard balls in motion on it. It might be the case that because of the initial conditions, these two balls would never collide. If so, then the statement that these balls would never collide would be lawlike inasmuch as it would always be true, but yet it would be different from, say, the laws of reflection which hold for the collisions of the balls with the cushions. These latter laws could be called *basic* or *necessary,* the former *structure dependent* or *contingent.*

Here we have a view which is nearer to Kneale's than to Sellars'. For it might seem that such basic laws would be 'principles of natural necessitation' to use Kneale's phrase. That is, we would almost seem to have here a concept of metaphysical necessity instead of the concept of pragmatic necessity discussed by Sellars.

The difficulty about this distinction is that of applying it to our actual universe. For until, like God, we know all the secrets of the universe, it will be impossible to know which laws are structure dependent and which are not. If, for example, the Mach-Einstein program were carried through, then some of the basic laws of mechanics would be shown to be structure dependent. A second possible criticism of this distinction is that its persuasiveness is only superficial. Examples like the billiard table make the distinction appear important because experience in our actual universe has taught us that the laws of collision hold for all billiard tables and for other collisions as well, not just for the one table; the fact that the balls never meet we know would never hold for every table. That is, in considering the simple billiard table universe we tacitly bring in our knowledge of our actual multi-billiard-table universe. If—to formulate a rather outrageous counterfactual—I were one of the two billiard balls (and still possessed means of making observations), I might distinguish the two types of law but neither would seem more *basic* or more *necessary* than the other. "The laws of collision might have been different, the initial conditions might have been different—but both are what they are."

Finally it should be noticed that even if one did accept this distinction, it would not coincide with the distinction Sellars wants to draw between his second and third class of general statements. For it is, I think, clear that psychological laws, sociological laws, biological laws, and possibly, as we have seen, even the laws of mechanics are structure dependent. There cannot be many basic laws remaining. If it were only these basic (and probably as yet unknown) laws that give rise to counterfactuals, then counterfactuals would be confined to the pages of the most recondite physics books, instead of being a pervasive feature of our ordinary *language.*

C. Thus the word *necessary* (and its cognates), when included in a general statement in the object language or used in the metalanguage to describe an object-language generalization, is ambiguous. When included in the statement it signals a claim about the way in which the generalization is justified; in the other cases when used about the generalization, it says something about the role the generalization plays in scientific theory. And, if this explanation is correct, we need no longer look for ties of necessary connection or even for statements (other than tautologies) true *ex vi terminorum,* or for object-language expressions of logically prior material rules of inference.

Actually some of the plausibility of Sellars' argument is due to the misleading use of the phrase *in point of fact.* We are left with the impression that the only alternative to his view of generalizations as expressions of material rules of inference is to take them as generalizations which are 'in point of fact true.' But though in one sense an empiricist would accept this, it is unintentionally misleading because we often use the phrase *in point of fact* to distinguish between general statements which can only be validated by perfect induction—'all the coins in my pocket were minted after 1940'—and those which are reached by ampliative induction. Of course, any empiricist would admit that, if and when we find them, the basic laws of the universe will, in a sense, be true in point of fact. But this is only to deny that they are tautologies, 'principles of natural necessitation' (whatever that may mean), or rules. It is not to deny that these laws fit into a complex theoretical system and that they are such that they are not easily refutable by direct empirical evidence. And, more important, it is not to assert that they only *happen* to be true or are true *by chance*—for this notion of chance only applies *within* a universe in which either the basic laws are indeterministic or else are so complex that men do not know them; the notion cannot meaningfully be applied to the universe as a whole. When one realizes this, one sees that there is no need to accept Sellars' third category of generalizations. Empirical generalizations of differing scope and generality are sufficient for prediction, retrodiction, explanation, and the formation of counterfactual conditionals; what more could one ask for?

We have thus seen how unsatisfactory are most of the arguments that have been put forward in support of the thesis that material rules of inference are an indispensable feature of our language. The strongest argument is that which starts from the fact that we often argue directly from particular premise to particular conclusion without stating any major premise. If one wished to construct a logic that represented the way we explicitly argued, then one would have to recognize material rules of inference. But does the idea of such a logic make sense? We have seen that it would also have to be applied to what previously have been regarded as enthymemes of the second order. It is however clear that such a logic would have to be so complicated as to be completely unwieldy—do we not before using logic often 'put statements into logical form,' that is, say that statements of different grammatical form have the same logical form? A logic which considered all the different ways arguments might be put would be very different from any present logic, and it would almost certainly be a logic which would be of little philosophical use.

If then one rejects the idea of such a logic, very little remains to be said in favor of material rules of inference, inference licenses, or inference tickets. In fact, I fail to see that these terms are of any use in a philosophical investigation of the way in which we talk, either in science or everyday discourse.

NOTES

1. J. S. Mill, *System of Logic,* Book II, Chapter III. G. Ryle, "If', 'So', and 'Because'" in Philosophical Analysis, ed. M. Black, pp. 323–340. M. Black, *Problems of Analysis,* pp. 191ff.
2. But see the paper by M. Scriven in this volume where he argues that rules *can* be true or false.
3. I owe the substance of this paragraph to a remark made in discussion by Dr. W. Rozeboom.
4. R. B. Braithwaite, *Scientific Explanation,* Chapter IX. Cambridge: Cambridge Univ. Press, 1953. Max Black, *Problems in Analysis,* pp. 191–208. Ithaca, N.Y.: Cornell Univ. Press, 1954.
5. Cf. *Westminster Confession of Faith,* Chapter 1.
6. "Should We Attempt to Justify Induction?" *Philosophical Studies,* April 1957.
7. P. 66. New York: Macmillan, 1947.
8. "Necessary Truths," forthcoming in *Mind.*
9. S.E. Toulmin, *Introduction to the Philosophy of Science.* * London: Hutchinson's Univ. Libr.; New York; Longmans, Green and Co., 1953.
10. This has been pointed out by reviewers of Toulmin's book, e.g., M. Scriven in *Philosophical Review,* 64:124–128 (1955) and E. Nagel in *Mind,* 63:403–412 (1954).
11. *Mind,* 62:313–338 (1953). Professor Sellars' paper in this volume completes, and also to some extent modifies, the theory outlined in his earlier article. Since I wrote the present paper before his paper for this volume was completed, I have discussed in the text only his earlier article. Thus some of my criticisms are not applicable to his present views, and some of them have been implicitly countered by some of his arguments.
12. Professor Sellars now admits that his 1953 paper did not take into account these singular counterfactuals, and that it would be misleading to say that generalizations corresponding to these singular counterfactuals were true *ex vi terminorum.* His paper in this volume makes it clear that he is willing to say such generalizations are in a way necessary and that he therefore does not need to have recourse to either of the arguments I discuss in the two succeeding paragraphs.

SUGGESTIONS FOR FURTHER READING

Brodbeck, M. (ed.) *Readings in the Philosophy of the Social Sciences.* New York: The Macmillan Co., 1968, pp. 239–336.

Campbell, N. *What Is Science?* New York: Dover Publications, Inc., 1952, pp. 37–57.

Goodman, N. *Fact, Fiction and Forecast.* London: The Athlone Press, 1954, pp. 13–36.

Kaplan, A. *The Conduct of Inquiry.* San Francisco: Chandler Publishing Co., 1964, pp. 84–125.

Krimerman, L.I. (ed.) *The Nature and Scope of Social Science.* New York: Appleton Century-Crofts, 1969, pp. 205–350.

Nagel, E. *The Structure of Science.* New York: Harcourt, Brace and World, Inc., 1961, pp. 47–76.

Pap, A. *An Introduction to the Philosophy of Science.* New York: The Free Press, 1962, pp. 289–306.

Rescher, N. *Hypothetical Reasoning.* Amsterdam: North-Holland Publishing Co., 1964.

———. "Lawfulness as Mind-dependent." *Essays in Honor of Carl G. Hempel.* N. Rescher (ed.) *et al.* Dordrecht, Holland: D. Reidel Publishing Co., 1971, pp. 178–197.

Smart, J.J.C. *Between Science and Philosophy.* New York: Random House, Inc., 1968, pp. 121–174.

Toulmin, S. *The Philosophy of Science.* London: Hutchinson University Library, 1953, pp. 57–104.

SECTION V

Scientific Theories

The first article in this section purports to "convey the experiences behind the term 'theory' " in the social and the exact sciences, where such conveyance is understood as telling us "what is meant by a theory" in these two kinds of sciences. By now several lights should be flashing in your head—or bells should be ringing, if you prefer that metaphor. This author, Rapoport, seems to assume that sciences may be distinguished on the basis of exactness of *something* and that sciences that are somehow inexact are also (and *only*?) social. Moreover, he seems to want to give us a *report* of what the term "theory" means, rather than a *proposal* of what it should mean. So you will want to be sure that he does report and not stipulate as he examines four senses of "theory."

The second article attempts to answer the question "What is a theory?" and to offer a critique of four allegedly inadequate answers to this question. Achinstein begins by asking first, "What does it mean to say that someone has a theory?" and then he suggests at least six things that it is supposed to mean. The qualifier "at least" is very important here, because in fact Achinstein has an enormous number of different claims packed into his six numbered paragraphs. For example, consider the "second" condition he cites for stating that someone has a theory. It includes 1) "*A* does not know, nor does he believe, that *T* is false." That is two things. 2) "On the contrary, he believes that *T* is true or that it is plausible to think *T* is true." That is one *or* two other things. 3) "Moreover, he cannot immediately and readily come to know that *T* is false, or could not at the inception of his belief in *T*." That is one *or* two more. It may well be that every condition Achinstein suggests is necessary and important, but in order to decide whether or not they are, one must first identify all of them. You must not be misled into believing that because he *numbers* six conditions and claims to have only six conditions, therefore he *has* only six conditions. You will have to be a more careful logician than Achinstein in order to derive the full benefit of his analysis.

READING 12

Various Meanings of "Theory"

Anatol Rapoport

So many discussions go astray because the same words are used in different senses by adherents of different points of view that it seems imperative to start practically every discussion by clarifying the meanings of terms. Yet this problem is easier posed than solved. We in academic life owe understandable allegiance to erudition and to elegance of expression, and all too often we take a definition to be adequate (in the sense of clarifying meaning) if it sounds well. More is required, of course. Clarification of meaning (whether couched in formal definitions or in illustrative examples) takes place only if the terms defined are actually geared to the experience of the people concerned. This is a serious problem, because the experiences of people, although they overlap, can be widely disparate. Particularly among us in academic life the disparity may be quite wide. For our experience is very largely the experience of thinking, and thinking is tempered by language in the broadest sense, that is, by the way ideas are organized. And various ways of organizing ideas are imposed on us by our disciplines. Discipline means constraint. Discipline is essential for any organized activity. And so in academic disciplines, "discipline" means constraint on the mode of thought. It prescribes the repertoire of concepts, the patterns of classification, the rules of evidence, and the etiquette of discourse.

Cross-disciplinary endeavor, therefore, depends on the ability of the participants to think in terms of more than one language—a feat more difficult than the ability to think, say, both in English and French, because the languages of the disciplines vary not only in their vocabularies and grammars (as ordinary languuages do) but also in deeper aspects, whose meaning I hope can be conveyed by the phrase, "principles of organizing thought."

Reprinted by permission of the author and the *American Political Science Review, 52,* 1958, pp. 972–988.

I

Our concern here is the role of theory. I will first try to convey the experiences behind this term characteristic of the exact sciences (which I will define by their predominant thought patterns). Here I am on sure ground, because this is the field in which I myself have been "disciplined." Afterwards I will try to convey the experiences behind the term "theory" in other than exact sciences, particularly in the social sciences. Here I can only give my impressions. I am neither a political scientist nor a psychologist (to give two examples of "non-exact" sciences), and so I have not been properly disciplined to speak with authority on the way a term like "theory" is used there. My remarks are to be taken only as an account of how the thinking of social scientists and others about theory appears to someone who thinks of theory in terms by the discipline of exact science.

In the exact sciences, a theory is a collection of theorems. This concept of theory is also a partial definition of an exact science. It is the *entire* definition of any self-contained branch of mathematics. Some maintain, however, that mathematics is not a science, because it makes no assertions about the observable world. Whether one demands that a science must necessarily make assertions about the outside world is a matter of taste. I tend to accept this limitation on what is to be called a "science" and thus I agree to exclude mathematics from the sciences. What then is an exact science?

To begin with, it is a collection of theorems (to be presently defined); but the theorems have to be translatable into assertions about the tangible world and these assertions should be verifiable within certain limits of accuracy. The theorems are what makes a science exact; the accuracy of the assertions is what makes it successful.

Next we define a theorem. A theorem is a proposition which is a strict logical consequence of certain definitions and other propositions. The validity of a theorem, then, usually depends on the validity of other theorems. This tracing of antecedents goes on until the rock bottom is reached . . . assertions which are not proved but simply assumed, and terms which are not defined but simply listed. In a mathematical system it is unnecessary (in fact, impossible) either to prove these basic assumptions (the postulates) or to define the basic terms. This is what Bertrand Russell meant when he said that in mathematics we never know what we are talking about, nor whether what we are saying is true. In logic, the situation is exactly the same. Indeed logic is often taken to be a branch of mathematics or vice versa.

According to our criterion for science, however, we demand that some of the terms used in a science be related extensionally to referents and at least some of the assertions be empirically verifiable. I say some, not necessarily all, and this is an important point to which I will return.

Practically all the exact sciences we know are mathematical or at least highly mathematicized. This is by no means an indication of some supernatural power inherent in mathematics but rather of the propensity of mathematicians for preempting new territories and of adjusting and extending mathematical methods so as to be able to deal with different content areas. The central fact is that a necessary adjunct of an exact science is a set of completely rigid rules of deduction. It is the rigidity of these rules, not the accuracy of the assertions or precision of measurements, which makes an exact science. Now wherever there are such rules, a symbolism is invented as a purely mnemonic device. Wherever a symbolic notation occurs coupled with rules of deduction, *i.e.*, of manipulating the symbols, the mathematician steps in and assumes jurisdiction over the territory, or else the practitioners of the newly invented system of symbol manipulation are called mathematicians. This is what accounts for the all-pervasiveness of "mathematics" in the exact sciences.

II

I will now describe in greater detail by an illustrative example what is meant by a theory in an exact science. Specifically I will take a problem from mechanics, the earliest and one of the most successful of the exact sciences.

Consider a pendulum, that is, a weight supported by a string. The problem is to "explain" its motion. Immediately the question arises, what one means by "explain"? Obviously any explanation will contain statements introduced by "because" in reply to questions starting with "why." But what kind of questions are these? They are likely to be determined by what is observed about the pendulum. What is observed, in turn, will depend on what is singled out for observation. A question like "Why does the pendulum move around?" is so vague that it frustrates any attempt to answer it. Or, viewed in another way, it is trivially easy to answer, *because* it is so vague. One might say, for instance, that the pendulum moves around because there are forces acting on it.

The first task of an exact science, therefore, is to make the questions precise. The question "Why does the pendulum move *as it does?*" is more to the point. But the phrase "as it does" now lays the questioner open to a counter-question "What do you mean, 'as it does'?" This is a challenge to describe how in fact the pendulum does move, and this temporarily turns the attention away from an "explanation" toward description. To explain anything we must first circumscribe just what we are going to explain.

The first problem, then, is to *describe* the motion of the pendulum. This immediately introduces a motivation to simplify the situation. If the pendulum is constrained only by the string, its bobbing about will appear at first too chaotic to yield to a systematic description. Let us therefore constrain the pendulum to move in one plane, the way clock pendulums move. Now it moves just "back and forth."

But "back and forth" is still too crude a description. How can we make it more precise? Here is where the fundamental orientation of mechanics (of motion) as an exact science begins to direct one's methods of observation. This orientation prescribes what shall be of prime interest in any investigation of motion. If a moving object can be specified at any given moment by its position, the description of motion consists of associating a sequence of positions with a sequence of moments of time. In the simplest case, the position can be uniquely specified by a single number and, of course, time can also be so specified. The position of the pendulum bob, for example, can be specified by the angle of deflection (positive or negative) from the plumb line, or else equally well by the horizontal displacement of the center of the bob from the position of rest. A complete description of the motion, then, will be given by a table specifying the deflection at each moment of time. Such a table is called a mathematical function.

It can be demonstrated by experiment that the mathematical function $x = A \sin (mt)$ will very nearly, but not quite, describe the motion of the pendulum. Here x is the horizontal deflection of the bob, A is the maximum deflection, t is time, and m is a certain constant to be presently discussed. The qualification "not quite" is of paramount importance, as we shall see. First, however, let us see what we mean when we say that a mathematical function describes a set of data.

Actually a set of data in which values of two variables are related can be represented by a set of points in the plane, as many points as there have been readings. In particular, suppose one took only three readings of the pendulum position and got three points on a graph whose axes are displacement *vs.* time. A mathematical function will "describe" these data if the curve corresponding to the function can be passed through these points. Now a circle or a parabola can always be passed through any three (non-collinear) points, and so can many other kinds of curves. All of them, then, can be said to "describe" this limited set of data. Of course, if more readings are taken, the additional points may not lie on the same curve. However, given any number of points a great many curves can be made to fit them, which is to say that a great many equations can describe the data, indeed exactly, not approximately. Why, then, is the particular function $x = A \sin (mt)$ chosen, even though it does *not* describe the data exactly?

This function is chosen because it does more than describe the data. It *explains* the data in the sense of "explanation" as it is used in an exact science. And the discrepancies between it and the data are accounted for by the inadequacies of the fundamental assumptions in terms of which the explanation is made.

The fundamental assumptions concern the general laws which supposedly govern the motion of a body subjected to forces. In the case of the pendulum, we can from the geometry of the situation analyze the forces acting on the bob, hence *derive* the mathematical form of the motion. The equation above gives the derived (idealized) motion of an idealized body in an idealized environment. The discrepancy between the idealized and the actual state of affairs is supposed to account for the discrepancies between the prescribed and the idealized motion. Some of these idealizing assumptions are as follows:

1. The supporting string or rod is supposed to have no mass.
2. No friction or air resistance acts on the pendulum.
3. The bob is assumed to have mass but no extension.
4. The (small) horizontal displacement is supposed to be proportional to the angle of displacement.

Etc.

All of these assumptions are false. Yet the physicist continues to make them. Why? In return for sacrificing precision (precision must in any event be sacrificed wherever measurements are involved), he gains simplicity and, what is more important, he gets at the fundamentals (almost in the Platonic sense of the word) of the situation. This allows him to subsume a great many phenomena under a single scheme. For example, the assumed laws of motion and idealized properties of the pendulum allow the physicist to derive *from the same set of postulates* the following additional relations:

1. The period of the pendulum will be independent of the mass of the pendulum.
2. Within limits, it will be independent of the amplitude of oscillation.
3. For large amplitudes, the period will become dependent on the amplitude, and the precise nature of this dependence can be predicted.
4. The period will be directly proportional to the square root of the length of the pendulum.
5. The period will be inversely proportional to the square root of the acceleration of gravity. This relation explains the meaning of the constant m in the equation of the pendulum's motion. The constant involves the square root of the ratio of the acceleration of gravity to the length of the pendulum.
6. The horizontal velocity of the pendulum will be given by the function $v = mA \cos (mt)$.
7. The same scheme can be extended to the spherical pendulum, *i.e.*, one not constrained to a single plane. That is to say, from the same set of assumptions one can get many additional results, for instance, the precession of the plane of rotation of the spherical pendulum, etc. Indeed, all the observed motions of the heavenly bodies are derived from exactly the same three or four assumptions about motion which are supposed to underlie the swinging of a simple pendulum.

The story of the pendulum illustrates the power of the mathematical model. A mathematical model is much more than a description of events in terms of the mathematical relations among the variables. It is rather a set of assumptions often referring to a highly idealized situation, from which assumptions the relations to be observed are *derived,* to be compared with observations. Agreement with observations corroborates the model. Most important for corroboration is the prediction of *other* relationships, perhaps not hitherto observed. At any rate, the more are the relationships derived, and the fewer are the relationships assumed, the more powerful is the model. A trivial model does no more than lead to relationships which observations had suggested in the first place. Such a model penetrates no deeper than the observational level and is therefore purely descriptive, not explanatory. It simply restates in other terms what has been observed.

The method just outlined is applicable to many widely disparate exact sciences, though the content, of course, will be different in each case. In mechanics, for example, the fundamental assumptions have to do with the laws of motion. In mathematical genetics, another exact science in the sense defined, but not as successful as mechanics, the fundamental assumptions have to do with something quite different, namely the re-shuffling, segregation, and recombination in sexually reproducing organisms of entities supposed to be carriers of separate inherited traits. In this science, the laws governing these events are assumed to be not laws of mechanics or of electrodynamics or of thermodynamics but "laws of chance." For these too an exact mathematical theory exists. And on the basis of it, given certain combinations of genotypes, certain patterns of mating, certain linkages among the genes upon the chromosomes, certain selection pressures exerted by the environment, etc., the distributions of the genotypes and of the phenotypes in the succeeding generations can be computed. Here too discrepancies between prediction and observation are unavoidable, because of the idealizing assumptions which cannot be avoided. But the principle of investigation is the same as in theoretical mechanics. One derives a collection of theorems about how things should happen under idealized conditions. The discrepancies are attributed to the imperfections in the assumptions, and in the initial observations. The discrepancies provide the leverage for further refinements of the theory.

Economics may be viewed in the same light. The assumptions here have to do (in the classical picture) with the relations between price levels, supply, demand, maximization of profit (or expected profit) by the so-called "economic man," etc. The mathematical scheme being given, predictions on the basis of initial observations can be made. In mathematical economics agreement between theory and observation is not often good, and many are led to dismiss mathematical methods in economics on this ground. In the light of what has been said, such an attitude is not justifiable. The predictive power of mathematical meteorology, especially in its early stages, was also quite poor. Yet there was never any question that air currents, temperature gradients, and all the other conditions studied in meteorology are subject to the strict laws of physics. The low success of primitive meteorology is entirely attributable to the complexity of the phenomena compared with the drastically simplified assumptions which it was still possible to handle mathematically. Crude meteorology was merely a stage in the development of more refined and more successful meteorology. The original conceptual scheme was correct. Only the tools had to be sharpened.

Thus the bluntness of the available mathematical tools is not a sufficient ground for rejecting them in principle. It is an entirely different matter if the question is raised whether the tools applied are the right *kind* of tools. As is conceded by all who understand the mathematical method, its power is enormous where it can be applied. The big question is, where can it be applied? Clarification is needed here. For many, mathematics means classical mathematics, that is, mathematics of eighteenth-century physics, derived from the differential and integral calculus. As pointed out, the range of logical disciplines now called "mathematics" has enormously increased, both in techniques and in the variety of conceptualizations, so that any inadequacy of classical mathematics for dealing with problems of social science can by no means be taken as an indication of the inapplicability of mathematics in principle.

Nevertheless, let us first examine the conditions that must be fulfilled in order that classical mathematical methods may be utilized to advantage in theoretical social science.

III

First, there must be sharply defined, quantitative variables singled out for study. In the mechanics of motion, we have seen that position and time were fundamental. Actually there are three fundamental kinds of quantities in mechanics, from which all others are derived: length, time, and mass. Nor is there any question (in classical mechanics) how these quantities are to be measured. In other branches of physics, there are other quantities, for example, heat and temperature in thermodynamics, electric charge current and strength of electric and magnetic fields in electrodynamics, etc. Where probability theory is used, the fundamental quantities are frequencies of occurrence of events or of "types," and so on. In pursuing the question to what extent a quantitative variable can be sharply defined, a most important problem looms: the problem of recognition.

Note that the problem is trivial in mechanics. To determine the position of an object, we must, of course, recognize the object in all positions, but this is ordinarily so easy as to present no problem. As we pass from physics to chemistry, the problem of recognition becomes more important. For example, before quantities or concentrations of substances can be measured, the substances must be recognized. To be sure, there are unambiguous rules for recognizing substances by their "properties," and to the extent that the "properties" are defined in terms of classical physical measurements, the problem of recognition is solved. But it makes itself known.

As we pass to biology, the problem of recognition becomes even more serious. It now takes special training, sometimes quite prolonged, to tell one type of organism from another, one tissue from another, to interpret what is seen through a microscope, etc.

In the behavioral sciences, the problem of recognition becomes paramount. Since these sciences have only recently arisen from the humanities, their terms are derived largely from common sense and from intuitive notions at best, and from deeply rooted pre-scientific notions and prejudices at worst. Outside of science, no need may be felt to endow terms with operational meanings: one's intuitive meaning seems to suffice on the basis of the universal naive assumption that the other's perceptions are like one's own, or else something is wrong with the other's perceptions. Where there is no consensus on recognition, there can, of course, be no question of quantification or measurement, and so the first requirement of exact (or mathematical) science seems to be not fulfilled. Some workers in behavioral sciences feel that this difficulty precludes in principle the extension of mathematical methods to behavioral science.

The other condition usually assumed necessary for an exact science is this. Given a set of unambiguously measurable variables, one must be able to choose some assumptions about how they are related which reasonably reflect "reality." It is conceded, of course, that only an idealization of reality can be reflected in any finite set of assumptions, but it is maintained that the idealization should at least come close to reality. In mechanics, this condition is fulfilled. True, there is no such thing as frictionless motion, a perfectly rigid body, extensionless mass, etc., but these idealizations are in many instances well enough "approached" by reality. Once a set of assumptions is chosen, the mathematics required to derive theorems and conclusions from them must be amenable to being handled by the human mind. Again, it is maintained by some that even if unambiguous variables could be singled out (as they are in economics) and even if reasonably accurate assumptions could be made, the resulting mathematical system would be too unwieldy to be useful—there are too many relevant variables, and they are too intricately interwoven to permit treatment by existing mathematical techniques.

When it is proposed to simplify the situation by holding all but a few variables constant, it is pointed out (quite correctly) that, first, in many fields of investigation this is practically impossible: one cannot experiment with national economics or with real political systems; second, even where experimentation is possible, controlled conditions introduce distortions of such magnitude as to make extrapolations from controlled to natural situations, *i.e.*, from *in vitro* to *in situ* (to use the physiologist's terms), worthless.

Now I will develop what I want to say further along two lines. First I want to recognize to a certain extent the justice of the criticism of *premature* uses of mathematics in the social sciences and thereby the inevitability of an entirely different conception of "theory" and of the notion of a model, which have arisen in social science. But then I want to point out the limitations of this conception and its underlying assumption concerning the nature and function of mathematical methods and of mathematics itself.

IV

There are social scientists who understand the nature and the importance of the operational definition. Whether motivated by a hope of an ultimate possibility of mathematization of their disciplines (they are more likely to criticize premature mathematization, not mathematization in principle), or because they have been caught up by the spirit of positivism, which has become dominant in modern systematic thought, they undertook the task of creating a sound and consistent terminology of social science, particularly in sociology.

This task, however, would be trivial if it were confined to a compilation of a glossary, no matter how "operational" the definitions in it. For definitions are arbitrary—they are no more than agreements on how to use terms. Implicit in the work of Parsons and of Levy (to name two workers who recognize the importance of systematization) are not only attempts to map observable events on terms (which is what is done in operational definitions). Their main efforts are directed toward selecting events and combining them in such a fashion as to make the terms applied to these combinations fruitful in the development of a social theory which eventually is to become a collection of theorems—statements in "if so . . . then so" terms.

They ask, for example, in effect, "What is a social action?" Given a certain philosophic orientation, questions like this can be appallingly misunderstood Traditional philosophy is cluttered with questions of this type where the implicit assumption is that behind each word in use there must be a reality, and that the business of the philosopher is to discover it, so that making a "proper definition" is tantamount to establishing a truth. It has always been the curse of philosophy (until this curse was lifted by the logical positivists) to assume that entities called politics, society, power, welfare, tyranny, democracy, milieu, progress, etc., actually exist, just as cats, icebergs, coffee pots, and grains of wheat exist, and that each has an essence discoverable by proper application of reason and observation. I add observation, because I am speaking not only of the Platonists but also of the Aristotelians.

Now I certainly am not trying to say what is often said in vulgarized versions of the logical positivist position, namely that "concrete" objects certainly exist while "abstractions" don't. A "cat" is no less an abstraction than "progress," when you come to think of it. The problem is not one of existence but one of consensus. Not what *is* a cat, but what easily recognizable objects shall be *called* cats, is the first question. Because agreement is comparatively easy to reach on this question, we can pass immediately to the study of the cats themselves, their "nature," if you wish. But where agreement is not easy, that is, where one cannot immediately agree on an easily recognizable class of events which shall be subsumed under the term "democracy" or "status" or "power," it is futile to pass to the study of these supposed entities.

The systematizers understand in varying degrees the nature of this semantic problem, and they try to come to grips with it. They ask in effect, "What sort of thing shall we *call* a social action?" Consensus is not easy to reach, because the various definitions will presumably have different consequences. "Social action" once defined will presumably be a key term in some social science discipline. It will (hopefully) appear in the theorems of future theory. Therefore its particular definition serves to focus attention on the component events from which the definition is compounded. It may or may not be fruitful to focus attention on this or that combination of events. Hence the problem of definition becomes a "theoretical" problem, something which is often difficult for the natural scientist to recognize.

In the same spirit we can interpret the questions "What is a political act? "What is an economic act?" This search for primary, supposedly elemental, acts is itself inspired by the role of the atom concept in chemistry (as Easton points out). It is not so much a question of whether these "elementary particles" exist: just naming them does not confer existence. It is rather a question of whether our observations can be so organized that the *assumption* that they exist gives us a heuristic and predictive advantage. Incidentally, this is the only sense in which the so called "elementary particles" of physics can be said to exist.

We have, then, so far two distinct meanings of theory. For the natural, especially the physical scientist, theory, as we have said, is a collection of derived theorems tested in the process of predicting events from observed conditions. The physical scientist is able to address himself to problems of this sort, because for him the problems of recognition, of definition, and of meaningful classification either do not exist or have been largely solved. For the social scientist, all too often, the latter kinds of problems are central. The social scientist's aim, therefore, must be lower than that of the physicist.

It is often difficult to concede that one's aims are lower. So there are social scientists who will insist instead that their aims are "different." They are likely to say that they aim at "understanding" events, not at predicting them. For the physical scientist, however, "understanding" is synonymous with prediction. For this reason the physical scientist is likely to become impatient with the social scientist's distinction between the two. Moreover, the physical scientist often associates so called "understanding" (divorced from predictive ability) with ancient philosophical "explanations" which were unsurpassed in their vagueness or in tautological triviality. For this reason, too, the physical scientist often looks upon social science down his nose. And also for this reason, some investigators, usually those with physical science backgrounds, in an attempt to be constructive (that is, willing to extend to social science the power and respectability of the physical sciences) try to take the bull by the horns and to construct mathematical models of social behavior or of historical process wherever quantifiable variables can be found, not bothering too much with the question of whether these variables are germane to the sort of thing social science is trying to do.

The success of these attempts is spotty but by no means negligible. But because it is spotty, and because the relevance of the results to the important questions of social science is uncertain and (let's face it) because many social scientists do not read mathematics and for this reason develop defensive attitudes toward its methods, there is a reaction in the social science camp. This reaction is largely, I think, the source of the disclaimer to the effect that social science is not interested in prediction but only in "understanding." When asked what the proponents of "understanding" mean by it, they are in difficulties. It is as difficult to convey the meaning of "understanding" (in its subjective sense, as it is used here) as it is to convey the meaning of "appreciation" or of "perception." Yet these words are full of "meaning," of sorts. All of us "know" what they mean in the same sense that we "know" what vinegar smells like or how velvet feels. Pressing the issues of "meaning" of understanding is not fair in this situation. But it is fair to raise the question whether it is proper to give the name "science" to an activity which aims only at subjective understanding of this sort.

This is not a rhetorical question. I am not at all sure that the answer is categorically "no," although I suspect that I prefer "no" to "yes." Yet there is no denying that this intuitive organization of perception (akin to appreciation) is an important component in the psychology of science. Without it I doubt whether any but utilitarian motivation would exist for scientific activity, and I doubt whether science could get very far on utilitarian motives alone.

We have, then, a third meaning of "theory" in the attempts of social scientists (these attempts are no longer tolerated in physical science) to achieve and to impart intuitive understanding of social behavior, of the nature of institutions, of political systems, of cultures, and such matters. The language of such theory is largely metaphorical, although a great deal of factual material may be brought "in support." To "support" a theory in this context means at best to marshal factual material (historical and political events, case histories, etc.) in such a way that the reader who views this "evidence" through the metaphors, concepts, and definitions of which the "theory" is constructed will have the experience of "understanding." There is no need to say that even such attempts at concretization are often lacking in the writings of social scientists.

You may gather that as I mention concepts of theory farther and farther from those which enjoy hegemony in the physical science, I am becoming more and more skeptical about the scientific worth of such concepts. To a certain extent this is true, but I do not wish to draw a sharp line anywhere. The "worth" of a theory is not calculable by a set of cut and dried criteria any more than a man's worth as a member of the community is calculable in terms of how much he produces. In particular, metaphor and analogy, although they cannot be accepted as scientific "explanations," are sometimes important aids in the sense that they prepare the mind to make some precise investigations. It is in this sense that the so-called "models" of the non-exact sciences are to be appreciated. They are like the diagrams of geometry, neither necessary nor sufficient for the sort of proof that mathematical rigor demands, but often helpful for the eventual construction of such proofs.

There is also a branch of psychology which partakes in this sort of theorizing, namely, "depth psychology," to which the Freudian system also belongs. This branch of psychology is singularly poor in predictive capacity, either deterministic or statistical. Nor are there many attempts to make its terms operational, similar to the attempts of the systematizers in sociology. The aim is intuitive understanding of what makes up personality (another term which is only intuitively understood). It is strange for me, whose habitat is mathematics, to say this, but I think that depth psychology, particularly the contribution of Freud, is the richest area of behavioral science. I only regret that the disparity between the soft-heads and the hard-heads is so great that it is difficult for them to lay out a common program in which intuitive insights can be translated into strict deductions and verifiable generalizations.

V

There is still a fourth sense in which theory is used, in particular "political theory," namely in the normative, value-laden sense. In this sense, political theory would be concerned, for example, with the question of what is the best form of government. I have been specifically warned to avoid this issue on the ground that too much ink has already been shed over it. To some it seems that concern with what "ought to be" is farthest removed from science, which properly concerns itself with "what is." I will take serious issue with this position.

Whatever I know of the situation in political science is, of course, largely hearsay. In particular I take advantage of learning about the ideas of men whose books I have not read from the very few books on political science which I have read. Thus whatever I know about the ideas of James Bryce I have read in a book by Easton. Easton tells me that Bryce felt that generalization must be firmly rooted in "fact." He also knows how the idea is carried to extremes in some sections of American political science This "hyperfactualism," as Easton calls it, is, of course, quite understandable. The passage from undisciplined speculation, left over from the times when all science was rooted in philosophy, to militant empiricism has occurred in so many sciences as to suggest the operation of a law. In physical science hyperfactualism died on the vine. It might have proliferated if Francis Bacon's recommendations were ever carried out. But the greatest scientist who was Bacon's contemporary happened to be Galileo Galilei, and he chose a different path. If he had taken "facts" too seriously and too meticulously, he could never have enunciated the general law of falling bodies, because it would never account for the falling of leaves from trees, nor for the fall of rain drops, which between them account for probably 99 per cent of the falling that ever occurs on this planet. Neither leaves nor rain drops follow Galileo's law even approximately. Therefore his law is factually false. But it is true nevertheless, in a deeper sense. Without such ideally true and factually false laws, mathematical physics would have never left the ground.

Galileo's was, in a way, a normative theory. It described not how bodies fell but how they ought to fall under idealized conditions. In this sense one can well see how a theory can be normative and yet truly scientific. The idea of a truly scientific normative theory of action is not to pontificate about morality but to prescribe a correct course of action on the basis of a given desiderata, and in certain (usually idealized) conditions. Such a theory may not have any "practical value," because the idealized conditions may never obtain, but it may have immense heuristic value. In particular, it may through its underlying analysis of the fundamentals of the situation impart to the social scientist just the part of intuitive understanding of an area of investigation that he is seeking.

An example *par excellence* of this sort is game theory, a mathematical structure, which for the most part deals with situations which seem exceedingly remote from the subject matter of social science. However, centuries of scientific experience should have taught us that remoteness of a theory from a particular content area is no indication of its relevance or irrelevance. The fever which derives its name from "bad air" was not really understood until the events in the life of a certain mosquito became known. The harnessing of natural forces owes a tremendous lot to the ancients' curiosity about the anomalous behavior of amber and to one man's logical analysis of an experiment, which had been designed for no other reason than to determine the earth's motion relative to "absolute space."

The relevance of game theory to social science, particularly to political science (although originally the most direct applications were thought to be to economics) resides in the circumstance that game theory distills the logical essence of the situation which Catlin has termed the political act, namely a desire to fulfill an act of will in a context where conflict with others' desires to fulfill their acts will is to be expected.

Interpreted in a physical context, the metaphor "conflict of forces" calls for some sort of equilibrium theory. Such a theory can be and has been developed purely metaphorically. The concepts of "force," "pressure," "balance of power," "leverage," stability," "instability," are mostly terms borrowed from physics. Descriptions of conflict situations in these terms sound like descriptions of physical systems. But of course the analogy is a metaphorical, not a logical one, *i.e.,* the similarity is felt intuitively, not derived as a consequence or an isomorphism between two situations. Therefore metaphorical models of conflict, although they may be valuable for a variety of reasons, cannot be expected to yield logically compelling theorems, let alone theorems translatable into predictions.

There have been attempts to construct real mathematical models of conflict by utilizing the conceptual apparatus of classical mathematics. One such attempt, a very ambitious one, was undertaken by the late Lewis F. Richardson, who cast international rivalries in terms of differential equations and interpreted the stabilities and instabilities of the resulting systems of equations as the stabilities or instabilities of certain international situations. Since Richardson's theory is mathematical, its conclusions are definitive and compelling. In this they have an advantage over the metaphorical theories of the same sort. Its success as a predictive theory is, as would be expected, extremely limited. One set of data was fitted very well by the assumptions of the model, namely the arms expenditures of the two coalitions during the armament race preceding World War I. But even this extremely good fit is not impressive, since there are too many free parameters in the model and the points to be fitted are too few. One could at this point recall our previous argument that a mathematical theory always begins by treating an idealized situation and this beginning serves as a point of departure for greater refinements.

VI

The merits of game theory, however, are of quite another kind. It too deals with idealized situations; in particular it assumes "perfectly rational players." But it departs radically from earlier attempts to cast behavior into models of the mathematical physics type in one very important respect. The older models assumed the metaphysical basis of mechanics. That is, they described systems whose "states" were determined by a causality operating on the "here and now," the way physical states are determined. The state of any physical system is a consequence of the immediately preceding state in accordance with the laws governing infinitesimal changes of state.

In contrast, game theory is primarily a decision theory. It too casts situations into sequences of states. But each successive state is determined by a decision made by a rational being who foresees all possible outcomes and chooses a course of action, which in some way is likely to yield the best outcome under the circumstances. The phrase "under the circumstances" is crucial, for here game theory goes to the real heart of the matter. Each decision maker controls only a part of the situation. In making his decisions, he is aware that other decision makers whose "interests" may be opposed to his also make "rational decisions," and moreover take into account the decisions which *he* is likely to make. No physical theory treats of such situations.

Whereas the mathematical theories of behavior borrowed from the methods of physics (and often chemistry) depend for their success on the special assumptions concerning the interaction of variables and on the possibility of measuring certain key parameters, game theory is largely independent of special assumptions and measurements.

The independence from measurements is achieved by simply by-passing the problem. The only numerical variables are "utilities," *i.e.*, degrees of preference by the several decision makers for the various possible outcomes, and these are simply regarded as given. Independence from special assumptions obtains, because game theory is entirely "normative." It assumes "complete rationality" of the decision makers.

To some workers in game theory, particularly those who are oriented to social science applications, the most interesting results are paradoxically those which show up the inadequacies of game theory, *i.e.*, the indeterminacies of results based on the indeterminacy of the concept of "rationality" in all but the simplest of situations.

To illustrate, let us examine two game-like situations, in the first of which "rationality" can be satisfactorily defined, but not in the second. Suppose two decision makers have two courses of action each. Each of the four pairs of choices leads to an outcome which is denoted by a pair of "pay-offs," that is utilities accruing to each player. The situation can be represented by a 2×2 matrix. The first player chooses one of two rows; the second, one of two columns. The entries in the matrix are the pay-offs to the first and second players respectively, as in Figure 1.

$$\begin{bmatrix} 0,\ 0 & 3, -3 \\ 1, -1 & 2, -2 \end{bmatrix}$$

Figure 1

Note that the pay-offs in this case all add up to zero, *i.e.,* what the first player wins, the second loses. These games are called zero-sum.

This game has a "solution" which prescribes a "rational" choice to each player, namely for the first player to choose the second row, and for the second player the first column. This is so, because the first player can obviously guarantee for himself a win of 1, and the second player can *prevent* him from winning more than one, even if each choice is made without knowledge of the other's choice. Here a true "balance of power" exists, but note that this balance is not analogous to any physical balance of "opposing forces." It is a balance based on the *logic* of the situation, involving rational decisions.

The "solution" is much less definite in our next example, shown in Figure 2. Here the pay-offs do not add to zero. Such

$$\begin{bmatrix} 1,\ 1 & -2,\ 2 \\ 2, -2 & -1,\ -1 \end{bmatrix}$$

Figure 2

a game is called non-zero sum. If we follow the first player's "rational" reasoning, we must concede that he should choose the second row because *no matter which* column the second player chooses, he is better off by choosing the second row. By symmetry, the second column is the second player's "best" choice. But this pair of "best" choices results in a loss to both players (– 1, – 1), whereas their "worst" choice would have given both a win. Obviously "self-interest" is not a self-evident concept.

Agreement between the players to choose first row, first column would be considered "rational" in this case. But admitting agreements of this sort leads to questions of coalition formation, an immensely fertile field in game theory and a very difficult one, because of the ambiguities which plague the concepts of "power," "self-interest," "rationality," etc.—terms taken for granted in much discussion of individual and social behavior, but which game theory has undertaken to define with mathematical precision, and has led into formidable conceptual difficulties in consequence.

The central problem in game theory thus appears to be logical analysis, specifically logical analysis of situations in which common sense often fails. Where such logical analysis can be pushed to a definitive conclusion, the theory of games can be considered a normative theory. For example, the equilibrium solution of a two-person zero-sum game is essentially a prescription to two rational players how to choose their strategies.

But sometimes no such conclusion can be reached on the basis of existing concepts. The following extremely instructive example is taken not from game theory proper but from closely allied investigations. I shall cast it in terms likely to be of interest to political scientists.

Suppose three men, *A, B, C,* rank three courses of action, *X, Y, Z,* according to their preference 1, 2, 3. Let their rankings be displayed in the matrix shown in Figure 3.

	X	*Y*	*Z*
A	1	2	3
B	2	3	1
C	3	1	2

Figure 3

What is a fair compromise? Suppose we apply the majority rule to paired comparisons. Noting that *X* is preferred to *Y* by *A* and *B* (a majority) and *Y* to *Z* by *A* and *C* (again a majority), we are tempted to assign to *X, Y,* and *Z* the ranks, 1, 2, 3 respectively. However, this violates the majority rule in the case of *X* and *Z,* because *Z* is preferred to *X* by *B* and *C* (again a majority!).

Kenneth Arrow has shown that such impasses are *certain* to arise wherever more than 2 persons rank more than 2 alternatives; that moreover no decision rules except those which seem undesirable in a democracy, such as dictatorial or entirely arbitrary prescriptions, can be assigned which do not contain inherent contradictions of the sort just noted.

The methods of such analyses, like the methods of game theory, are those of exact science. The situations portrayed certainly remind us of political situations. They involve conflicts of interest, advantages of coalition, social decision rules, in other instances also arbitration schemes, calculation of power indices, etc. Here, then, is an exact science, seemingly applicable to politics (or perhaps, economics). But what does "applicable" mean? If it is taken to mean as involving the possibility of translating the theorems into predictions about human behavior, I am afraid it is stretching the imagination to call game theory "applicable."

The conditions of translatability are not met. Situations are seldom so clear-cut as to be describable in terms of a few "alternatives"; preferences are never so clear as to be measured in utilities; men are seldom rational. In this sense game theory fares no better than the classical mathematical models of behavior borrowed from physics and chemistry.

But in another sense, game theory and similarly oriented investigations are a genuine step forward. For they have burst through the framework of thought imposed by physical science on those who wished to apply mathematical methods to behavioral science. The floodgates have opened and a torrent of entirely new concepts has rushed by. I wish to note in passing that a flourishing terminology is more often a calamity in a rhetorically oriented discipline than an indication of its richness. Vagueness allows undisciplined proliferation and duplication, often providing a mask of erudition for lack of originality and insight. Not so in exact science. There concepts hold their own only if they provide points of anchorage for genuine theorems, not simply for rhetorical speculation.

But what good is it, the empiricist is bound to ask, which is his perfect right. No hidebound answer can be given. Only if one believes that far-reaching and deep-digging logical analysis is essential in any discipline which aspires to the status of science, will one be satisfied with the answer that game theory is an example of such an analysis, that it is of relevance to political science because its fundamental concepts are idealizations of what political science is about, namely decisions made amid partly conflicting and partly coincident interests of rational, calculating beings. Theory of this sort represents one of two poles. The other is meticulous empiricism, of the kind espoused by Bryce, by most historians, and by the positivists of Jerome Frank's persuasion in legal thought. The arguments of the empiricists are well known. They are worthy arguments. The arguments in favor of pure theory are much harder to present because social scientists do not often come into contact with really powerful pure theories, the kind that grow on mathematical soil. One can only say that their worth has been amply demonstrated elsewhere. The physicist might spend thousands of years studying the behavior of ocean waves on a beach in most meticulous detail; in the end he would be no wiser than before with regard to what is essential in wave motion.

The really profound understanding of waves is quite independent of observing any real waves. For that matter, the most important waves in our lives are not even observable directly—I am referring to the waves which underlie our entire telecommunication system and also subatomic events.

It is these kinds of essences which pure theory seeks. It goes without saying that *ultimately* the findings of theory must somehow be translated into real predictions and observations. But to demand this too soon is not wise. It would be like demanding cash payment in every business transaction. Such "hard-headedness," although aimed at security, would actually disrupt the system of credit on which any complex economy must depend to a great degree.

Theory, then, is like a system of credit. One has a right to demand that *somewhere* there are assets to back up the transactions. But, as often as not, these assets may be in the future, and the very act of questioning their existence may set in motion a chain reaction which will preclude their existence.

This is what I meant by the statement that not every conclusion of an exact theory has to be translatable into observation. In the case of the recent constructions of exact theories of presumed relevance to social science, we will do well to extend to them the most liberal credit. After all, in our society the thinker's time comes cheap.

READING 13

Theories

Peter Achinstein

One of the central aims of science is the construction of theories; and one of the central aims of the philosophy of science is the study of the concept "theory." Much of this concept has already been made in the preceding discussion on definitions and the interpretation of terms; further appeals to it will occur later when we turn to the subject of theoretical terms and to analogies and models. The word "theory" has both a narrow and a broad use. The former is illustrated by expressions such as "the Bohr theory of the atom" and "the kinetic theory of gases"; the latter, by "physical theory" and "nuclear theory." Most of this chapter will concern the narrower use; the broader one will be examined briefly in the final section.

The first question to be raised is simply: What is a theory? I think the same considerations are relevant in answering this question whether we are dealing with theories inside or outside science. And I shall compare the view to be developed with several other conceptions of theories found in the philosophy of science. I shall then concentrate on theories in science, especially those in physics, and consider what elements can be distinguished in their presentations, the different types of presentations that are or might be given, and the advantages and disadvantages of each.

1. CRITERIA FOR THEORIES

Let me approach the question "What is a theory?" indirectly by first asking: What does it mean to say that someone has a theory? Several conditions will be proposed as semantically relevant and

Reprinted by permission of the publisher from P. Achinstein, *Concepts of Science,* 1968, The Johns Hopkins University Press, pp. 121–137.

quite central for this, although possibly only one (the fourth) is logically necessary. The nonsatisfaction of each of the other conditions would in and of itself tend to count, to some extent, against saying that someone has a theory without necessarily precluding it. I shall then show how the concept of a theory can be defined by reference to the conditions for having a theory. Let us consider, then, what is typically the case when someone, call him *A*, has a theory *T*.

1) *A* does not know that *T* is true, although he believes that *T* is true or that it is plausible to think it is. Moreover, he cannot immediately and readily come to know the truth of *T*, or could not at the inception of his belief in *T*. He may think he knows that *T* is true. But *we* say he has a theory *T* when we impute to *A* a lack of knowledge concerning the truth of *T*. Of course, *A* may later come to know that *T* is true. If so, it is no longer appropriate for us to say that he has a *theory* *T* (or that he has *T* as a theory) unless we doubt he knows the truth of *T*. This condition represents the conjectural or speculative element associated with theories.

Consider, for example, the kinetic theory of Daniel Bernoulli. One of the reasons we are willing to say Bernoulli had the theory that gases are composed of an enormous number of tiny particles in rapid motion is because he did not know, and could not immediately and readily come to know, that this was so. He did give certain arguments to support the theory. For example, he showed how it afforded a quantitative explanation of the relationship between pressure, temperature, and density of a gas.[1] However, such arguments could not be said to demonstrate the truth of the theory; Bernoulli could not be described as having known, or having been in a position to come to know immediately and readily, that gases contain particles of the sort he postulated. He might be so described if, for example, a super-microscope had always been available by means of which such particles could readily have been observed. But then there would be some hesitation in speaking of this as his *theory*. Thus, one reason why the proposition "My pen is filled with ink" is not a theory I have is because I know it to be true; and even if I did not, I could readily come to know it in a simple and direct way.

I have spoken here of truth, but this is a bit of an oversimplification, since truth is not always clearly ascribable to a theory as a whole. A theory may contain a large set of assertions some of which are true and others of which are not, thus making the question of the truth of the entire set not clearly decidable. Moreover, there will be considerations of scope: how many sorts of items are subject to the principles of the theory? There will also be considerations of accuracy: within what limits does *T* describe what is actually the case? Usually the principles of the theory will be formulated without mention of scope, so that if items in a certain range are discovered that do not satisfy the theory, it may not be clear whether to say that the theory has been shown to be false or to be limited in scope. Certainly if very many such ranges are discovered, the former description might be given. Analogous remarks apply to accuracy.

Accordingly, when I speak here of knowing and believing that *T* is true, I mean to imply something complex, namely, knowing or believing either that *T* is true (where truth is clearly applicable to *T*) or that on the whole *T* says what is the case in a more or less accurate way and—where the question of scope is relevant—does so with respect to a reasonably extensive range of items.

2) *A* does not know, nor does he believe, that *T* is false. On the contrary, he believes that *T* is true or that it is plausible to think *T* is true. Moreover, he cannot immediately and readily come to know that *T* is false, or could not at the inception of his belief in *T*. Of course, *A* may later come to know or believe that *T* is false; but if so, we would no longer say that he has the theory *T*. For example, Kepler's first theory about the orbit of the planet Mars was that it was a perfect circle. However, this theory yielded consequences seriously at odds with what was observed, and he came to disbelieve it.[2] If anyone at that time had claimed that Kepler (still) had the "circular" theory, he would have been mistaken.

The same considerations regarding *truth* apply here as in the first condition. Accordingly, someone satisfies the present condition if he thinks it reasonable to suppose that *T* is more or less accurate for a fairly extensive range of items. (This is one feature that distinguishes a theory from what in Chapter Seven will be called a "theoretical model." As will be explained in that chapter, someone may propose a theoretical model knowing full well that it is applicable only within a very limited range and not very accurate even within that range.)

3) *A* believes that *T* provides, or will eventually provide, some (or a better) understanding of something and that this is or will be one of the main functions of *T*. By providing an understanding I mean something quite broad that can be done in a number of related ways, for example, by explaining, interpreting, removing a puzzle, showing why something is not surprising, indicating a cause or causes, supplying reasons, analyzing something into

simpler, more familiar, or more integrated components. Also, what is to be understood may be of many different sorts. Some authors write as if a theory always provides an understanding of one thing, namely, some surprising phenomenon that has been observed. For example, Hanson suggests that whenever a theory is proposed the following inference is involved:

Some surprising phenomenon *P* is observed.
P would be explicable as a matter of course if *H* were true.
Hence there is reason to think that *H* is true.[3]

No doubt very often when one has a theory he believes that it provides a better understanding of a surprising phenomenon that has been observed. But one may have a theory and believe that it provides a better understanding of something which could not be classified as surprising, or even as a phenomenon, and indeed may never have been observed. *A* may propose a particular theory about the atom because he believes that it provides a better understanding of its nature. Of course this may indirectly provide a better understanding of certain phenomena observed in cloud chambers, for example, but the latter need not be the only sort of thing for which *A* believes his theory provides understanding. Nor need it be the central thing as far as *A* is concerned. Indeed *A* could have proposed his theory without knowing of any observable phenomena for which his theory provides understanding.

I said that when *A* has a theory *T* he believes that one of *T*'s main functions is to explain, interpret, remove puzzles, and so forth. Normally *A* will be able to indicate how his theory purports to do some of these things, but this is not absolutely required. *A* might be said to have a theory *T* even though he is unable to supply actual explanations, interpretations, puzzle-removers, and so forth. What will suffice, I think, is that he believes these are possible and central for the theory and that he be able to identify at least some of the sorts of items for which his theory might be able to provide understanding. In short, *A* believes that *T* is at least, and importantly, a *potential* explainer, interpreter, remover of puzzles, analyzer, and so forth. The less central these potential roles are for *A*, the less willing are we to speak of what *A* has as a theory.

There are cases in which the term "theory" is used where such roles may seem minimal or even nonexistent, as when someone is said to have the theory that Senator Claghorn will not be running in the next election. We might characterize such a use as stretched or even careless and say that, strictly speaking, this is not a theory but a conjecture, speculation, or guess. There is another possibility. In such a case, when we call this a theory we may be thinking of it in connection with a larger set of assertions that does more properly constitute a theory, for example, that the Republicans believe that they will be meeting stiff competition in the next election and so will want to nominate a more colorful, dynamic, and youthful personality. One of the important functions of the latter may indeed be to explain, analyze, provide reasons, and so forth.

4) *T* consists of propositions purporting to assert what is the case. For *A* to have a theory he need not actually have formulated such propositions in a language. And if he does formulate them, he may do so in several different ways, that is, use different sentences to express them. Moreover, contrary to what some instrumentalist philosophers suggest, *T* is not a set of *rules,* literally speaking, although it may contain rules expressed propositionally. For example, it may contain the proposition that if an electric current flows through a straight portion of wire and if we place the thumb of our right hand along the wire pointing in the direction of the current, the other fingers will point in the direction of the magnetic field (the so-called right-hand rule).

Which propositions *are* the theory? Normally when one speaks of a theory what is being referred to is a set containing at least its central and distinctive assumptions. By *assumptions* I mean those propositions of the theory that are not treated as being derived from others in the theory. By *central* I mean those regarded as expressing the most important ideas of the theory. By *distinctive* I mean those that more than others serve to identify it as that particular theory and distinguish it from others. Typically the central assumptions will be the distinctive ones, and vice versa.

A theory may have associated with it (use, presuppose, have as consequences) many propositions not in the set of central and distinctive assumptions. In referring to the Bohr theory of the atom one may be referring to assumptions concerning the quantization of angular momentum of the election in the atom and the quantization of energy radiated or absorbed by the atom. Yet, in addition, the theory uses Coulomb's law, Newton's second law of motion, and the principle of conservation of mechanical energy in order to derive the desired consequences. These additional laws and principles might be spoken of as part of the theory *in the sense that* they must be used to derive the desired consequences. Similarly, when one refers to Newton's theory of mechanics one may be referring to his three laws of motion and the law of gravitation. Yet there are principles that are derivable from these laws together with special assumptions about specific systems, for example, principles governing unsupported bodies, oscillator motion, and accelerated reference systems. Such derived principles might be spoken of as part of the theory *in the sense that* they are derivable from its central and distinctive assumptions when further assumptions are made about systems of particular sorts.

With respect to the question "Which propositions are the theory?" two extreme answers might be given. One is that only the central and distinctive assumptions are the theory (only Bohr's two assumptions, only Newton's four laws). The other is that everything associated with the theory (including additional assumptions and consequences) is the theory. Both answers are somewhat arbitrary. I think the best procedure is simply to distinguish the central and distinctive assumptions of a theory from the additional principles (possibly from other theories) it may use or presuppose, and these in turn from consequences of the theory. These are all "associated with" the theory, but in different ways. In short, there is no simple and nonarbitrary answer to the question "Which propositions are the theory?"

Finally, there is a sense in which items other than propositions, for example, concepts and definitions, may be said to belong to a theory; but to speak of *A*'s theory, in the sense I am now considering, is to speak of such items primarily in connection with propositions; as being part of them, in the case of concepts, as presupposed by them, in the case of definitions. Concepts and definitions may be cited and explained in a presentation of *A*'s theory, though the theory proper is a set of propositions with which they are associated.

5) *A* does not know of any more fundamental (theory) *T′* from which he knows that the set of central and distinctive assumptions of *T* can be simply and directly derived, where *A* satisfies all the other conditions with respect to *T′*. Newton, for example, in Book I of the *Principia,* formulated the principle of conservation of momentum (Corollary III) and the principle of inertia for the center of mass of a system of bodies (Corollary IV). Yet in doing so he did not formulate a theory that he had, but only certain propositions that follow, and that he knew to follow (simply and directly), from such a theory. Of course a scientist may have a theory that he applies to some relatively specific system *S*; and what he says about *S* may also be described as a theory that he has. For example, in Book III of the *Principia,* Newton applied his theory of mechanics to the tides on the earth, and what he said might be described as his theory of tides. But to ob-

tain this theory he needed to go beyond the laws of dynamics and gravitation and introduce special assumptions and observations about the character of the earth. Moreover, the derivations are neither simple nor direct.

In short, *A* may know that some of the principles of a theory that he has are (or are simply and directly derivable from) principles of some more fundamental theory that he also has. However, he does not know this about the entire set of his central and distinctive assumptions.

6) *A* believes that each of the assumptions in the set regarded as distinctive and central will be helpful, together with others associated with the theory, in providing an understanding of those items for which *A* believes the theory may provide an understanding. He need not believe that each such assumption is required for this purpose; he may, indeed, believe that some are redundant. Nor must he believe that each such assumption will help provide an understanding of *every* item for which the theory provides some understanding, though he should believe that there will be items that all or most of these assumptions will help to illuminate; that is, he should believe in a *joint* effort on the part of his central and distinctive assumptions.

This condition is intended to capture what I regard as a weak "working together" requirement on theories. It precludes conjoining the central and distinctive assumptions of what are viewed as wholly unrelated theories and calling the result a theory. It also precludes stringing together wholly unrelated propositions, where each is intended to illuminate something different, and calling the result a theory. Even if someone were committed to Malthus' theory of population and to Bohr's theory of the atom, he would not have a theory consisting of the conjunction of these; nor would he have one consisting of a conjunction formed by taking one principle from Malthus' theory, one from Bohr's, one from Newton's, and so on.

I have cited six conditions for having a theory. Can these be used in saying what it is to be a theory?[4] One might be tempted to define a theory as follows: Where *T* is a set of propositions, *T* is a theory if and only if there is some person *A* who, with respect to *T,* satisfies the six conditions specified above. But this would be too stringent, for in some situations we might want to count *T* as a theory even if no one now satisfies these conditions, that is, even if no one now has *T*. Let us try a weaker definition: *T* is a theory if and only if there is some person *A* who, with respect to *T,* now satisfies these conditions or did so at some previous time. However, this would preclude counting as a theory some *T* that no one ever had, has now, or even will have—but that it might be plausible to imagine someone as having (had). In some situations a *T* of the latter sort will count as a theory; for example, we are considering various possible structures that might be imputed to the atomic nucleus. In other situations such a *T* will not be counted as a theory; for example, we are considering only theories of the atomic nucleus that someone has actually had, or only theories that someone has actually had and are at present live options in physics: in such situations if a *T* was mentioned that no one ever had, the reply might be that there is so such theory, or that it does not count as a theory for present purposes. This reply would be inappropriate in a situation of the former sort.

What is classifiable as a theory, then, depends in part on the context of classification: this can vary so that in one situation what is called for is a *T* that someone has had; in another, a *T* that may not have been had by anyone but is still something that is reasonable to imagine someone as having; in another, a *T* had by leading scientists of today; and so forth. Accordingly, from the fact that there is a person *A* who satisfies the six conditions with respect to some *T*, it does not follow that *T* can always be counted as a theory. Moreover, I have said that these conditions are semantically relevant for having a theory but (with the possible exception of the fourth) are not logically necessary. So what we can say about the conditions is this: if an *A* does satisfy (all or most of) them, then he has what in appropriate contexts can be classified as a theory; and if he has what in some context is classifiable as a theory, then he satisfies (all or most of) them. In short, these conditions, together with contextual considerations, are semantically relevant for having a theory.

The following characterization of a theory can now be offered. To say that some *T* is a theory is, depending on the context, to say one or more of the following: that (at least) some person *A* now satisfies conditions (1) through (6) with respect to *T;* or that some person at a previous time satisfied these conditions with respect to *T*; or that some person now satisfies these conditions with respect to *T*, and *T* is a live option today; or that it is plausible to imagine some person as having satisfied these conditions with respect to *T;* and so forth. Another way of putting this, although truncated and rough, is as follows: *T* is a theory, relative to the context, if and only if *T* is a set of propositions that (depending on the context) is (was, might have been, and so forth) not known to be true or to be false but believed to be somewhat plausible, potentially explanatory, relatively fundamental, and somewhat integrated.

2. OTHER CONCEPTIONS OF THEORIES

Let me now contrast the conception of a theory just developed with four others to be found in the philosophy of science.

The first I shall call the *axiomatic* account. It is held by those defending the Positivist position on interpretation discussed in Chapter Three, but I shall formulate it in a more general way that will be neutral on topics such as partial interpretation and the theoretical-nontheoretical distinction. On the axiomatic account a theory is no more and no less than a set of sentences ("axioms") stated in a specified vocabulary (containing primitive and defined terms) together with all the consequences ("theorems") of this set plus the proofs of these consequences. In short, a theory is a hypothetico-deductive system. This conception is usually embellished with further requirements that concern the *adequacy* of the theory rather than whether or not it is a theory; that is, for a theory to be an adequate one, on this account, it must contain at least some general statements, it must be simple, confirmable and confirmed by observations, and capable of generating new consequences.

What is involved in having a theory? Champions of the axiomatic account do not address themselves directly to this question. Some might want to say that for *A* to have a theory (though not necessarily an adequate one) is simply for *A* to have formulated or to be willing to formulate (in a specified vocabulary) a set of sentences as axioms, another as theorems, and proofs of the latter. I shall call this the weaker version of the axiomatic account. However, in the light of considerations in the previous section, others might want to say that for *A* to have a theory ***T***, it is not sufficient simply for *A* to have formulated or be willing to formulate axioms, theorems, and proofs. *A* must also not know or be readily able to know that *T* is true; and he must not know or believe *T* to be false, but, on the contrary, must think *T* reasonable or plausible in some measure at least. I shall call this the stronger version.

Let us consider this account in the light of the conception I have suggested. On the weaker version, one who has a theory *T* can know that *T* is true or false, as the case may be, thus ignoring my first and second conditions relevant for having a theory; the stronger version was of course constructed so as not to violate these conditions. Since, on the weaker version, for *A* to have a theory is simply for *A* to have formulated or to be willing to formulate a set containing axioms, theorems, and proofs, it is not required or relevant that *A* believe that one of the main functions of these is (or will be) to provide some understanding of something, thus violating my third condition. Nor is this condition satisfied even if, following the stronger version, we add requirements concerning *A*'s knowledge and beliefs regarding the truth and falsity of *T*. Proponents of the axiomatic account may reply, however, that an *adequate* theory, that is, one that satisfies the further conditions they specify, does provide explanations. For something is explained, on this view, if a statement describing it is deduced from an adequate theory.[5] The deductive model of explanation assumed by the axiomatic account is a complex topic which cannot be explored here. However, there are, I think, numerous examples of "adequate" theories, in the axiomatic sense, that could not justifiably be said to explain something described in the theorems of such theories. Let me cite just one example.

When an electric discharge is passed through hydrogen, light is given off that, when analyzed with a spectroscope, is shown to consist of a series of sharp lines of definite wavelengths. This has been known since 1860. In 1885, Balmer proposed the following simple relationship between the wavelengths of the lines:

$$\frac{1}{\omega} = R\left(\frac{1}{2^2} - \frac{1}{\tau^2}\right), \tau = 3,\ 4,\ 5,\ \ldots,$$

where ω is the wavelength of the line and R is a constant for hydrogen. From this "axiom," together with the further "axiom" that $R = 109{,}677 cm^{-1}$, the wavelength of each of the known lines can be derived. Moreover (as it actually happened), by substituting $\tau = 7,\ 8,\ \ldots$ in the above formula, and using the value for R, the existence of new lines was predicted and later observed. This then, is, or can be put in the form of, an axiomatic system (indeed one that satisfies the criteria for an acceptable theory indicated by proponents of the axiomatic account: it is general, empirically testable and confirmed, simple, and capable of being used to make new predictions). However, it provides no explanation whatever of the existence of the observed lines, of why the wavelengths are what they are, of their discreteness, and so forth. At most, it organizes the known data in such a way that we are in a better position to seek a theory to explain the initial phenomena.

On either version of the axiomatic account, one who has a theory T may have some more fundamental theory from which he knows T to be simply and directly derivable. In short, any subset of propositions in (or implied by) a theory is itself a theory; and this ignores my fifth condition relevant for having a theory. On either version of the axiomatic account there is no "working together" requirement, so that any unrelated propositions one may string together can constitute a theory, and the conjunction of any two theories can itself be a theory—thus violating my sixth condition. Finally, on either version of the axiomatic account something counts as a theory independently of the context. Any set T consisting of axioms, theorems, and proofs is a theory irrespective of whether anyone has (had, might have had, and so forth) T; indeed, it is a theory irrespective of whether anyone ever contemplated or could have contemplated such a set. This ignores contextual considerations relevant in deciding what is to count as a theory.

Supporters of the axiomatic account may reply that they are not concerned with how the concept of a theory is normally employed. They want to replace it with another concept that they think will be more satisfactory for certain purposes. This replacement of one concept by another regarded as more satisfactory is what Carnap calls explication.[6] By Carnap's own admission, one explicates a concept, that is, replaces concept C by C' only where C leads to certain difficulties (inconsistencies, unclarities, paradoxes). But those defending the axiomatic account never even try to show what difficulties are generated by using what they call the "pre-analytic" concept of a theory. Indeed, they never consider this concept at all. I have tried to point out the differences between their concept and what I regard as one standard concept of a theory employed with (as well as outside) science.

A second view about theories has achieved recent popularity through the writings of Feyerabend.[7] Unlike the axiomatic account, it distinguishes theories from mere empirical generalizations. The latter are considerably restricted in their application, arrived at by observing instances, and refutable in a simple and direct way by conflicting evidence. Theories, on the other hand, are very general unrestricted assumptions. They are ones to which we are deeply committed and would abandon not simply in the light of conflicting evidence but only when we become deeply committed to other such assumptions. They are not arrived at merely by observing instances. Indeed observations are made in the light of a theory; terms used in reporting observations depend for their very meanings on the theory. Feyerabend includes in the set of theories not only those recognized as such in science (Newton's theory of mechanics, the special theory of relativity), but also certain "ordinary beliefs (e.g. the belief in the existence of material objects), myths (e.g., the myth of eternal recurrence), religious beliefs, etc."[8]

Part of this view, having to do with the meaning-dependence of terms, was discussed in the previous chapter. What I want to note here is that this conception of a theory ignores a number of the conditions in section 1. For example, it makes no requirement that one who has a theory not know that it is true; that is, it omits the speculative or conjectural element. Nor is there a requirement that one who has such a theory believe that it does or will provide understanding of something and that this is one of its main functions. Any deep general commitment refutable only by another deep general commitment would count as a theory on this view, whether or not such a commitment is supposed to provide understanding. Furthermore, classifying any such commitment as a theory ignores contextual considerations of the sort noted earlier. The account also fails to provide a necessary condition. To have a theory *T*, it is not required that *T* be criticizable only in the light of other theories and that *T* be incapable of refutation by conflicting evidence, independently of alternative theories.

However, as with the axiomatic account, the claim may not be that this represents a standard use of the term "theory," even in science. So I want to note two proposals about theories that do appear to be making such a claim.

The first is that of Hanson, who developed it in part in reaction to the axiomatic approach.[9] Hanson rejects the latter on the ground that scientists do not start from hypotheses and then deduce the data; they begin with the data and then search for hypotheses to explain them. Thus, Kepler did not begin with the hypothesis of an elliptical orbit for Mars. He began with observations recorded by Brahe and "struggled back from these, first to one hypothesis, then to another, then to another, and ultimately to the hypothesis of the elliptical orbit."[10] Now a theory is something that organizes initially puzzling observed data into an intelligible pattern, so that these data are "explicable as a matter of course." Hanson writes:

> What is it to supply a theory? It is to offer an intelligible, systematic, conceptual pattern for the observed data. The value of this pattern lies in its capacity to unite phenomena which, without the theory, are either surprising, anomalous, or wholly unnoticed.[11]

To compare Hanson's account with mine, let us suppose that one who has a theory is now supplying it for some purpose. If, as seems quite possible, Hanson means to be suggesting a sufficient condition for supplying a theory, then his account is too liberal. For example, one condition that I suggested as relevant for having a theory but that seems irrelevant for Hanson is that *A* does not know, and cannot (or could not at first) readily come to know, that *T* is true. Suppose that *A* and *B* are walking in the woods and *B* observes something that he finds very puzzling moving on a distant tree. Once this is called to his attention, *A* informs *B* that it is a bear climbing a tree, though all that is visible from the present vantage point are the bear's paws clinging to the tree.[12] Here *A* has organized what *B* observes into an "intelligible, systematic, conceptual pattern" so that what *B* observes is no longer "surprising" or "anomalous." *A*, however, is not (at least in my example) supplying some *theory* that he has, since he knows that it is a bear, and even if he did not, he could (have) come to know this in a simple and direct way. Nor, it seems to me, would we speak here of *A*'s theory, though what *A* says does explain what *B* sees, does remove *B*'s puzzlement, and so forth.

If, on the other hand, Hanson means to be suggesting only a necessary condition for supplying a theory, then, one is prompted to ask, What else is relevant? The text does not say. Even so, the condition he does propose is too stringent, and the reason why was suggested earlier.

His account implies that the supplier of a theory must have observed or be aware of certain phenomena for which the theory offers an explanation. This, of course, will frequently happen, but what a theory explains, or is supposed to explain, need not be any phenomena that have been observed but may be items of other sorts, even ones not yet observed. Indeed there might be cases in which the proponent of a theory is unable to produce actual explanations of anything (observed phenomena or otherwise), though he is committed to believing that his theory can be used in explanations (or at least to promote some sort of understanding) of various things and that this will be one of its main functions.

The final conception of a theory I want to discuss is more complex than any of the others. It is that of Sylvain Bromberger.[13] He first introduces the notion of a *p*-predicament ("*p*" for puzzled or perplexed) for which he gives this definition: "*A* is in a *p*-predicament with regard to [question] *Q* if and only if, on *A*'s views, *Q* admits of a right answer, but *A* can think of no answer to which, on *A*'s views, there are no decisive objections."[14] For example,

Bromberger informs us that he happens to be in a *p*-predicament with regard to the question, "Why do teakettles emit a humming noise just before the water begins to boil?" but not with regard to the question, "What is the height of Mt. Kilimanjaro?", although he knows the answer to neither. The point is that he can think of nothing as an answer to the first that is not ruled out by his views about the world, whereas he can think of several possible answers to the second (for example, 15,000 feet, 8,000 feet), no one of which is precluded by his views.

Using the notion of a *p*-predicament, Bromberger provides the following definition of a theory. For *T* to be a theory, it is necessary and sufficient that *T* be a known proposition (or set of propositions) but not known to be true and not known to be false; and that *T* provide an answer to a question *Q* with regard to which it is in principle possible to be in a *p*-predicament; or that *T* provide one among many such answers which are mutually exclusive, where each such answer, in the absence of the others, would contain the only answer to *Q* not precluded by the conditions set by the question; or else that *T* be the contrary of some presupposition of *Q*.[15]

One problem with this conception stems from its requirement that for *T* to be a theory, it must not be known to be false. This would rule out theories now known to be false (for example, the phlogiston theory). Recognizing this to be an unfortunate aspect of his definition, Bromberger proposes that we think of the definition as specifying only what it means to be an "acceptable theory." But he fails to indicate the respect or respects in which he is considering acceptability. Not every set of propositions meeting his criteria would count as an acceptable theory in all relevant respects. On his conception, a theory not known to be false, but for which a considerable amount of disquieting evidence exists, could be an acceptable theory; or similarly, a theory could be an acceptable one, on this conception, even though the answers that it provides to questions, although not known to be false, are generally regarded as implausible, much too complex, and so forth. Furthermore, the requirement that the theory not be known to be true is too stringent. A set of propositions whose truth was once not known, but now is, might legitimately be classified as a theory.

Nor is Bromberger's proposal sufficient for a theory. Suppose *B* in 1965 raises the question, "Who will be the Republican Presidential nominee in 1968?"; and suppose that he happens to be in a *p*-predicament with respect to this question because he can think of no answer to which, on his views, there are no decisive objections. *A*, who hears *B*'s question but does not know of his *p*-predicament, proposes the answer "Charles Percy," someone unknown to *B* but not decisively precluded by his views (and where it is not known whether this answer is true or false). According to Bromberger's definition, in proposing this answer, *A* would be proposing a theory. However, if my conception of a theory is correct, then this need not be so. *A* may not be serious in his answer: even though he may not know whether it is true or false he may believe it to be false. Also, not knowing of *B*'s *p*-predicament, *A* did not propose the answer for the purpose of explaining, removing puzzles, and so forth; nor did he think of this as potentially one of the main functions of his answer.

Indeed, if Bromberger's conditions are taken to be sufficient for being a theory, then almost any known proposition not known to be true or to be false is a theory. According to his proposal, for *T* to be a theory, it is sufficient that *T* be a known proposition not known to be true or to be false and that a question *Q* be formulatable with regard to which it is in principle possible for someone to be in a *p*-predicament, where *T* provides an answer to *Q* (or one among several such answers); it is also sufficient that *T* be the contrary of some presupposition of *Q*, where *T* is not known to be true or to be false. Suppose I say, "The cat is on the mat," where this is not known to be true or to be false. A question can now be formulated ("Where is the cat?") such that it is in principle possible to describe someone as being in a *p*-predicament with regard to that question. (For example, such a person believes that there is no mat on which the cat could be and, having looked everywhere else without success, can think of no place the cat could be.) Since my original proposition ("The cat is on the mat") provides an answer to the question "Where is the cat?" with regard to which it is in principle possible for someone to be in a *p*-predicament, and since we are supposing that this proposition is not known to be true or to be false, it is a theory on Bromberger's account. This is not to deny that it could be a theory in some contexts (provided other conditions of the sort mentioned in the previous section are satisfied). The point is simply that, on Bromberger's account, in any context, and no matter what other conditions obtain, if this statement is made and is not known to be true or to be false, it is classifiable as a theory.

Should Bromberger not agree that the question "Where is the cat?" could, even in principle, generate a *p*-predicament, we can choose another question, namely, "How come the cat is on the mat?" Surely it is in principle possible for someone (who, for example, believes he has just put the cat on the mat and now believes the cat is no longer there) to be in a *p*-predicament with regard to this question. But the proposition "the cat is on the mat" is a contrary of a presupposition of the question just formulated; so, on Bromberger's definition, it is a theory (if it is not known to be true or to be false).

Suppose, then, we try to strengthen Bromberger's requirements as follows. We require not that in principle it be possible to formulate a question to which the theory is an answer (or is the contrary of a presupposition of the question), but that such a question actually have been formulated by the proponent of the theory or by others and that the theory be proposed to answer this question, among others. Furthermore, let us require not that in principle it be possible for someone to be in a *p*-predicament with respect to that question, but that some people actually be in that *p*-predicament. Such conditions would obviously not be satisfied by virtually every known proposition not known to be true or to be false.

However, these requirements would be much too stringent. They would not be necessary for a theory. Suppose that in answer to the question "How was the solar system formed?" a group of leading astronomers pondering the latest evidence unanimously agree on three possibilities no one of which is definitely precluded by their views. Then the Astronomer Royal arrives, carefully examines all the evidence, and proposes a fourth quite different answer to this question. I think it is perfectly clear that the Astronomer Royal might have proposed a theory even though no one is in a *p*-predicament with respect to the question "How was the solar system formed?" and even though he knows that no one is in such a predicament. In general, a person *A* may propose a theory *T* in answer to a question *Q* even though those for whom he is proposing this theory can generate a number of possible answers none of which is precluded by their views and even though *A* knows this to be so.

NOTES

1. Daniel Bernoulli, "On the Properties and Motions of Elastic Fluids, Especially Air," *Kinetic Theory,* ed. S. G. Brush (Oxford, 1965), pp. 57–65.
2. For an illuminating discussion, see N. R. Hanson, *Patterns of Discovery* (Cambridge, England, 1958). Chap. 4.
3. *Ibid.,* p. 86.
4. In the discussion that follows I am indebted to Stephen Barker.
5. See Carl G. Hempel and Paul Oppenheim, "Studies in the Logic of Explanation," *Philosophy of Science,* 15 (1948), 135–75.

6. See Chap. Three, sec. 4, where this is explained.
7. "Explanation, Reduction, and Empiricism," *Minnesota Studies in the Philosophy of Science,* ed. H. Feigl and G. Maxwell (Minneapolis. 1962), III, 28–97. See also Thomas Kuhn, *The Structure of Scientific Revolutions* (Chicago, 1962).
8. 'Problems of Empiricism," *Beyond the Edge of Certainty,* ed. R. G. Colodny (Englewood Clilffs, N.J., 1965), p. 219, note 3.
9. *Patterns of Discovery,* Chap. 4.
10. *Ibid.,* p. 72.
11. *Ibid.,* p. 121.
12. This bear example is used by Hanson himself in illustrating what he means by "seeing the pattern" in the phenomena. *Ibid.,* p. 12.
13. Sylvain Bromberger, "A Theory about the Theory of Theory and about the Theory of Theories," *Philosophy of Sciences The Delaware Seminar,* ed. Bernard Baumrin (New York, 1963), II.
14. Bromberger, "An Approach to Explanation," *Analytical Philosophy,* ed. R. J. Butler (Oxford, 1965), II, 82.
15. "A Theory about the Theory of Theory," p. 102.

SUGGESTIONS FOR FURTHER READING

Ackerman, R. *The Philosophy of Science*. New York: Western Publishing Co., Inc., 1970, pp. 64–70, 87–100 and 103–114.

Baumrin, B. (ed.) *Philosophy of Science, The Delaware Seminar*: Vol. 2 *1962-1963*. New York: Interscience Publishers, 1963, pp. 3–138.

Brodbeck, M. (ed.) *Readings in the Philosophy of the Social Sciences*. New York: The Macmillan Co., 1968, pp. 457–572.

Brody, B.A. (ed.) *Readings in the Philosophy of Science*. Englewood Cliffs, N.J.: Prentice-Hall, Inc., 1970, pp. 190–222, 252–293.

Feigl, H. and Brodbeck, M. (eds.) *Readings in the Philosophy of Science*. New York: Appleton-Century-Crofts, Inc., 1953, pp. 235–308.

Kaplan, A. *The Conduct of Inquiry*. San Francisco: Chandler Publishing Co., 1964, pp. 294–326.

Krimerman, L.I. (ed.) *The Nature and Scope of Social Science*. New York: Appleton-Century-Crofts, 1969, pp. 205–350.

Nagel, E. *The Structure of Science*. New York: Harcourt, Brace and World, Inc., 1961, pp. 79–152.

Smart, J.J.C. *Between Science and Philosophy*. New York: Random House, Inc., 1968, pp. 121–174.

Toulmin, S. *The Philosophy of Science*. London: Hutchinson University Library, 1953, pp. 105–139.

SECTION VI

Observation and the Meaning of Scientific Terms

The first article in this section presents several proposed criteria of empirical meaningfulness and attempts to elucidate their strengths and weaknesses. Philosophers who have been called "logical empiricists" or "logical positivists" have generally held the view that meaningful assertions may be divided into two sorts, namely, those that are cognitively meaningful and those that are emotively meaningful. The latter are supposed to be emotively appealing, sentimentally suggestive and poetically pleasing, if you like, but basically *mush,* i.e., they really don't assert anything, so strictly speaking they can't be true or false. For example, the sentences "Martin Luther King, Jr. was a good man," "The Columbia Ice Fields are beautiful" and "Archie Bunker is an ass" are allegedly only emotively meaningful. They are expressions of feelings, not truth-claims.

Cognitively meaningful claims are supposed to be either logically or else empirically true or false. If they are logically true or false then they are said to be analytic or self-contradictory, respectively. For example, "A horse is a horse" is analytic, and "A horse is not a horse" is self-contradictory. The question that remains, then, is simply this: How should one characterize empirically meaningful sentences, or, more precisely, what criterion should be used to distinguish empirically meaningful sentences (or assertions, statements or propositions) from those are not empirically meaningful? Since many, if not all and only, scientific claims must be empirical = about the world in which we live, the construction of a criterion of empirical meaningfulness is of fundamental importance for the philosophy of science and technology. Unfortunately, as Hempel shows, so far (and things haven't changed much on this score since 1950, when his essay first appeared) no one has been able to produce a viable criterion.

The second article criticizes the arguments of those who hold that "what is perceived depends upon what is believed," and defends the roughly contrary view that "scientific facts are neutral." Kordig takes great pains to lay out the bare bones of the arguments he attacks, and to appreciate his strategy, the reader must match his pains with careful study. The idea behind the view of those whom Kordig opposes is perfectly straightforward. If, for example, you and I have different beliefs about what this figure is, we will "see" different things.

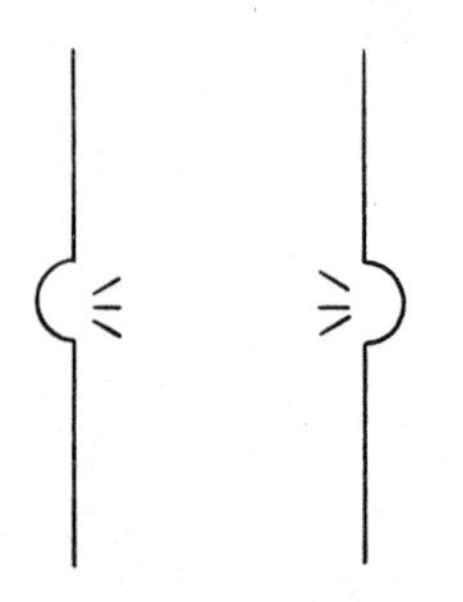

What is that? A man with two warts on his arm? A blown-up picture of Popeye's knee? A bear hanging on to a tree from the other side? It seems that what we "see" depends on the conceptual net we throw on what's "out there." (When I asked my three-year-old daughter what that was, she answered, "Two lines with bumps.") But Kordig tries to show that this line of reasoning is too fast, and if he is right, the idea of more or less independent, neutral facts in the world obtains a new lease on life.

READING 14

Problems and Changes in the Empiricist Criterion of Meaning

Carl G. Hempel

1. INTRODUCTION

The fundamental tenet of modern empiricism is the view that all non-analytic knowledge is based on experience. Let us call this thesis the principle of empiricism.[1] Contemporary logical empiricism has added [2] to it the maxim that a sentence makes a cognitively meaningful assertion, and thus can be said to be either true or false, only if it is either (1) analytic or self-contradictory or (2) capable, at least in principle, of experimental test. According to this so-called *empiricist criterion of cognitive meaning, or of cognitive significance,* many of the formulations of traditional metaphysics and large parts of epistemology are devoid of cognitive significance—however rich some of them may be in non-cognitive import by virtue of their emotive appeal or the moral inspiration they offer. Similarly certain doctrines which have been, at one time or another, formulated within empirical science or its border disciplines are so contrived as to be incapable of test by any conceivable evidence; they are therefor qualified as pseudo-hypotheses, which assert nothing, and which therefore have no explanatory or predictive force whatever. This verdict applies, for example, to the neo-vitalist speculations about entelechies or vital forces, and to the "telefinalist hypothesis" propounded by Lecomte du Noüy.[3]

The preceding formulations of the principle of empiricism and of the empiricist meaning criterion provide no more, however, than a general and rather vague characterization of a basic point of view,

Reprinted by permission of the author and *Revue internationale de philosophie,* Bruxelles, *4,* 1950, pp. 41–63.

and they need therefore to be elucidated and amplified. And while in the earlier phases of its development, logical empiricism was to a large extent preoccupied with a critique of philosophic and scientific formulations by means of those fundamental principles, there has been in recent years an increasing concern with the positive tasks of analyzing in detail the logic and methodology of empirical science and of clarifying and restating the basic ideas of empiricism in the light of the insights thus obtained. In the present article, I propose to discuss some of the problems this search has raised and some of the results it seems to have established.

2. CHANGES IN THE TESTABILITY CRITERION OF EMPIRICAL MEANING

As our formulation shows, the empiricist meaning criterion lays down the requirement of experiential testability for those among the cognitively meaningful sentences which are neither analytic nor contradictory; let us call them sentences with empirical meaning, or empirical significance. The concept of testability, which is to render precise the vague notion of being based—or rather baseable—on experience, has undergone several modifications which reflect an increasingly refined analysis of the structure of empirical knowledge. In the present section, let us examine the major stages of this development.

For convenience of exposition, we first introduce three auxiliary concepts, namely those of observable characteristic, of observation predicate, and of observation sentence. A property or a relation of physical objects will be called an *observable characteristic* if, under suitable circumstances, its presence or absence in a given instance can be ascertained though direct observation. Thus, the terms "green," "soft," liquid," "longer than," designate observable characteristics, while "bivalent," "radioactive," "better electric conductor," and "introvert" do not. Terms which designate observable characteristics will be called *observation predicates.* Finally, by an *observation sentence* we shall understand any sentence which—correctly or incorrectly—asserts of one or more specifically named objects that they have, or that they lack, some specified observable characteristic. The following sentences, for example, meet this condition: "The Eiffel Tower is taller than the buildings in its vicinity," "The pointer of this instrument does not cover the point marked '3' on the scale," and even, "The largest dinosaur on exhibit in New York's Museum of Natural History had a blue tongue"; for this last sentence assigns to a specified object a characteristic—having a blue tongue—which is of such a kind that under suitable circumstances (e.g., in the case of my Chow dog) its presence or absence can be ascertained by direct observation. Our concept of observation sentence is intended to provide a precise interpretation of the vague idea of a sentence asserting something that is "in principle" ascertainable by direct observation, even though it may happen to be actually incapable of being observed by myself, perhaps also by my contemporaries, and possibly even by any human being who ever lived or will live. Any evidence that might be adduced in the test of an empirical hypothesis may now be thought of as being expressed in observation sentences of this kind.[4]

We now turn to the changes in the conception of testability, and thus of empirical meaning. In the early days of the Vienna Circle, a sentence was said to have empirical meaning if it was capable, at least in principle, of complete verification by observational evidence; i.e., if observational evidence could be described which, if actually obtained, would conclusively establish the truth of the sentence.[5] With the help of the concept of observation sentence, we can restate this requirement as follows: A sentence S has empirical meaning if and only if it is possible to indicate a finite set of observation sentences, O_1, O_2, . . . , O_n, such that if these are true, then S is necessarily true, too. As stated, however, this condition is satisfied also if S is an analytic sentence or if the given observation sentences are logically incompatible with each other. By the following formulation, we rule these cases out and at the same time express the intended criterion more precisely:

(2.1) *Requirement of complete verifiability in principle:* A sentence has empirical meaning if and only if it is not analytic and follows logically from some finite and logically consistent class of observation sentences.[6]

This criterion, however, has several serious defects. The first of those here to be mentioned has been pointed out by various writers:

(*a*) The verifiability requirement rules out all sentences of universal form and thus all statements purporting to express general laws; for these cannot be conclusively verified by any finite set of observational data. And since sentences of this type constitute an integral part of scientific theories, the verifiability requirement must be regarded as overly restrictive in this respect. Similarly, the criterion disqualifies all sentences such as "For any substance there exists some solvent," which contain both universal and existential quantifiers (i.e., occurrences of the terms "all" and "some" or their equivalents); for no sentences of this kind can be logically deduced from any finite set of observation sentences.

Two further defects of the verifiability requirement do not seem to have been widely noticed:

(*b*) Suppose that S is a sentence which satisfies the proposed criterion, whereas N is a sentence such as "The absolute is perfect," to which the criterion attributes no empirical meaning. Then the alternation SvN (i.e., the expression obtained by connecting the two sentences by the word "or"), likewise satisfies the criterion; for if S is a consequence of some finite class of observation sentences, then trivially SvN is a consequence of the same class. But clearly, the empiricist criterion of meaning is not intended to countenance sentences of this sort. In this respect, therefore, the requirement of complete verifiability is too inclusive.

(*c*) Let "P" be an observation predicate. Then the purely existential sentence "(Ex)P(x)" ("There exists at least one thing that has the property P") is completely verifiable, for it follows from any observation sentence asserting of some particular object that it has the property P. But its denial, being equivalent to the universal sentence "(x) $\sim$ P(x)" ("Nothing has the property P") is clearly not completely verifiable, as follows from comment (*a*) above. Hence, under the criterion (2.1), the denials of certain empirically—and thus cognitively—significant sentences are empirically meaningless; and as they are neither analytic nor contradictory, they are cognitively meaningless. But however we may delimit the domain of significant discourse, we shall have to insist that if a sentence falls within that domain, then so must its denial. To put the matter more explicitly: The sentences to be qualified as cognitively meaningful are precisely those which can be significantly said to be either true or false. But then, adherence to (2.1) would engender a serious dilemma, as is shown by the consequence just mentioned. We would either have to give up the fundamental logical principle that if a sentence is true or false, then its denial is false or true, respectively (and thus cognitively significant); or else, we must deny, in a manner reminiscent of the intuitionistic conception of logic and mathematics, that "(x) $\sim$ P(x)" is logically equivalent to the negation of "(Ex) P (x)." Clearly, the criterion (2.1), which has disqualified itself on several other counts, does not warrant such drastic measures for its preservation; hence, it has to be abandoned.[7]

Strictly analogous considerations apply to an alternative criterion, which makes complete falsifiability in principle the defining characteristic of empirical significance. Let us formulate this criterion as follows: A sentence has empirical meaning if and only if it is capable, in principle, of complete refutation by a finite number of observational data; or, more precisely:

(2.2) *Requirement of complete falsifiability in principle:* A sentence has empirical meaning if and only if its denial is not analytic and follows logically from some finite logically consistent class of observation sentences.[8]

This criterion qualifies a sentence as empirically meaningful if its denial satisfies the requirement of complete verifiability; as is to be expected, it is herefore inadequate on similar grounds as the latter:

(*a*) It rules out purely existential hypotheses, such as "There exists at least one unicorn," and all sentences whose formulation calls for mixed—i.e., universal and existential—quantification; for none of these can possibly be conclusively falsified by a finite number of observation sentences.

(*b*) If a sentence S is completely falsifiable whereas N is a sentence which is not, then their conjunction, S.N. (i.e., the expression obtained by connecting the two sentences by the word "and") is completely falsifiable; for if the denial of S is entailed by some class of observation sentences, then the denial of S.N. is, *a fortiori*, entailed by the same class. Thus, the criterion allows empirical significance to many sentences which an adequate empiricist criterion should rule out, such as, say "All swans are white and the absolute is perfect."

(*c*) If "P" is an observation predicate, then the assertion that all things have the property P is qualified as significant, but its denial, being equivalent to a purely existential hypothesis, is disqualified (cf. (a)). Hence, criterion (2.2) gives rise to the same dilemma as (2.1).

In sum, then, interpretations of the testability criterion in terms of complete verifiabilty or of complete falsifiability are inadequate because they are overly restrictive in one direction and overly inclusive in another, and because both of them require incisive changes in the fundamental principles of logic.

Several attempts have been made to avoid these difficulties by construing the testability criterion as demanding merely a partial and possibly indirect confirmability of empirical hypothesis by observational evidence.

(2.3) A formulation suggested by Ayer[9] is characteristic of these attempts to set up a clear and sufficiently comprehensive criterion of confirmability. It states, in effect, that a sentence S has empirical import if from S in conjunction with suitable subsidiary hypotheses it is possible to derive observation sentences which are not derivable from the subsidiary hypotheses alone.

This condition is suggested by a closer consideration of the logical structure of scientific testing; but it is much too liberal as it stands. Indeed, as Ayer himself has pointed out in the second edition of his book, *Language, Truth and Logic,*[10] his criterion allows empirical import to any sentence whatever. Thus, e.g., if S is the sentence "The absolute is perfect," it suffices to choose as a subsidiary hypothesis the sentence "If the absolute is perfect then this apple is red" in order to make possible the deduction of the observation sentence "This apple is red," which clearly does not follow from the subsidiary hypothesis alone.[11]

(2.4) To meet this objection, Ayer has recently proposed a modified version of his testability criterion. The modification restricts, in effect, the subsidiary hypothesis mentioned in (2.3) to sentences which are either analytic or can independently be shown to be testable in the sense of the modified criterion.[12]

But it can readily be shown that this new criterion, like the requirement of complete falsifiability, allows empirical significance to any conjunction S.N., where S satisfies Ayer's criterion while N is a sentence such as "The absolute is perfect," which is to be disqualified by the criterion. Indeed, whatever consequences can be deduced from S with the help of permissible subsidiary hypotheses can also be deduced from S.N. by means of the same subsidiary hypotheses, and as Ayer's new criterion is formulated essentially in terms of the deducibility of a certain type of consequence from the given sentence, it countenances S.N. together with S. Another difficulty has been pointed out by Professor A. Church, who has shown [13] that if there are any three observation sentences none of which alone entails any of the others, then it follows for any sentence S whatsoever that either it or its denial has empirical import according to Ayer's revised criterion.

3. TRANSLATABILITY INTO AN EMPIRICIST LANGUAGE AS A NEW CRITERION OF COGNITIVE MEANING

I think it is useless to continue the search for an adequate criterion of testability in terms of deductive relationships to observation sentences. The past development of this research—of which we have considered the major stages—seems to warrant the expectation that as long as we try to set up a criterion of testability for individual sentences in a natural language, in terms of logical relationship to observation sentences, the result will be either too restrictive or too inclusive, or both. In particular it appears likely that such criteria would allow empirical import, in the manner of (2.1)(*b*) or of (2.2)(*b*), either to any alternation or to any conjunction of two sentences of which at least one is qualified as empirically meaningful; and this peculiarity has undesirable consequences because the liberal grammatical rules of English as of any other natural language countenance as sentences certain expressions ("The absolute is perfect" was our illustration) which even by the most liberal empiricist standards make no assertion whatever; and these would then have to be permitted as components of empirically significant statements.

The predicament would not arise, of course, in an artificial language whose vocabulary and grammar were so chosen as to preclude altogether the possibility of forming sentences of any kind which the empiricist meaning criterion is intended to rule out. Let us call any such language an *empiricist language.* This reflection suggests an entirely different approach to our problem: Give a general characterization of the kind of language that would qualify as empiricist, and then lay down the following:

(3.1) *Translatability criterion of cognitive meaning:* A sentence has cognitive meaning if and only if it is translatable into an empiricist language.

This conception of cognitive import, while perhaps not explicitly stated, seems to underlie much of the more recent work done by empiricist writers; as far as I can see it has its origin in Carnap's essay, *Testability and Meaning* (especially part IV).

As any language, so also any empiricist language can be characterized by indicating its vocabulary and the rules determining its logic; the latter include the syntactical rules according to which sentences may be formed by means of the given vocabulary. In effect, therefore, the translatability criterion proposes to characterize the cognitively meaningful sentences by the vocabulary out of which they may be constructed, and by the syntactical principles governing their construction. What sentences are singled out as cognitively significant will depend, accordingly, on the choice of the vocabulary and of the construction rules. Let us consider a specific possibility:

(3.2) We might qualify a language L as empiricist if it satisfies the following conditions:

(a) *The vocabulary of L* contains:

(1) The customary locutions of logic which are used in the formulation of sentences; including in particular the expressions "not," "and," "or," "if . . . then . . . ," "all," "some," "the class of all things such that . . . ," ". . . is an element of class . . .";

(2) Certain *observation predicates.* These will be said to constitute the basic empirical vocabulary of L;

(3) Any expression definable by means of those referred to under (1) and (2).

(b) *The rules of sentence formation for L* are those laid down in some contemporary logical system such as *Principia Mathematica.*

Since all defined terms can be eliminated in favor of primitives, these rules stipulate in effect that a language L is empiricist if all its sentences are expressible, with the help of the usual logical locutions, in terms of observable characteristics of physical objects. Let us call any language of this sort a thing-language in the narrower sense. Alternatively, the basic empirical vocabulary of an empiricist language might be construed as consisting of phenomenalistic terms, each of them referring to some aspect of the phenomena of perception or sensation. The construction of adequate phenomenalistic languages, however, presents considerable difficulties,[14] and in recent empiricism, attention has been focussed primarily on the potentialities of languages whose basic empirical vocabulary consists of observation predicates; for the latter lend themselves more directly to the description of that type of intersubjective evidence which is invoked in the test of scientific hypotheses.

If we construe empiricist languages in the sense of (3.2), then the translatability criterion (3.1) avoids all the shortcomings pointed out in our discussion of earlier forms of the testability criterion:

(*a*) Our characterization of empiricist languages makes explicit provision for universal and existential quantification, i.e., for the use of the terms "all" and "some"; hence, no type of quantified statement is generally excluded from the realm of cognitively significant discourse;

(*b*) Sentences such as "The absolute is perfect" cannot be formulated in an empiricist language (cf. (*d*) below); hence there is no danger that a conjunction or alternation containing a sentence of that kind as a component might be qualified as cognitively significant;

(*c*) In a language L with syntactical rules conforming to *Principia Mathematica,* the denial of a sentence is always again a sentence of L. Hence, the translatability criterion does not lead to the consequence, which is entailed by both (2.1) and (2.2), that the denials of certain significant sentences are non-significant.

(*d*) Despite its comprehensiveness, the new criterion does not attribute cognitive meaning to *all* sentences; thus, e.g., the sentences "The absolute is perfect" and "Nothingness nothings" cannot be translated into an empiricist language because their key terms are not definable by means of purely logical expressions and observation terms.

4. THE PROBLEM OF DISPOSITION TERMS AND OF THEORETICAL CONSTRUCTS

Yet, the new criterion is still too restrictive—as are, incidentally, also its predecessors—in an important respect which now calls for consideration. If empiricist languages are defined in accordance with (3.2), then, as was noted above, the translatability criterion (3.1) allows cognitive import to a sentence only if its constitutive empirical terms are explicitly definable by means of observation predicates. But as we shall argue presently, many terms even of the physical sciences are not so definable; hence the criterion would oblige us to reject, as devoid of cognitive import, all scientific hypotheses containing such terms—an altogether intolerable consequence.

The concept of temperature is a case in point. At first glance, it seems as though the phrase "Object x has a temperature of c degrees centigrade," or briefly "$T(x) = c$" could be defined by the following sentence, (D): $T(x) = c$ if and only if the following condition is satisfied: If a thermometer is in contact with x, then it registers c degrees on its scale.

Disregarding niceties, it may be granted that the definiens given here is formulated entirely in reference to observables. However, it has one highly questionable aspect. In *Principia Mathematica* and similar systems, the phrase "if p then q" is construed as being synonymous with "not p or q"; and under this so-called material interpretation of the conditional, a statement of the form "if p then q" is obviously true if (though not only if) the sentence standing in the place of "p" is false. If, therefore, the meaning of "if . . . then . . ." in the definiens of (D) is understood in the material sense, then that definiens is true if (though not only if) x is an object not in contact with a thermometer—no matter what numerical value we may give to c. And since the definiendum would be true under the same circumstances, the definition (D) would qualify as true the assignment of any temperature value whatsoever to any object not in contact with a thermometer! Analogous considerations apply to such terms as "electrically charged," "magnetic," "intelligent," "electric resistance," etc., in short to all disposition terms, i.e., terms which express the disposition of one or more objects to react in a determinate way under specified circumstances. A definition of such terms by means of observation predicates cannot be effected in the manner of (D), however natural and obvious a mode of definition this may at first seem to be.[15]

There are two main directions in which a resolution of the

difficulty might be sought. On the one hand, it could be argued that the definition of disposition terms in the manner of (D) is perfectly adequate provided that the phrase "if . . . then . . ." in the definiens is construed in the sense it is obviously intended to have, namely as implying, in the case of (D), that even if x is not actually in contact with a thermometer, still if it *were* in such contact, then the thermometer *would* register c degrees. In sentences such as this, the phrase "if . . . then . . ." is said to be used counterfactually; and it is in this "strong" sense, which implies a counterfactual conditional, that the definiens of (D) would have to be construed. This suggestion would provide an answer to the problem of defining disposition terms if it were not for the fact that no entirely satisfactory account of the exact meaning of counterfactual conditionals seems to be available at present. Thus, the first way out of the difficulty has the status of a program rather than that of a solution. The lack of an adequate theory of counterfactual conditionals is all the more deplorable as such a theory is needed also for the analysis of the concept of general law in empirical science and of certain related ideas. A clarification of this cluster of problems constitutes at present one of the urgent desiderata in the logic and methodology of science.[16]

An alternative way of dealing with the definitional problems raised by disposition terms was suggested, and developed in detail, by Carnap. It consists in permitting the introduction of new terms, within an empiricist language, by means of so-called reduction sentences, which have the character of partial or conditional definitions.[17] Thus, e.g., the concept of temperature in our last illustration might be introduced by means of the following reduction sentence, (R): If a thermometer is in contact with an object x, then $T(x) = c$ if and only if the thermometer registers c degrees.

This rule, in which the conditional may be construed in the material sense, specifies the meaning of "temperature," i.e., of statements of the form "$T(x) = c$," only partially, namely in regard to those objects which are in contact with a thermometer; for all other objects, it simply leaves the meaning of "$T(x) = c$" undetermined. The specification of the meaning of "temperature" may then be gradually extended to cases not covered in (R) by laying down further reduction sentences, which reflect the measurement of temperature by devices other than thermometers.

Reduction sentences thus provide a means for the precise formulation of what is commonly referred to as operational definitions.[18] At the same time, they show that the latter are not definitions in the strict sense of the word, but rather partial specifications of meaning.

The preceding considerations suggest that in our characterization (3.2) of empiricist languages we broaden the provision *a* (3) by permitting in the vocabulary of L all those terms whose meaning can be specified in terms of the basic empirical vocabulary by means of definitions or reduction sentences. Languages satisfying this more inclusive criterion will be referred to as thing-languages in the wider sense.

If the concept of empiricist language is broadened in this manner, then the translatability criterion (3.1) covers—as is should—also all those statements whose constituent empirical terms include "empirical constructs," i.e., terms which do not designate observables, but which can be introduced by reduction sentences on the basis of observation predicates.

Even in this generalized version, however, our criterion of cognitive meaning may not do justice to advanced scientific theories, which are formulated in terms of "theoretical constructs," such as the terms "absolute temperature," "gravitational potential," "electric field," "ψ function," etc. There are reasons to think that neither definitions nor reduction sentences are adequate to introduce these terms on the basis of observation predicates. Thus, e.g., if a system of reduction sentences for the concept of electric field were available, then—to oversimplify the point a little—it would be possible to describe, in terms of observable characteristics, some necessary and some sufficient conditions for the presence, in a given region, of an electric field of any mathematical description, however complex. Actually, however, such criteria can at best be given only for some sufficiently simple kinds of fields.

Now theories of the advanced type here referred to may be considered as hypothetico-deductive systems in which all statements are logical consequences of a set of fundamental assumptions. Fundamental as well as derived statements in such a system are formulated either in terms of certain theoretical constructs which are not defined within the system and thus play the role of primitives, or in terms of expressions defined by means of the latter. Thus, in their logical structure such systems equal the axiomatized uninterpreted systems studied in mathematics and logic. They acquire applicability to empirical subject matter, and thus the status of theories of empirical science, by virtue of an empirical interpretation. The latter is effected by a translation of some of the sentences of the theory—often derived rather than fundamental ones—into an empiricist language, which may contain both observation predicates and empirical constructs. And since the sentences which are thus given empirical meaning are logical consequences of the fundamental hypotheses of the theory, that translation effects, indirectly, a partial interpretation of the latter and of the constructs in terms of which they are formulated.[19]

In order to make translatability into an empiricist language an adequate criterion of cognitive import, we broaden therefore the concept of empiricist language so as to include things-languages in the narrower and in the wider sense as well as all interpreted theoretical systems of the kind just referred to.[20] With this understanding, (3.1) may finally serve as a general criterion of cognitive meaning.

5. ON "THE MEANING" OF AN EMPIRICAL STATEMENT

In effect, the criterion thus arrived at qualifies a sentence as cognitively meaningful if its non-logical constituents refer, directly or in certain specified indirect ways, to observables, but it does not make any pronouncement on what "the meaning" of a cognitively significant sentence is, and in particular it neither says nor implies that that meaning can be exhaustively characterized by what the totality of possible tests would reveal in terms of observable phenomena. Indeed, *the content of a statement with empirical import cannot, in general, be exhaustively expressed by means of any class of observation sentences.*

For consider first, among the statements permitted by our criterion, any purely existential hypothesis or any statement involving mixed quantification. As was pointed out earlier, under (2.2) (*a*), statements of these kinds entail no observation sentences whatever; hence their content cannot be expressed by means of a class of observation sentences.

And secondly, even most statements of purely universal form (such as "All flamingoes are pink") entail observation sentences (such as "That thing is pink") only when combined with suitable other observation sentences (such as "That thing is a flamingo").

This last remark can be generalized. The use of empirical hypotheses for the prediction of observable phenomena requires, in practically all cases, the use of subsidiary empirical hypotheses.[21] Thus, e.g., the hypothesis that the agent of tuberculosis is rod-shaped does not by itself entail the consequence that upon looking at a tubercular sputum specimen through a microscope, rod-like shapes will be observed: a large number of subsidiary hypotheses, including the theory of the microscope, have to be used as additional premises in deducing that prediction.

Hence, what is sweepingly referred to as "the (cognitive) meaning" of a given scientific hypothesis cannot be adequately characterized in terms of potential observational evidence alone, nor can it be specified for the hypothesis taken in isolation. In order to understand "the meaning" of a hypothesis within an empiricist language, we have to know not merely what observation sentences it entails alone or in conjunction with subsidiary hypotheses, but also what other, non-observational, empirical sentences are entailed by it, what sentences in the given language would confirm or disconfirm it, and for what other hypotheses the given one would be confirmatory or disconfirmatory. In other words, the cognitive meaning of a statement in an empiricist language is reflected in the totality of its logical relationships to all other statements in that language and not to the observation sentences alone. In this sense, the statements of empirical science have a surplus meaning over and above what can be expressed in terms of relevant observation sentences.[22]

6. THE LOGICAL STATUS OF THE EMPIRICIST CRITERION OF MEANING

What kind of a sentence, it has often been asked, is the empiricist meaning criterion itself? Plainly it is not an empirical hypothesis; but it is not analytic or self-contradictory either; hence, when judged by its own standard, is it not devoid of cognitive meaning? In that case, what claim of soundness or validity could possibly be made for it?

One might think of construing the criterion as a definition which indicates what empiricists propose to understand by a cognitively significant sentence; thus understood, it would not have the character of an assertion and would be neither true nor false. But this conception would attribute to the criterion a measure of arbitrariness which cannot be reconciled with the heated controversies it has engendered and even less with the fact, repeatedly illustrated in the present article, that the changes in its specific content have always been determined by the objective of making the criterion a more adequate index of cognitive import. And this very objective illuminates the character of the empiricist criterion of meaning: It is intended to provide a clarification and *explication* of the idea of a sentence which makes an intelligible assertion.[23] This idea is admittedly vague and it is the task of philosophic explication to replace it by a more precise concept. In view of this difference of precision we cannot demand, of course, that the "new" concept, the explicatum, be strictly synonymous with the old one, the explicandum.[24] How, then, are we to judge the adequacy of a proposed explication, as expressed in some specific criterion of cognitive meaning?

First of all there exists a large class of sentences which are rather generally recognized as making intelligible assertions, and another large class of which this is more or less generally denied. We shall have to demand of an adequate explication that it take into account these spheres of common usage; hence an explication which, let us say, denies cognitive import to descriptions of past events or to generalizations expressed in terms of observables has to be rejected as inadequate. As we have seen, this first requirement of adequacy has played an important rôle in the development of the empiricist meaning criterion.

But an adequate explication of the concept of cognitively significant statement must satisfy yet another, even more important, requirement: together with the explication of certain other concepts, such as those of confirmation and of probability, it has to provide the framework for a general theoretical account of the structure and the foundations of scientific knowledge. Explication, as here understood, is not a mere description of the accepted usages of the terms under consideration: it has to go beyond the limitations, ambiguities and inconsistencies of common usage and has to show how we had better construe the meanings of those terms if we wish to arrive at a consistent and comprehensive theory of knowledge. This type of consideration, which has been largely influenced by a study of the structure of scientific theories, has prompted the more recent extensions of the empiricist meaning criterion. These extensions are designed to include in the realm of cognitive significance various types of sentences which might occur in advanced scientific theories, or which have to be admitted simply for the sake of systematic simplicity and uniformity,[25] but on whose cognitive significance or non-significance a study of what the term "intelligible assertion" means in everyday discourse could hardly shed any light at all.

As a consequence, the empiricist criterion of meaning, like the result of any other explication, represents a linguistic proposal which itself is neither true nor false, but for which adequacy is claimed in two respects: first in the sense that the explication provides a reasonably close *analysis* of the commonly accepted meaning of explicandum—and this claim implies an empirical assertion; and secondly in the sense that the explication achieves a *"rational reconstruction"* of the explicandum, i.e., that it provides, together perhaps with other explications, a general conceptual framework which permits a consistent and precise restatement and theoretical systematization of the contexts in which the explicandum is used—and this claim implies at least an assertion of a logical character.

Though a proposal in form, the empiricist criterion of meaning is therefore far from being an arbitrary definition; it is subject to revision if a violation of the requirements of adequacy, or even a way of satisfying those requirements more fully, should be discovered. Indeed, it is to be hoped that before long some of the open problems encountered in the analysis of cognitive significance will be clarified and that then our last version of the empiricist meaning criterion will be replaced by another, more adequate one.

NOTES

1. This term is used by Benjamin (2) in an examination of the foundations of empiricism. For a recent discussion of the basic ideas of empiricism see Russell (27), Part Six.
2. In his stimulating article, "Positivism," W. T. Stace argues, in effect, that the testability criterion of meaning is not logically entailed by the principle of empiricism. (See (29), especially section 11.) This is correct: according to the latter, a sentence expresses knowledge only if it is analytic or corroborated by empirical evidence; the former goes further and identifies the domain of cognitively significant discourse with that of potential knowledge; i.e., it grants cognitive import only to sentences for which—unless they are either analytic or contradictory—a test by empirical evidence is conceivable.
3. Cf. (19), Ch. XVI.
4. Observation sentences of this kind belong to what Carnap has called the thing-language (cf., e.g., (7), pp. 52-53). That they are adequate to formulate the data which serve as the basis for empirical tests is clear in particular for the intersubjective testing procedures used in science as in large areas of empirical inquiry on the common-sense level. In epistemological discussions, it is frequently assumed that the ultimate evidence for beliefs about empirical matters consists in perceptions and sensations whose description calls for a phenomenalistic type of language. The specific problems connected with the phenomenalistic approach cannot be discussed here; but it should be mentioned that at any rate all the critical considerations presented in this article in regard to the testability criterion are applicable, *mutatis mutandis,* to the case of a phenomenalistic basis as well.
5. Originally, the permissible evidence was meant to be restricted to what is observable by the speaker and perhaps his fellow-beings during their lifetimes. Thus construed, the criterion rules out, as cognitively meaningless, all statements about the distant future or the remote past, as has been pointed out, among others, by Ayer in (1), Chapter I; by Pap in (21), Chapter 13, esp. pp. 333 ff.; and by Russell in (27), pp. 445-47. This difficulty is avoided, however, if we permit the evidence to consist of any finite set of "logically possible observation data," each of them formulated in an observation sentence. Thus, e. g., the sentence S_1, "The tongue of the largest dinosaur in New York's Museum of Natural History was blue or black" is completely verifiable in our sense; for it is a logical consequence of the Sentence S_2, "The tongue of the largest dinosaur in New York's Museum of Natural History was blue"; and this is an observation sentence, as has been shown above.

And if the concept of *verifiability in principle* and the more general concept of *confirmability in principle,* which will be considered later, are construed as referring to *logically possible evidence* as expressed by observation sentences, then if follows similarly that the class of statements which are verifiable, or at least confirmable, in principle includes such assertions as that the planet Neptune and the Antarctic Continent existed before they were discovered, and that atomic warfare, if not checked, may lead to the extermination of this planet. The objections which Russell (cf. (27), pp. 445 and 447) raises against the verifiability criterion by reference to those examples do not apply therefore if the criterion is understood in the manner here suggested. Incidentally, statements of the kind mentioned by Russell, which are not actually verifiable by any human being, were explicity recognized as cognitively significant already by Schlick (in (28), Part V), who argued that the impossibility of verifying them was "merely empirical." The characterization of verifiability with the help of the concept of observation sentence as suggested here might serve as a more explicit and rigorous statement of that conception.

6. As has frequently been emphasized in empiricist literature, the term "verifiability" is to indicate, of course, the conceivability, or better, the logical possibility of evidence of an observational kind which, if actually encountered, would constitute conclusive evidence for the given sentence; it is not intended to mean the technical possibility of performing the tests needed to obtain such evidence, and even less does it mean the possibility of actually finding directly observable phenomena which constitute conclusive evidence for that sentence—which would be tantamount to the actual existence of such evidence and would thus imply the truth of the given sentence. Analogous remarks apply to the terms "falsifiability" and "confirmability." This point has been disregarded in some recent critical discussions of the verifiability criterion. Thus, e.g., Russell (cf. (27), p. 448) construes verifiability as the actual existence of a set of conclusively verifying occurrences. This conception, which has never been advocated by any logical empiricist, must naturally turn out to be inadequate since according to it the empirical meaningfulness of a sentence could not be established without gathering empirical evidence, and moreover enough of it to permit a conclusive proof of the sentences in question! It is not surprising, therefore, that his extraordinary interpretation of verifiability leads Russell to the conclusion: "In fact, that a proposition is verifiable is itself not verifiable" *(l. c.)*. Actually, under the empiricist interpretation of complete verifiability, any statement asserting the verifiability of some sentence S whose text is quoted, is either analytic or contradictory; for the decision whether there exists a class of observation sentences which entail S, i.e., whether such observation sentences can be formulated, no matter whether they are true or false—that decision is a matter of pure logic and requires no factual information whatever.

A similar misunderstanding is in evidence in the following passage in which W. H. Werkmeister claims to characterize a view held by logical positivists: "A proposition is said to be 'true' when it is 'verifiable in principle'; i. e., when we know the conditions which, when realized, will make 'verification' possible (cf. Ayer)." (cf. (31), p. 145). The quoted thesis, which, again, was never held by any logical positivist, including Ayer, is in fact logically absurd. For we can readily describe conditions which, if realized, would verify the sentence "The outside of the Chrysler Building is painted a bright yellow"; but similarly, we can describe verifying conditions for its denial; hence, according to the quoted principle, both the sentence and its denial would have to be considered true. Incidentally, the passage under discussion does not accord with Werkmeister's perfectly correct observation, *l. c.*, p. 40, that verifiability is intended to characterize the meaning of a sentence—which shows that verifiability is meant to be a criterion of cognitive significance rather than of truth.

7. The arguments here adduced against the verifiability criterion also prove the inadequacy of a view closely related to it, namely that two sentences have the same cognitive significance if any set of observation sentences which would verify one of them would also verify the other, and conversely. Thus, e.g., under this criterion, any two general laws would have to be assigned the same cognitive significance, for no general law is verified by any set of observation sentences.

The view just referred to must be clearly distinguished from a position which Russell examines in his critical discussion of the positivistic criterion. It is "the theory that two propositions whose verified consequences are identical have the same significance" ((27), p. 448). This view is untenable indeed, for what consequences of a statement have actually been verified at a given time is obviously a matter of historical accident which cannot possibly serve to establish identity of cognitive significance. But I am not aware that any logical positivist ever subscribed to that "theory."

8. The idea of using theoretical falsifiability by observational evidence as the "criterion of demarcation" separating empirical science from mathematics and logic on the one hand and from metaphysics on the other is due to K. Popper (cf. (22), section 1-7 and 19-24; also see (23), vol. II, pp. 282-285). Whether Popper would subscribe to the proposed restatement of the falsifiability criterion, I do not know.

9. (1), Ch. I.—The case against the requirements of verifiability and of falsifiability, and favor of a requirement of partial confirmability and disconfirmability is very clearly presented also by Pap in (21), Chapter 13.

10. (1), 2d ed., pp. 11-12.

11. According to Stace (cf. (29). p. 218), the criterion of partial and indirect testability, which he calls the positivist principle, presupposes (and thus logically entails) another principle, which he terms the *Principle of Observable Kinds*: "A sentence, in order to be significant, must assert or deny facts which are of a kind or class such that it is logically possible directly to observe some facts which are instances of that class or kind. And if a sentence purports to assert or deny facts which are of a class or kind such that it would be logically impossible directly to observe any instance of that class or kind, then the sentence is non-significant." I think the argument Stace offers to prove that this principle is entailed by the requirement of testability is inconclusive (mainly because of the incorrect tacit assumption that "on the transformation view of deduction," the premises of a valid deductive argument must be necessary conditions for the conclusion (*l. c.*, p.225)). Without pressing this point any further, I should like to add here a remark on the principle of observable kinds itself. Professor Stace does not say how we are to determine what "facts" a given sentence asserts or denies, or indeed whether it asserts or denies any "facts" at all. Hence, the exact import of the principle remains unclear. No matter, however, how one might choose the criteria for the factual reference of sentences, this much seems certain: If a sentence expresses any fact at all, say *f*, then it satisfies the requirement laid down in the first sentences of the principle; for we can always form a class containing *f* together with the fact expressed by some observation sentence of our choice, which makes *f* a member of a class of facts at least one of which is capable, in principle, of direct observation. The first part of the principle of observable kinds is therefore all-inclusive, somewhat like Ayer's original formulation of the empiricist meaning criterion.

12. This restriction is expressed in recursive form and involves no vicious circle. For the full statement of Ayer's criterion, see (1), second edition, p. 13.

13. Church (11).

14. Important contributions to the problem have been made by Carnap (5) and by Goodman (15).

15. This difficulty in the definition of disposition terms was first pointed out and analyzed by Carnap (in (6); see esp. section 7).

16. The concept of strict implication as introduced by C. I. Lewis would be of no avail for the interpretation of the strong "if . . . then . . ." as here understood, for it refers to a purely logical relationship of entailment, whereas the concept under consideration will, in general, represent a nomological relationship, i.e., one based on empirical laws. For recent discussions of the problems of counterfactuals and laws, see Langford (18); Lewis (20), pp. 210-230; Chisholm (10); Goodman (14); Reichenbach (26), Chapter VIII; Hempel and Oppenheim (16), Part III; Popper (24).

17. Cf. Carnap (6); a brief elementary exposition of the central idea may be

found in Carnap (7), Part III. The partial definition (R) formulated above for the expression "$T(x) = c$" illustrates only the simplest type of reduction sentence, the so-called bilateral reduction sentence.
18. On the concept of operational definition, which was developed by Bridgman, see, for example, Bridgman (3, 4) and Feigl (12).
19. The distinction between a formal deductive system and the empirical theory resulting from it by an interpretation has been elaborated in detail by Reichenbach in his penetrating studies of the relations between pure and physical geometry; cf., e. g., Reichenbach (25). The method by means of which a formal system is given empirical content is characterized by Reichenbach as "coordinating definition" of the primitives in the theory by means of specific empirical concepts. As is suggested by our discussion of reduction and the interpretation of theoretical constructs, however, the process in question may have to be construed as a partial interpretation of the non-logical terms of the system rather than as a complete definition of the latter in terms of the concepts of a thing-language.
20. These systems have not been characterized here as fully and as precisely as would be desirable. Indeed, the exact character of the empirical interpretation of theoretical constructs and of the theories in which they function is in need of further investigation. Some problems which arise in this connection—such as whether, or in what sense, theoretical constructs may be said to denote—are obviously also of considerable epistemological interest. Some suggestions as to the interpretation of theoretical constructs may be found in Carnap (8), section 24, and in Kaplan (17); for an excellent discussion of the epistemological aspects of the problem, see Feigl (13).
21. This point is clearly taken into consideration in Ayer's criteria of cognitive significance, which were discussed in section 2.
22. For a fuller discussion of the issues here involved cf. Feigl (13) and the comments on Feigl's position which will be published together with that article.
23. In the preface to the second edition of his book, Ayer takes a very similar position: he holds that the testability criterion is a definition which, however, is not entirely arbitrary, because a sentence which did not satisfy the criterion "would not be capable of being understood in the sense in which either scientific hypotheses or commonsense statements are habitually understood" ((1), p. 16).
24. Cf. Carnap's characterization of explication in his article (9), which examines in outline the explication of the concept of probability. The Frege-Russell definition of integers as classes of equivalent classes, and the semantical definition of truth—cf. Tarski (30)—are outstanding examples of explication. For a lucid discussion of various aspects of logical analysis see Pap (21), Chapter 17.
25. Thus, e. g., our criterion qualifies as significant certain statements containing, say, thousands of existential or universal quantifiers—even though such sentences may never occur in everyday nor perhaps even in scientific discourse. For indeed, from a systematic point of view it would be arbitrary and unjustifiable to limit the class of significant statements to those containing no more than some fixed number of quantifiers. For further discussion of this point, cf. Carnap (6), sections 17, 24, 25.

SUGGESTIONS FOR FURTHER READING

1. Ayer, A. J., *Language, Truth and Logic,* Gollancz, London, 1936; 2nd ed., 1946.
2. Benjamin, A. C., "Is Empiricism Self-refuting?" (*Journal of Philos.,* Vol. 38, 1941).

3. Bridgman, P. W., *The Logic of Modern Physics,* The Macmillan Co., New York, 1927.
4. Bridgman, P. W., "Operational Analysis" (*Philos. of Science,* Vol. 5, 1938).
5. Carnap, R., *Der logische Aufbau der Welt,* Berlin, 1928.
6. Carnap, R., "Testability and Meaning" (*Philos. of Science,* Vol. 3, 1936, and Vol. 4, 1937).
7. Carnap, R., *Logical Foundations of the Unity of Science,* In: *Internat. Encyclopedia of Unified Science,* I, 1; Univ. of Chicago Press, 1938.
8. Carnap, R., *Foundations of Logic and Mathematics. Internat. Encyclopedia of Unified Science,* I, 3; Univ. of Chicago Press, 1939.
9. Carnap, R., "The Two Concepts of Probability" (*Philos. and Phenom. Research,* Vol. 5, 1945).
10. Chisholm, R. M., "The Contrary-to-Fact Conditional" (*Mind,* Vol. 55, 1946).
11. Church, A., Review of (1), 2nd. ed. (*The Journal of Symb. Logic,* Vol. 14, 1949, pp. 52–53).
12. Feigl, H., "Operationism and Scientific Method" (*Psychol. Review,* Vol. 52, 1945). (Also reprinted in Feigl and Sellars, *Readings in Philosophical Analysis,* New York, 1949.)
13. Feigl, H., "Existential Hypotheses; Realistic vs. Phenomenalistic Interpretations," (*Philos. of Science,* Vol. 17, 1950).
14. Goodman, N., "The Problem of Counterfactual Conditionals" (*Journal of Philos.,* Vol. 44, 1947).
15. Goodman, N., *The Structure of Appearance.* Harvard University Press, 1951.
16. Hempel, C. G., and Oppenheim, P., "Studies in the Logic of Explanation" (*Philos. of Science,* Vol. 15, 1948).
17. Kaplan, A., "Definition and Specification of Meaning" (*Journal of Philos.,* Vol. 43, 1946).
18. Langford, C. H., Review in *The Journal of Symb. Logic,* Vol. 6 (1941), pp. 67–68.
19. Lecomte du Noüy, *Human Destiny,* New York, London, Toronto, 1947.
20. Lewis, C. I., *An Analysis of Knowledge and Valuation,* Open Court Publ., La Salle, Ill., 1946.
21. Pap, A., *Elements of Analytic Philosophy,* The Macmillan Co., New York, 1949.
22. Popper, K., *Logik der Forschung,* Springer, Vienna, 1935.
23. Popper, K., *The Open Society and Its Enemies,* 2 vols., Routleldge, London, 1945.
24. Popper, K., "A Note on Natural Laws and So-called 'Contrary-to-Fact Conditionals' " (*Mind,* Vol. 58, 1949).
25. Reichenbach, H., *Philosophie der Raum-Zeit-Lehre,* Berlin, 1928.
26. Reichenbach, H., *Elements of Symbolic Logic,* The Macmillan Co., New York, 1947.
27. Russell, B., *Human Knowledge,* Simon and Schuster, New York, 1948.
28. Schlick, M., "Meaning and Verification" (*Philos. Review,* Vol. 45, 1936). (Also reprinted in Feigl and Sellars, *Readings in Philosophical Analysis,* New York 1949).
29. Stace, W. T. "Positivism" (*Mind,* Vol. 53, 1944).
30. Tarski, A., "The Semantic Conception of Truth and the Foundations of Semantics" (*Philos. and Phenom. Research,* Vol. 4, 1944). (Also

reprinted in Feigl and Sellars, *Readings in Philosophical Analysis,* New York, 1949.)
31. Werkmeister, W. H., *The Basis and Structure of Knowledge,* Harper, New York and London, 1948.
32. Whitehead, A. N., and Russell, B., *Principia Mathematica,* 3 vols., 2nd ed., Cambridge 1925-1927.

READING 15

The Theory-Ladenness of Observation

Carl R. Kordig

Revolutionary new views concerning science have recently been advanced by Feyerabend, Hanson, Kuhn, Toulmin, and others. The claim that there are pervasive presuppositions fundamental to scientific investigations seems to be essential to the views of these men. Each would further hold that transitions from one scientific tradition to another force radical changes in what is observed within each tradition. I wish here to discuss and evaluate these views.[1]

I

Feyerabend claims that what is perceived depends upon what is believed ([5], pp. 220–221); and he maintains that among really efficient alternative theories (for the purpose of mutual criticism) "each theory will possess its own experience, and there will be no overlap between these experiences" ([5], p. 214). According to Feyerabend "scientific theories are ways of looking at the world; and their adoption affects our general beliefs and expectations, and thereby also our experiences . . ." ([3], p. 29). Toulmin, Hanson, and Kuhn concur with this view. Toulmin claims that men who accept different "ideals" and "paradigms" will see different phenomena. He thinks theories not only give significance to facts, but also determine what facts are for us at all ([21], pp. 57, 95). Like Feyerabend, Toulmin asserts that "we see the world through" our fundamental concepts of science (e.g., inertial motion) "to such an extent that we forget what it would look like without them" ([21], p. 101). Indeed, both Feyerabend and Toulmin would have previous thinkers "live in an observational world very different from our own" ([5], p. 221; cf. also, [21], p. 103). Kuhn expresses quite similar views. He feels that,

> ...during revolutions scientists see new and different things when looking with familiar instruments in places they have looked before. It is rather as if the professional community had been suddenly transported to another planet [for] ... paradigm changes do cause scientists to see the world of their research engagement differently. In so far as their only recourse to that world is through what they see and do, we may want to say that after a revolution scientists are responding to a different world. ([13], p. 110)

This "world of their research-engagement" is similar to the objects Hanson claims scientists see in the sense of 'see' relevant to science. For Kuhn, quite literally, different "paradigms" transform observations and experience ([13], pp. 6, 110–111). "Paradigms determine large areas of experience at the same time" ([13], p. 128). He believes that after a revolution scientists work in a different world ([13], pp. 110, 117, 119, 134, 149). And one thing he means by this is that the data themselves change ([13], p. 134). This is quite different from a more traditional view. Many philosophers of science would surely want to say that what alters when a "paradigm" changes is only the scientist's interpretation of fixed data. On their view, Priestly and Lavoisier—contrary to Kuhn—both saw oxygen, but they interpreted their observations differently; Aristotle and Galileo both saw pendulums, but they differed in their interpretations of what they both had seen. For Kuhn, however, "the process by which either the individual or the community makes the transition from constrained fall to the pendulum or from dephlogisticated air to oxygen is not one that resembles interpretation" because of "the absence of fixed data for the scientist to interpret" ([13], pp. 120–121). "Normal," not "revolutionary," science utilizes interpretation because it is a deliberative enterprise "that aims to refine, extend, and articulate a paradigm that is already in existence" ([13], p. 121). During a revolution when the scientist embraces a new paradigm, he does not *interpret*. Rather, according to Kuhn, he experiences a Gestalt shift. Gestalt shifts are "sudden and unstructured," interpretation is not. Presumably there could be a time when one is half-through interpreting something, but there could not be a time when one is half-through experiencing a Gestalt shift (cf. Hanson, [8], p. 10). Kuhn maintains, along with Hanson, that the Gestalt sense of seeing is the sense relevant to revolutions in science. Normal science, on Kuhn's view, leads to the recognition of anomalies which

> . . . are terminated not by deliberation and interpretation but by a relatively sudden and unstructured event like the gesalt (*sic*) switch. ([13], p. 121)
>
> What were ducks in the scientist's world before the revolution are rabbits afterwards. ([13], p. 110)

Kuhn thus prefers, in analyzing scientific revolutions, to employ a conceptual scheme which has a "sudden and unstructured event" play the central role.[2]

II

Hanson's remarks about observation in science are indeed quite provocative. Unlike Kant he maintains that even *particular* theories determine what is seen. He admits, however, that scientists in different traditions see the same thing in one sense of 'see' ([8], pp. 5, 7, 8, 18, 20). For example, *something* about the visual experiences of Johannes Kepler and Tycho Brahe, when on a hill watching the dawn, is the same for both. Namely, both have a visual experience of a brilliant yellow-white disc centered between green and blue color patches ([8], pp. 8, 18). And for both, the distance between this disc and the horizon is increasing ([8], p. 182, note 6).

Hanson, however, claims that this type of seeing does not exhaust the concept. He feels, with Kuhn, that there is also a sense in which two observers do not see the same thing, do not begin from the same data, even though they are visually aware of the same object ([8], p. 18). He presents us with several Gestalt figures to illustrate this second sense of seeing—"seeing-as" ([8], pp. 8–18).

Well, just what have Gestalt examples to do with science? One might agree that Gestalt examples provide genuine examples of *seeing*; yet one might still doubt that observations relevant and important to science and scientific disputes are of this kind. Hanson wants to remove this doubt. He wants to claim that examples of *seeing-as,* analogous to his Gestalt examples, have occurred within the history of science; moreover, he wants to claim that they were more important than neutral observations for understanding scientific change and controversy.

There are then for Hanson these two senses of 'see.' In one sense (the neutral one) scientists, but presumably not *qua* scientists, see the same thing. In the other they do not. And Hanson, with Kuhn, urges it is in the latter sense that we get *scientific* "data" ([8], pp. 4–5, 17; [13], p. 134). Thus he concludes that Tycho and Kepler do not begin their inquiries from the same data, do not make the same observations, do not even see the same thing ([8], p. 5). Together on a hill Tycho and Kepler, according to Hanson, see (in the non-neutral sense of 'see') different things in the east at dawn. How could it be otherwise? "Practicing in different worlds, the two . . . scientists see different things when they look from the same point in the same direction" (Kuhn, [13], p. 149). "Kepler and Tycho are to the sun as we are to fig. 4,

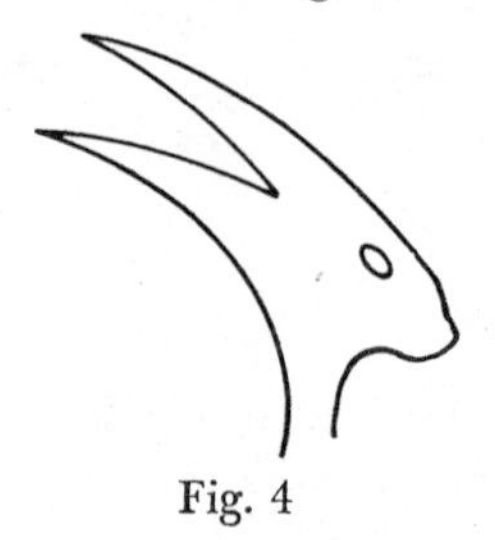

Fig. 4

when I see the bird and you see only the antelope" ([8], p. 18).

A difficulty arises here. Hanson says you and I both see his *fig. 4,* although we don't both see a bird. If then the sun is to be equated with *fig. 4,* as Hanson has just suggested, *both* Tycho and Kepler should see it. Hanson's analogy, if accurate, seems to undercut his own position: he maintains that it is the sense in which Tycho and Kepler do *not* observe the same thing which must be grasped if one is to understand fundamental disagreements within science.

Let us, however proceed to examine Hanson's argument that Tycho and Kepler see different things in a sense important and relevant to science.[3] The argument starts from two virtually incontestable premisses,

(1) m_0 (Tycho) sees X_0 (the sun Tycho sees).

(2) m_1 (Kepler) sees X_1 (the sun Kepler sees).

Now, for Hanson to *see* an object X is to *see* that if $A_1, \ldots A_n$, were done to X then $B_1, \ldots B_n$ respectively would result ([8], pp. 20–21, 22–23, 29–30, 58–59, 97).[4] In short, it is to see that X behaves in certain essential ways, i.e., has certain deep dispositional properties. For one to see Tycho's sun (X_0) is for him to at least see that from some celestial vantage point, the sun is such that it could be watched circling our fixed earth ([8], p. 23). That is, he would see a sun which is essentially mobile ([8], pp. 17, 23–24, 182). "Watching the sun at dawn through Tychonic spectacles would be to see it in something like this way" ([8], p. 23). Hanson, therefore, adds (3) as a further premiss to his argument.

(3) If anyone sees X_0 then he sees that $P_0(X_0)$.

where 'P_0' stands for 'is mobile,' a predicate in Tycho's system which holds essentially of the sun.

The situation is however, different for Kepler's sun. For one to see Kepler's sun (X_1) is for him at least to see that the sun does not behave in the above "Tychonic" way (cf. [8], p. 23). Why? Well, Kepler sees a static sun, whereas Tycho sees a mobile sun ([8], pp. 17, 23–23, 182). When Kepler sees the sun, he sees the horizon dipping or turning way from it. When Tycho sees the sun, he sees it ascending or rising ([8], pp. 23, 182). "The shift from sunrise to horizon-turn is analogous to the shift-of-aspect phenomena [viz., the Gestalt examples] already considered . . ." ([8], pp. 23–24). That is, Hanson adds another premiss, (4), to his argument.

(4) If anyone sees X_1 then he sees that $P_1(X_1)$.

where 'P_1' stands for 'is static', a predicate in Kepler's system which holds essentially of the sun.

From (1) and (3), Hanson validly infers (5),

(5) m_0 sees that $P_0(X_0)$.

And from (2) and (4), he similarly infers (6),

(6) m_1 sees that $P_1(X_1)$.

Now, m_0 and m_1 presumably know that anything that is static is not mobile. Therefore, Hanson obtains (7) from (6),

(7) m_1 sees that $\sim P_0(X_1)$.

From (5) and (7) he then concludes (8),

(8) $X_0 \neq X_1$.

(8) says Tycho and Kepler do not see the same sun. But the validity of the inference from (5) and (7) to (8) is questionable. It depends on the meaning of 'sees that.' We will examine this shortly. Let us for the moment assume it to be valid. Then one link remains in Hanson's argument. Why is this sense of 'see,' and not the neutral sense, important and relevant to Kepler's and Tycho's observations? Simply because the differences in what they see amounts to the difference between an essentially geocentric and essentially heliocentric universe. And this *is* the essential and deep difference between Tycho's and Kepler's physical theories.

The preceding is Hanson's argument. I think it is fallacious. Hanson implicitly uses the phrase 'sees that' to mean either 'knows that' (cf., e.g., [8], pp. 18, 20–22) or 'believes that' (cf., e.g., [8], pp. 23–24). In either case his argument is unacceptable.

B

Consider the case in which 'sees that' means 'knows that.' In this case Hanson's argument is valid but leads to absurd consequences; and we shall discover that this is because its premisses are false. From (5) and the fact that we are taking 'sees that' to mean 'knows that' we can infer (9),

(9) m_0 knows that $P_0(X_0)$.

Now, whatever else it may be, what is known is at least true. That is, if m_0 knows that $P_0(X_0)$, then we can conclude that $P_0(X_0)$ is the case. Therefore (10) follows,

(10) $P_0(X_0)$.

From (7) we can similarly infer (11),

(11) $\sim P_0(X_1)$.

To show Hanson's argument is valid we will show that (8), his conclusion, follows from (10) and (11). And this is readily done: for if $X_0 = X_1$, by (10), we would have $P_0(X_1)$ which contradicts (11). Thus, Hanson's argument is valid.

It is nevertheless, fallacious. The careful reader may have already noted the oddity of (10). Yet (10) was needed to make Hanson's argument valid; and it was indeed obtainable on the supposition that 'sees that' means 'knows that.' (10) is strange. Tycho's theory is false. He was mistaken about the status of the sun in the solar system. He mistakenly thought it was mobile. Either we hold (10) or not. If we hold (10), we will have to give up the widely accepted view that Tycho was wrong in thinking his sun revolved around the earth. On the other hand, if we give up (10), we will be forced to give up either (1) or (3). They were the premisses which entailed (10). One could not, therefore, rationally hold them and deny (10). Now, the truth of (1) is, I think, beyond question. To deny it amounts to the claim that Tycho didn't see the sun he saw, which is absurd. If 'sees that' means 'knows that' Hanson's premiss (3) should, therefore, be abandoned if we deny (10).

Hanson would, given his philosophical position, have to maintain that a geocentric world picture really isn't wrong after all. However, Hanson's own point of view is also that of modern science. Thus, he should say that a geocentric world picture is wrong, and indeed was wrong before Kepler. He should therefore maintain that a geocentric point of view is both wrong and not wrong if his own point of view includes both modern science and his own philosophical position. At the expense of the viewpoint of modern science he might maintain that his philosophical position is the single, preferred, or absolute point of view. But then he would be open to the charge that, in spite of his explicit disclaimers, he is not engaging in a *descriptive* philosophy of science; he would be, rather, philosophically legislating to science—telling scientists that what they regard as correct scientific beliefs (e.g., that the solar system has never been geocentric) are in fact, wrong. And this charge *Hanson* would have to take seriously. He thinks that a principal merit of his own approach is its descriptivism and its avoidance of the normative excesses of positivism.

Further, there are reasons which tend to suggest the falsity of many claims that, *prima facie* at least, are similar and analogous to (3) and (4). The Hansonian enterprise is not unique. Many philosophers have tried to thread knowledge into what they came to call 'sense data'. In the philosophical literature on perception and knowledge, sense data of various sorts have been proposed in order to function as the basic perceptual objects. The tendency in sense data accounts is to attempt to find or to postulate some type of object, the perception of which is *sufficient* for knowledge of the *seeing-that* or *knowing-that* kind. Hanson's search for the basic perceptual objects of science, the perception of which is sufficient for knowledge of the *seeing-that* or *knowing-that* kind is thus, in spite of his claims to the contrary ([8], p. 22), not so different from the approach of the sense data theorists. Similarly, Hanson thinks that " 'Seeing that' threads knowledge into our seeing" ([8], p. 22), and that it is "the logical element which connects observing with our knowledge" ([8], p. 20). Daniels ([2]) and Sellars ([19]) have argued with respect to the sense data positions that none of the things customarily labelled 'sense data' guarantee that the person who sees them will know that or see that they have attributes or qualities of a particular sort. Daniels, for instance, claims that one can still veridically see any of the customary types of sense data and be mistaken about what color it is, what shape it is, etc. ([2], pp. 4–5). Daniels' argument hinges on the fact that *seeing-that* and *knowing-that* are for the most part intentional whereas *seeing things or events* is for the most part not. I can see that a thing is a lamp without seeing that it is a collection of atoms and without seeing that it is my maiden aunt's favorite possession. *Seeing-that usually* has more to do with *knowing-that,* I think, than does seeing things. One cannot see that a thing is a lamp without knowing that it is. But one can see one's maiden aunt's favorite possession without knowing that it is one's maiden aunt's favorite possession. Veridical seeings of things often should not, therefore, be taken as sufficient conditions of *seeing-that* or *knowing-that.*

This might be thought puzzling in view of the fact that we often answer such questions as "How do you know that the tallest girl in Asylum High School was at the play?" by saying "I saw her there." But suppose that the tallest girl in Asylum High School also happens to be the youngest Smith girl and that one does see the tallest girl in Asylum High School there. If this is so, one also sees the youngest Smith girl there. How, then, by seeing the youngest Smith girl there, does one end up knowing that the tallest girl in Asylum High School was at the play, but not knowing that the youngest Smith girl was? This would, I think, be indeed paradoxical if one takes veridical seeings of things cited in such answers as "I saw her there," as sufficient conditions of *seeing-that* or *knowing-that.* The proper conclusion to draw, therefore, seems to be that veridical seeings of things, or events, or sense data of the sorts discussed by Daniels, often are not sufficient conditions of *seeing-that* or of *knowing-that.*

Although this conclusion does not *establish* the falsity of Hanson's premisses (3) and (4), it does tend to make them *somewhat* implausible, for each of his premisses also claims that *seeing* is a sufficient condition for *seeing-that* or *knowing-that.* One might here object by claiming that (3) and (4) are special and that their truth instead hinges on the fact that P_0 and P_1 are *essential* properties of X_0 and X_1, respectively. This, however, will not do and not merely because of Quinean reasons. At any given time an object may have many essential properties which are unknown. Yet this cannot reasonably be said to prevent us from seeing the object. That we know an object's essential properties is not, therefore, a necessary condition of our seeing the object.

It is interesting to note that what we have said here is not opposed to a Kantian position. Kant would *not* have held that we cannot experience things unless we *know that* or *see that* that they presuppose *his* categories; otherwise there would have been no point to his writing the *Critique.*

But if the role of such answers as "I saw her there" is usually not one of logical sufficiency for *seeing-that* or *knowing-that,* then what is its usual relation to these latter concepts? J. L. Austin perhaps provides an answer: by saying that one sees a thing or event, one indicates to one's hearer that one is in a position here and now *to see that* it has attributes of a particular sort (to see that the tallest girl in Asylum High School is in the audience) ([1], pp. 47–50). But the fact that one can see the youngest Smith girl in the audience and not see that the youngest Smith girl is in the audience shows that much more than seeing x is relevant to one's seeing that F(x).

C

Let us now consider the case where 'sees that' means 'believes that.' In this case Hanson's argument again is fallacious. (5) and (7) become (12) and (13) respectively.

(12) m_0 believes that $P_0(X_0)$.

(13) m_1 believes that $\sim P_0(X_1)$.

But to conclude (8) from (12) and (13) is invalid. m_0 believes that P_0 holds essentially of X_0. m_1 believes that $\sim P_0$ holds essentially of X_1. But this does not imply (8). A counter-example arises from noting the consistency of (14), (15), and (16):

(14) Mohammed X, a Black Muslim, believes that Mohammed Ali is good and, essentially so.

(15) Adolph Übermensch, a Nazi, believes that Cassius Clay is not good, and essentially so.

(16) Yet, Mohammed Ali = Cassius Clay.

Hence, two people can believe that contradictory properties hold essentially of the *same* object. Thus, when 'sees that' means 'believes that' Hanson's argument is invalid.

In addition (3) and (4) become implausible. They would, if true, yield consequences which are both strange and inconsistent with other parts of Hanson's approach. If (3) were true, Tycho could not rationally revise his essential beliefs about the sun he sees and still be able to see it. Let us assume that Tycho sees the sun X_0 at time t_0. Therefore, by (3) with 'believes that' substituted in place of 'sees that', Tycho at t_0 believes that $P_0(X_0)$. That is, Tycho at t_0 believes that the sun is essentially mobile. The moment t_1 (where t_1 is later than t_0) Tycho rationally believed that $\sim P_0(X_0)$ it would follow that he would no longer be able to see X_0 if (3) were true: assuming he did see X_0 at t_1 then, by (3), he would believe that $P_0(X_0)$; and this is contrary to our previous assumption that he rationally believed that $\sim P_0(X_0)$. To believe both that $P_0(X_0)$ and that $\sim P_0(X_0)$ is not rational. Tycho could not, therefore revise his beliefs about X_0. (4) results in an analogous consequence. In short, it would be impossible for (say) Tycho, in a rational way, to believe or learn that the sun he saw before is not mobile if he was still able to see it. This is paradoxical, for in an important sense, scientific beliefs are about matters-of-fact (data, phenomena, etc.). Hanson would not disagree with this; he too would maintain that scientific beliefs express the connections thought to obtain between matters-of-fact. What then about scientific disputes and disagreements? We can grant that scientific disputes are usually not disputes over facts in the sense that courtroom disputes often are (cf.: Wilson: 'Yes, I saw Jones kill Smith'; Benson: 'No, Jones was having dinner with me when Smith was killed.'). Yet scientific disputes are disputes over facts in another sense. They involve scientists' adherence to incompatible beliefs *about* facts. Such beliefs express the different connections that each thinks obtains between matters-of-fact. This point too Hanson has noted well:

> The difference is not about what the facts are, but it may very well be about how the facts hang together. ([8], p. 118)

Given that Hanson (correctly, in my opinion) maintains this—and it is a central tendency in his approach—it is hard to see how the rest of his view would make it possible for scientific disputants to revise their beliefs as to the connections they think obtain between objects which they continue to experience. In short, it would be impossible for Tycho, in a rational way, to believe, or learn, that a sun which he could continue to see throughout the time interval t_1-t_0 was not mobile. If Tycho, in a rational way, believes that $\sim P_0(X_0)$ at t_1, then Tycho could not see X_0 at t_1. Granted, if he were to revise his beliefs in this way, he would see another sun X_1 (where $X_1 \neq X_0$). At t_1, however, his new and revised belief $\sim P_0(X_0)$ would not be about X_1. '$\sim p_0(X_0)$' would be about X_0. Generalizing, this would imply that one's *revised* scientific beliefs are at any time not *about* what one can see at *that time*.

This is paradoxical. It is also, as we noted, opposed to the tendency in Hanson's position which regards scientific beliefs as expressions of the connections thought to obtain between the elements of possible experience and which regards scientific disagreements as differences "about how the facts hang together."

There is another reason which would tend to suggest the implausibility of (3) and (4) when 'sees that' means 'believes that.' *Believing-that* locutions are for the most part intentional; *seeing things or events* locutions are for the most part not. I can believe that a thing is a lamp without believing that it is a collection of atoms and without believing that it is my maiden aunt's favorite possession. But, if I see a lamp, I see a collection of atoms; and if the lamp is my maiden aunt's favorite possession, I see my maiden aunt's favorite possession. Thus, one can see a collection of atoms without believing that it is a collection of atoms; one can see one's maiden aunt's favorite possession without believing that it is her favorite possession. Veridical seeings of things should not, therefore, usually be taken as sufficient conditions of *believing-that's.*

The situation is even worse. The very conclusion of Hanson's argument leads to untoward consequences. (8) tells us Tycho and Kepler each observe different things in a sense. And Hanson's conclusion is that this sense is the one relevant to their scientific disagreements. But given this, it becomes impossible for them to have scientific disagreements about the sun; they are not talking about the same sun because in the sense relevant to science they don't see the same sun. And each is talking about what he sees. Their beliefs about the sun are not, therefore, *rival* beliefs. If I observe a bird and claim it can fly while you observe an antelope and claim it cannot, we are not expressing *rival* beliefs. We are seeing different things; thus, our respective claims about them are not *rival* claims. Just so, if Tycho sees something different than Kepler, then the beliefs they express about what they see are not *rival* beliefs. In short, the sense in which Tycho and Kepler do not see the same thing is *not* relevant to their scientific disagreements if these disputes are indeed disputes "about how the facts hang together."

D

Hanson suggests that observations are theory-laden, that different scientists see different things because they espouse different theories, that they could not have observed what they did if they did not hold the theory that they did. This, however, is incompatible with a principal tendency of the rest of his approach, namely, the stress he places on the importance of "retrodictive inferences" to science. Indeed, his view as to observation would preclude the possibility of such inferences.

Hanson maintains that retrodictive inference is central to science. He thinks its form is the following:

1. Some surprising phenomenon P is observed.
2. P would be explicable as a matter of course if II were true.
3. Hence there is reason to think that II is true ([8], p. 86).

Hanson thinks that:

> From the observed properties of phenomena the physicist reasons his way toward a keystone idea from which the properties are explicable as a matter of course. ([8], p. 90)

Hanson's position on observation, if correct, however, would make it impossible for the scientist to reason in this way. We could not begin with data and search for the keystone principles which explained them if, as Hanson maintains, we could not observe this data unless we already *saw that* these keystone principles were correct (cf. "The knowledge is there in the seeing," [8], p. 22). One could not begin with an observed phenomenon P and search for the as yet unknown "keystone idea" or "hypothesis" which would explain it. For if we could, this would imply that we do not *see that* the keystone idea or hypothesis H is correct at the time we initially observe the phenomenon P. Thus, we could not have observed P to begin with for the requisite knowledge (H or the keystone idea) would *ex hypothesi not* have been "there in the seeing" at the outset ([8], p. 22; cf. also [8], pp. 20–21, 26, along with (3) and (4) above). Retrodictive inferences would therefore be impossible. Consider Kepler. Initially he did observe a sun P_0. But *initially* he did *not* see that this sun was static (H). Only later did he reason his way towards this conclusion and it was a conclusion in part based on his *(and Brahe's)* observations of the sun. But if (4) is correct, then the second he saw that the sun is static his observation of it would have changed ("the shift-of-aspect phenomena . . . is occasioned by differences between what Tycho and Kepler think they know") ([8], p. 24). And it is this new sun P_1 which he sees that is supposed to be explicable by, H. H patterns, structures, and explicates P_1, not P_0. It is the sun Kepler sees at the time he *sees that* H which is supposed to be explicable by means of H. Thus, Kepler really would not have found an explanation for his initial observation P_0. His inference therefore would not have been "retrodictive" given Hanson's account of such inferences.

III

Let us next turn to Professor Feyerabend. Within the "logical empiricist" tradition, accounts of experience usually maintain that neutral observations are possible and that observational results can be stated and verified independently of the theories investigated (cf. Hempel, [11], pp. 20–22, 43–44). Feyerabend, however, thinks that:

> In these accounts it is taken for granted that observational results can be stated and verified independently, at least independently of the theories investigated. This is nothing but an expression, in the formal mode of speech, of the common belief that experience contains a factual core that is independent of theories. . . . ([5], p. 151).

Feyerabend seems to maintain, with Kant, that there are no data without concepts. For him, the given is a myth. Further, he would seem to argue, with Kant, that *even if there were such data,* "Intuition without concepts is blind." That is, they would exhibit a radical poverty of content which would preclude their epistemic relevance to the testing of scientific theories ([5], pp. 194–197, 202–215). These claims, however, do not imply, as he seems to think, that neutral observations are not possible or that observational results cannot "be stated and verified independently of the theories investigated." Affirming the possibility of neutral observations and the possibility of observational results being stated and verified independently of the theories investigated is logically weaker than affirming an unconceptualized given; it is not the case that the former are "nothing but an expression in the formal mode of speech of the common belief that experience contains a factual *core* (my italics). . . ." Professor Scheffler has recently seemed to argue just this ([18], Chapter 2). He argues persuasively against C. I. Lewis' notion of the given ([5], esp. Chap. 2) and would agree with Feyerabend that pure data are a myth. However, Scheffler maintains that testing of theories does *not* presuppose unconceptualized data. He thinks that observation is conceptualized. But this, he claims, is only to say that concepts (or categories) are employed to classify and individuate phenomena. Hypotheses, on the other hand, are formulated concerning the distribution of phenomena in the categories employed. The correctness of those hypothesis is tested by observation.

Feyerabend, *et al.* might object that when scientists have formulated incompatible theories concerning a particular realm of phenomena, observation cannot provide an independent basis for deciding between the theories. If observations are structured by concepts, then scientists advocating distinct theories will structure their observations differently. Thus, each scientist will make observations relevant only to his own theory.

Scheffler thinks his distinction between categories and hypotheses would provide an answer to such an objection ([18], p. 43). Using one basic set of categories, any number of alternative hypotheses may be formulated. Competing scientific theories are often alternative hypotheses formulated, on Scheffler's view, using one basic set of categories. This point is, of course, moot and one of the very questions at issue. In addition, however, Scheffler believes, correctly in my opinion, that *even if* two scientists employ "non-overlapping category systems" (presumably, systems in which no category of one is identical to any category of the other) their respective categories may "have in common certain identical items, although they group them differently" ([18], p. 41). His example of two non-identical categories, "giraffe" and "animal" presumably supports this. Thus, he maintains that "Two category systems, as wholes, may even have all their items in common, though classifying them in different ways" ([18], p. 41). For these reasons Scheffler holds that scientists advocating alternative theories often can invoke the same observations as relevant to testing both (or all) the competitors and as providing a basis for deciding which is more adequate.

Feyerabend, *et al.* could perhaps respond to this argument in a manner not unsimilar to Professor Quine's thoroughgoing pragmatic approach. Feyerabend might well deny the validity of the distinction between (a) concepts, definitions, and analytic truths, which determine the basic scheme of categories to be employed, and (b) the hypotheses or general empirical truths. He might grant that in a scientific theory or language we can make an abstract distinction, for example, between (a) general conceptual truths and (b) general empirical statements (hypotheses); yet he would perhaps deny that we can, in studying the basis for rejecting or modifying scientific theories, distinguish between those considerations which lead us simply to reject a hypothesis and those considerations which lead us to reject a system of categories. He might hold that the total theory consisting of the system of empirical hypotheses *and* the "analytic" statements which formulate concepts faces the test of observation as a loosely integrated unit. In adjusting the theory to accommodate observations, he might continue, there is a great deal of flexibility in regard to which hypotheses and statements are allowed to be altered. There is, Feyerabend might contend, no valid distinction within statements of a theory between those which formulate concepts or categories and those which express empirical hypotheses.

Such a reply is, it seems to me, unsatisfactory. It would at this point perhaps be illuminating to briefly note a *prima facie* similarity and a *prima facie* difference between Quine's and Feyerabend's approach. Both would seem to deny there is an analytic-synthetic distinction, and not merely in the sense that there is no sharp distinction. As Gilbert Harman points out ([10], pp. 125–127), for Quine this denial takes the form of a claim that nothing is analytically true, that all truth or falsity is synthetic; for Quine, the analytic-synthetic distinction is "meaningless" in that one cannot make sense of it in any way such that there turn out to be analytic truths. But for Feyerabend, given his radical meaning variance position, this denial should take the form of a claim that nothing is synthetically true, that all truth or falsity is linguistic; for Feyerabend, the analytic-synthetic distinction is "meaningless" in that one cannot make sense of it in any way such that there turn out to be synthetic truths; the "truths of language—truths of fact" distinction resembles the nonwitch-witch distinction, which fails to distinguish anything since there are no witches.

The basic conflict, however, between the defender of the possibility of common observations and his critic need not, at this point, hinge either on the distinction between concepts (categories) and empirical hypotheses or on the Duhem-Quine Thesis. We might agree that both categorical schemata and particular empirical hypotheses in the form of expectations and beliefs influence both observational mode and content. Nevertheless, this does not compel us to assert that observational mode and content cannot be *shared* by holders of different empirical hypotheses. The fact that what is observed is influenced by belief does not imply that what is observed cannot be shared by holders of different beliefs; not every influence need be a one-to-one influence. Further, there may exist much in common between the two sets of beliefs and this may make common observations possible. The gap in scientific transitions is much exaggerated, as it always is in revolutionary changes; if, in any given "scientific revolution," a list were made of the agreements of the two sides, it would be enormously large. It is a historical truism that revolutions change much less than they seem to on the surface, and this may be applied to the problems we are discussing.[5] In a conflict between competing scientific theories, there are normally many basic principles held in common by advocates of each of the competing views.[6] Such common principles might even be said to form a system of concepts within which the conflicting hypotheses can be formulated.

IV

Professor Kuhn's overall view of observation is quite similar to Professor Feyerabend's and Professor Hanson's, as we have already indicated (I). Let us, however, examine some of the specifics of his position.

A

Kuhn feels that Sir William Herschel's discovery of the planet Uranus provides an example of a celestial body that was seen differently ([13], pp. 114–115). He thinks that this discovery was a prime example of a transformation in the scientists' field of vision, one quite analogous to a Gestalt switch. His evidence for this conclusion is the following: On at least seventeen different occasions between 1690 and 1781, several of Europe's most eminent observers "had seen a star in positions that we now suppose must have been occupied at the time by Uranus" ([13], p. 114). "One of the best observers in this group had actually seen the star on four successive nights . . ." ([13], p. 114). Herschel "observed the same object" but after further investigation "announced that he had seen a new comet!" ([13], p. 114). Only later did Lexell suggest that the orbit was probably planetary From this evidence Kuhn concludes,

> When that suggestion [Lexell's] was accepted, there were several fewer stars and one more planet in the world of the professional astronomer. A celestial body that had been observed off and on for almost a century was seen differently. . . . ([13], pp. 114–115)

This same type of argument is used again and again by Professor Kuhn. For example,

> . . . after the assimilation of Franklin's paradigm, the electrician *looking* at a Leyden jar *saw* [my italics] something different from what he had seen before. The device *had become* [my italics] a condenser. . . . ([13], p. 117)

Again,

> Lavoisier . . . *saw* oxygen where Priestly had *seen* [*my italics*] dephlogisticated air and where others had seen nothing at all. ([13], p. 117)

Pendulums too are described as involving a "shift of vision" ([13], p. 118). Kuhn claims,

> Pendulums were *brought* into *existence* [my italics] by something very like a paradigm-induced Gestalt switch. ([13], p. 119)

Let us now evaluate these remarks. The historical "evidence" to the effect that the discovery of the planet Uranus was an instance of a Gestalt-type shift is dubious. It would make much better sense to say that the astronomer-observers mistakenly believed that what they had seen in the sky was a star, whereas it really was a planet. The fact is that Herschel didn't claim Uranus was a comet until *after* some further scrutiny ([13], p. 114); after this scrutiny he "announced that he had seen a new comet" ([13], p. 114). But *announcing after* investigation, that what one had previously seen is a comet is surely different from *actually seeing* something different. What Herschel did was simply announce that he had some justification for believing that what he saw was a comet. That is, Herschel announced that he had some justification for believing that what he saw possessed the dispositional properties that comets possess.

Lexell suggested that the orbit was "probably planetary" "after fruitless attempts to fit the observed motion to a cometary orbit" ([13], p. 114). Surely, what this suggests is *not* that Lexell now *saw* a planet ([13], p. 114) or now perceived differently ([13], pp. 110, 119); rather it suggests that *because of* theoretical considerations Lexell now *believed that* the "observed motion" was that of a planet. That is, Lexell now rationally believed that the same object everyone had *seen* possessed the property of being a planet. He believed that it behaved in planetary ways.[7] Not *everyone,* of course, believed that what they saw was a planet. Some were mistaken about the dispositional properties possessed by what they saw. Some believed that it was a star. And believing that it is a star is quite different from actually seeing or perceiving a star (cf. section II-C above). If "there were several fewer stars and one more planet in the world of the professional astronomer" ([13], p. 114) *after* Lexell's suggestion was accepted, as Kuhn claims, then something else must also follow. It must follow that *before* Lexell there were more stars in the world of the professional astronomer. But if this is true one can no longer say that these astronomers before Lexell were *mistaken* about the number of stars. Even if one held that to say whether they were mistaken is to say something about their viewpoint, Kuhn would have to hold they were not mistaken. This is because from Kuhn's viewpoint there really were more stars in their world. It is not just that they believed there were more stars in their world. According to Kuhn there *really* was this number of stars ([13], p. 114). Thus these astronomers were not wrong after all. However, Kuhn's own point of view is also that of modern science. Thus, he should also say that the pre-Lexell framework is wrong, and indeed was wrong before Lexell. He should therefore maintain that the pre-Lexell framework is both wrong and not wrong if his own point of view includes both modern science and his own philosophical position. Kuhn would thus end up espousing a contradiction unless he were to give up either his philosophical point of view or his scientific point of view. At the expense of the viewpoint of modern science he might maintain that his philosophical position is the single, preferred, or absolute point of view. But then he would be open to the charge that, in spite of his explicit disclaimers ([13], pp. 93, 161, 165, 169), he is not engaging in a *descriptive* philosophy of science; he would be, rather, philosophically legislating to science—telling scientists that what they regard as correct scientific beliefs are, in fact, wrong. And this charge Kuhn would have to take seriously. He feel that a principal merit of his own approach is its descriptivism and its avoidance of the normative excess of "logical empiricism."

The same holds for Kuhn's other examples. After the assimilation of Franklin's paradigm, the electrician looking at a Leyden jar did *not* see something different. The device did not *become* a condenser. Rather, it was now *believed* (or known) to possess new properties, condenser-properties. Different features of what was seen were now held or believed to be more important. Their importance was due to their now more central relation to the new conceptual framework. Shifts of vision *are* shifts of *vision,* not belief. No profit results from an inordinate intertwining of these concepts. Sight is one thing, belief another. The same holds for dephlogisticated air versus oxygen, and falling stones versus pendulums.

These paradigm shifts involved, and were due to, shifts of belief, not vision. (Hanson argues that different interpretations are not the cause of rival scientific accounts. He argues rather that disparities arise because the fundamental visual data have changed ([8], pp. 8–11). Some of these arguments are irrelevant. They are directed at *Gestalt* examples. And this is the very question at issue. Are Gestalt shifts central to what goes on during *scientific* change?)

B

Further, Kuhn's view, as does Hanson's (cf. II-C), implies rival traditions are not really *rival.* This is because they are about different subject matters. We will soon argue this in general for Kuhn, Toulmin, and Feyerabend. Now, however, let us examine one of Kuhn's examples. In describing a famous debate between the French chemists Proust and Berthollet, Kuhn says,

> The first claimed that all chemical reactions occurred in fixed proportion, the latter that they did not. Each collected impressive experimental evidence for his view. Nevertheless the two men *necessarily* [my italics] talked through each other, and their debate was entirely inconclusive. Where Berthollet saw a compound that could vary in proportion. Proust saw only a physical mixture. To that issue neither experiment nor a change of definitional convention *could* [my italics] be relevant. The two men were as fundamentally at cross-purpose as Galileo and Aristotle had been. ([13], p. 131)

Notice how Kuhn's position entails that these two men *necessarily* talked through each other, that they were *necessarily* at cross purposes. Perhaps they were, in fact, at cross purposes. Perhaps they did, in fact, talk through each other. But why necessarily? Kuhn's methodology and radically idealist theory of perception alone is what forces us to this untoward conclusion. *His* view indeed implies that they were necessarily at cross purposes, that they necessarily talked through each other. For to the extent they were brought up and trained in a given, particular tradition, they had no control over what they saw. Given that one is practicing *within* a given tradition, one either sees things or one doesn't. Given then that the visual worlds of these scientists were different, as Kuhn maintains, then the claims they make about their respective worlds are not *rival* claims. The claims, instead, will be at cross purposes. When they make the claims, they will indeed be talking through one another. And necessarily so, given that they espouse different paradigms. But just what leads to these undesirable conclusions? I believe it is in part Kuhn's strange use of words, his strange and misleading locution, that is responsible. Language can mislead. And it in good part is responsible for the odd results we have just noted. Precisely what is responsible is the following claim which, in my opinion, is either false or misleading: [8]

(17) "Where Berthollet saw a compound that could vary in proportion, Proust saw only a physical mixture." ([13], p. 131)

Why not simply assert (18)?

(18) Berthollet believed that chemical reactions involved compounds that could vary in proportion, whereas Proust believed that chemical reactions involved only physical mixtures.

Any "talking through" one another at "cross purposes" that might have gone on would, if we hold (18) instead of (17), be due to the men themselves, and not to the nature of *science, qua* science, as Kuhn would have it. Men do have control over what they believe about what they see; at least more control than over what they see. And if *these* men didn't, then it was due to their own dogmatism, not to any defect of, or constraint on, their vision.

V

More general, yet analogous, objections hold against Feyerabend, Hanson, Toulmin, and Kuhn. Each of their views (cf. I), I shall argue, prevents scientists within a tradition from revising their beliefs as to the "essential" properties of experience. Progress could not, therefore, be achieved. Second, I will maintain that each of their views prevents the scientific theory after a revolution from being a rival or alternative to the scientific theory before the revolution. Third, I will argue that their views are beset by problems as to how a theory interacts with "the common environment" to produce the world. Fourth, I will argue that if their views were correct, then no theory could be tested or falsified by any observations; observations and observation reports could not lead to the rational rejection of the particular scientific theory which is employed; nor could they lead to the rational acceptance of a new and revolutionary scientific theory. Each of these four conclusions constitues some grounds for doubting the positions which lead to them.

A

Let us proceed with the first point. Roughly put, essential revisions of beliefs about experience could not be made in a rational way if Feyerabend, Hanson, Toulmin, or Kuhn were correct. This is analogous to the problem that we have already noted with regard to the details of Hanson's position (cf. II-C). There we argued that (3) and (4) were implausible (cf. pp. 458–459). If they were true, Tycho and Kepler could not rationally revise their essential beliefs about the sun they see and continue to be able to see it. Generalizing, Hanson's position as to observation would imply that a scientist's revised, essential (or "important" if one prefers to eschew essentialism) beliefs are not at any time about what he could see *at that time.* The concept of scientific progress would therefore be attenuated.

The same type of argument can readily be extended to Feyerabend, Toulmin, and Kuhn. Feyerabend says that among really efficient [9] alternative theories "each theory will possess its own experience, and there will be no overlap between these experiences" ([5], p. 214). Toulmin's view is similar as we have already noticed (cf. I). He thinks men who accept different ideals and paradigms see different phenomena ([21], p. 5); theories determine what are facts for us at all ([21], p. 95). Kuhn too believes that after a revolution scientists work in a different observational world; what they experience has changed ([13], pp. 110, 117, 119, 134, 194). The general problem is the following: Assume we accept a scientific theory, T_1. What we experience, E_1, will now be different from what we would experience if we accepted any significantly different "alternative" theory. How, therefore, could we, in a rational way, ever significantly *revise* our theories and beliefs so that they would be about what we would still experience? On their view, the moment we changed our beliefs regarding the essential properties of experience, experience itself would change. In short, it would be impossible for us to revise our essential beliefs about what we would still be able to experience. The second we revise them, they would no longer be about what we could still experience. This would imply that a scientist's revised essential beliefs about experience are at any time not about what he could experience at that time (cf. pp. 458–459). And this is absurd. If these beliefs were scientific, then they would be confirmable or falsifiable and would purport to express truths about what we could experience. But if there were nothing that we *could* experience which would satisfy or falsify these beliefs at the time we hold them (in virtue of our very holding of them), then, indeed, these beliefs would be neither confirmable nor falsifiable for us. Thus, in particular, they would not be *scientific* and would not express truths about what we could experience.

B

So much for the first point. Now for the second, which is quite similar. Feyerabend's, Hanson's Toulmin's, and Kuhn's views, I maintain, prevent the scientific theory after a revolution, T_2, from being a *rival,* or *alternative,* to the scientific theory, T_1, before the revolution. On their view, T_1 determines experience E_1 which is different than E_2, the experience determined by T_2. But given this it becomes difficult, if not impossible, for the scientist accepting T_1 to have professional disagreement about the experience E_1 with the scientist accepting T_2; this is because they are not talking about the same experience or world; in the sense relevant to science it is claimed that they don't experience the same things; $E_1 \neq E_2$. And each scientist is talking about what he experiences. Their beliefs about experience and the world are not, therefore, *rival* beliefs. Nor are T_1 and T_2 *alternatives.* What is T_1 an *alternative to?* It is not an alternative to T_2, for on their view, T_2 is talking about something different (E_2) than T_1 is (E_1). T_1 and T_2 are not alternative views of the same world. On their view, the world has radically changed. Nor are they alternative views of experience, for this is radically different. If I accept a current version of quantum mechanics and you accept a current version of sociology, we are not thereby disagreeing. We are not thereby expressing rival or alternative beliefs. Just so with any T_1 and T_2 on the above view. If I observe a duck and claim it can fly while you observe a rabbit and claim it can't fly, we are not expressing rival or alternative beliefs. We are seeing different things; thus, our respective claims about them are not *rival* or *alternative* claims. Just so, if one scientist experiences different things than another, then the beliefs they express about their objects of experience are not rival or alternative beliefs.[10]

This consequence is further directly opposed to one of the principal methodological considerations that motivates Feyerabend to hold the radical observational variance position in the first place, namely, the "principle of proliferation: Invent, and elaborate theories which are *inconsistent* [my italics] with the accepted point of view, even if the latter should happen to be highly confirmed and generally accepted" ([7], pp. 223–224). This principle is central to Feyerabend's positive methodology and is a "main consequence" of his "abstract model for the acquisition of knowledge" ([7], p. 223). This principle, under the name of 'theoretical pluralism', "is assumed to be an *essential feature* [his italics] of all knowledge," for "Criticism must use alternatives" ([4], pp. 6–7). Yet it is precisely his radical observational variance position which, as we have seen, precludes two different theories from contradicting each other and from being inconsistent. Rather than being "an *essential* feature of all knowledge," the principle of proliferation would instead be an *impossible feature* of all knowledge. And on a normative interpretation it would, in effect, call for the impossible, given the doctrine of radical observational variance. Prescriptions for the impossible, though not strictly contradictory, are somewhat awkward and become especially suspect employed within a scientific methodology. Given all this, I, therefore, find it quite ironic that Feyerabend puts forward precisely this methodological principle as a *reason* for his espousal of radical observational variance (cf., e.g. [4], pp. 7–8).

C

A third problem which besets their views is one concerning the range of fact, and of interaction between theory and fact.

Kuhn, for example, denies that during a scientific revolution there exist fixed data which the scientist interprets ([13], p. 121 top). Yet he also claims that the scientists' world is determined "jointly by the environment and the particular normal-scientific tradition" ([13], p. 11, cf. also p. 122). Here Kuhn presumes that the "environment" is itself unaffected by theory; it is rather the "world" which is so affected. The environment combines with the theory to form the world. In this sense the environment may itself be said to be "fixed." Kuhn gives no reasons why scientists should not study the nature and properties of this evironment. Indeed its investigation would be empirical and would be, for this reason, worthy of scientists' attention. Yet it is presumed fixed and unaffected by theory.

It would therefore be neutrally available to scientists even during scientific revolutions when different theories are supposedly combining with it to form different worlds. But it is just the availability of fixed data during scientific revolutions which Kuhn, as we have noted, would deny.

There is, in addition, an interaction problem with the radical observational variance view. Several important and related questions naturally arise given this view; and *prima facie,* at least, they are legitimate questions. *How* does the "normal-scientific tradition" operate on, and change, this environment to form the scientists' world? Just how do paradigms or theories interact with the "common environment" to produce "data" ([13], pp. 122, 134; [21], pp. 95–96)? What is the nature of this "manufacturing process" ([3], pp. 50–51)? What happens when new data arise? Where is the *locus* of this interaction? If "However construed, the construing is there in the seeing" ([8], p. 23), then we may ask, "Precisely *where* in the seeing is it?" It does not help whatsoever to say that what happens is a "sudden and unstructured event" ([13], p. 121). This "answer" is nothing more than an expression of a basic unintelligibility of their view.

More traditional approaches to the philosophy of science would have scientific activity consist in the interpretation of neutral and fixed observational data. And a virtue of such approaches is that these questions simply do not arise. *Prima facie,* at least, they would not be legitimate questions within the conceptual framework of a more traditional philosophy of science. The problem is that answers to such questions are not explicitly provided by either Hanson, Kuhn, Toulmin, or Feyerabend. Nor do the answers, in an obvious way, follow from, or inhere implicitly in, the conceptual scheme they employ to analyze science; yet it is this very scheme which gives rise to such questions. In short, these writers, in my opinion, lack a well-thought-out and developed epistemology. Is there any reason why they cannot use some accepted epistemological theory to deal with, and perhaps answer, these questions? This is, of course, an open and difficult question. Many different general epistemological theories have made their appearance in the history of thought and the approach of Feyerabend, Kuhn, Hanson, and Toulmin is perhaps not incompatible with all of them. Their approach is, however, incompatible with either a Platonistic or Kantian epistemology. Consider a Platonistic approach. Here objects of experience and observation would be constituted through participation in *invariant* Forms which exist whether or not we believe that they exist. Such objects would thus be constituted in a way which is independent of our theoretical *beliefs.* Platonistic answers to the above questions would therefore be inappropriate for they would invoke a theory which is incompatible with the radical observational variance view. On a Kantian approach, objects of experience would be constituted by our Categories and Forms of Intuition each of which is essentially invariant from person to person. Objects of experience for Kant are objective, invariant, and theory-neutral because they are constituted, not by particular scientific theories, but by general and invariant Categories and Forms of Intuition. Such objects would therefore be constituted in a way which is independent of our particular scientific beliefs. They would also be independent of our general philosophical beliefs as to (say) *which* Categories and Forms of Intuition experience must presuppose; one of the purposes Kant had in writing the *Critique of Pure Reason* was to convince his disbelieving philosophical opponents that their experiential world already depended on Kantian Categories and Forms of Intuition—whether or not these opponents initially believed that Kantian Categories and Forms of Intuition were correct. Kantian answers to the above questions would therefore be in appropriate. They would invoke an epistemological theory which is incompatible with the radical observational variance view.

D

The fourth problem which besets their views is that if they were true it would follow that no theory could be tested or falsified by any observations. According to Hanson, Feyerabend, Kuhn, and Toulmin, observations presuppose, and are laden with, the particular scientific traditions of the time. Therefore, I will argue that observations and observation reports could not lead to the rational rejection of the scientific tradition. Nor could they lead to the rational acceptance of a new and revolutionary tradition.

Professor Hanson, unlike Kant, indeed feels that what is observed presupposes and is laden with the particular theory of the time ([8], pp. 10, 23, 149, 157; [9], p. 38):

> [the uncertainty principle] is built into the outlook of the quantum physicist, into every observation of every fruitful experiment since 1925. The facts recorded in the last thirty years of physics are unintelligible except against this conceptual backdrop. ([8], p. 149)
> The observations and the experiments are infused with the concepts; they are loaded with the theories. ([8], p. 157)

Feyerabend also feels that observational results cannot be arrived at, stated, or verified, independently of the particular theories investigated ([5], pp. 151, 194–197, 214, 220–221).

Kuhn ([13], pp. 6, 110–111, 117, 119, 128, 134, 149) and Toulmin ([21], pp. 57, 95, 101, 103: [23], p. 463) concur. Indeed, these views may be part of what prompts Professor Shapere to remark:

> The view that, fundamental to scientific investigation and development, there are certain very pervasive sorts of *presuppositions* [my italics], is the chief substantive characteristic of what I have called the new revolution in the philosophy of science.... ([20], p. 48)

Such a view on their part is, in my opinion, implausible for two reasons.

1. Consider the first. Let us assume that observations made within a particular scientific tradition presuppose, and are "loaded with," the scientific theory T employed at that time. Any observation report 0 of these observations will then presuppose T. On Professor Van Fraassen's account of the concept of presupposition ([24], esp. pp. 137–138), the only developed account of this notion known to me, it then follows that 0 is neither true nor false unless T is true. And from this it follows that if 0 is true, then T is true and if 0 is false, then T is true. Thus, we cannot maintain that 0 is true and T is false; nor can we maintain that 0 is false and T is false. The latter result is rather remarkable; it implies that false predictions cannot be used as a basis for denying T. The situation, however, is still worse. No observation report 0, whether true or false, can be used to reject T; if 0 presupposes T, then, whenever 0 is true, T is true, and whenever 0 is false, T is true. The only way out of this dilemma is to maintain that reports of observations are neither true nor false. But this is tantamount to denying that science is an empirical and cognitive enterprise. In any case, we would have to regard as neither true nor false *all* reports of observations made during former times, if we regarded as false the theory employed during those times. In particular, the observation reports of Brahe, Kepler, Newton, Leverrier, Priestly, Gibbs, and Michelson-Morley would have to be regarded as neither true nor false.

2. Consider the second point. No true or false observation report which presupposes a theory T could serve as a basis for the rational acceptance of a new theory T_1 which is mutually inconsistent with T. If we accept T_1, we would have to deny T, given T and T_1 are mutually inconsistent. If we deny T, we would have to maintain that 0 is neither true nor false. If we maintain that 0 is neither true nor false, it is hard to see how 0 could serve as a *basis* for the rational acceptance of T_1. In particular, if we accept relativistic physics (T_1), we would have to deny classical physics (T) since these two theories are mutually inconsistent. If we deny classical physics, then we would have to maintain that the observa-

tion reports (0) of the Michelson-Morley experiment, of Leverrier's observations of Mercury's rotation round the sun, etc., were neither true nor false. Being neither true nor false, it is hard to see how they could have served as a basis for the rational acceptance of relativistic physics over classical physics. These observation reports, however, *are* generally considered by scientists to be correct and true, and are part of what led to the rational acceptance of relativistic physics over classical physics (cf. e.g., [16], pp. 542–547).

VI

We have, therefore, found some methodological justification for assuming that, in a non-trivial sense, experience is neutral with respect to alternative scientific theories. At best, the contradictory view implies that theory change is not rational and without justifiation. At worst, the contradictory view leads to unintelligible and absurd consequences. In addition to such methodological justification, the historical examples that we discussed (esp. in IV) have suggested that scientific observations are neutral.

There are, then, difficulties in the position of Feyerabend, Hanson, Kuhn, and Toulmin as to observations in science. I have tried to show that an alternative to their view is preferable. Nevertheless, their position does suggest something valuable. Scientific revolutions did not consist merely in the discovery of new facts or merely in a closer attention to facts already known. Not everything a scientist observes has equal potential for testing or confirming his theory. Some things a scientist experiences may be relatively irrelevant to his theory. In this sense, we may say that different theories define different realms of experience; that is, the potential of the experiences for confirmation and test have changed with change of theory. In this sense, we may say that observation is theory-laden. Some observations change in importance when theories change. An observation, or type of observation, that used to be important with respect to a theory, T_1, may no longer be as important when we abandon T_1 and accept, and use, another theory, T_2. And this entails a sounder view of the history of science. Past scientists, as Feyerabend, Hanson, Kuhn, and Toulmin recognize well, need not be blamed for lack of attention to observational detail which we attend to. They need not be blamed for overlooking things we do not overlook; this is because they were sometimes looking for different things—the things that would (a) resolve the particular problems and anomalies engendered by *their* theory and (b) confirm and test their theory to a high degree ([a] and [b] are not unrelated).

Past scientists need not be blamed for not performing the experiments that present scientists perform; and this even if they did have the experimental apparatus which would have enabled them to do so. As Toulmin aptly puts it ". . . the questions we ask nature inevitably depend on prior theoretical considerations" ([21], p. 101). And different theories may ask different questions, yield different types of predictions (some more central than others) with differing logical interconnections in need of different types of observational test. Hanson is also sensitive to this point. He suggests that differences in their conceptual organization more or less guaranteed that Mach and Hertz would not have the same research problems, would not perform the same experiments ([8], p. 118). The direction of each's inquiry was somewhat different; and this was most likely due to a differing internal and logical, or conceptual, structure between the two theories. Tentative theories or hypotheses are needed to give direction to a scientific investigation. They determine what data should be collected at a given point in a scientific investigation. Data are highly relevant to a tentative theory which is advanced for consideration when either their occurrence or non-occurrence can in some sense be centrally inferred from this theory. And only these data—the relevant data—should be collected. Some thinkers have unfortunately denied this. As Feyerabend puts it,

> What *is* to be critized is the attempt *without* much guidance from thought to collect as many useless facts as possible from as many domains as possible, and to expect that science will one day miraculously profit from the collection thus assembled. ([5], p. 156)

Past scientists did, in fact, perform different experiments from present scientists. Moreover, they should have. They were testing a different theory than present-day scientists are testing. What a scientist ends up seeing and observing partly depends on what he is looking for. Scientists who are looking for new things obviously end up observing new things, if they find what they are looking for. (Note that this does not imply that they could not still see the old things if they looked). In this sense, also, observation is "theory-laden." But, as we argued earlier, any stronger sense is implausible. Nevertheless, the sense in which observation *is* theory-laden is not an uninteresting sense. Different theories highlight different features of the world. The scientist's attention is directed to those parts of experience which, in Kuhn's words, "promise opportunity for the fruitful elaboration of an accepted paradigm" ([13], p. 125). Scientists do, and should, devote little attention to processes which are, in Feyerabend's words, "situated at the periphery of their enterprise" ([5], p. 152). "Science," as Kuhn claims, indeed "does not deal in all possible laboratory manipulations" ([13], p. 125). It selects only those relevant for testing a theory currently under consideration. "As a result, scientists with different paradigms engage in different concrete laboratory manipulations." "After a scientific revolution many old measurements and manipulations become irrelevant and are replaced by others instead" ([13], pp. 125, 128). There is nothing pernicious about the theory-ladenness of observations in *this* sense. After a revolution, theory T_2 still accounts for the observations and the phenomena that T_1, the previous theory, accounted for. But T_2 does more. It accounts for what T_1 didn't account for. T_2 also generates some new predictions. And keeping in mind that T_2 accounts for the phenomena that T_1 accounted for, scientists then direct their attention to test the outcome of these new predictions. They may, in carrying out these new tests and experiments, observe some new types of experience. But this does not imply they no longer see what they saw before. They still see the old things. But they don't bother with them after their occurrence has been accounted for by the new theory. Their attention becomes refocussed. They look for additional things. They see and observe additional things. The realm of experience has *widened* with the acceptance of T_2 to include these new observations.[11] But this does not imply that previous observations, made using T_1, are somehow either annihilated, changed, or no longer accounted for by T_2. Widening is not annihilating. Explaining is not explaining away.

Return to an earlier example for a moment (II-C). Mohammed X, the black Muslim, possesses a quite different set of beliefs than Adolf Übermensch, the Nazi. Each would interrogate the young heavyweight (Cassius Clay = Mohammed Ali) somewhat differently. Each would ask him somewhat different questions before deciding whether he should be stripped of his championship and imprisoned. And they might differ as to his guilt or innocence. Yet, in spite of their widely differing beliefs they would have asked him many of the same questions. Enough of these would ensure that they were each talking about the same man. Namely, each would inquire as to his place of birth, the number of his professional fights, his height, his weight, etc. Just so for scientific theories. The fact that scientists ask different questions of nature does not imply that they are not talking about the same things. It does not imply that there is not a large overlap between their two realms of experience; it does not imply one of the realms is not an extension of the other.

CONCLUSION

Recognizing something valuable in Feyerabend's, Hanson's, Kuhn's, and Toulmin's position on the observation, we must still reject much of it for the reasons given throughout this paper; their whole position, taken as is, destroys the possibility of comparing and judging theories by reference to experience. We have urged that the use of Gestalt examples as the paradigmatic case for science tends to generate more problems than it solves. Hanson urges that after a student has received scientific training, he no longer sees the same "glass and metal instrument" ([8], p. 15) he saw before. He now "sees the instrument in terms of electrical circuit theory, thermodynamic theory, . . ., etc." ([8], p. 15). We have urged, instead, that such statements mean nothing other than that the student now has new *beliefs,* perhaps expressed by means of new concepts, about the instrument, about what it does and is, and about how it works. We have urged that the claim that such a scientist *sees* something different is unjustified. Believing is not seeing.

In fact, it has not been shown that facts are not neutral. We have suggested that scientific facts are neutral. Moreover, we have suggested that it is methodologically desirable that they be so: for the purposes of being able to compare, judge, and revise alternative and rival theories by reference to experience; for the purpose of there being able to *be* alternative and rival theories at all; for the purpose of avoiding an interaction problem between theory and environment; for the purpose of being able to test or falsify a theory by reference to observations. And none of what we have maintained is inconsistent with our claim that different theories redirect and refocus the scientist's attention or with our claim that different theories lead the scientist to *look for* different things. New theories, if they are better theories, *do, and should,* cause us to examine and observe new and additional aspects of sensory experience; they should cause us to attend to more and more of this experience. In this sense, we may say that what scientists observe does change; it increases. In this sense, we may say that observations are theory-laden. This sense is not pernicious, nor is it inconsistent with what we have maintained above; what a scientist observes, and therefore sees, before he accepts a new theory, is still the same; it remains recorded; and he could see it again if he looked. A new theory usually *does,* and moreover *should,* account for previous observations made when using a previous, *rival* theory. But the new theory *also* accounts for more. It accounts for observations the previous theory positively failed to account for. And it accounts for observations the older theory didn't even talk about. In this sense experience has widened. New theories are more *comprehensive* than old theories. Science is, therefore, a cumulative and expanding enterprise.

NOTES

1. I must acknowledge my debt to R. H. Thomason, my dissertation advisor, for his many helpful comments on earlier drafts of this chapter. I am also indebted to A. I. Fine, H. Margenau, and the late N. R. Hanson for stimulating discussions on these topics. For a more complete account of these matters, cf. my *The Justification of Scientific Change* (Dordrecht, Holland: D. Reidel, 1971, forthcoming).
2. Some might here object that this should preclude Kuhn from describing the *structure* of scientific revolutions except in the sense that they have no structure. I do not feel that such an objection is entirely forceless.
3. Not only do Tycho and Kepler see different things, but laymen are compared to infants and idiots in that they are held to all be literally blind to what the physicist sees ([8], pp. 17, 20, 22).
4. Hanson implicitly claims "seeing X as Y" is the same as "seeing an object X." Namely, seeing X as Y is to see that if $A_1, \ldots A_n$ were done to X then $B_1, \ldots B_n$ respectively would result (*cf.* [8], p. 21, where Hanson discusses fig. 1). This is, unfortunately, incompatible with Hanson's explicit claim that he is not identifying *seeing* with *seeing-as* "I do not mean to identify seeing with seeing as" ([8], p. 19). Since 'sees that' has the same sense in both claims, so would Hanson's two types of 'seeing.'
5. Professor Toulmin notes this well ([22], pp. 83–85); and Professor Kuhn has very recently amended his "normal-revolutionary" distinction to one in which theoretical *micro*-revolutions are going on continously [14].
6. "$a = F/m$" is one such neutral principle in both classical mechanics and relativistic mechanics.
7. Cf. "After investigating the matter further, I now agree with you. I did not, as I had previously thought, see John commit the crime at all. As you say, John definitely was in Japan at the time. It was, after all, Charles whom I saw."
8. It is my opinion that many of Kuhn's claims in [13] are either false or else provide additional examples of what Professor Ryle has termed "systematically misleading expressions" in his now classic article, [17]. Cf.: "When an expression is of such a syntactical form that it is improper to the fact recorded, it is systematically misleading in that it naturally suggests to some people—though not to 'ordinary' people—that the state of affairs recorded is quite a different sort of state of affairs from that which it in fact is" ([17], p. 16).
9. They will be "efficient," according to Feyerabend, if they fulfill in good measure the methodological aim of mutual critism. Cf. [4], pp. 7–8.
10. Perhaps there may be forms of cognitive conflict other than contradiction. Even if this may be so—although it is not at all clear to me how it could—my above argument does hold for a large class of conflicts; it shows that cognitive conflict in the sense of contradiction would be precluded.
11. By suggesting the unification of terrestrial and celestial laws, the Copernican revolution made the projectile a legitimate source of information about planetary motions. Cf. [12], p. 230.

REFERENCES

1. Austin, J. L., "Other Minds," in *Philosophical Papers* (Oxford: Oxford University Press, 1961).
2. Daniels, C. B., "Colors and Sensations: Or How to Define a Pain Ostensively," *American Philosophical* Quarterly, 4 (1967), pp. 231–237.
3. Feyerabend, P. K., "Explanation, Reduction and Empiricism," in *Minnesota Studies in the Philosophy of Science,* eds. H. Feigl and G. Maxwell, *Scientific Explanation, Space and Time,* vol. III, pp. 28–97 (Minneapolis: University of Minnesota Press, 1962).
4. Feyerabend, P. K., "How to Be a Good Empiricist: A Plea for Tolerance in Matters Epistemological," in B. Baumrin, ed., *The Delaware Seminar in Philosophy of Science,* vol. 2, pp. 3-39 (New York: Interscience, 1963).
5. Feyerabend, P. K., "Problems of Empiricism," in *Beyond the Edge of Certainty,* ed. R. Colodny, pp. 145–260 (Englewood Cliffs: Prentice-Hall, 1965).
6. Feyerabend, P. K., "On the 'Meaning' of Scientific Terms," *The Journal of Philosophy,* 62 (1965), pp. 266–274.
7. Feyerabend, P. K., "Reply to Criticism," in *Boston Studies in the Philosophy of Science,* ed. by R. S. Cohen and M. W. Wartofsky, vol. 2, pp. 223–261 (New York: Humanities Press, 1965).
8. Hanson, N. R., *Patterns of Discovery* (Cambridge: The University Press, 1958).
9. Hanson, N. R., *The Concept of the Position* (Cambridge: The University Press, 1963).
10. Harman, G., "Quine on Meaning and Existence, I," *The Review of Metaphysics,* 21 (1967), pp. 124–151.
11. Hempel, C. G., *Fundamentals of Concept Formation in Empirical Science* (Chicago: University of Chicago Press, 1952).
12. Kuhn, T. S. *The Copernican Revolution* (New York: Vintage Books, 1957).
13. Kuhn, T. S. *The Structure of Scientific Revolutions* (Chicago: University of Chicago Press, 1962).
14. Kuhn, T. S., "Logic of Discovery or Psychology of Research?," to be published in *The Philosophy of K. R. Popper,* ed. P. A. Schilpp.
16. Mason, S. F., *A History of the Sciences* (New York: Collier Books, 1962).
17. Ryle, G., "Systematically Misleading Expressions," in *Logic and Language (First and Second Series),* ed., by A. Flew (Garden City: Anchor Books, 1965).
18. Scheffler, I., *Science and Subjectivity* (Indianapolis: The Bobbs-Merrill Company Inc., 1967).
19. Sellars, W., "Empiricism and the Philosophy of Mind," in *Science, Perception and Reality,* pp. 127–196 (London: Routledge and Kegan Paul, 1963).
20. Shapere, D., "Meaning and Scientific Change," in *Mind and Cosmos,* ed. R. Colodny, pp. 41–85 (Pittsburgh: University of Pittsburgh Press, 1966).
21. Toulmin, S., *Foresight and Understanding* (New York: Harper and Row, 1961).

22. Toulmin, S., "Conceptual Revolutions in Science," *Synthese,* 17 (1967), pp. 75–91.
23. Toulmin, S., "Reply," *Synthese,* 18 (1968), pp. 462–463.
24. Van Fraassen, B. C., "Presupposition, Implication, and Self-reference," *Journal of Philosophy,* 65 (1968), pp. 136–152.

SUGGESTIONS FOR FURTHER READING

Achinstein, P. *Concepts of Science.* Baltimore: The Johns Hopkins Press, 1968, pp. 157–202.

Braithwaite, R.B. *Scientific Explanation.* Cambridge: Cambridge University Press, 1955, pp. 22–87.

Hanson, N.R. *Patterns of Discovery.* Cambridge: Cambridge University Press, 1958, pp. 4–49.

Kaplan, A. *The Conduct of Inquiry.* San Francisco: Chandler Publishing Co., 1964, pp. 34–83.

Kyburg, H.E. *Philosophy of Science: A Formal Approach.* New York: The Macmillan Co., 1968, pp. 87–115.

Martin, M. *Concepts of Science Education.* Glenview, Ill.: Scott, Foresman and Co., 1972, pp. 103–131.

Pap, A. *An Introduction to the Philosophy of Science.* New York: The Free Press, 1962, pp. 3–76.

Scheffler, I. *The Anatomy of Inquiry.* New York: Alfred A. Knopf, 1963, pp. 127–224.

Wartofsky, M.W. *Conceptual Foundations of Scientific Thought.* New York: The Macmillan Co., 1968, pp. 99–122.

SECTION VII

Induction, Probability and the Appraisal of Hypotheses

The first article in this section attempts to prove that the five most popular arguments designed to show that "a jusification of induction is either impossible or unnecessary or both" are "inconclusive." The problem of justifying induction or inductive inference is usually referred to as Hume's Problem, after the eighteenth century Scottish philosopher who first posed it. There are two apparently plausible ways to distinguish inductive from deductive arguments, neither of which is entirely satisfactory. Before turning you loose on this first article, then, it will be worthwhile to take a careful look at these alternatives.

On what we may call the *formalist account,* an argument is said to be deductive if and only if its premises imply its conclusion. Such arguments are also said to be truth-preserving, since whatever truth is in their premises is passed on to their conclusions. On the contrary, the premises of inductive arguments do not imply their conclusions, but merely make them more or less probable, well-supported or plausible. In other words, inductive arguments are not truth-preserving; the truth of their premises is not necessarily passed on to their conclusions.

The main virtue of the formalist account is that it allows one to describe Hume's problem with considerable facility. If the premises of deductive arguments imply their conclusions, then one is guilty of self-contradiction when he accepts the premises but not the conclusions. That provides one with the strongest possible warrant for accepting the conclusions of deductive arguments. But, as Hume argued, such a warrant is entirely lacking for inductive arguments; and that is precisely where Hume's problem begins.

The main disadvantage of the formalist account is that it makes it logically impossible to distinguish *valid* deductive arguments from deductive arguments *simpliciter.* On this account there simply are no invalid deductive arguments. Invalid arguments, by any reasonable definition, do not imply their conclusions. They are not truth-preserving. Hence, an invalid deductive argument in the formalist view must be a nontruth-preserving *and* truth-preserving argument, which is impossible. Thus, all deductive arguments on this account are valid arguments. There are no formally incorrect deductions because formal incorrectness is sufficient to destroy deductiveness itself on this account. What has happened here is precisely analogous to what happened to Plato's account of a mathematician. One is only a mathematician, he claimed, insofar as one does not make a mistake. The result was that he was suddenly left without any poor mathematicians, indeed, without any less-than-perfect mathematicians.

In order to avoid the infelicity outlined in the preceding paragraph, some philosophers (myself included) adopt what may be called a *suppositional account.* In this view, an argument is said to be deductive if it is supposed to imply its conclusion and inductive if it is supposed to make its conclusion more or less probable or acceptable. Thus, when an argument is characterized as deductive or inductive in this view, one is not evaluating it in any way. To say that an argument is deductive or inductive is merely to say what it is supposed to do, whether or not it does.

The main virtue of the suppositional account is that it allows one to distinguish valid deductive arguments from deductive arguments *simpliciter.* *Valid* deductive arguments do precisely what deductive arguments are supposed to do; namely, they imply their conclusions. They *are,* as they are supposed to be, truth-preserving. Formally incorrect deductive arguments, like poor mathematicians, do not do what they are supposed to do. But they are nonetheless deductive arguments, just as less-than-perfect mathematicians are nonetheless mathematicians. (What other kinds of mathematicians are there?) Similar remarks apply *mutatis mutandis* to inductive arguments.

But—and here is the crunch—there is also an infelicity involved in the suppositional account. For in this view, Hume's problem cannot be put as sharply as before. Now we have to say that the conclusions of formally correct = valid deductive arguments must be accepted to avoid self-contradiction, but the conclusions of formally correct = (inductively) valid inductive arguments must *also* be accepted to avoid another contradiction. That is, it would be self-contradictory to say that the premises of some inductive argument make its conclusion more or less probable or acceptable, *but* the conclusion is not acceptable, probable, or whatever. If an argument must do such and such to be correct or to be a good example of its kind, then when it does precisely that, it must be regarded as correct or good. In sum, on the suppositional account, Hume's problem tends to disappear, since the same sort of warrant now stands behind formally correct deductive *and* inductive arguments.

There is a further difficulty with the suppositional account that should be mentioned. Using the formalist definition, one may state immediately that any argument having the form of *modus ponens* is deductive and any argument having the form of a so-called *statistical syllogism* is inductive. Here are the two forms:

Modus Ponens Form	*Statistical Syllogism Form*
If p then q	Most A's are B.
p	x is an A
Therefore, q.	So, probably x is a B.

Inspection of the *forms* of arguments in the light of our definitions is all that is required to determine which arguments are deductive and which are inductive. On the other hand, this is not possible with the suppositional account. It may be supposed, for example, that an argument patterned after *modus ponens* makes its conclusion more or less probable. After all, the conclusions of such arguments do have logical probabilities of one given their premises. So the supposition is perfectly well founded. But someone may balk at the prospect of regarding *modus ponens* as an inductive argument form, and he may balk even more when an argument patterned after the form of a statistical syllogism is regarded as deductive!

The primary aim of the second article in this section is to examine three interpretations of the term "probability," seven theories of measurement for probability-values, and the strengths and weaknesses of these interpretations and theories. It is not likely that I can say anything better here than I said there, so I won't try.

The final article is intended to "assess the weight of simplicity in the construction and acceptance of scientific theory." One of the most frequently offered reasons for proposing a particular scientific hypothesis is that it is, after all, *simpler* than its alternatives. For a few years it seemed to many people as if a satisfactory account of simplicity would solve Hume's Problem and several others. However, to Bunge, it seemed that a closer inspection of scientific practice itself and various senses of "simplicity" revealed quite a different story. Besides elucidating various meanings of "simplicity" and the role simplicity in its various forms plays and/or might play in the assessment of scientific theories, he introduces other criteria of evaluation such as "empirical interpretability," "depth," "scrutability" and so on. So far no one has given all these notions the attention they deserve. Are they as clear as they might be? Are they an exhaustive and/or an exclusive set of criteria?

READING 16

Should We Attempt to Justify Induction?

Wesley C. Salmon

In the broadest sense, an inductive inference is any non-demonstrative inference to a matter of fact. An inductive rule, then, would be any non-deductive rule of inference for drawing matter of fact conclusions, provided that such a rule does not sanction drawing self-contradictory conclusions from any consistent set of premises (including the null set). I regard the problem of justifying a choice from among the wide variety of possible inductive rules. The question whether past experience is to be a guide to the future is included in the problem thus formulated, for among the possible rules are some which render evidence about the past irrelevant to predictions of the future.[1]

In recent years a rather large number of philosophers have argued that the attempt to justify induction ought to be abandoned. They have supported this claim by arguments designed to show that a justification of induction is either impossible or unnecessary or both. Within this paper I shall call such philosophers "anti-warrantists"; those who believe it worthwhile to persist in attempting to find a justification of induction will be called "warrantists." The anti-warrantists have frequently charged that there is no genuine problem of justifying induction—if there appears to be a problem it is because of a misconception of the nature of induction or justification, or because of a similar kind of confusion. To whatever extent there is a "problem," it is solved by exposing the confusions which led us to demand a justification in the first place, not by producing a justification. It will be the purpose of this paper to examine the arguments of the anti-warrantists to see whether they have, indeed, disposed of the problem

Reprinted by permission of University of Minnesota Press, from *Philosophical Studies, 8,* 1957, pp. 33–48.

of induction. The arguments of the anti-warrantists fall into two major groups. The first contains arguments designed to show that a justification of induction is impossible; the second contains those designed to show that it is unnecessary.

1.1. The anti-warrantist's most common argument is that Hume has shown the impossibility of proving that inductions will ever again be successful; hence, the warrantist is bound to meet with failure in his attempt to justify induction. However, certain warrantists can reply that they are not attempting this impossible task. They propose a different kind of justification—one not open to the kind of objection Hume raised. In particular, authors like Feigl, Kneale, and Reichenbach—called "practicalists" by Max Black (2a)—have held that, because of certain deductive relations between inductive rules and the aims of inquiry, some rules are superior to others *for the purposes inductive inference is designed to serve.*

But, it might be objected, getting true conclusions from true premises as often as possible is the aim of induction. How could any rule be shown to be suited to this aim without at the same time showing that this rule will, in fact, produce true conclusions? The practicalist attempts to answer this objection by proving that, although we cannot guarantee the success of any method, still we can demonstrate that some rules are bound to lead to true conclusions if any rules will. If we have a set of possible rules $\{R_i\}$, and if it can be shown that the situations on which any of these rules will produce true conclusions is a subset of the situations in which R_1 will yield true conclusions, then R_1 is justified, for it is as good or better than any of the others for drawing true conclusions. This is precisely the way in which a practicalist such as Reichenbach tries to justify a standard inductive method such as his *Rule of Induction* (7, p. 444).

In his essay "'Pragmatic' Justifications of Induction" (2a), Max Black presents a detailed criticism of the practicalist theory of induction. He takes issue with four contentions which he associates with practicalism (2a, pp. 157–58):

i. Some experts have held that inductive policies are bound to be applicable in all possible worlds.[2]
ii. Inductive policies have often been praised for being "self-correcting."
iii. It has been urged that the inductive policies at any rate satisfy the *necessary* conditions for prediction and generalization; so that anybody following them can be sure of having done everything in his power to discover factual truth, although he can have no guarantee of success.

iv. Sometimes it is conceded that there are methods alternative to induction (which the last argument in effect denied), but it is added that all such methods ("clairvoyance," etc.) must in turn be tested inductively, and are therefore to be regarded as pseudo-inductive methods.

i. The first thesis—that induction must be applicable in every possible world—is, I believe, mistaken, but it is no necessary part of the practicalist position. The practicalist does not hold that induction *must* work. He holds that if any method works, induction does. With Black, we may reject the first thesis as incorrect, but this does no damage to practicalism. We might add that Black in his essay "How Difficult Might Induction Be?" has pointed out that there are possible worlds in which prediction is impossible (2c). In order to refute the practicalist, however, it would be necessary to describe a possible world in which induction would not work but some other method would.

ii. According to the second thesis, induction is self-correcting. Black objects that the only sense in which induction is self-correcting is that additional data bring about *revisions* in the conclusions. But there is no guarantee that these revisions bring one closer to the correct conclusions; they may take us further from the truth. There are any number of methods which merely revise their conclusions as more data accumulate, but there is no particular virtue in this characteristic unless the revision constitutes an improvement. Black illustrates this point by contrasting two widely different methods. The first is a standard inductive method which directs us to use the relative frequency in the observed sample as our estimate of the limit of the relative frequency. Black describes a second method which he calls the "counter-inductive method." According to this method, the smaller the observed relative frequency of an event, the greater is our estimate of the limit of the relative frequency. Unfortunately, Black's method leads to contradiction (8); however, there are consistent methods which have the same general feature and which will serve our purposes as well as Black's counter-inductive method.[3] Hereinafter, when we speak of a counter-inductive method we will be referring to a consistent analogue of Black's method. Black points out that each of these methods, the standard method and the counter-inductive method, leads to revised estimates of the limit of the relative frequency as it is applied to new samples with different constitutions. Hence, Black concludes, the two methods are self-correcting in exactly the same sense.

However, Black has overlooked an important sense in which the standard method is self-correcting and in which the counter-inductive method is not. If the relative frequency in a sequence converges to some limit, the standard method will eventually lead to estimates of that limit which are accurate within any desired degree of approximation, however small. This statement is demonstrably true. Hence, persistent use of the standard method in such a sequence must sooner or later yield estimates which are close to the correct value of the limit. On the other hand, the counter-inductive method leads, in general, to estimates which always differ by at least some positive quantity from the observed relative frequency. The counter-inductive method may produce some accurate estimates of the limit, but there is no guarantee that it will ever do so. Furthermore, it can be shown that, if the sequence has a limit (other than one half), persistent use of the counter-inductive method will lead to estimates of the limit which are consistently inaccurate. This is the sense in which the standard inductive method is self-correcting and the counter-inductive method is not.

iii. The third thesis—that inductive policies satisfy the necessary conditions for prediction and generalization—is a crucial one. Unfortunately, Black's discussion of this thesis involves a major misinterpretation and a related major oversight. As Black construes the thesis it means that the use of induction is a necessary condition of successful prediction of the future—that the standard inductive method is the only method that will produce correct predictions of the future. This is not what the practicalist maintains. The practicalist holds that the ability of the standard inductive method to predict the future is a necessary condition of the predictability of the future. The practicalist does not claim that the inductive method is the only method which can correctly predict; he claims that the inductive method can predict successfully if any other method can. Hence, the demonstration that some other method might predict correctly is no refutation of the practicalist position. The practicalist maintains that, if the inductive method fails, so will every other method. In order to refute the practicalist, therefore, it is necessary to show that some other method could succeed where the inductive method would fail.

As a result of this misunderstanding, Black claims that the practicalists have unduly restricted their definition of the aims of cognitive inquiry in order to prove that the use of induction is a necessary condition of successful prediction. In particular, he criticizes Reichenbach for characterizing these aims as the ascertainment of limits of relative frequencies. He quotes the following statement from Reichenbach (7, p. 474): "Scientific method pursues the aim of predicting the future; in order to construct a precise formulation for this aim we interpret it as meaning that scientific method is intended to find limits to the frequency."

Then Black comments (2a, p. 175): "He is narrowing the aim, not just formulating it more precisely. This narrowing of the aim makes it possible for Reichenbach to view the search for limits of the frequency as a necessary condition for the success of inductive method. But anybody who says he wants to predict the future but is not interested in finding the limits of relative frequencies of occurrence of characters in infinite series is not contradicting himself. Reichenbach is not analyzing scientific method but redefining it for his own purposes."

Black might be right in maintaining that not every problem of prediction can be analyzed in terms of the problem of finding limits or relative frequencies, but he offers very little reason for his contention. Reichenbach, on the other hand, has argued at great length that all problems of prediction can be reduced to this form. Since the argument is long and technical we can only briefly indicate a few of its main points. First, Reichenbach argues that the theorems of the calculus of probability are tautological when the single non-logical primitive, the probability implication, is interpreted as the limit of the relative frequency. From this it follows that all probability derivations can be construed as derivations from premises about the limits of relative frequencies. In addition, Reichenbach, following Venn, gives an interpretation of probability of single events in which this concept is defined in terms of probability referred to infinite sequences. And finally, the probability of scientific theories is interpreted in terms of Bayes's Theorem, showing that the assignment of a probability to a theory may be regarded as a derivation from premises which deal with limits of relative frequencies. Reichenbach's argument may not be completely adequate, but in the light of this argument it is not as obvious as Black seems to feel that Reichenbach has merely redefined scientific method for his own purposes.

iv. The fourth thesis which Black attributes to the practicalists —the thesis that methods other than the standard inductive methods must be tested by induction—is again a case of a misunderstanding on Black's part. The practicalist need not, and does not, assert that alternative methods must be based upon or tested by induction. The practicalist has two theses about alternative methods. First, he requires that a method of inference be justified if it is to be considered methodologically sound. Assuming induction to be justified, inductive testing of these alternative methods would be reasonable. But this thesis has nothing in particular to do with the justification of induction. Second, the practicalist maintains that alternative methods *can,* not *must,* be tested inductively. This seems to be a correct doctrine, for we can investigate inductively the causal or statistical relations between predictions made according to a given method and the events predicted. But the significance of our ability to test alternative methods inductively is this. If someone says, "I can conceive of a world in which clairvoyance is a consistently successful method of prediction, but in which induction cannot predict successfully," we can deny this possibility on the ground that we could learn inductively about the success of clairvoyance; hence the inductive method would be successful in such a world. This point is usually introduced in support of the contention discussed above, that induction will work if any method will.

Black makes the point that the alternative methods may just as well be utilized to evaluate the inductive method as vice versa. Someone might use clairvoyance to determine whether induction is going to be successful in the future. But if induction is independently justifiable, as the practicalist tries to show, while clairvoyance is not, then, although one could use clairvoyance to test the standard inductive method, it would be utterly pointless. It would be a case of testing a justifiable method by one which was unjustified.

This examination of Black's arguments against the practicalists shows, I believe, that his objections do not hold. He has failed to show that the practicalist approach cannot succeed; indeed, he has not even shown that all practicalist attempts thus far are insufficient. I do not wish to maintain that an adequate justification of induction has yet been given (9). My point is that we have not been given reason for giving up the search.

1.2 Let us now examine a second argument of the anti-warrantist for the impossibility of justifying induction. According to this argument, justification consists in showing that whatever is to be justified conforms to certain already accepted principles or rules. In particular, an inference is justified if it can be shown to conform to the relevant rules of inference. Sometimes these rules can, in turn, be justified by reference to other rules or principles. But to ask for a justification of *all* rules of inference is without sense, for no rules or principles are available in terms of which a justification could possibly be given. When we have called into question so much that there no longer remain any rules or principles to which a justification could be referred, then we have reached the limits of justifiability. Thus, to question any particular inductive inference is legitimate, for it can be justified or refuted in terms of the general canons of induction, whereas, to question induction in general leaves no canons in terms of which the justification can occur. This view is held by Strawson (10).

If the foregoing theory is correct, empirical knowledge is, at bottom, a matter of convention. We choose, quite arbitrarily it would seem, some basic canons of induction; there is no possibility of justifying the choice. They are arbitrary in the sense that cognitive considerations do not force their acceptance. It is perfectly conceivable that someone else might select a different set of inductive canons, and if so, there would be no way of showing that one set was better than another for purposes of gaining factual knowledge. Yet, such a person would regard certain inferences as justified which we would regard as unjustified. He would hold certain conclusions to be well established while we would hold the same conclusions to be disconfirmed. This is the sense in which conventionalism follows from the Strawson theory.

Herbert Feigl has given an answer to this contention of Strawson, and it consists in providing a clear sense for the question of the justification of induction in general (4, 5, 6). Feigl distinguishes two kinds of justification. He calls the first of these "validation"; it is the kind of justification Strawson describes. An inference is validated by showing that it is governed by an accepted rule. A rule of inference is validated by showing that it can be derived from other accepted rules or principles. There is, however, a second form of justification called "vindication." This kind of justification consists in showing that a given decision, policy, or act is well adapted to achieving a certain end. Translated into Feigl's terminology, Strawson's thesis becomes the innocuous claim that it is impossible to validate induction in general; only particular inductive rules and inferences can be validated. However, the warrantist is not attempting to validate the basic inductive canons; he seeks to vindicate them. The warrantist intentionally goes beyond the limits of validation, but he does not go beyond the limits of justification. To maintain that he transgresses the limits of justification would be tantamount to a denial that vindication is a kind of justification. It is difficult to imagine any argument that could possibly support such a denial.

The appeal to vindication requires, obviously, some aims or goals in terms of which a vindication can be given. It is at this point that one of the main controversies in the whole philosophy of induction occurs. The practicalist wants to vindicate induction by reference to the aim of attaining correct predictions and true conclusions. The critic will immediately point out that it is impossible to prove that induction will ever achieve this goal. It might therefore be concluded that there is no possibility of ever vindicating induction.

There are two major alternatives at this point. On the one hand, we may revise our conception of the aim of induction in an attempt to escape the necessity of proving that induction is well suited to the aforementioned purpose of arriving at true conclusions. On the other hand, we may hold, as the practicalist does, that it is possible to show that some inductive rules are better suited than others to the purpose of arriving at true results, even though it is impossible to prove that one will be successful while another will not. Let us consider the first of these alternatives. A large number of authors have suggested that we might justify induction as a tool for establishing *reasonable* beliefs, since it is impossible to show that induction will lead to *true* beliefs. According to this view, induction could be vindicated as leading, not necessarily to true conclusions, but rather to reasonable ones. Strawson, not really content with a view which implies sheer conventionalism, argues for this kind of justification when he is not busy arguing that no justification is needed. The argument is based chiefly upon an analysis of the meaning of "reasonable" which purports to establish that reasonable beliefs, by definition, are beliefs which have good inductive support. Strawson says (10, p. 249), "'to call a particular belief reasonable or unreasonable is to apply inductive standards . . ." A little later, he further comments (10, p. 257): "to ask whether it is reasonable to place reliance on inductive procedures is like asking whether it is reasonable to proportion the degree of one's convictions to the strength of the evidence. Doing this is what 'being reasonable' *means* in such a context."

It seems to me that there are fatal objections to this approach. The term "reasonable" is, after all, virtually a synonym of "justifiable." To have reasonable beliefs is to have beliefs that are well grounded by justifiable methods. "Reasonable," then, partakes of the same ambiguity as "justifiable"—one sense referring to validation, the other to vindication. Thus, believing reasonably in one sense means holding beliefs which are sanctioned by inductive and deductive canons. In this sense, reasonable beliefs are beliefs which have been arrived at by methods which can be *validated* by reference to the accepted principles of inductive and deductive inference. In the second sense, "reasonable" means the adoption of methods and techniques which will most efficiently bring about one's ends and goals. This sense of "reasonable" corresponds to *vindication*. It is clear that using inductive methods is reasonable in the sense of "reasonable" which corresponds to *validation*. Now the problem of the justification of induction assumes the form "Is there any justification for being reasonable?" It will not do to reply that this question has the obvious tautological answer "It is reasonable to be reasonable." In view of the two distinct meanings of "reasonable" this answer may be no tautology at all, for it may contain an equivocation on the term "reasonable." Therefore, we must not lightly dismiss the question about a justification for being reasonable.

If we ask, "Why be reasonable?" construing "reasonable" in the sense related to vindication, the answer is easy to find. Being reasonable, in this sense, means adopting methods which are best suited to the attainment of our ends. Since we are motivated to achieve our ends, the realization that a method is reasonable constitutes a sufficient reason for adopting that method. To be unreasonable, in this sense, is to invite frustration. If, however, we shift to the sense of "reasonable" which is associated with validation, the answer to the question "Why be reasonable?" is much less clear. Presumably, the answer would be that to be reasonable is to be scientific and to use methods which have worked well for us. To be unreasonable would be to hold beliefs which are ill grounded and which run great danger of being false. But in so saying, have we not begged the very question which is at issue in the problem of induction? Surely there is no particular intrinsic value in being scientific or proceeding in accord with the standard inductive methods. We adopt these methods because we regard them as the best methods for establishing matter of fact conclusions. But when the problem of induction is raised, the question at issue is whether the standard inductive methods are, in fact, well suited to the purpose of establishing these factual conclusions.

It may be that the two senses of "reasonable" which we have distinguished are extensionally equivalent—that procedures are reasonable in the one sense if and only if they are reasonable in the other. But it would be a mistake merely to assume that this is the case. When a term has two distinct definitions it is not permissible to assume that the two definitions are equivalent; if there is such equivalence it must be shown. This is especially true when there are arguments which indicate that the supposed equivalence may not hold. Hume's arguments are just such arguments. If we try to show that such equivalence does hold—if we try to show that the standard inductive methods are those best suited to the purpose of arriving at correct beliefs—we are undertaking the task of the warrantist.

In accord with the philosophic fashion of the times one may be tempted to ask what is the ordinary meaning of "reasonable." Perhaps the ordinary sense of "reasonable" ensures that proceeding according to the standard inductive rules is reasonable. This is probably true of the ordinary sense. But this only shows that ordinary usage is established by people who are unaware of Hume's arguments. To say that ordinary people are untroubled by Humean doubts about induction may simply mean that ordinary people are philosophically ignorant. They assume that the two senses of reasonable distinguished above are equivalent partly because they have never thought of the distinction and partly because, had they thought of the distinction, they would have been unaware of any considerations which would lead to the conclusion that possibly the two senses are not equivalent. To cite ordinary use in this context, then, does not solve the philosophic question. It sanctions neglect of the philosophic question by virtue of an equivocation.

The attempt to vindicate inductive methods by showing that they lead to reasonable belief is a failure. If we assume that inductive beliefs are reasonable in the sense of being based on justifiable methods of inference, we are begging the question. If we regard beliefs as reasonable simply because they are arrived at inductively, we still have the problem of showing that reasonable beliefs are valuable. This is the problem of induction stated in new words. If we regard beliefs as reasonable simply because they are arrived at inductively and we hold that reasonable beliefs are valuable for their own sake, it appears that we have elevated inductive method to the place of an intrinsic good. On this latter alternative it would seem that we use inductive methods, not because they enable us to make correct predictions or arrive at true explanations, but simply because we like to use them. It sounds very much as if the whole argument (that reasonable beliefs are, by definition, beliefs which are inductively supported) has the function of transferring to the word "inductive" all of the honorific connotations of the word "reasonable," quite apart from whether induction is good for anything. The resulting justification of induction amounts to this: If you use inductive procedures you can call yourself "reasonable"—*and isn't that nice!*

1.3. The third argument which might be used to prove the impossibility of carrying out the warrantist program amounts, essentially, to a denial that there are any fundamental inductive rules or principles with which to begin the process of justification. Black seems to hold this view (2b, pp. 195, 208). The absence of fundamental inductive principles might be accounted for in either of two ways:

i. Suppose we raise a question about the correctness of some inductive rule R_1. Suppose further that R_1 can be validated by reference to an accepted rule R_2. Suppose in addition that R_2 can be validated by reference to R_3, etc. In short, suppose we can continue indefinitely justifying one inductive method by a different inductive method without ever exhausting the supply. In that case there would be an infinite sequence of inductive rules which does not end in any supreme rule or principle; hence, there would be no place to start in the attempt to give a justification of induction in general. This state of affairs is suggested by Black when he says (1, p. 88), "Every inductive principle can be justified—but not all at the same time."

To avoid confusion on this point it is important to distinguish carefully between particular inductive inferences and rules of inference. It is probably true that one can go on indefinitely supporting the conclusion of one inductive inference by another inductive inference, but frequently all these inferences will be governed by one or a very few rules. It is a very different matter to maintain that there are inexhaustible supplies of rules which make it possible to justify one by another without end. If we are going to test an inductive rule inductively we will ask what its frequency of success has been—this is what an inductive test amounts to. But in so doing we soon arrive at some form of induction by enumeration, and it is hard to imagine what other inductive rule we could bring into play to criticize it. The claim that there could be an infinite sequence of different rules as well as different inferences seems implausible in the extreme.

ii. Black seems to feel that the search for fundamental rules of inductive inference takes us so far in the direction of abstractness that, when we have found any rule which looks as though it is basic, it turns out to be so hopelessly abstract that it is useless. Take the rule of simple enumeration. According to Black, from the mere fact that all A's so far have been B, without any additional information, it is impossible to make any reasonable judgment as to whether the remaining A's will be B. Only if we know what kind of things A's are and in what conditions they have been observed can we decide whether the fact that all of them have been B is evidence that the others will also be B. Black points out that from the time we first learn how to draw conclusions we are taught the circumstances in which we can safely generalize and those in which we cannot. Whenever we make inferences we do so in the light of a good deal of auxiliary knowledge.

This is probably a true factual statement about all our inductive inferences, but it ought not to be confused with logical analysis. When we judge that it is safe to generalize in some circumstances and not in others we are making use of inductive inferences. To be sure, we may have been taught, as infants, when to apply a certain rule and when not to. But in so doing we are applying criteria which have been established inductively, if not by ourselves, then by others. There seems to be no reason to suppose we cannot investigate the evidence which supports such criteria and the inductive methods by which they can be established. Again, it would seem probable that the inductive methods involved soon reduce to some form of induction by enumeration.

It is strange that Black calls it an "assumption" to suppose that there might be a "single supreme principle of induction" (2b, p. 208). Reichenbach, for example, has offered extended arguments to show that induction by enumeration is such a principle. We took note of this argument in another connection when we considered whether the aims of prediction could be correctly characterized as the search for limits of frequencies. Since induction by enumeration is the method Reichenbach proposes for the determination of limits, the same argument supports the contention that every induction can be analyzed in terms of induction by enumeration. This argument, incidentally, precedes Reichenbach's attempt to justify induction and does not presuppose it. It seems that Black is remiss in not giving more serious attention to this important argument against the view he supports.

This concludes our consideration of arguments designed to prove that it is impossible to justify induction. We shall now discuss the group of arguments designed to show that no justification of induction is necessary. Arguments of this second type ordinarily follow arguments of the first type. After we have been told we cannot have a justification of induction the attempt is made to convince us that we never really wanted one in the first place.

2.1 Anti-warrantists have sometimes held that arguments such as Hume's only prove that induction and deduction are distinct and different forms of inference and that inductive inferences do not possess deductive validity. When some of the more skeptical writers have said that inductive conclusions are unjustified and without support, the anti-warrantists continue, all they are really showing is that such conclusions do not have *deductive* support. But, they do have *inductive* support. There are two kinds of support, inductive and deductive. There is no reason to cast aspersions on one of these kinds of support and deplore the fact that it is different from the other. When we see this point, the apparent need for a justification of induction disappears. It is recognized as an irrational demand that inductive inferences be transformed into deductive inferences (1, pp. 61–88; 11).

We agree that it would be useless to demand that inductive support be transformed into deductive support, but we must protest that the warrantist—at any rate, the practicalist—is not attempting any such feat. Rather, the warrantist proceeds on the principle that justified modes of inference may be used to justify other modes. Deductive systems are regarded as purely formal and a priori; hence, in principle they can be established independent of and prior to the establishment of any empirical knowledge. Deductive inference is the tool which is available at the beginning of the task of attempting a rational reconstruction of empirical knowledge. A consistent logic may be utilized because it cannot produce false conclusions from true premises; its justification consists in just this. Hence, when the warrantist attempts to find a deductive justification of inductive inference, he is simply using a justified system of inference in his attempt to justify a type of inference which is as yet—within his philosophical investigation—unjustified. He is not relegating induction to an inferior position; he is simply taking up problems of justification in a logical order.

If we refer back to the statement of the problem of justification as it was set out at the beginning of this paper, we realize how inappropriate are the charges of the anti-warrantists on this count. We might perfectly well agree that there is such a thing as inductive support and that this is quite distinct from deductive support. But the problem is, which of the vast range of possible inductive rules is it that defines this kind of support? This kind of question does not involve any confusion of induction with deduction or any desire to transform induction into deduction.

2.2. Finally, let us examine Max Black's contention that it is possible to provide inductive support for inductive rules (2b). This argument is classed as an argument against the necessity of justification of induction because Black uses it to show that no general justification of induction is needed. A given inductive rule can be established by a self-supporting argument, according to Black. He offers two examples of self-supporting inductive arguments. But, since he admits that one of these examples has an obviously false premise, we shall confine our attention to the other. It runs (2b, p. 197):

> In most instances of the use of R_2 in arguments with true premises examined in a wide variety of conditions, R_2 has been successful.
>
> Hence (probably):
>
> In the next instance to be encountered of the use of R_2 in an argument with a true premise, R_2 will be successful.

The rule R_2 itself reads:

> To argue from *Most instances of A's examined in a wide variety of conditions have been B* to (probably) *The next A to be encountered will be B.*

This is clearly a case of argument which conforms to the Rule R_2 and has as its conclusion that R_2 will be successful.

Black holds that this kind of self-supporting argument is neither circular nor trivial. Suppose someone had noticed that in several cases in which R_2 had been used it had led to a true prediction. His immediate reaction might be to regard R_2 as a correct inductive rule. But, wanting to be cautious and not accept rules uncritically, he could look around for many other instances of the application of R_2. If he does so and finds that R_2 has produced a true conclusion in the majority of newly examined situations in which it has been applied, Black says, is he not justified in regarding R_2 as better substantiated than before? If there is any doubt, consider what would have happened if the new instances he examined had all turned out to be cases of unsuccessful inference. This would have tended to show R_2 as unreliable. Now, if negative cases can tend to disconfirm, positive cases must tend to confirm. Hence, the further investifiation of R_2 does, in fact, help to support the reliability of R_2.

To evaluate Black's analysis, let us examine another "self-supporting" argument. Earlier in the discussion we introduced a counter-inductive method. Let us (somewhat inexactly) formulate a rule which corresponds to this method. We will call this rule "R_3"; it will read: To argue from *Most instances of A's examined in a wide variety of conditions have not been B* to (probably) *The next A to be encountered will be B*. The "self-supporting" argument which is to correspond to R_3 will be:

> In most instances of the use of R_3 in arguments with true premises examined in a wide variety of conditions, R_3 has been *un*successful.

Hence (probably):
In the next instance to be encountered of the use of R_2 in an argument with a true premise, R_3 will be successful.

It is to be noted, first, that this argument is governed by R_3 and, second, that if R_2 and R_3 are both applied in the same situations, the premises of both "self-supporting "arguments will be true but the conclusion of at least one must be false. R_2 and R_3 are conflicting rules in the sense that from the same premises they will almost always produce contrary conclusions. But, to a person who holds R_2 exactly the same evidence will support R_2 as will support R_3 for a person who holds R_3. This indicates that neither argument is genuinely self-supporting. For if one were to raise seriously the question "Which is the better of these two rules for making predictions?" he would be unable to get any answer from such "self-supporting" arguments.

The reason these arguments are not genuinely self-supporting is that the main question is begged by the use of the term "support" or its cognates such as "evidence." For an argument to establish its conclusion, either inductively or deductively, two conditions must be fulfilled. First, the premises must be true, and second, the rules of inference used by the argument must be correct. To be sure, the rule of inference does not have to be stated explicitly as part of the argument, nor does a principle which corresponds to the rule have to be incorporated in the argument as a premise. But unless we are justified in accepting the premises as true and in accepting the rules of inference as correct, the argument is inconclusive. The so-called self-supporting arguments are therefore circular in the following precise sense: the conclusiveness of the argument cannot be established without assuming the truth of the conclusion. It happens, in this case, that the assumption of the truth of the conclusion is required to establish the correctness of the rules of inference used rather than the truth of the premises, but that makes the argument no less viciously circular. The circularity lies in regarding the facts stated in the premises as *evidence for* the conclusion, rather than as evidence against the conclusion or as no evidence either positive or negative. To regard the facts in the premises as evidence for the conclusion is to assume that the rule of inference used in the argument is a correct one. And this is precisely what is to be proved. If the conclusion is denied, then the facts stated in the premises are no longer evidence for the conclusion. Someone who had adopted a skeptical view of R_2 might regard the fact that R_2 had been successful most of the time in the past as evidence that it won't be successful in the future because, say, it is 'running out of luck."

Black concludes his essay "Inductive Support of Inductive Rules" with the statement (2b, p. 208): "Any philosopher who seriously questions the admissibility of induction will have equally grave doubts about *any* induction. Thus an inductive inference to the reliability of some inductive rule will still leave such a skeptical philosopher a prey to his skeptical doubts." Near the beginning of the same essay, Black comments (2b, pp. 193–94): "It is to be presumed that the reader can recognize some instances of inductive correctness; if not, this essay will be of no value to him."

I think it *is* to be presumed that most readers can recognize some instances of inductive correctness, and when we ask for a justification of induction we do not presume otherwise. We merely ask for the grounds of such recognition. Black seems to feel this request is improper. He has failed to show, however, why it is improper, beyond showing that it is difficult to fulfill. If there is such a thing as inductive correctness and if instances of it can be recognized, it seems likely that there are criteria of inductive correctness which can be found and vindicated. We have, at any rate, failed to find reasons for condemning the search.

The five foregoing arguments are, in my opinion, the most important arguments of the anti-warrantists. Each is inconclusive. Therefore, in answer to the major question of this paper—should we attempt to justify induction?—two comments will be sufficient. First, we have not been given any good reason for abandoning the attempt. Second, important questions hang on the justifiability of induction. If induction cannot be justified, inductive beliefs become conventional (as explained under 1.2); if induction can be justified, this conventionalism can be circumvented. Since we need hardly argue the philosophical significance of the doctrine of conventionalism, an affirmative answer to the original question seems inescapable.

NOTES

1. The "dagger-method" described in (3) is an example.
2. In his discussions, Black uses "inductive policies" to mean standard inductive methods such as Reichenbach's *Rule of Induction*.
3. The "diamond-method" described in (3) is one such.

REFERENCES

1. Black, Max. *Language and Philosophy.* Ithaca: Cornell University Press, 1949.
2. Black, Max. *Problems of Analysis.* Ithaca: Cornell University Press, 1954.
   a. "'Pragmatic' Justifications of Induction," pp. 157–90.
   b. "Inductive Support of Inductive Rules," pp. 191–208.
   c. "How Difficult Might Induction Be?" pp. 209–25.
3. Burks, Arthur. "The Presupposition Theory of Induction," *Philosophy of Science,* 20:177–97 (1953).
4. Feigl, Herbert. "De Principiis Non Disputandum . . ." in Max Black, ed., *Philosophical Analysis.* Ithaca: Cornell University Press, 1950.
5. Feigl, Herbert. "Scientific Method without Metaphysical Presuppositions," *Philosophical Studies,* 5:17–28 (1954).
6. Feigl, Herbert. "Some Major Issues and Developments in the Philosophy of Science of Logical Empiricism," in Herbert Feigl and Michael Scriven, eds., Minnesota Studies in the Philosophy of Science, Vol. I, *The Foundations of Science and the Concepts of Psychology and Psychoanalysis.* Minneapolis: University of Minnesota Press, 1956.
7. Reichenbach, Hans. *The Theory of Probability.* Berkeley: University of California Press, 1949.
8. Salmon, Wesley C. "Regular Rules of Induction," *Philosophical Review,* 1956.
9. Salmon, Wesley C. "The Predictive Inference," *Philosophy of Science,* forthcoming.
10. Strawson, P.F. *Introduction to Logical Theory.* New York: John Wiley & Sons, 1952.
11. Will, Frederick L. "Generalization and Evidence," in Max Black, ed., *Philosophical Analysis.* Ithaca: Cornell University Press, 1950.

READING 17

Meaning and Measures of Probability

Alex C. Michalos

THE LOGICAL INTERPRETATION

a) The Problem of Meaning

As we suggested earlier, according to the logical interpretation, "probability" is interpreted as a *logical relation* between sentences (or propositions, or statements). For example, relative to the sentence

Either a head or a tail will turn up.

the sentence

A head will turn up.

may be assigned a probability of $\frac{1}{2}$. The relation between the given sentence(s) and the sentence whose probability is to be determined is considered *similar to* but *weaker than* the relation between the premisses and conclusion of a valid demonstrative argument. In the latter case, we say the premisses *imply* the conclusion. Here we say the premisses *partially imply* the conclusion. The question of the existence, comparative strength, or degree of the relation of *partial implication* between two or more sentences is a purely logical one. That is, the answers to such questions must be determined by logical analysis, *not* by empirical investigation. Thus, probability sentences are *logically true* if they are true at all, and *logically false* (i.e., self-contradictory) if they are false at all. While certain experiences or empirical investigations may persuade us to abandon certain probability sentences which were determined *a priori,* such

Reprinted by permission of the author and publisher from Alex C. Michalos, *Principles of Logic,* © 1969, Prentice-Hall, Inc., Englewood Cliffs, N.J.

experience can not falsify these statements any more than they could verify them. Since the sentences are analytic or self-contradictory, strictly speaking, experience is irrelevant to them. (We shall consider this peculiarity in detail later.)

b) The Problem of Measurement

Granting, then, that probabilities are to be determined by and based on logical considerations alone, the question is, *How do we obtain initial numerical probabilities?* A rough provisional answer to this question is this: *Count* logical possibilities and determine the ratios of certain numbers of these to certain numbers of others. We shall consider two theories of measuring probabilities within the logical interpretation which might be roughly characterized by this answer, namely, the *classical theory* and the *logical range theory.*

i) The Classical Theory. According to the **classical theory** *of measuring probabilities* which was created by Pierre Simon de Laplace in the latter half of the eighteenth century, we measure the probability of an event of a certain kind in three steps: 1. Count the total number of equally possible events of a certain kind. 2. Count the number of cases of these which are "favorable" to the event whose probability is to be measured. 3. The ratio of the number of "favorable" cases to the total number of equally possible events is the required numerical value; i.e.,

$$\text{(A)} \quad \text{Measure of probability} = \frac{\text{number of favorable cases}}{\text{total number of equally possible events.}}$$

Since events may be described by *sentences,* we may rewrite (A) thus:

$$\text{Measure of probability} = \frac{\text{number of favorable sentences}}{\text{total number of equally possible sentences.}}$$

Consider some examples.

Example 1

What is the probability of tossing a head with a fair coin? (A coin is fair if its design is such that its two faces have an equal chance of turning up on any toss.)

Solution. The sentence whose probability we wish to measure is

A head turns up.

There are two equally possible sentences (describing two equally possible events), namely,

A head turns up.
A tail turns up.

Hence, the denominator of our fraction is 2. There is one "favorable" sentence (describing one "favorable" event), namely,

A head turns up.

So the numerator of our fraction is 1. Hence, the probability of tossing a head with a fair coin is $\frac{1}{2}$, which is also the probability of the sentence describing that event. Dividing the numerator by the denominator, we may convert this ordinary fraction into a decimal fraction, namely, 0.5. Either expression '$\frac{1}{2}$' or '0.5' will do.

We have repeatedly emphasized the fact that probabilities are always *relative* to something. We shall call the set or collection of events (or sentences, or hypotheses), relative to which the probability of some event (or sentence, or hypothesis) is determined, a **sample space.** It is, after all, a "space" full of *samples* of a whole population (viz., the whole collection). Occasionally we shall speak of activities engaged in to obtain sample spaces as **experiments;** e.g., counting the total number of cards in a deck, tossing a coin, rolling a die, etc. Then the various events in such sample spaces will be called **outcomes** (e.g., a king, a queen, an ace, etc.; a head; a six; etc.) and the sample itself will be called an **outcome space.** The sample or outcome space which happens to contain the *total number* of events under discussion, *whether or not* they are *equally possible,* will be called the **universe of discourse** or **universal set.** When the sample or outcome space contains *some* but not all of the events in the universal set, we sometimes refer to it as a **subset** of the universal set. *Every* probability value is relative to *some* or a *part* of some universe of discourse.

EXAMPLE 2

Suppose a fair *die* is tossed. What is the probability of the hypothesis that a face with six dots will turn up?

Solution. A die is a small cube with six different faces, namely,

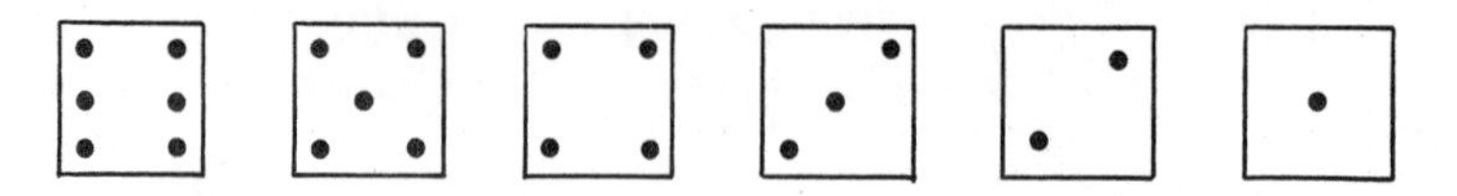

Our *sample space,* then, contains six events and is identical with our *universe of discourse.* The hypothesis whose probability we wish to measure is

A face with six dots will turn up.

There are six *equally possible* hypotheses, one for each face. Hence, the denominator of our fraction is 6. There is one "favorable" hypothesis. So the numerator of our fraction is 1. Hence, the probability of the hypothesis that a six will turn up is 1/6.

This value 1/6 is called the **initial** or **absolute** probability of the hypothesis that a six will turn up; it is the probability of the hypothesis relative to the *initially given universe of discourse or universal set*. It is contrasted with **conditional** or **relative** probability, which is the probability of a hypothesis relative to a *subset* of the universe of discourse. That is, although *all* probabilities are relative to some sample space, only those which are relative to a sample space which is a subset (part) of the (whole) universe of discourse are called 'relative' or 'conditional probabilities'. Consider the following example of relative or conditional probability.

EXAMPLE 3

Given the information that a die has been tossed and that a face with an *even* number of dots has turned up, what is the probability of the hypothesis that a six has turned up?

Solution. Because we know a face with an even number of dots has turned up, we can not use the universal set as our sample space. Instead, we must use only a subset of it, namely, the *three equally possible even-numbered* faces.

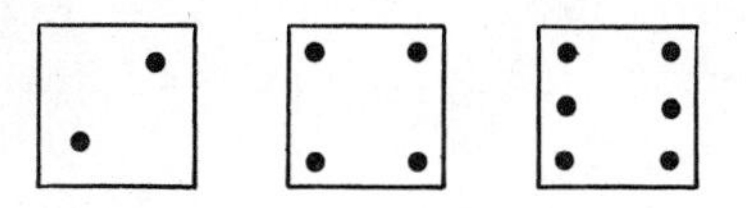

So the denominator of our fraction is 3. We still have only one "favorable" hypothesis; i.e., that a face with six dots has turned up. Hence, the numerator of our fraction is 1. Thus, the *relative* or *conditional* probability of the hypothesis that a six has turned up, *given* the information that a face with an even number of dots has turned up, is $\frac{1}{3}$.

The classical theory of measuring probabilities is useful *provided that* we have a set of *equally possible* events (or hypotheses, or sentences) to begin with. Therefore, it is imperative that we devise some criterion for determining exactly which events are equally possible and which are not. The criterion employed by Laplace was called **The Principle of Nonsufficient Reason;** i.e., in the absence of any reason for considering certain events to be *not* equally possible may be considered equally possible. John Maynard Keynes called this **The Principle of Indifference** and we shall use this term too, because it is shorter. Since "equally possible" events are events with equal initial probabilities, the principle prescribes the assignment of *equal numerical values* for all initial probabilities. This prescription, however, is *ambiguous* and leads directly to contradictions. For suppose we are interested in the probability of throwing a six with a fair die. Since the die is fair, we have *no reason* to believe that any of the six sides is more probable than any other. Hence, according to the principle, we must assign a probability of 1/6 to the hypothesis that a face with six dots will turn up. But we also have *no reason* to expect a six rather than a nonsix. So, according to the principle, we must assign a probability of $\frac{1}{2}$ to the hypothesis that a face with six dots will turn up. But *this contradicts* our original assignment. Furthermore, we have no reason to expect a five rather than a nonfive. So the probability of the hypothesis that a face with five dots will turn up is $\frac{1}{2}$. But applying the *Special Addition Principle* of the calculus of probability (which is explained in Sec. 6.2), we find that the probability of the hypothesis that *either* the face with five dots *or* the face with six dots will turn up is *one*. That is, it is *certain* that either a five or a six will turn up! Evidently the principle is taking too much for granted.

When contradictions and extravagant probabilities are obtained by applying the principle as we did in the last paragraph, most people suspect that it is *not* the principle which is at fault, but the application of that principle. Most people are unwilling to admit that we have no reason to expect a nonsix rather than a six. After all, they reason, nonsix *includes* or *is made up of* five other events, and what better reason could we want for *not* assigning six and nonsix equal initial probabilities? They suggest, then, that the principle might be modified somehow to block such misapplications or misunderstanding as we have just seen. Let us consider the possibility more carefully.

To begin with, we should distinguish divisible events from indivisible ones. **Indivisible events** are events which for *some* reason (e.g., logical, physical, technological, or practical) cannot be broken up into component events. They are *ultimate alternatives* or *simple* events in the sense that they are not composed (or compounded) of others, but all others are composed (or compounded) of them. For example, each of the faces of a die; the faces of a coin; births, deaths; (for some purposes) apples, men, automobiles; etc. Anything which for one reason or another we might wish to treat as an indivisible element or unit could serve as an example. Events which are *not* indivisible, then are **divisible.** They are compound events, i.e., events which are composed of others; e.g., nonsix; even-numbered face; (for some purposes) bald men, bitter apples, blue automobiles; etc. Given this distinction between divisible and indivisible events, we may modify the Principle of Indifference thus: In the absence of any reason for considering certain *indivisible* events to be *not* equally possible, they may be considered equally possible. This modified principle is strong enough to withstand the *sort* of attacks we considered above (though *not all* attacks) *provided that* we can reach some agreement about which events are indivisible and which are not in any given experiment. Usually (and fortunately), as in the case of six and nonsix, nothing more than common sense or intuition is required for investigators to reach such an agreement. Hereafter when we refer to the Principle of Indifference, we shall always mean the modified version. . . .

ii) The Logical Range Theory. The most thorough and systematic investigation of the theory of measuring probabilities by logical ranges has been given by the contemporary philosopher Rudolf Carnap. To begin with, we must imagine that we have some very simple world to describe. It is a world with only one primitive property, say the property of being red, and only two individuals, say Arf and Barf. The property is **primitive** in the sense that it is not to be defined in terms of any other properties in the little world, but another property (that of being nonred) is defined in terms of it. There are only four possible states of affairs in this world, namely,

S1 Arf is red and Barf is red.
S2 Arf is red and Barf is not red.
S3 Arf is not red and Barf is red.
S4 Arf is not red and Barf is not red.

(Notice the similarity between this table and ordinary truth-tables for two sentences with two possible truth-values.) Instead of talking about states of affairs, Carnap introduces the technical term "state description." A **state description** is a sentence which indicates, for *every* individual in the universe of discourse and for *every* primitive property, whether or not the individual has that property. Sometimes the term 'individual distributions' is used instead of 'state description.' The former is appropriate since the thing designated is a sentence describing the way in which *all* the properties in the universe of discourse are *distributed* among *all* the *individuals.* When there are n individuals in the universe and r primitive properties, there are 2^{nr} state descriptions. In the simple little world we are considering, we have $n = 2$ and $r = 1$; hence we have $2^{nr} = 2^{2 \times 1} = 2^2 = 4$ state descriptions S1, S2, S3, and S4.

The **logical range** of any *sentence* is the class consisting of all those state descriptions in which the sentence is true. Thus, for example, the *logical range* of the sentence

Arf is red.

is the class consisting of S1 and S2. These two state descriptions describe the only logically possible states of affairs in our little world in which the given sentence is true. That is, if the state of the world is such that either

S1 *Arf is red* and Barf is red.
or
S2 *Arf is red* and Barf is not red.

adequately describes it, then the given sentence is true. Again, the *logical range* of the sentence

Arf and Barf are red.

is the class consisting of the single state description

S1 *Arf is red* and *Barf is red.*

Here S1 describes the only logically possible state of affairs in our little world in which the given sentence is true. The *logical range* of the sentence

Either Arf is red or Barf is red.

is the class consisting of the three state descriptions

S1 *Arf is red* and *Barf is red.*
S2 *Arf is red* and Barf is not red.
S3 Arf is not red and *Barf is red.*

If any one of these state descriptions adequately describes the present state of our world, then the given sentence is true. Moreover, S1, S2, and S3 describe the only logically possible states of affairs in which the given sentence *may* be true in this world.

Notice, now, that the middle two state descriptions S2 and S3 are related to each other in a rather peculiar way, a way in which neither of them is related to either of the other state descriptions (S1 and S4) or in which neither of these (S1 and S4) are related to each other. If we substitute 'Barf' for 'Arf' and 'Arf' for 'Barf' in the middle two state descriptions, then S2 becomes S3 and S3 becomes S2. That is,

S2 Arf is red and Barf is not red.

becomes

S3 Barf is red and Arf is not red.

and

S3 Arf is not red and Barf is red.

becomes

S2 Barf is not red and Arf is red.

When state descriptions are related in such a way that one may be obtained from another by substituting the names of certain individuals for others (in the manner illustrated above), they are **isomorphic.** Thus, S2 and S3 are isomorphic state descriptions.

A **structure description** is a disjunction of all the state descriptions which are isomorphic to a given one. In the little world of our example there are three structure descriptions, namely,

St 1 Arf is red and Barf is red.

which is identical with S1;

St 2 Either Arf is red and Barf is not red, or Arf is not red and Barf is red.

which is the *disjunction* of S2 and S3; and

St 3 Arf is not red and Barf is not red.

which is identical with S4. While we happen to know *exactly which* individuals in our little world have which properties, all that is required for the complete characterization of a structure description is that we know *how many* individuals have which properties; e.g., we might characterize the second structure description St 2 by indicating that in it there is one red individual and one that is not red. It does not matter which individuals are or are not red. Sometimes the term "statistical distribution" is used instead of "structure description." The former is appropriate since the thing designated is a sentence giving us a certain *statistic* about the *distribution* of properties in our world, namely, the number of individuals in this world which have *each* of the available properties. A census taker, for example, would only be interested in statistical distributions such as the number of men, women, boys, and girls living in a certain house; the number of automobiles purchased in a year; the number of deaths; etc.

According to the **logical range theory** *of measuring probabilities* as it was developed by Carnap, then, we measure the probability (Carnap would say the "degree of confirmation") of a hypothesis in three steps: a) Apply the Principle of Indifference to the structure descriptions of some universe of discourse. b) Apply the principle *again* to the isomorphic state descriptions which make up every structure description. The result of these two steps is that each *state description* has been assigned an *initial probability*. c) The measure of the initial probability of any hypothesis then is the *sum* of the initial probabilities of all the state descriptions in the *range* of that hypothesis. That is,

Measure of initial probability of hypothesis = sum of initial probabilities of state descriptions in range of that hypothesis.

Consider some examples.

EXAMPLE 4

Given the little world of Arf and Barf, what is the probability of the hypothesis that Barf is not red?

Solution. Let us begin by abbreviating our description of this world. Let 'a' and 'b' be short for the proper names 'Arf' and 'Barf', respectively; and let 'R' and '$\overline{R}$' be short for the predicates 'is red' and 'is not red', respectively. Using familiar logical notation, the abbreviated description of our world looks like this.

S1	$Ra \cdot Rb$
S2	$Ra \cdot \overline{R}b$
S3	$\overline{R}a \cdot Rb$
S4	$\overline{R}a \cdot \overline{R}b$

Following step a), we assign equal probabilities to the structure descriptions.

Structure Descriptions

St 1	$Ra \cdot Rb$	$\frac{1}{3}$
St 2	$\left\{ \begin{array}{l} Ra \cdot \overline{R}b \\ \overline{R}a \cdot Rb \end{array} \right\}$	$\frac{1}{3}$
St 3	$\overline{R}a \cdot \overline{R}b$	$\frac{1}{3}$

Next, according to step b), we assign equal probabilities to each of the state descriptions which make up every structure description.

	Structure Descriptions		*State Descriptions*
St 1	$Ra \cdot Rb$	$\frac{1}{3}$	$\frac{1}{3}$ (S1)
St 2	$\left\{ \begin{array}{l} Ra \cdot Rb \\ Ra \cdot \overline{R}b \end{array} \right\}$	$\frac{1}{3}$	$\left\{ \begin{array}{l} \frac{1}{6} \text{ (S2)} \\ \frac{1}{6} \text{ (S3)} \end{array} \right.$
St 3	$\overline{R}a \cdot \overline{R}b$	$\frac{1}{3}$	$\frac{1}{3}$ (S4)

Table 5.21

The *range* of the hypothesis that Barf is not red, '$\overline{R}b$', is the class of those state descriptions in which it is true, namely, S2 and S4. The sum of the probabilities of these two state descriptions is $\frac{1}{6} + \frac{1}{3} = \frac{1}{2}$. Hence, the probability of the hypothesis that Barf is not red is $\frac{1}{2}$.

It should be emphasized that the *double application* of the Principle of Indifference is a somewhat problematic novelty of Carnap's logical range theory. Carnap's predecessors, Ludwig Wittgenstein and Johannes von Kries, and most people apply the principle directly to state descriptions. The result of such an application looks like this.

	State Descriptions	
S1	$Ra \cdot Rb$	$\frac{1}{4}$
S2	$Ra \cdot \overline{R}b$	$\frac{1}{4}$
S3	$\overline{R}a \cdot Rb$	$\frac{1}{4}$
S4	$\overline{R}a \cdot \overline{R}b$	$\frac{1}{4}$

In Sec. c) on page 385 we shall consider Carnap's argument in defense of the double application. We are presenting Carnap's position rather than that of, say, one of his predecessors because Carnap's position has been more thoroughly developed and discussed.

EXAMPLE 5

Suppose three fair coins are tossed. What is the probability of the hypothesis that *at least two* of them will turn up heads?

Solution. We have a little world with two properties, that of being a head and that of being a tail. There are three individuals, the three coins. If a coin turns up heads (tails) [i.e., has the property of being a head (tail)], then it is logically impossible for it to turn up tails (heads) at the same time. The two properties exclude one another and are, therefore, called 'mutually exclusive.' We might, then, consider the property of being a tail as simply the *lack* of the *primitive* property of being a head. Thus, when we determine the number of state descriptions in this world, we may assume we have $r = 1$ primitive property and $n = 3$ individuals. So we shall need $2^{nr} = 2^{3 \times 1} = 2^3 = 8$ state descriptions. Let us abbreviate 'is a head' by 'H', 'is a tail' by '$\overline{H}$', 'the first coin' by 'a', 'the second coin' by 'b', and 'the third coin' by 'c'. Then our abbreviated table of state and structure descriptions will look like this.

		Structure Descriptions	*State Descriptions*
St 1	$Ha \cdot Hb \cdot Hc$	$\frac{1}{4}$	$\frac{1}{4}$ (S1)
St 2	$Ha \cdot Hb \cdot \overline{Hc}$		1/12 (S2)
	$Ha \cdot \overline{Hb} \cdot Hc$	$\frac{1}{4}$	1/12 (S3)
	$\overline{Ha} \cdot Hb \cdot Hc$		1/12 (S4)
St 3	$Ha \cdot \overline{Hb} \cdot \overline{Hc}$		1/12 (S5)
	$\overline{Ha} \cdot Hb \cdot \overline{Hc}$	$\frac{1}{4}$	1/12 (S6)
	$\overline{Ha} \cdot \overline{Hb} \cdot Hc$		1/12 (S7)
St 4	$\overline{Ha} \cdot \overline{Hb} \cdot \overline{Hc}$	$\frac{1}{4}$	$\frac{1}{4}$ (S8)

Table 5.22

(Notice the similarity between this table and an ordinary truth-table for three sentences with two possible truth-values.) We have four structure descriptions since none, one, two, or all three coins might have the property of being a head. Applying the Principle of Indifference to these structure descriptions [according to step a)], each one is assigned a probability of $\frac{1}{4}$. Applying the principle *next* to the state descriptions which make up these structure descriptions, we find the values listed under 'state descriptions' in Table 5.22. The *range* of the hypothesis that *at least two coins* will turn up heads is the class consisting of those state descriptions in which it is true, namely, S1—S4. The sum of the probabilities of these four state descriptions is $1/4 + 1/12 + 1/12 + 1/12 = 1/2$. Hence, the probability of the hypothesis that at least two coins will turn up heads is 1/2.

Again, it should be noted that Carnap's predecessors and many other people would apply the Principle of Indifference directly to state descriptions. As a result, each state description in Table 5.22 would be assigned an initial probability of 1/8.

So far we have seen how *initial* probability values are obtained with the logical range theory. Little more needs to be said to explain how *conditional* probabilities are obtained. We measure the (conditional) probability of a hypothesis represented by '*H*', *given* some evidence '*E*', by the fraction whose numerator is the sum of the initial probabilities for all the state descriptions in the range of '*H* and *E*' and whose denominator is the sum of the initial probabilities for all the state descriptions in the range of '*E*'; i.e.,

$$\text{Measure of probability of '}H \text{ given } E\text{'} = \frac{\text{Sum of initial probabilities of state descriptions in range of '}H \text{ and } E\text{'}}{\text{Sum of initial probabilities of state descriptions in range of '}E\text{'.}}$$

(We are using the capital letters 'H' and 'E' as sentential variables here because they seem to be more suggestive than, say, 'p' and 'q'.) Consider the following example.

EXAMPLE 6

Given the information that three coins have been tossed and that the first two have turned up heads, what is the probability that the third coin is a head?

Solution. We may use Table 5.22 as a description of our universe of discourse.' The evidence 'E' is interpreted as

(Coin) a is *a* head and *b* is a head.

That is,

$$Ha \cdot Hb.$$

The *range* of 'E' is the class consisting of S1 and S2. The sum of the initial probabilities of S1 and S2 is $1/4 + 1/12 = 1/3$. So $1/3$ is the denominator of our fraction. The numerator is the sum of the initial probabilities in the range of '*H and E*'; i.e.,

(The third coin) *c* is a head *and* (coin) *a* is a head *and* *b* is a head.

This is just state description 1, which has an initial probability of $\frac{1}{4}$. Hence, the numerator of our fraction is $\frac{1}{4}$. Therefore, the (conditional) probability of the hypothesis that (the third coin) *c* is a head given the evidence that the first two are heads is $(\frac{1}{4})/(\frac{1}{3}) = \frac{3}{4}$.

c) Some Advantages and Disadvantages of the Logical Interpretation

There are at least five advantages to the logical interpretation of 'probability' and probability sentences when either the classical or the logical range theories (with a *single* rather than a *double* application of the Principle of Indifference) are used to obtain numerical values. In the first place, the interpretation usually allows us to obtain probabilities which agree with our *intuition*. Given, for example, a lottery ticket with 1 chance in 1000 of winning, most people do not *expect* to win. There are, after all, 999 "unfavorable" cases and only 1 "favorable." On the other hand, if someone had the 999 tickets, he would feel very confident about his chances of winning the lottery. Of course, given 1 ticket or 999, no one could be *certain* of the outcome. Nevertheless, in either case most people would have very little doubt about it. The logical interpretation is usually in complete agreement with such *intuitions*. In the *second* place, the interpretation frequently allows us to obtain probabilities which agree with our *observations*. Suppose, for example, that someone repeatedly played lotteries in either of the ways suggested above (i.e., always buying only 1 or else 999 of a total possible 1000), he would win with roughly the same frequency. That is, in the long run he would win just about *once* or else nearly 999 times in every 1000. We would observe, then, a *frequency* of wins which corresponds fairly closely to the probabilities obtained with the logical interpretation. Coins usually *do* turn up heads about half of the time in long-run experiments and each of the sides of a die *does* turn up roughly $\frac{1}{6}$ of the time. The logical interpretation agrees fairly well with such *observations*.

This second advantage leads directly to a *third*, namely, that we may use sentences about logical probability (or probabilities based only on logical considerations) to *appraise* predictions and actual frequencies *prior to observation*. To see how this is possible, we must be very clear about the meaning of probability sentences on the logical interpretation. We have said that sentences about logical probability are logically true (analytic) or logically false (self-contradictory); e.g.,

> The probability of the hypothesis that a head will turn up is $\frac{1}{2}$.

is logically true if it is true at all or it is logically false. This is tantamount to saying that such probability sentences have no *predictive force*. They are not *about* what will or will not happen. Rather, as their name suggests, they are *about probability conceived as a logical relation;* e.g., the example above is an analytic sentence about the logical relation (probability) of a particular hypothesis, namely,

> A head will turn up.

to some evidence (sentences) about the shape of a coin, etc. The latter hypothesis *has* predictive force; i.e., it is a prediction *about* what will or will not happen. It is an empirical sentence *about* the *appearance* of a particular face of a coin. Now, the fact that probability sentences (on the logical interpretation) are not predictions does *not* prevent us from using them to assess predictions *prior* to the observation of outcomes. If, for instance, Otto Optimist buys 1 of 1000 lottery tickets and *predicts* that he will win the lottery, we know just about what this prediction is worth. We know just about what it is worth *because* we have the (logical) probability sentence

> The probability of the hypothesis that a ticket numbered such and such will be drawn is $\frac{1}{1000}$.

The probability sentence has no predictive force, but it may be used to *assess* predictions.

Again, suppose Otto has always bought only 1 ticket for each lottery *and* that he has won roughly 50 percent of the time. Given our probability sentence, such success is incredible—unless, of course, the lotteries have been "fixed". Given Otto's purchases, his success, and our probability sentence, we *suspect* that there has been foul play. Nobody is *that* lucky. What we are doing, then, is using our probability sentence to appraise the actual frequency of success which Otto has been experiencing. *Before* we observe that frequency, we use the logical interpretation to *estimate* it. The more that frequency disagrees with our logical probability estimate, the more we suspect that chance alone is not responsible for Otto's success. Hence, while the probability sentence has no predictive force, it may be used to *assess actual frequencies.*

The *fourth* advantage of the logical interpretation is that it is always *applicable.* There is no particular problem applying the notion to such unique events as; e.g., the death of your Uncle Stanley on the last day of October 1980. Finally, we must list the comparatively *easy access* to probability values as the *fifth* advantage of this interpretation. We do not have to engage in any lengthy investigation of some part of the physical world or examine anyone's psychological characteristics; i.e., we do not have to consider the physical universe or anyone's psychology. We merely have to consider relevant logical possibilities and determine certain ratios among various numbers of them.

There are at least two *disadvantages* of the logical interpretation. In the *first* place, there is the problem of the *proper application* of the Principle of Indifference to obtain initial probabilities. According ot Carnap's logical range theory, the principle should be applied first to *structure* descriptions and then to state descriptions. If it is applied this way, however, we may obtain very *inaccurate estimates* of actual frequencies. Suppose, for example, that we toss 10 fair coins all at once and ask for the probability of the hypothesis that all 10 will turn up heads. We have, then, a world with one primitive property (that of being a head) and 10 individuals. Hence, we have 11 possible structure descriptions, namely, all 10 individuals (coins) are heads, exactly 9 are heads, 8 are heads, . . . , none are heads. Hence, applying the principle to structure descriptions, the initial probability of the hypothesis that all 10 coins will turn up heads is 1/11. But the actual frequency of the event described by this hypothesis is (or *would be* if we ran some experiments) *much closer* to 1/1024. The application of the Principle of Indifference to structure descriptions, then, may yield very inaccurate estimates.

It *seems* that the way out of this problem is already before us, for if the principle is applied directly to *state* descriptions, as suggested by Wittgenstein and von Kries, we obtain an initial probability of 1/1024 for the hypothesis in question. This is basically the approach taken by the classical theory of measuring probabilities. Carnap has insisted, however, that our problems cannot be eliminated in this fashion. His argument runs as follows: According to a generally accepted **Principle of Learning from Experience,** "other things being equal, a future event is to be regarded as the more probable, the greater the relative frequency of similar events observed so far under similar circumstances." But if the Principle of Indifference is applied directly to state descriptions, then we may be forced to violate the Principle of Learning from Experience. We may be forced to assign conditional probabilities to hypotheses which are *always identical* to their initial probabilities, i.e., we may be unable to alter probabilities by altering our evidence. Hence, we cannot apply the Principle of Indifference directly to state descriptions.

With the help of Example 5 in our discussion of the logical range theory, we may illustrate the problem Carnap has suggested. (Our illustration is the same in all important respects to one that Carnap constructed using black and white balls in an urn.) If the Principle of Indifference is applied directly to the state descriptions in Table 5.22, each state description will have an initial probability of $\frac{1}{8}$. If we know that three coins have been tossed and that the first two have turned up heads, then the probability of the hypothesis that the third coin is a head is *not* $\frac{3}{4}$ as before, but $\frac{1}{2}$. That is, now we have the probability of the evidence '*E*'

$$Ha \cdot Hb$$

is $\frac{1}{4}$, the sum of the probabilities of S1 and S2 (i.e., the range of '*E*'). The probability of '*H and E*'

$$Ha \cdot Hb \cdot Hc$$

is now $\frac{1}{8}$, the probability of S1. Hence, the probability of '*H* given *E*' is $(\frac{1}{8})/(\frac{1}{4}) = \frac{1}{2}$. Suppose, now, that *instead* of the evidence '*E*', we are given '*E*′'

Coin *a* is not a head and *b* is a head.

That is, we know three coins have been tossed and

$$\overline{H}a \cdot Hb$$

What is the probability of the hypothesis that *c* is a head? It is *still* $\frac{1}{2}$. The range of '*E*′' is S4 and S6. (See Table 5.22.) Hence, the probability of 'E' is $\frac{1}{8} + \frac{1}{8} = \frac{1}{4}$. '$H$ *and* E' is interpreted as

$$\overline{H}a \cdot Hb \cdot Hc$$

which is just S4. Hence, its probability is $\frac{1}{8}$. Dividing $\frac{1}{8}$ by $\frac{1}{4}$ we have $\frac{1}{2}$, the probability of the given hypothesis. Hence, given

$$Ha \cdot Hb$$

or

$$\overline{H}a \cdot Hb$$

the probability of 'Hc' is $\frac{1}{2}$. But the sum of the probabilities for the state descriptions in the range of 'Hc' *alone* is *also* $\frac{1}{2}$. Thus, the conditional probability of this hypothesis is *identical* to its initial probability; i.e., we are unable to alter the probability of the hypothesis by altering the evidence. Hence, the Principle of Learning from Experience is violated if the Principle of Indifference is applied directly to state descriptions. Therefore, according to Carnap, it would be a mistake to assume that the problem of the proper application of the Principle of Indifference is solved by applying the principle to state descriptions instead of structure descriptions.

The *second* disadvantage of the logical interpretation has been suggested by the philosopher Alfred Ayer. Ayer claims that regardless of the amount of evidence gathered, every correctly determined probability sentence is as good as every other. A probability sentence on this interpretation is *logically* true if it is true at all. Among equally analytically true sentences, we cannot select better or worse, more or less correct ones. They are all equally correct and, therefore, equally acceptable. It follows, then, that there is no reason for ever trying to *increase* the amount of *evidence* gathered for any hypothesis. Once we have any properly determined probability sentence, we should halt all investigation and the sooner the better. Consider the following example.

There is going to be a football game Saturday between the professional teams of Atlanta and Boston. We are interested in determining the probability of the hypothesis 'H' that Atlanta will win. If we assume that we know nothing about either team, we should say the probability of our hypothesis is $\frac{1}{2}$. *A priori* Atlanta's winning is as probable as not. Suppose, however, that we have three bits of information.

A: Six of Atlanta's better players are ill.
B: The game is going to be played in Boston's stadium.
C: Atlanta has lost seven games in a row.

Furthermore, suppose we know that the initial probability of 'H' is *higher* than the probability of 'H' relative to the initial evidence

'*E and A*'; that relative to '*E and A and B*', the probability of '*H*' is even *lower* than it is relative to '*E and A*'; and that relative to '*E and A and B and C*', the probability of '*H*' is *still lower*. That is, as we increase our evidence, the probability of the hypothesis that Atlanta will win is *decreased*. It is becoming less and less probable that Atlanta will win. Now, most people would say that the probability of a hypothesis should be determined on *all* available relevant evidence. Carnap calls this the **Principle of Total Evidence.** They would say that the most *accurate* probability sentence about the outcome of the given game is that which is based on *all* our evidence '*E and A and B and C*'. But there is *no warrant* at all for such claims on the logical interpretation. Each of the probability sentences suggested above would be as accurate as every other, since each one is necessarily true. Thus, according to Ayer, there is no reason to prefer the sentence involving *all* our evidence over that involving only the *initially* given evidence. Any resources spent gathering evidence beyond the initially given evidence are wasted. . . .

THE PHYSICAL INTERPRETATION

a) The Problem of Meaning

As we suggested earlier, according to the physical interpretation, 'probability' is interpreted as a *physical attribute* of certain physical systems. The systems may be physical ranges, experimental arrangements, or sequences of events. Thus, for example, the probability sentence

> The probability of throwing a six with this fair die is $\frac{1}{6}$.

may be given any of the following (rough) physical interpretations.

> The proportion of the *physical range* of the event of a six turning up to the *physical range* of the event of tossing a fair die is $\frac{1}{6}$.
>
> The experimental arrangement of tossing a fair die has the *propensity* of turning up about $\frac{1}{6}$ of the time.
>
> The *relative frequency* (in the long run) of tossing a six with a fair die is $\frac{1}{6}$.

The proportions of physical ranges, the propensities, and the relative frequencies are all physical attributes of certain physical systems. The question of the existence, comparative amount, or degree of the physical attributes possessed by any systems is an *empirical* or *factual one*. That is, the answers to such questions must be determined by empirical investigation, *not* by logical analysis. Thus, probability sentences are *empirically true* if they are true at all and *empirically false* if they are false at all. Their truth or falsity is determined *a posteriori* if it is determined at all. Hence, experience is highly relevant to these sentences.

b) The Problem of Measurement

Granting, then, that probabilities are to be determined by and based on physical considerations alone, the question again arises, *How do we obtain initial numerical probabilities?* A rough provisional answer to this question is this: *Count* physical attributes and determine the ratios of certain numbers of these to certain numbers of others. We shall consider two theories of measuring probabilities within the physical interpretation which might be roughly characterized by this answer, namely, the *physical range theory* and *frequency theories.*

i) The Physical Range Theory. The physical range theory of measuring probabilities, as it has been developed by the philosopher William Kneale, is similar to Carnap's logical range theory in two important respects. The notions of a *range* and a *Principle of Indifference* are fundamental for both theories. But it is precisely in these two fundamental ideas that the main *differences* between the two theories lie. In the next few paragraphs we shall explain these differences. We shall *not* present the physical range theory from the bottom up as it were. Rather we shall *assume* that the physical range theory is *exactly the same* as the logical range theory already discussed, and we shall merery make *two* changes in the latter.

To begin with, you recall that the *logical* range of a sentence is the class of those state descriptions in which the sentence is true and that we find out which state descriptions are in this class *a priori* by logical analysis. Now, we may say that the ***physical range*** of an *event* is the class of those states of affairs which contain that event and that we find out which states of affairs contain the event in question *a posteriori* by empirical investigation. Consider, for example, the little world described in Table 5.21. The *physically* possible states of affairs are described by the four state descriptions. To say that these states of affairs are **physically possible** is to say that given the existing laws of nature (e.g., the laws of physics, chemistry, biology, etc.), any of these states of affairs *may* occur. These are, to use Kneale's phrase, "real possibilities." The events *contained in* these states of affairs or the events which *constitute* these states of affairs *are* contained in or *do* constitute these states of affairs *as a matter of fact,* not as a matter of logic. The sentence

The *physical* range of the event of Arf's being red is the class consisting of the states of affairs described by S1 and S2 (in Table 5.21).

is a factual (or empirical, or synthetic) one. We must examine the little world to find out whether or not the event in question is *physically* possible given those states of affairs. We know the event is *logically* possible because we know the sentence describing that event is not self-contradictory. But we must investigate the world to reach a decision about physical possibility. This, then, is the *first* fundamental difference between the physical and logical range theories: In the *physical* range theory, the range of an *event* is a *physical entity* (a physical possibility) which either exists as a matter of *fact* or does not. In the *logical* range theory, on the other hand, the range of a *sentence* is a *logical entity* (a logical possibility) which exists as a matter of *logic* or does not.

You recall that the Principle of Indifference is: In the *absence of any reason* for considering certain indivisible events (or sentences) to be *not* equally possible, they may be considered equally possible. The **Physical Principle of Indifference** suggested by Kneale is: In the *presence of a reason* for considering certain indivisible events to be equally possible, they must be considered equally possible. If we conduct a thorough investigation of certain events which, for one reason or another, we are treating as *indivisible* and find good reasons for considering them as equally possible, then we *should* assign them equal initial probabilities. Notice that these initial probabilities are assigned (perhaps *discovered* would be more appropriate) *a posteriori*. We *know* certain events are equally probable; that is why we say they are. This contrasts sharply with the principle of the logical theorists who say that certain events may be regarded as equally probable *because* they do not have contrary evidence. The physical theorist bases his initial probabilities on the *information* he has obtained by empirical investigation. The logical theorist bases his initial probabilities on his *lack* of information. While the particular values obtained by each theory might be identical, the *bases* of the assignments are quite distinct.

Given the Physical Principle of Indifference and physical ranges, then, we would proceed to obtain initial probabilities by applying the principle directly to *state descriptions*. Since this procedure and its problems are similar to those discussed under the logical range theory, we shall not take them up again here.

Before we leave the physical range theory, it is worthwhile and convenient to mention the so-called **propensity theory** which was introduced by the philosopher Karl Popper. Popper claims that when we utter such sentences as, say,

The probability of tossing a head with this fair coin is 0.5.

we mean that a particular part of the physical universe, namely our *experimental arrangement,* is such that it has a *tendency, disposition,* or *propensity* to produce the event of a head turning up about one-half of the time.

If we think of an experimental arrangement as a set of physical ranges with certain propensities, then we have a rough intuitive idea of the propensity theory. Instead of talking about ranges of physical *possibilities,* we would talk about ranges of physical *propensities.* Instead of measuring ranges of possibilities, we would measure ranges of propensities. The only difference, then, between a propensity theory of measuring probabilities and a physical range theory would be the difference between physical *propensities* and physical *possibilities—whatever that difference is.*

Alternatively, we might ignore the problem of what physical propensities *are,* or are *like,* and concentrate on the actual *frequencies* of the occurrences of certain events. The frequencies would be regarded as direct manifestations of propensities or tendencies in certain experimental arrangements. With this view (which seems to be nearer to Popper's), probabilities in the sense of propensities are measured by relative frequencies. Whether or not there is a *significant* difference between this view of the propensity interpretation and the frequency theories we are going to discuss is an open question.

ii) ***Frequency Theories.*** We shall consider two types of frequency theories, one involving sequences with a finite number of elements and the other involving sequences with an infinite number of elements.

iia) According to the **finite frequency theory** of measuring probabilities which has been advocated by Bertrand Russell, we measure the probability of an event in three steps: 1. Select a relevant sequence of repetitive events and count its members. 2. Count the number of cases in that sequence which are "favorable" to the event whose probability is to be measured. 3. The ratio of the number of "favorable" cases to the total number of events in the selected sequence is the required numerical value. This ratio is the *relative frequency* of the occurrence of "favorable" cases in some sequences of events. It is the frequency of "favorable" cases relative to the selected sequence.

There are two main differences between this theory and the classical theory. The first already been emphasized. Here we obtain numerical values *not* by considering mere logical possibilities but by *observing* some physical system. The second important difference is the introduction of the idea of a *sequence* of *repetitive events.* An event is called **repetitive** if it is repeatable an unlimited number of times; e.g., the appearance of a head on the toss of a certain coin, producing a defective machine, drawing a ball from a barrel (with replacement), catching a fish, shooting a duck, hitting a baseball, etc. Strictly speaking, we should say that repetitive events are nothing more than *kinds, types,* or *sorts* of events. It is the *kind* or *type* that recurs, not any particular event. You cannot, for instance, make *the very same hit* on your second turn at bat that you made on your first turn. But you can make *another hit;* i.e., you can repeat the *type* of performance but not the *particular* performance. A particular performance is an event which is not repeatable and it may, therefore, be called a **unique** event; e.g., the appearance of a head on the *next* toss of a certain coin, the birth of Fabian, drawing the very *first* ball from a barrel, shooting the *third* duck to fly over your barn, hitting three homeruns in one game for the first time, the death of Teddy Roosevelt, etc. The distinguishing characteristic of *unique* events is that they occur *only* once. The distinguishing characteristic of *repetitive* (strictly, *kinds, types,* or *sorts* of) events is that they *may* occur more than once. A **sequence** of repetitive events is an aggregate of such events. As far as the finite frequency theory is concerned, only *finite* sequences are considered.

Example 1

What is the probability of tossing a three with a fair die?

Solution. To measure the probability of tossing a three with a fair die, we begin by *tossing the die* a number of times. Each toss of the die is a member of a *sequence* of tosses which is called a **reference class;** the reference class is nothing more than a special kind of *sample space.* It always has a *finite* number of members. The *original* or *initially given* reference classes are still called 'universes of discourse'. If, for example, a die is tossed six times then we have a sequence (or a reference class, or a universe of discourse) with six events. Each time the face with three dots turns up, we have a "favorable" case; e.g., if we toss only one three in six times, we have one "favorable" case. The measure of the probability of throwing a three with a fair die, then, is simply the ratio of the number of "favorable" cases to the total number of events in the reference class. In the present case, we might have a probabilility of $\frac{1}{6}$.

Example 2

Rodney Rotgut produced 100 jugs of homemade whiskey. He

sold 90 of them at $1.00 a jug and the others at $.50. What is the probability of selling a jug of Rotgut's whiskey for $1.00?

Solution. Our reference class consists of the sequence of 100 *sales*. There are 90 "favorable" cases in this class; i.e., 90 jugs were sold at $1.00 each. Hence 90/100 or 90 percent of the jugs sold at $1.00. Thus, the probability of selling a jug of Rotgut's whiskey for $1.00 is 0.9.

EXAMPLE 3

Suppose you are picking apples from a tree in the dark. After your basket is full, you bring it into the barn and examine its contents. Table 5.31 is a summary of what you find.

	Wormy	Nonwormy
Red apples	20	30
Green apples	10	40

Before venturing up the tree again, you are interested in finding the following probabilities:

a) The probability of picking a red apple.
b) The probability of picking a wormy apple.
c) The probability that any green apple picked will be wormy.
d) The probability that any nonwormy apple picked will be red.

Solution. To begin with, you have a sequence of 100 *picks*. To determine the number of times you picked a red apple, find the *sum* of the entries on the *line* following 'red apples'. You picked 50 red apples. Hence, the answer to a) is: The probability of picking a red apple is 50/100 or 0.5. To determine the number of wormy apples picked, find the *sum* of entries in the *column* below 'wormy'. You picked 30 wormy apples. Hence, the answer to b) is: The probability of picking a wormy apple is 30/100 or 0.3.

Here c) and d) both involve *conditional* probabilities. In c) we are given the special condition that a *green* apple has been picked, and we want to know the probability that it is wormy. Our *new sample space*, then, is the sequence consisting of only 50 *green* apples. To determine the number of green apples which are wormy, simply read the entry in the lower left corner of the table. You picked 10 green apples with worms. Hence, the ratio of the number of green *and* wormy apples to the number of green apples is 10/50; i.e., 10/50 or 20 percent of all the green apples picked are wormy.

Thus, the answer to c) is: The probability that any green apple picked will be wormy is 0.2.

To determine the probability that any nonwormy apple picked will be red, begin by finding the total number of nonwormy apples; i.e., find the sum of the entries in the column below 'nonwormy'. Our *conditional sample space* contains 70 nonwormy apples. Exactly 30 of these are red. Hence, the ratio of the number of nonwormy *and* red apples to the total number of nonwormy apples is 3/7 or about 0.43.

iib) According to the **limiting frequency theory** of measuring probabilities which has been developed by Richard von Mises, we measure the probability of an event in three steps: 1. Select a relevant *random* sequence of events. 2. Determine the *limit* of the relative frequency of "favorable" cases in this sequence. 3. That *limit* is the required numerical value.

There are two main differences between this *limiting* frequency theory and the *finite* frequency theory. First, here we are concerned with sequences which are infinitely long. If we tried to *count* all the members in such sequences, we would never finish counting. Second, we are concerned with *limiting* values of relative frequencies in such sequences, not merely with a finite number of events. The first of these novelties requires no explanation, but the second does. More precisely, we must begin with an explanation of the ideas of a *limit* and a *random* sequence.

Suppose a coin is tossed 10 times with the following outcomes.

$$H \quad H \quad T \quad H \quad T \quad H \quad T \quad H \quad T \quad T$$

The '*H*'s' and '*T*'s' are short for 'heads' and 'tails', respectively. After each toss we may determine the *proportion* of heads up to that point. The fractions below each '*H*' and '*T*' indicate these proportions.

H	*H*	*T*	*H*	*T*	*H*	*T*	*H*	*T*	*T*
1/1	2/2	2/3	3/4	3/5	4/6	4/7	5/8	5/9	5/10

After four tosses, $\frac{3}{4}$ of the events in the sequence are "favorable". After seven tosses, 4/7 of the events are "favorable". At the end of our experiment, the ratio of "favorable" cases to the total number of events in the sequence is 5/10. As we increased the number of members in this sequence, the proportion of heads approached $\frac{1}{2}$. If the proportion of heads continues to approach $\frac{1}{2}$ as the length of the sequence is increased indefinitely, then we say that the *limit* of the relative frequency of heads in this sequence is $\frac{1}{2}$. Similarly, if the length of the sequence is increased indefinitely and the proportion of heads approaches $\frac{3}{4}$ or $\frac{5}{6}$, then we say that the *limit* of the relative frequency of heads in this sequence is $\frac{3}{4}$ or $\frac{5}{6}$, respectively. In either of these latter two cases, we would say that the coin was not fair. Quite generally, then, we may define the limit of a sequence of such fractions as follows. The **limit** of a sequence of fractions is a number which is such that after a certain place in the sequence has been reached, all the fractions in the sequence are very close to it. To make this phrase "very close" more precise, we may say that given *any* small number *e*, it is possible to find a place in the sequence after which all the fractions lie between the limit *L* and *e* (i.e., between $L - e$ *and* $L + e$). Any sequence which has a limit is called **convergent. Divergent** sequences have no limits. If a sequence *is* convergent then "in the long run" the procedure we have just described will find it.

The second notion that must be explained is that of *randomness.* Suppose the following sequence is continued indefinitely.

(A) *H T H T H T H T H T* . . .

The limit of the relative frequency of '*H*'s' in this sequence is $\frac{1}{2}$. In *this* respect, there may be no difference between this sequence (A) and the following.

(B) *T H H H T H T H T T* . . .

However, while the *limit* of the relative frequency of heads in both of these sequences may be the same, the *order* of the sequences is quite different. The '*H*'s' and '*T*'s' are distributed in a *regular* fashion in (A); i.e., every '*H*' is followed by a '*T*' and vice versa. But in (B) the '*H*'s' and '*T*'s' are distributed in an *irregular* fashion; i.e., sometimes an '*H*' is followed by a '*T*', sometimes it is followed by two '*T*'s', sometimes it is followed by another '*H*', etc. The '*H*'s', and '*T*'s' in the latter sequence are supposed to illustrate the beginning of a sequence of randomly distributed '*H*'s' *and* '*T*'s'; i.e., a sequence of '*H*'s' and '*T*'s' such that *no one knows or could know what is coming next.*

Two conditions have been used to define random distributions. A sufficient condition of randomness called 'insensitivity to place selection' was introduced by von Mises. A necessary condition of randomness called 'freedom from aftereffect' was introduced by Karl Popper. We shall use the former stronger condition in our definition of 'random sequences'.

Insensitivity to Place Selection: Each of the two sequences (A) and (B) illustrated above contains the first 10 members of an infinitely long sequence. In (A), the first member of the sequence is a head, the second is a tail, the third a head, etc. In other words, we could describe the order of (A) by saying that the *first place* in the sequence is occupied by a head, the second place is occupied by a tail, the third place by a head, etc. Now, if we selected every member of the sequence (A) occupying an *even* place, we would have a sequence of *tails* made up of the tail in the second place, the tail in the fourth place, the tail in the sixth place, etc. That is to say, we would have constructed a *subsequence* according to the following *rule of place selection:* Select all even-placed members of the original sequence. The limit of the relative frequency of '*H*'s' in the subsequence is 0. Applying this same rule to (B), we have the following subsequence.

(B′)	*H*	*H*	*H*	*H*	*T*	. . .
	1/1	2/2	3/3	4/4	4/5	

As the fractions beneath the letters show, the limit of the relative frequency of '*H*'s' in this subsequence is *not* $\frac{1}{2}$, but $\frac{4}{5}$. Other rules of place selection might select every odd-placed member of the original sequence, every third place, every fourth place, etc.

If a sequence is such that events are distributed in it in such a way that the limit of the relative frequency of the events in a *subsequence* constructed according to *any* rule of place selection (subject to three conditions mentioned below) is the *same* as the limit of the relative frequency of the events in the original sequence, then the later is said to be **insensitive to place selection.** Sequences which are insensitive to place selection are **random** sequences; i.e., they are sequences of randomly distributed events.

The three conditions put on rules of place selection are that a) they must make no reference to any event or attribute whose probability is being determined; b) they must be applicable to a whole sequence no matter how long it is extended; and c) they must be specified *before* observing the sequence. Thus, if we happened to be interested in deciding whether or not '*H*'s' are randomly distributed in a sequence of '*H*'s' and '*T*'s', then, according to a), acceptable rules of place selection would not make *any* references to '*H*'s'. A rule which selected only the *first two* members of a sequence would be rejected by b) because it does not apply to the whole sequence. Again, if we made up our rules *after* observing the sequence, then if we had the time, energy, and ingenuity, we could always make up a rule to construct a *subsequence* in which the limit of the relative frequency of certain events in it *differed* from the limit in the *original* sequence. Hence, according to c), only rules which are specified *before* observing the sequence are acceptable.

Freedom from Aftereffect: Consider the following sequence (C).

(C) *H H H T H T T T H T H H H T T H T T T H* . . .

The limit of the relative frequency of '*H*'s' in this sequence is $\frac{1}{2}$. If we formed a subsequence by selecting every member of (C) which is *immediately preceded* by a '*T*', we would find the following.

(C1) *H T T H H T H T T H . . .*

The limit of the relative frequency of '*H*' in (C1) is also $\frac{1}{2}$. Again, if we formed a subsequence by selecting every member of (C) which is immediately preceded by '*H T H*', we would obtain the following.

(C2) *T H*

The limit of the relative frequency of '*H*' in (C2) is $\frac{1}{2}$. Similarly, selecting every member which is immediately preceded by '*H H H T*', we find a subsequence in which the limit of the relative frequency of '*H*' is $\frac{1}{2}$.

If a sequence is such that events are distributed in it in such a way that the limit of the relative frequency of the events in a *subsequence* constructed by selecting the events immediately following *any* group of predecessors is the *same* as the limit of the relative frequency of the events in the original sequence, then the latter is said to be **free from aftereffect.** According to Popper, then, sequences which are free from aftereffect may be called 'random sequences'; i.e., they are sequences of randomly distributed events. Since, as we mentioned above and as G. H. von Wright has demonstrated, *freedom from aftereffect* is only a *necessary* condition of randomness, we are not using it in our definition of 'random sequences'. The fact that it is a *weaker* condition than *insensitivity to place selection* may be seen by noticing that a rule for the selection of members with certain predecessors is nothing more than a *special sort* of rule of place selection; i.e., Popper's condition is included in von Mises' condition. Consider two examples of probability measurements with the limiting frequency theory.

Example 4

The Hangitall Company produces about 4000 coat hangers a week. The quality control department claims that about 5 percent of the hangers made every week are defective. What is the probability that any hanger produced by the Hangitall Company is defective?

Solution. We have a theoretically unlimited sequence of *hangers produced* by Hangitall, and we know that week after week about 5 percent of them are defective. We suppose the defective hangers are randomly distributed and that the limit of the relative frequency of defective hangers produced by Hangitall is 0.05. Hence, the probability that any hanger produced by the Hangitall Company is defective is 0.05.

EXAMPLE 5

A certain baseball player has a lifetime batting average of 0.282. What is the probability that he will make a hit?

Solution. We have a theoretically unlimited sequence of *times at bat.* We suppose that the hits are randomly distributed in this sequence and that the limit of the relative frequency of hits is roughly 0.28. Hence, the probability that the given baseball player will make a hit is 0.28.

c) Some Advantages and Disadvantages of the Physical Interpretation

There are two main advantages of the physical interpretation. The first one reveals an immediate superiority *over* the logical interpretation. The *first* advantage of the physical interpretation is the resolution of Alfred Ayer's problem of the equal correctness of all nonself-contradictory probability sentences. On the logical interpretation, such sentences are analytic and, therefore, one is as good as another. Hence, there is no reason to prefer probability sentences involving *all* available relevant evidence over those involving only *initially* given evidence. However, the physical interpretation provides quite different options. On this interpretation, probability sentences are never logically true (analytic) or logically false (self-contradictory). They are factual (or empirical, or synthetic) sentences which may be *corrected* or *improved* by further investigation. Hence, there is an excellent reason to prefer probability sentences involving *all* available relevant evidence over those involving only initially given evidence; namely, the former *could* be more accurate.

The *second* advantage of the physical interpretation is that it satisfies the *intuitive belief* of many people that probability sentences are *about* physical phenomena. Many people believe that when we utter a probability sentence, we *mean* to be saying something about some part of the physical universe. Hence, any interpretation which does not meet this requirement is intuitively unsatisfactory for these people. Certainly such sentences as

The probability of tossing a head with a fair coin is 0.5.

seem to be analogous to

The temperature of the room is 70°F.
The weight of the stone is 14 pounds.
The length of the bridge is 100 feet.

Each of these sentences seems to be *about* some property of the physical universe. Most people would reject any interpretation of the latter three sentences which had the result that the sentences were not *about* physical phenomena but *about* logical possibilities or psychological attiudes. Similarly, there are many people who would reject any interpretation of probability sentences which have the same result. The fact, then, that the physical interpretation does not have this result contributes to its intuitive satisfactoriness and must be considered a major advantage.

There are two main disadvantages of the physical interpretation. The *first* disadvantage is that sentences about physical probabilities are *not available prior to observation* to be used to appraise predictions and actual frequencies. For example, recall the case of the lottery which Otto believed he would win though he had a ticket with an *a priori* probability of 1/1000 of being drawn. Our appraisal of Otto's prediction *seemed* quite legitimate, but if the logical interpretation is given up in favor of the physical one, such an appraisal would be completely without foundation. On the physical view, *prior* to observation we have no reason at all to assign any event any probability. Hence, on this view, we must abandon *a priori* appraisals of predictions. Again, suppose Otto won roughly 50 percent of the lotteries in which he bought only one ticket. Given this information and the physical interpretation, we have no good reason to *suspect* foul play. We have certain events occurring with a certain frequency and, for all we know, that frequency is just as it should be; i.e., for all we know, the lotteries are not "fixed". Again, then, while the *prior* appraisal of this frequency *seemed* legitimate, such appraisals must be abandoned when the physical interpretation is adopted. On the physical interpretation, logical possibilities are simply useless for the assessment of predictions and actual frequencies.

The *second* disadvantage of the physical interpretation is that either a problem of the proper application of the Physical Principle of Indifference *or* a problem of the proper selection of reference classes must be faced. If either a physical range or (one view of) a propensity theory is used to measure probabilities, then one must decide whether the physical principle is to be applied to state descriptions or to structure descriptions. This is basically the same problem that logical theorists face. On the other hand, if a frequency theory is used to measure probabilities, then one must decide how to select proper reference classes. This problem is frequently overlooked, but it is as serious as the other. Consider the following example.

Suppose it is given that

1) The probability of '*H, given A*' is x.
2) The probability of '*H, given A and B*' *is* y. ($x \neq y$)

Each of these sentences gives the probability of '*H*' relative to a different reference class. But which sentence should we accept? Which is the more accurate? Most people would say that we should prefer sentence 2) because that involves all our available evidence or, as Hans Reichenbach says, "the narrowest class for which reliable statistics can be compiled". Unfortunately this answer is not (as Reichenbach recognized) always helpful. There are cases in which 1) would be prefered to 2). For example, suppose it is known that

1′) The probability that Roger will receive an 'F' on a math exam *given* his past performance on such exams is x.
2′) The probability that Roger will receive an 'F' on a math exam, *given* his past performance on exams covering math *and* music is y. ($x \neq y$)

This time 1′) would be accepted as a more accurate or reliable estimate of Roger's performance on a forthcoming math exam than 2′). It is clear, then, that the problem of the proper selection of reference classes is not solved by the "narrowest class" rule. Moreover, in the absence of some plausible solution to this problem, the choice of proper reference classes remains somewhat arbitrary.

Before closing this discussion of the physical interpretation, we must mention the problem which relative frequency theories have with *unique* events. You recall that such theories require sequences of events to begin with; i.e., the universe of discourse for such theories is always a sequence of *repetitive* events. Hence, in the absence of such a sequence, it is logically impossible to measure the probability of any event. Now, unique events are by definition events which are not repetitive. Hence, a crucial question arises for those who wish to measure probabilities using some sort of frequency theory; namely, How can we measure the probability of unique events? For example, how can we measure the probability of the death of George Washington in 1799? There are two alternatives. In the first place, the frequentist may simply deny that his theory is *applicable* to unique events and ignore them completely. Whether or not they have measurable probabilities on some other theory, he may say that they surely do not have measurable probabilities in any frequency sense. Hence, they are beside the point. On the other hand, the frequentist may consider the problem of assigning probabilities to unique events a serious challenge and attempt to meet it somehow. Hans Reichenbach's theory of the "weight of hypotheses" is just such an attempt and it will be discussed in detail in Chapter 8. Whatever the outcome of this problem of unique events, it must be emphasized that it is *not* a problem for the physical interpretation in general. It is a problem for those who wish to *measure* probabilities using some sort of frequency theory. Advocates of the physical interpretation who employ some sort of physical range or propensity theory of measuring probabilities do not have any particular problem with unique events. (Or, if they do have this problem, then it is shared by *every* other theory given *any* interpretation.) . . .

THE PSYCHOLOGICAL INTERPRETATION

a) The Problem of Meaning

As we suggested earlier, according to the psychological interpretation, 'probability' is interpreted as a *psychological attitude* toward the occurrence or nonoccurrence of certain events. Probability sentences are about actual *degrees* of *belief* or *credibility*. According to those who defend this interpretation, people are supposed to have various discernible *belief-states* ranging from complete certainty to total skepticism. Sometimes we say we *know* such and such is the case; i.e., we *feel certain* that it is the case. Sometimes we say we *believe very strongly* that such and such is the case. Sometimes our belief is not quite so strong. Sometimes we have serious doubts, etc. For example, we know (if we know *anything*) that we were born; we believe very strongly that we shall survive our fiftieth birthday; we believe somewhat less strongly that we shall survive our seventieth birthday; we have serious doubts about surviving our ninetieth birthday; and we do not believe we shall be around to celebrate our one hundred and tenth birthday. Although you might believe more or less strongly about any of the events mentioned here (e.g., you may have serious doubts about surviving your seventieth birthday), none of these *belief* sentences sounds peculiar *as* a belief sentence. People do, as a matter of fact, express themselves *as if* they had various *discernible belief-states* or *actual degrees of belief* or *degrees of credibility*.

Those who adopt the psychological interpretation, then, assume that probability sentences are *about* the same things that belief sentences are about. Hence, to say that a certain event is probable is to say that someone believes it will occur. To say that it is more probable that Mr. Clean is right than that Mr. Dirty is right is to say that someone would sooner believe Mr. Clean than Mr. Dirty. To say that it is highly probable that it will rain is to say that someone strongly believes that it will rain. No attempt is made to answer the question, '*Why* does someone have this or that belief?' Logical theorists generally insist that if an answer to this question were sought, the *logical* foundations of the interpretation would be revealed; and physical theorists usually claim, on the contrary, that its *physical* foundations would show up. While it is surely an important and interesting question to raise, it may be ignored entirely without disastrous consequences. There is no logical absurdity involved in the view that probability sentences are about actual belief-states *whatever* the causes of the latter or whatever sort of evidence led to the latter. At any rate, this is where the psychological theorist begins.

b) The Problem of Measurement

Granting, then, that probabilities are to be determined by and based on psychological considerations alone, the question again arises, *How do we obtain initial numerical probabilities?* According to the **personal odds theory** introduced by Frank P. Ramsey, we measure the probability of an event in two steps: 1. Determine the highest odds a rational person would be willing to offer in a bet on the event in question. 2. If the odds in favor of the event are m to n ($m{:}n$), then the probability of the event is $m/(m+n)$. The numerical values obtained by this method are usually called **personal probabilities.** The reason for the name 'personal odds theory' is apparent; i.e., the numerical values are derived from personal odds. Because the fraction $m/(m+n)$ is called a 'betting-quotient', this method of measuring probabilities is often referred to as 'the betting-quotient theory.'

Two assumptions must be noted immediately. In the first place, it is assumed that a person's *betting behavior* is a direct reflection or indication of his degrees of *belief*. If, for example, a person is willing to give *very high odds* in a bet on the occurrence of some event, then it is assumed that he *believes very strongly* that the event will occur. In the second place, contemporary defenders (e.g., Leonard J. Savage) of this or more sophisticated personal odds theories assume that the rational person referred to in step 1. should be a *slightly* "idealized" person; i.e., he is a person who never bets foolishly or capriciously. In particular, he does not place his bets in such a way that it is impossible for him to win. This is about as far as our "idealization" may be pushed, since it is *actual* degrees of belief we are interested in measuring, *not ideal* ones. John G.

Kemeny has shown that if a person's betting behavior is such that the odds he is willing to give yield probabilities which are consistent with the principles of the calculus of probability, then it *is* possible for him to win; i.e., he is a sufficiently "rational" bettor and his betting-quotients are "fair".

EXAMPLE 1

A certain rational bettor is willing to give odds of 3 to 1 in favor of the Red Sox in their next game. What is the probability that the Red Sox will win?

Solution. The odds in favor of the Red Sox are $m{:}n$, which is 3:1. Hence, the probability that the Red Sox will win is $m/(m + n) = 3/(3 + 1) = \frac{3}{4}$.

EXAMPLE 2

The owner of a certain racetrack is willing to give you odds of 100 to 1 *against* your horse coming in first. What probability is he assigning to the event of your horse coming in first?

Solution. If the owner of the track is willing to give you odds of 100 to 1 *against* your horse coming in first, then, as far as he is concerned, the probability that your horse will *not* come in first is $m/(m + n) = 100/(100 + 1) = 100/101$. We may consider the two events, that of your horse coming in first and that of your horse not coming in first, as the only two members of our universe of discourse. The *sum* of probabilities of all the events in a universe of discourse is 1. Hence, to find the probability that your horse *will* come in first, we simply subtract 100/101 from 1; i.e., the probability that the owner is assigning to the event of your horse coming in first is 1/101.

EXAMPLE 3

Suppose Clarence is only willing to make an *even* bet on some event. What is the probability of that event?

Solution. Clarence is not giving any odds at all. Hence, $m{:}n$ equals 1:1. So the probability of the event in question is $1/1(1 + 1) = \frac{1}{2}$.

c) Some Advantages and Disadvantages of the Psychological Interpretation

There are four main advantages of the psychological interpretation. *First,* no Principle of Indifference is required; *second,* probability sentences are empirically testable. The first point needs no explanation. With respect to the second, it must only be emphasized that the testability of probability sentences is contingent upon the assumption that degrees of belief are in complete agreement with betting behavior. While the latter is publicly observable, the former is not. Hence, the assumption is crucial for this second advantage.

The third advantage of this interpretation is identical to the second advantage of the logical interpretation and it reveals a superiority of the psychological interpretation over the physical one. The *third* advantage is that sentences about psychological probabilities are available *prior to observation* to be used in the appraisal of predictions and actual frequencies. Enough has already been said about the use of probability sentences for both purposes. All that must be added is that, *unlike* probability sentences in the logical interpretation, probability sentences in the psychological interpretation *have* predictive force. For example, on the latter interpretation

The probability that a six will turn up is $\frac{1}{6}$.

is a claim about the odds a person will or will not be willing to give in a bet on the occurrence of a six. Hence, both kinds of probability sentences may be used in appraisals, but only the latter has predictive force.

The *fourth* advantage of the psychological interpretation is its agreement with ordinary ways of talking about probabilities. We have already seen that people express themselves as if they had various recognizable belief-states and that it is generally assumed that there is a close relation between these states and probability. The psychological interpretation is only a short step from these familiar facts of life. The assumed *proximity* between belief-states and probability becomes an *identity*. To the man on the street, this must be considered a major advantage.

There are two main disadvantages of the psychological interpretation. In the *first* place, according to the psychological interpretation, the probability of an event may be radically altered if someone simply *changes his mind*. For example, if some rational bettor feels extremely confident that a coin is going to turn up heads, we would have to say the probability of tossing a head is, say, 0.9. If he suddenly changes his mind, the probability of this event might drop considerably, say, to 0.2. As long as he does not violate any of the conditions of the probability calculus and does not bet in such a way that he *cannot* win, we would have to accept his initial probability assignments. Hence, this interpretation permits much more variation of initial probabilities than most people would be willing to grant. Furthermore, this extreme variability implies a *second* disadvantage: There might be an unalterable disagreement about the probability of some hypothesis (say, some scientific theory or law) either because the initial probability assigned to it by one rational bettor differs radically from that assigned to it by another or because the same person vacillates among a number of possible values. Hence, although this interpretation provides an objective basis (i.e., a publicly testable one), it *might not* provide a *stable* one.

READING 18

The Weight of Simplicity in the Construction and Assaying of Scientific Theories

Mario Bunge

One of the most difficult and interesting problems of rational decision is the choice among possible diverging paths in theory construction and among competing scientific theories—i.e., systems of accurate testable hypotheses. This task involves many beliefs—some warranted and others not as warranted—and marks decisive crossroads. Suffice to recall the current conflict between the general theory of relativity and alternative theories of gravitation (e.g., Whitehead's) that account for the same empirical evidence, the rivalry among different interpretations of quantum mechanics (e.g., Bohr-Heisenberg's, de Broglie-Bohm's, and Landé's), and the variety of cosmological theories (e.g., Tolman's cyclical model and the steady-state theory). They all account for the same observed facts although they may predict different kinds of as yet unknown facts; they are consequently, up to now, *empirically equivalent* theories even though they are conceptually different and may even involve different philosophical views—i.e., they are *conceptually inequivalent*.

In effect, empirically equivalent theories may differ in many respects: in the kinds of entities and properties they postulate; in their logical organization and in their explanatory and predictive power; in their empirical testability and in their conformity with the bulk of scientific knowledge and with certain philosophical prin-

Reprinted by permission of the author and *Philosophy of Science, 28,* 1961, pp. 120–149.

ciples. These and other characteristics are dealt with by certain metascientific criteria that will be investigated in the following.

The set of metascientific criteria dealing with the various traits of acceptable scientific theories is what guides the choice among competing courses in theory construction and among the products of this activity. Now, simplicity is often listed among the requirements that scientific theories are supposed to satisfy, and is correspondingly offered as a, and sometimes as the, criterion for making a rational decision of choice among empirically equivalent theories.

Yet, simplicity is not a single kind but, on the contrary, is a complex compound; furthermore, simplicity is not a characteristic isolated from other properties of scientific systems, and it often competes with further desiderata, such as accuracy. Therefore, in order to assess the weight of simplicity in the construction and acceptance of scientific theory, we must discuss the kinds of simplicity and their relevance to the main characteristics of scientific theory (sec. 1), as well as their relevance to the truth of scientific theory (sec. 2) and to the acceptance of scientific theories in actual practice (sec. 3).

1. SPECIES OF SIMPLICITY AND THEIR RELEVANCE TO SYSTEMATICITY, ACCURACY AND TESTABILITY

1.1 Kinds of Simplicity

Although the question whether reality itself is simple or not is a genuine issue in ontology and in science—as may be certified by any worker in microphysics—we are here concerned with the simplicity of the theories about sections of reality, so that we may disregard the ontological problem of the complexity of reality. Complex states of affairs may be accounted for by theories with a comparatively simple basis (e.g., classical mechanics), and on the other hand there will always be room for pedants capable of expressing simple situations (or, rather, situations requiring simple descriptions) in an unnecessarily complex way: as the Viennese joke goes, "*Warum denn einfach, wenn es auch kompliziert geht?*"

Now, a system of signs, such as a theory, may be complex (or simple) in various ways: [1] syntactically, semantically, epistemologically, or pragmatically. When speaking of the simplicity of sign systems we must therefore specify the kind of simplicity we have in mind. It will not do—save as a rough indication—to say that we mean *overall simplicity,* because owing to the extreme heterogeneity of its various components it may well turn out that the degrees of complexity in the various respects are not additive; just think of the syntactical complexity of a proposition, which depends among other things on the number of places of the predicates occurring in it, and on its epistemological complexity, or degree of abstractness (in the epistemological sense), which is such a hazy notion. Even if correct simplicity measures of theories were available, the problem of the metricization of their overall simplicity would have to be solved. Let us then carefully distinguish the various ways in which a system of meaningful signs (such as a scientific theory) can be said to be simple.

Syntactical simplicity (economy of forms) depends on (1) the number and structure (e.g., the degree) of the specific primitive concepts (basic extralogical predicates); (2) the number and structure of independent postulates, and (3) the rules of statement transformation. Syntactical simplicity is desirable because it is a factor of cohesiveness and, in a certain sense (but not in another), of testability—as will be seen shortly. *Semantical simplicity* (economy of presuppositions) depends on the number of specifiers of meaning of the basic predicates. Semantical simplicity is valued within limits because it facilitates both interpretation of signs and fresh starts. *Epistemological simplicity* (economy of transcendent terms) depends on closeness to sense-data. Epistemological simplicity is not desirable in and of itself, because it conflicts with logical simplicity and with depth. Finally, *pragmatical simplicity* (economy of work) may be analyzed into (1) psychological simplicity (intelligibility), (2) notational simplicity (economy and suggestive power of symbols), (3) algorithmic simplicity (ease of computation), (4) experimental simplicity (feasibility of design and interpretation of empirical tests), and (5) technical simplicity (ease of application to practical problems). Pragmatical simplicity is, of course, valued for practical reasons.

No dependable measure of any of the four above-mentioned kinds of simplicity of sign systems is known at present. Even the gauges of the syntactical simplicity of predicate bases proposed so far [2] do not do justice to metrical predicates, such as 'age' and 'distance', which are in a sense "infinitely" more complex than classificatory concepts (presence/absence predicates) such as 'liquid'. And the proposal of measuring the structural complexity of equations by the number of adjustable parameters in them [3] is insufficient, since other formal properties are relevant as well, and because it involves a confusion of formal complexity with difficulty of test, with generality, and with derivativeness (as opposed to fundamentality).[4] At any rate, none of these proposals deals with systems of propositions and none of them account for the various kinds of simplicity, whence they are inadequate to cope with our problem.

Moreover, *the various kinds of simplicity are not all compatible with one another and with certain desiderata of science.* Thus, a syntactical oversimplification of the basis (e.g., a drastic reduction of the number of primitives and principles) may entail both difficulties of interpretation and lengthy deductions. A semantical oversimplification may involve the severance of the given theory with the remaining body of knowledge—i.e., a loss of systematicity in the sum total of science. An epistemological simplification, such as the elimination of transcendent (transempirical) terms is not only a guarantee of superficiality but also of an infinite complication of the postulate basis.[5] Finally, a pragmatical oversimplification may involve a loss of insight. Consequently, it would be unwise to recommend overall simplicity even if we had an accurate concept of overall simplicity.

Truth, however difficult may be its philosophical elucidation, is the target of scientific research; hence to truth all other desiderata—including some simplicities—must be subordinated. Now, truth is not obviously related to simplicity, but to complexity. The syntactical, semantical, epistemological, and pragmatical complexity of scientific theories usually increases with their scope, accuracy, and depth, until a point is reached where complexity of some kind becomes uncontrollable and obstrusive to further progress, and simplification in some respects and within bounds is called for.

But only those simplifications will be admitted in science which render the theory more manageable, more coherent, or better testable: no simplification will be accepted if it severely cuts down either those characteristics or the depth, the explanatory power, or the predictive power of the theory. The complexity of the task of truth-preserving simplifications—which are possible only in advanced stages of theory construction [6]—can be estimated if it is recalled that economy, not poverty, is wanted. That is, we do not want mere parsimony—which is best achieved by abstaining from theorizing—but minimization of the means/ends ratio.[7] Not a simple-minded elimination of complexities is required, but a cautious reduction of redundancies, a sophisticated simplification in some respects, under the condition that it does not detract from truth.

Let us inquire what contribution, if any, logical simplicity makes to the coherence, accuracy, and testability of scientific theory, since these are three necessary conditions for something to be a scientific theory, even before it can be regarded as approximately true.

1.2 Relevance of logical simplicity to systematicity

Theories are *systems* of hypotheses (corrigible propositions) containing extralogical concepts that range over a specified universe, i.e., that refer to a definite subject matter. Systems are, of course, sets of interrelated units, and the cohesion of scientific theories—in contrast with the looseness of the heaps of conjectures and data we so often find in nonscience and in underdeveloped science—is ensured by (1) exact formulation, (2) distribution of the basic concepts among the various basic propositions (axioms), and (3) economy of basic concepts. Let us be more explicit.

(1) Logical Neatness, or Exact Formulation of the Postulates. Loosely stated propositions can only loosely be tied together. No definite deductions are possible from vaguely worded basic assumptions; no neat distinction between axioms and observable consequences can then be made, whence no empirical datum will be strictly relevant to either. Syntactical accuracy, a prerequisite of empirical meaning and testability, is automatically achieved by mathematical formulation (this being a major and seldom noticed reason why mathematical models are sought); and semantical exactness is improved—though probably never ensured in a complete way—by the explicit and accurate statements of the rules of meaning. Where ambiguity and vagueness reign an army of scholiasts is invited to start a scholastic movement, and a manifold of theories instead of a single system readily emerges.

(2) Conceptual Connectedness, or Sharing of Basic Concepts among Postulates. An instance of an extremely unsystematic set of mutually independent postulates would be

(1) $\ldots C_1 \ldots, \quad \ldots C_2 \ldots, \quad \ldots, \quad \ldots C_6 \ldots,$

in which none of the basic or primitive predicates C_1 through C_6 occurs in more than one axiom. A slightly more organized system would be

(2) $C_1\text{–}C_2, \quad C_3\text{–}C_4, \quad C_5\text{–}C_6,$

where '–' stands for ordinary and logical words tying the basic concepts together An even better organized system would be the chain-like set

(3) $C_1\text{–}C_2, \quad C_2\text{–}C_3, \quad C_3\text{–}C_4, \quad C_4\text{–}C_5, \quad C_5\text{–}C_6.$

An equivalent connectedness would be provided by the single postulate

(4) $$C_1\text{–}C_2\text{–}C_3\text{–}C_4\text{–}C_5\text{–}C_6;$$

but, of course, such a unification at the propositional level is not always possible: it may not correspond to actual fact.

In the four cases the axioms are mutually independent on condition that the basic predicates themselves be mutually independent (as tested, for example, by Padoa's method).[8] But in the first case we have a loose aggregate of postulates, no matter how precisely they are formulated, and in the second case we have a partial connectedness of primitives, whereas in the last two cases the tightness of the conceptual connection is obvious.

Notice that increase in conceptual connectedness need not result in simplification of the postulational basis: (3) and (4) are equally coherent at the conceptual level; only postulational economy is gained by (4). In general, postulational simplification, if it preserves the predicate basis, is sufficient but not necessary for achieving conceptual cohesiveness, which is in turn necessary for having system. But postulational simplification is not a mechanical procedure: its feasibility depends on the nature of the case, i.e. on whether there is in fact a direct connection among all the properties denoted by the predicates involved.

*(3) **Simplicity of the Predicate Basis.*** The less the number of primitive concepts of the theory, the larger the number of bridges among them and the derived concepts (definitions and theorems) will have to be; as a consequence, the greater will be the conceptual and propositional connectedness of the theory. (This is one reason for adopting a variational principle as the sole postulate of many physical theories: it effects maximum conceptual unification, although its interpretation and status is far from simple.) In short, economy of the predicate basis improves systematicity.[9]

Notice, however, that formal simplicity of the basis is just *one* of three *means* for achieving the desideratum of systematicity. In the second place, the simplification of the predicate basis of factual theories has a limit which is rooted to the real net of properties; thus, e.g., at the present time, at least the following properties of fundamental particles are regarded as mutually irreducible (though connected), hence as not interdefinable: location in space-time, mass, electric charge, spin, and parity. In the third place, a large number of basic concepts does not prevent an exact treatment, since mathematical techniques enable us to handle as large a number of variables as desired; moreover, it is often desirable to *increase* the number of variables to infinity in order to attain a deeper level of analysis (by handling, e.g., the Fourier transforms of the original variables, as is done in field theories). What is important is not to minimize the number of predicates—as required by phenomenalism ever since Kirchhoff—but *to keep them under control.*

In short, simplicity of the predicate basis is *sufficient but not necessary* for systematicity; moreover, simplification of the predicate bases of factual theories is limited by the richness of reality and by pragmatical (e.g., methodological) considerations.

1.3. Relevance of logical simplicity to accuracy and testability

Testability, a second outstanding feature of scientific theory, depends on systematicity. In effect, the latter is not merely a question of economy and elegance: a theory, whether formal or factual, has to be a tightly knit set of propositions if it is to be testable *as such,* i.e. as a unit. A dough of vague assumptions all standing on the same logical level, without strong logical relations of deducibility occurring in its body, cannot be tested the way genuine theories are: since all of the propositions of the pseudotheory are loosely related to one another, every one of them will face separately the trials of logic and/or experience. How could we test the axioms of a factual theory if we cannot spot their logical consequences? A chaotic mass of conjectures lacking logical organization—as is the case with psychoanalysis [10]—cannot be subjected to the test of experience as a whole: experience may at most confirm some of the loosely related conjectures of the pseudotheory, but no evidence will ever conclusively refute the whole set of vaguely stated *ad hoc* hypotheses—especially if they are mutually shielding. And a theory which stands no matter what experience may say, is not an empirical theory.

Logical neatness and conceptual connectedness are then not luxuries but *means for ensuring testability,* which in turn is a necessary—but, of course, not a sufficient—prerequisite for attaining approximate truth. Notice that simplicity of the predicate basis is favorably relevant to testability to the extent to which it is propitious to systematicity; but recall that this kind of simplicity, though sufficient, is not necessary for attaining systematicity, as the same goal can be attained by means of conceptual connectedness.

Again, systematicity is necessary but not sufficient to ensure testability: *accuracy* and *scrutability* of the basic predicates are necessary as well. The more exact a statement is, the easier it will be to dispose of it; vagueness and ambiguity—the secret of the success of fortune-tellers and politicians—are the best protections against refutation. Now, *accuracy demands complexity,* both formal and semantic: suffice to compare the simplicity of presystematic, ordinary, discourse with the complexity of scientific discourse; compare 'small' with 'of the order of one atomic diameter', and '$x > a$' with '$x = a$'. Not just the simplest but the simplest among *equally precise* propositions and systems is to be preferred, both because accuracy is an independent desideratum of science, and because it favors testability.

Scrutability of the basic predicates is a further, obvious, condition for testability. The basic predicates of a scientific theory need not be observable or measurable in a direct way (few of them are). Only, they must be open to public scrutiny by the method of science, and for this it is necessary and sufficient that the theory establishes exact relations among its basic predicates and observable predicates. Terms such as 'élan vital,' 'infantile sexuality', 'absolute space', and the like, do not make up testable sentences, whence they must be dropped.

If desired, this norm of scrutability may be called the principle of methodological simplicity—on condition that it be realized that it is not necessarily related to other kinds of simplicity, such as formal economy of the predicate basis. A theory containing a large number of scrutable predicates will be preferable to another theory containing fewer predicates but all or part of them inscrutable, if only because the former theory will be testable, unlike the latter. The *methodological status* of the predicate basis is far more important than its logical structure and number. Thus, 'electrically charged' is both syntactically and semantically more complex than 'providential', yet it is scrutable and may consequently occur in scientific theory, whereas the latter cannot. In short, accuracy and scrutability may be consistent with logical complexity. When this is the case we are ready to sacrifice simplicity.

On the other hand, an excessive logical complexity may obstruct testability, and particularly refutability,[11] this being the reason logical simplicity is desirable as long as it does not involve loss of accuracy, scope, and depth. Irrefutability may be achieved through the mutual protection of hypotheses containing inscrutable predicates. This can be performed in a commonsensical or in a technical way. An example of the former is the theory of extrasensory perception, in which every instance unfavorable to the hypothesis of telepathic transmission can be regarded as favorable to the hypothesis of precognition, or to the hypothesis that the subject has become tired by exerting his supernatural powers. An instance of achievement of irrefutability with more impressive means is any phenomenological theory containing a number of adjustable parameters, and designed to account for phenomena *ex post facto* without risking any assumption on the mechanism involved. (Thus, e.g., the phenomenological theory of nuclear forces is allowed to introduce a number of parameters which are not independently measurable and which can be freely varied within generous limits; moreover, the observable consequences of the theory are largely insensitive to qualitative variations in the shapes and depths of the potential wells. This is one of the reasons for preferring, as a description of reality, the meson theory of nuclear forces, which involves a definite mechanism.)

The requirement of testability leads in the long run either to shave off the mutually shielding hypotheses, or to an entirely fresh start. In the former case a simplification is performed—but, then, few confirming instances may remain; in the latter case the theory ensuing from the new look may be simpler or more complex, but in any case it will be more detailed and consequently more daring than the timorous phenomenological theory (which, if empirically validated, will be useful as a control of new, deeper, theories). At any rate, the falsity of simple theories is usually easier to expose than the falsity of complex theories, on condition that they are falsifiable at all. Parsimony in the number of empirically adjustable parameters is not the seal of truth, but the abortive of falsity.

1.4. Simplicity, likelihood, and truth

The simpler theories are easier tested both by experience and by further theories, i.e. by inclusion in or fitting with contiguous systems. Syntactical and semantical simplification are, then, sufficient to improve testability even though they are not strictly necessary to secure it. Yet, there is as great a distance between *testable* and *tested,* as there is between a promise and its fulfillment. Syntactical and semantical simplicity are relevant to the likelihood of scientific theories in so far as they are factors of both systematicity and testability. But the assessment of the degree of *likelihood* of a theory is one thing, and the estimate of its degree of *corroboration* is another: the latter is done *a posteriori,* after certain tests have been given—and these include empirical corroboration, checking of compatibility with the bulk of relevant knowledge, and checking of explanatory power. It is only in the *prior* estimate of the likelihood of a theory that considerations of simplicity can legitimately arise, and this in an indirect way, namely, through the contribution of simplicity to systematicity and testability.

Once a theory has been accepted as the truest available, we do not care much for its simplicity. It will not do to argue that this is because simplicity has already been built into the theory during its construction: as we have seen, epistemological simplicity is inconsistent with depth and with formal simplicity, and the latter is inconsistent with accuracy, which is not only a desideratum in itself but also a condition for testability.

Nor will probability save the thesis that simplicity is necessary for truth, as is held by the theory according to which the simpler theories are the more probable because the basis of every theory consists in the conjunction of a number of axioms, and the less the number of members that occur in the conjunction the greater will be its total probability (equal to the product of the probabilities of the single axioms). The inadequacy of this theory is patent: (1) it does not apply to theories containing at least one strictly universal law statement, since the probability of universal laws is exactly zero; (2) it is not the simplest but the more complex hypotheses which are the easier to fit with empirical data: think of a wavy line passing through or near all the dots representing empirical data on a coordinate plane, as contrasted with a syntactically simpler curve, such as a straight line; it is unlikely that a large number of empirical "points" lie on a simple curve. It is the more complex hypotheses—especially if devised *ex post facto* and *ad hoc*—which are *a priori* the more probable.[12] In short, simplicity is incompatible with a high *a priori* probability.

In summary, syntactical and semantical simplicity are, within limits, favorably relevant to systematicity and testability—not to accuracy and truth; yet, they are not necessary conditions of systematicity and testability.

Now, any number of testable systems can be invented to cope with a given set of empirical data; the question is to hit on the *truest* one—a scientific problem—and to *recognize the signs* of approximate truth—a metascientific problem. For, indeed, truth is not the unveiling of what had been occult, as the pre-Socratics and Heidegger have claimed: truth is made, not found, and to diagnose truth is as hard as to diagnose virtue. We have a working theory of the complete (not the approximate) truth of sentences involving only observational predicates,[13] but we have no satisfactory theory of the *approximate* truth of *theories*. To say that a factual theory is true if and only if its observable consequences are true and none is false, is inadequate not only because the theory may contain untestable assumptions and yet be consistent with observable facts, but also because there is no means of exhaustively testing the infinity of consequences (theorems) of quantitative scientific theories, and because the notion of approximate truth is involved in them.

Furthermore, we should know by now that all factual theories are, strictly speaking, false: that they are more or less approximately true. No decision procedure for recognizing the approximate truth of factual theories is available, but there are *symptoms* of truth, and the expert employs these signs in the evaluation of theories. Let us review these symptoms of truth and find out what simplicities, if any, are relevant to them.

2. DESIDERATA OF SCIENTIFIC THEORY, OR SYMPTOMS OF TRUTH

At least five groups of symptoms of the truth of factual theories can be distinguished: they may be called syntactical, semantical, epistemological, methodological, and philosophical. Each symptom gives rise to a criterion, or norm, occurring in the actual practice of weighing factual theories before and after their empirical test, in order to ascertain whether they constitute an improvement on competing theories, if any. We shall call them *assaying criteria.* They are the following twenty.

2.1. Syntactical requirements

(*1*) ***Syntactical Correctness.*** The propositions of the theory must be well-formed and mutually consistent if they are to be processed with the help of logic, if the theory is to be meaningful, and if it is to refer to a definite domain of facts. Syntactically crippled sets of signs, on the other hand, cannot be logically handled; they cannot be unambiguously interpreted either, and if they contain internal contradictions they may lead to a sterile multiplicity of irrelevant statements. Yet, every theory is somewhat muddled in its preliminary stages; therefore, *rough* syntactical correctness, and definite *possibility* of formal improvement, are more realistic criteria than final formal neatness—which may anyhow not be attainable.

Simplicity is obviously not a factor of syntactical correctness; on the other hand, simplicity facilitates the *test* of syntactical correctness.

(*2*) ***Systematicity or Conceptual Unity.*** The theory must be a unified conceptual system (i.e., its concepts must "hang together") if it is to be called a theory at all, and if it is to face empirical and theoretical tests as a whole—i.e., if the test of any of its parts is to be relevant to the remainder of the theory, in such a way that a judgment can eventually be passed about the corroboration or falsification of the theory as a whole.

As we saw before (sec. 1.2), simplification of the predicate basis of the theory is sufficient to improve systematicity, but is not necessary for attaining it and cannot be forced beyond bounds which are partly set by the theory's referent (e.g., an aspect of nature). Furthermore, the historical trend of science has not been the shrinking but the expansion of the predicate bases, along with the establishment of more and more connections—mainly by way of law statements—among the various predicates. A progressive conceptual enrichment coped with an increasing logical cohesion or integration is the tendency of science—not a unification by impoverishment.[14]

2.2. Semantical requirements

(3) ***Linguistic Exactness.*** The ambiguity, vagueness, and obscurity of the specific terms must be minimal to ensure the empirical interpretability and the applicability of the theory. This requirement disqualifies theories in which terms such as 'big', 'hot', 'psychical energy', or 'historical necessity', occur essentially.

Now, the elimination of such undesirables has little to do with simplification. Clarification is more often accompanied by complication or, at least, by showing an actual complexity underneath an apparent simplicity. Hence, simplicity is unfavorable relevant to linguistic exactness, or at most irrelevant to it.

(4) ***Empirical Interpretability.*** It must be possible to derive from the theory's assumptions—in conjunction with bits of specific information—statements that could be compared with observational statements, so as to decide the theory's conformity with fact.

Simplicity is clearly unfavorably relevant to this desideratum, since an abstract theory is simpler than an interpreted system.

(5) ***Representativeness.*** It is desirable that the theory represents or, rather, reconstructs actual events and processes, and not merely describes them and predicts their observable gross effects. In order to be representational—as opposed to phenomenological—a theory need not be pictorial, visualizable, or intuitable (although these characteristics warrant representativeness). It is sufficient that some of the symbols occurring in the postulates of the theory be assigned a literal meaning by being correlated with actual and essential (diaphenomenal) properties of the theory's referent. In other words, for a theory to be representational it is sufficient that some of its basic predicates be assumed to represent real and fundamental —not merely external—traits of actual entities.

In the course of the growth of science phenomenological or non-representational theories have been replaced or at least supplemented by representational theories, which attempt to offer descriptions and explanations in agreement with reality (Einstein's *Realbeschreibung*). Thus, theories of action at a distance were replaced by field theories, thermodynamics was supplemented by statistical mechanics, circuit theory by electron theory, synoptic by dynamic meteorology, simple evolution theories by the theory of evolution through natural selection.

There are various reasons for preferring representational to phenomenological theories: (*a*) a major aim of investigators is not just "to save appearances" in an economical way (conventionalism, phenomenalism, pragmatism), but to attain a deep understanding of facts, both observed and unobserved—and this purpose is served better by representational than by phenomenological theories; (*b*) representational theories satisfy better the requirement of external consistency, whereas phenomenological theories are *ad hoc;* (*c*) representational theories, not being limited to the empirical data at hand, are more apt to predict facts of an unknown, otherwise unexpected, kind; (*d*) representational theories take more risks than phenomenological theories: by saying more they comply better with the requirement of refutability.

Now, a strict adherence to the rules of logical and epistemological simplicity would require us to dispense with representational theories, since these usually involve not only the predicates of the related phenomenological systems but further, more abstract, predicates of their own. We would have to drop hundreds of working theories, among them the shell model of the atomic nucleus, the spin theory of ferromagnetism, and the chromosomic theory of heredity. Here, again, simplicity is not welcome.

(6) Semantical Simplicity. It is desirable, up to a point, to economize presuppositions; in this way empirical statements can be made and tested without presupposing the whole of science. This requirement is imposed in a moderate way and on pragmatical rather than on theoretical grounds, since it amounts to the possibility of approaching the new without having to master the old in its entirety. But external consistency, which is even weightier, competes with semantical simplicity. Thus conventional biology, which is methodologically "mechanistic", complies with external consistency and, by the same token, it is semantically complex since it presupposes physics and chemistry; on the other hand, vitalistic biology is semantically simpler but fails to be continuous with physics and chemistry.

The theoretical value of semantical simplicity lies in that it suggests the existence of objective levels of organization of reality. Thus, the mere possibility of talking meaningfully about some aspects of life, psyche, and culture, without expressly dealing with their material bases, shows that levels are to some extent autonomous; but the requirement of depth will always end by forcing us to discover the links of events on one level to events on contiguous levels, and particularly on the lower ones.[15]

Semantical simplicity is, in short, an ambiguous rule: it may render the handling (e.g., the test) of the theory possible, but it may also be a symptom of superficiality.

2.3. Epistemological requirements

(*7*) ***External Consistency.*** The theory must be consistent with the bulk of accepted knowledge if it is to have the support of more than just its instances, and if it is to be regarded as an addition to knowledge and not as an extraneous body. Revolutionary theories—in contradistinction to deviant or crackpot theories—are inconsistent with only a part of scientific knowledge, for the very criticism of old theories and the construction of new ones is performed on the basis of definite knowledge and in the light of more or less explicitly stated norms. Isolated heterodoxies do not imperil the bulk of established (yet provisional) knowledge; quite on the contrary, we question isolated theories in the light of accepted knowledge and rules of procedure.

External consistency was the strongest argument Copernicus advanced in support of his theory of planetary motions; he pointed out that, unlike Ptolemy's theory, his own accorded with the axioms of the prevailing physical theory (Aristotle's), which ruled that celestial bodies moved in circular paths.[16] The remarkable contradiction of *ESP* and other supernatural theories with the bulk of science is also—along with methodological reasons—a major ground for rejecting them.[17]

Simplicity is clearly unfavorable to external consistency, since the latter imposes a growing multiplicity of connections among the various chapters of science.

(*8*) ***Explanatory power.*** The theory must solve the problems set by the explanation of the facts and the empirical generalizations, if any, of a given domain, and must do it in the most exact possible way. To put it briefly, *Explanatory power = Range × Accuracy.* But the range of a theory cannot increase beyond every bound; a scientific theory cannot pretend to solve every problem, under penalty of becoming irrefutable.[18] In particular, a scientific theory must be single-edged, that is, it must not be capable of supporting contrary (e.g., contradictory) hypotheses or proposals, and it must not be consistent with contrary pieces of evidence. (A hypothesis, if self-consistent, and exactly formulated, cannot be compatible with contrary pieces of evidence; a pseudotheory can, provided its hypotheses are mutually shielding.) Both the theory of predestination and pyschoanalysis, which offer explanations for everything human and are never embarrassed by contrary evidence, violate this condition. As regards range, the explanatory power of scientific theories is intermediate between the explanatory power of pseudoscientific theories and that of commonsensical theories.

It is clear that simplicity is unfavorable to explanatory power, because a wide range is a class of numerous subclasses, each intensionally characterized by a set of properties, and because accuracy, the second factor of explanatory power, requires complication as well (cf. 1.3). Thus, inequalities are simpler than equalities: they are simpler to define, to establish, and to test. Yet we are often ready to sacrifice simplicity for accuracy, as shown by the fact that numerical and functional equations are the more abundant the greater the severity of the standards of accuracy and testability become. In short, the demand for simplicity is incompatible with the demand for explanatory power.

(9) Predictive Power. The theory must predict at least those facts which it can explain after the event. But, if possible, the theory should also predict new, unsuspected, facts and relations: otherwise it will be supported only by the past. In other words, predictive power can be analyzed into the sum of the capacity to predict a known class of facts, and the power to forecast new "effects", i.e. facts of a kind unexpected on alternative theories. The former may be called *forecast power,* the latter *serendipic power.*[19] To put it in a nutshell, *Predictive power* = *(Old range* + *New range)* × *Accuracy* = *Forecast power* + *Serendipic power.* (Of course, the serendipic power of a theory cannot in turn be predicted, even after completing its construction, since we neither know in advance all the logical consequences of the theory's axioms, nor the range of unknown facts).

(Although the logical structure of prediction is the same as that of explanation—namely, deduction of singular sentences from general laws conjoined with specific informations—explanatory power is not the same as predictive power. Pseudoscience is prolific in explanation *post factum* but barren in prediction. The theories of nuclear, atomic, and molecular physics can explain singular phenomena—or classes of possible single phenomena—but can predict collective actual phenomena only—or, alternatively, they can only predict the probabilities of singular facts. Historical theories—such as those of geology, evolution, and human society—have a high explanatory power but a small predictive power, even counting retrodictions. Besides, predictions are usually of facts and very seldom of laws, whereas explanations can be either of facts or of laws. Finally, predictions are made with the help of the lowest level theorems of a theory—the closest to experience—whereas explanations may take place on any level. These are some of the reasons for counting predictive power separately from explanatory power.) [20]

Simplicity is unfavorable to predictive power for the same reason that it is incompatible with explanatory power.

(10) Depth. It is desirable, but by no means necessary, that theories explain essentials and reach deep in the level structure of reality. No scientific theory is just a summary of observations, if only because every generalization involves a bet on unobserved kindred facts. But, whereas some theories just account for appearances, others introduce diaphenomenal (but scrutable) entities and properties by which they explain the observable in terms of the unobservable: it is in this sense that wave optics is deeper than phenomenological (geometrical) optics, and reflexology deeper than behavioristics.

The requirement of depth does not, of course, eliminate the less deep theories: they may well be retained along with the deeper ones if they contain useful concepts that somehow correspond to actual entities or properties. Wave optics does not eliminate the concept of light ray but elucidates it in terms of interference, and neurophysiology is expected to elucidate behavior patterns, and not to explain them away. The requirement of depth functions as a stimulus in theory construction (e.g., the present feeling of dissatisfaction with the superficiality of dispersion relations and other phenomenological theories in physics), and in the scientific reconstruction of deep prescientific theories (e.g., Marxist sociology and psychoanalysis, both rich in deep concepts and hints, but marred by a muddled logic and a complacent methodology).

Since depth involves epistemological sophistication, it is inconsistent with epistemological and pragmatical simplicity.

(11) Extensibility or possibility of expansion to encompass new domains.[21] Thus, Hamilton's formulation of dynamics is preferable to Newton's because it can cope with a wider class of dynamical problems and because it can be extended beyond dynamics (e.g., into field theory); yet, it is logically and epistemologically more complex than Newton's version of mechanics: it contains twice as many equations of motion, and the concepts of generalized coordinates and momenta. The same is true of Maxwell's theory of the electromagnetic field, which it was possible to extend to optics, in relation with its rivals.

The capacity for linking or unifying hitherto unrelated domains is connected with both external consistency and serendipic power, and it depends on the depth of the concepts and laws peculiar to the theory. Hence simplicity, which is unfavorably relevant to these characteristics, is also unfavorably relevant to extensibility. On the other hand, the actual expansion of a theory produces a methodological unification, in the sense that a single method can be employed to attack problems belonging to formerly disjoint sets. But a considerable syntactical, semantical, and epistemological complexity must first be swallowed: methodological simplification is not a prerequisite but a reward for the willingness to accept certain complexities.

(12) Fertility. The theory must have exploratory power: it must be able to guide new research and suggest new ideas, experiments, and problems, in the same or in allied fields. In the case of adequate theories, fertility overlaps with extensibility and serendipic power. But altogether inadequate theories may be stimulating, either because they contain some utilizable concepts and hypotheses (as was the case with the caloric theory of heat and the aether theories), or because they elicit new theories and experiments designed to refute them. On the other hand, virtuous theories may be barren because nobody takes an interest in them—e.g., because they are shallow, as is the case with those theories which are little more than summaries of empirical data. This is why fertility should be counted on its own.

Here, again, simplicity is either irrelevant or unfavorably relevant.

(13) Originality. It is desirable that the theory be new relative to rival systems. Theories made of bits of existing theories, or strongly resembling available systems, or lacking new concepts, are unavoidable and may be safe to the point of being uninteresting. The most influential theories are not the safest but those which are the more thought-provoking and, particularly, those which inaugurate new ways of thinking; and these are all deep, representational, and extensible theories, such as Newtonian mechanics, field theory, quantum theory, and evolutionism. As a noted physicist [22] says, "For any

speculation which does not at first glance look crazy, there is no hope".

Now, the rules of simplicity evidently prohibit or at least discourage the framing of new, bold, constructs: the trite path is the simplest. This is particularly so in the case where theories are available which have been empirically confirmed, but which are unsatisfactory for some reason—e.g., because they are phenomenological. The policy of simplicity will in this case disavow new approaches and will thereby stop the advancement of science.

2.4. Methodological requirements

*(14) **Scrutability.*** Not only the predicates occurring in the theory must be open to empirical scrutiny by the public and self-correcting method of science (sec. 1.3), but also the methodological presuppositions of the theory must be controllable. This requirement leads to rendering suspect (*a*) evidences of a kind that only the given theory would accept, and (*b*) techniques, tests, and alleged modes of knowing which—like sympathetic understanding and essence intuition—cannot be controlled by alternative means and do not lead to intersubjectively valid conclusions, or at the very least to arguable ones.

Again, this requirement conflicts with simplicity of certain kinds, since the logically simplest theories are those speculative systems that do not care for tests. If someone should insist in introducing the term 'simplicity' in this connection, allow him to call this the requirement of methodological simplicity, but remind him that this phrase should not be construed as imposing a simplification in method in the sense of a relaxation of the standard of rigor, or of a reduction in the variety of tests, but as simplifying the task of rigorously testing the theory and the tests. Otherwise, the methodologically simplest theory would be the one validated by the "method" of navel contemplation. But, of course, no rule of simplicity enlightens us as to whether a given construct (e.g., a quantum-mechanical operator) can be regarded as representing an observable property, or not. Criteria of scrutability of predicates are not simple, and they are often controvertible.[23]

In summary, implicitly is ambiguously relevant to scrutability (see sec. 1.3).

*(15) **Refutability.*** It must be possible to imagine cases or circumstances that could refute the theory.[24] Otherwise no genuine tests

could be designed, and the theory could be regarded as logically true, i.e., as true come what may—hence as empirically void.

A scientific theory may certainly contain an irrefutable premiss among its postulates, such as an existential hypothesis of the form 'There is at least one *x* such that *x* is an *F*' (without specifying precise location in either space or time), or a statistical law of the form 'In the long run, *f* approaches *p*'. The theory may even presuppose irrefutable metascientific principles, like "Every fact is explainable in the long run".[25] But all such irrefutable statements should be confirmable and somehow supported by the bulk of knowledge; furthermore, all the remaining premisses of the theory should be refutable and none of them should be exempt from being indicted by evidence by the interposition of shielding hypotheses; finally, none of the lower-level consequences of the theory should be indifferent to experience. In particular, no secure, uncorrigible data which "resist the solvent influence of critical reflection" (Russell) are to enter science, which is essentially corrigible knowledge.

Clearly, semantical, epistemological, and experimental simplicity are favorable to refutability. But syntactical simplicity is ambiguously relevant to it: on the other hand, refutability requires accuracy, which in turn involves complexity (see sec. 1.3); on the other hand, the fewer the predicates involved, and the simpler the relations assumed to hold among them, the easier it will be to refute the theory. But what if facts, indifferent as they are to our labors, stubbornly refuse to lend themselves to logical simplification? Forced simplification will lead to actual refutation rather than just securing testability.

(16) Confirmability. The theory must have particular consequences that may be found to agree with observation (to within technically reasonable limits), And, of course, actual confirmation to a large extent will be required for the acceptance of every theory. The insistence on confirmation as the sole assaying criterion (inductivism) opens the door to theories fraught with vague and inscrutable predicates (gypsy theories). Plenty of confirmation is not a guarantee of truth, since after all the evidence may all be selected, or conveniently interpreted, or else the theory may never have been subjected to severe tests. But, of course, even if insufficient, confirmation is necessary for the acceptance of theories.[26]

Now, a theory may be complicated ex profeso so as to increase its degree of confirmation; hence, simplicity is unfavorably relevant to confirmation.

(*17*) ***Methodological Simplicity.*** It must be technically possible to subject the theory to empirical tests. The theory may lead to the formulation of predictions that are too difficult, or even impossible, to test empirically at the moment, yet it may be a valuable theory that may stimulate the improvement of technical media. An unpredictable number of years will pass before a single empirical test of any of the quantum theories of the gravitational field may be supplied, but the mere proliferation of theories of this kind may stimulate the design of empirical tests.

In short, methodological simplicity to a moderate extent must be required, particularly from theories designed to elude or postpone *sine die* the trial of experience; if required too sternly, it may be obstrusive.

2.5. Philosophical requirements

(*18*) ***Level-parsimony.*** The theory must be parsimonious in its references to sections of reality other than those directly involved. The higher levels (actual or imaginary) should not be appealed to if the lower are enough, and distant levels should not be introduced without the intermediate ones. This requirement is, of course, violated by animistic theories of matter and by mechanistic theories of mind.

The rule of simplicity is ambiguous in this connection as it is in others. In effect, level-parsimony may be regarded as an instance of the rule; yet, what can be simpler than reductionism—down, as in the case of mechanism, or up, as in the case of idealism—which violates the rule of level-parsimony?

(*19*) ***Metascientific Soundness.*** The theory must be compatible with fertile metascientific principles, such as the postulates of lawfulness and rationality, and the relevant metanomological statements [27] (such as general covariance).

Simplicity is, in the best of cases, irrelevant to metascientific soundness—unless it is arbitrarily included among the symptoms of such a soundness, despite its ambiguous relevance to the other desiderata of scientific theory.

(*20*) ***World-view Compatibility.*** It is desirable that the theory be consistent with the common core of the *Weltanschauugen* prevailing in scientific circles—world-views which, anyway, do mold the very construction and acceptance of scientific theories. This requirement functions as a stabilizer: on the one hand, it leads us—along with external consistency, of which it is an extension—to reject crack. pot theories; on the other hand, it may delay or even prevent revolutions in our world-view, if the latter makes no room for its own change. Remember the cool reception given to field theory and to Darwin's theory in France one century ago.) The criterion of world-view compatibility must therefore be used with care. Anyway, it does intervene in theory assaying, and it is better to realize this than to be inadvertently dominated, in our valuation of theories, by some unscientific world-view. World-views and scientific theories should control and enrich each other.

Simplicity is, of course, as inconsistent with world-view compatibility as it was found to be with external consistency.

2.6. Further criteria?

Alternative criteria have from time to time been proposed, such as intelligibility (psychological simplicity), elegance, practical usefulness, operational character ("definability" of all concepts in terms of effective operations), high probability, and causality. As a matter of fact these criteria often influence our valuation of theories, but they can be shown to be inadequate.

In effect, intelligibility or intuitability is beside the point, because it is largely a subjective characteristic entirely independent of truth.[28] Elegance or beauty is not an independent but a derived characteristic of some theories: a theory arouses an aesthetic feeling in us if it is logically well organized, accurate, deep, wide, and original—and if we are deeply interested in the subject. Practical applicability is rather irrelevant to truth, as shown by the heap of pseudosciences that serve a purpose beneficial to their entrepreneurs and even, occasionally, to their victims. Operational character cannot be satisfied if metrical and/or transcendent (theoretical) predicates are allowed [29]—as they must if the theory is to be exact. A high *a priori* probability is inconsistent with accuracy and universality. And causality, unless understood in a very liberal sense—as general determinism, which is committed only to the postulate "Everything is determined in accordance with law by something else"—would be as mutilating to science as overall simplicity.[30]

Further legitimate requirements may, of course, appear with the progress of metascience and with the advancement of science itself. Do not standards of rigor become more and more exacting? But the twenty above-listed criteria constitute a complex enough set—particularly for the sake of ascertaining the weight of simplicity.

3. THE ACCEPTANCE OF SCIENTIFIC THEORIES: FIVE CASE HISTORIES

Let us illustrate the functioning of the assaying criteria listed above with a few celebrated cases. We shall concede that all the theories to be examined in the following are logically consistent and, to some extent, compatible with empirical information.

3.1. Theory of the planetary system

The geocentric and the heliocentric models of our planetary system are regarded by conventionalists as empirically equivalent and even as equivalent modes of speech; and it has been claimed and repeated that the sole reason for preferring the heliocentric system is its simplicity relative to the geocentric image, since—so the contention runs—there is really no reason to single out one system of reference (the Copernican) to another (the Ptolemaic). Both affirmations are false: the Copernicus-Kepler system accounts for a far greater set of phenomena than Ptolemy's does, and it was not adopted because of its greater simplicity—which it does not possess in all respects—but because it is supposed to be a truer image of facts, as suggested among other reasons by its fitting with contiguous theories, whereas the geocentric system is an *ad hoc,* isolated, theory.

Specifically, the Copernicus-Kepler system satisfies the following assaying criteria (cf. sec. 2) to an extent its rival could never dream of: (*a*) *external consistency:* compatibility with dynamics, gravitation theory, and cosmology. No system of dynamics employs Ptolemy's noninertial axes (which alone could yield Ptolemy's orbits); the paths of the planets both in Newtonian and in Einsteinian theory are essentially determined by the sun; [31] and every cosmogonical theory entails the hypothesis that the earth was formed some billion years ago along with the other planets and without any special privilege (in other words, the geocentric system is inconsistent with the theory of stellar evolution). Terrestrial axes are as good as the Copernican system of reference from a *geometrical* point of view only—i.e., as regards the shape of the orbits—but they are definitely inadequate from *kinematical* and *dynamical* points of view, among other reasons because the apparent velocities of celestial bodies may take on any value (being proportional to the distance from the earth), even beyond the speed of light, and because, Ptolemy's axes being noninertial, it is not possible to apply to them the principle of relativity: they are not relativistically equivalent to the Copernican axes (they are not related by means of a Lorentz transformation);[32] (*b*) *explanatory and predictive powers:* the heliocentric system accounts for the phases of the planets (predicted and discovered by Galileo in the case Venus), for the aberration of light (which enables us to determine both the velocity of the earth and the earth-sun distance), for the Doppler shift of the star's spectra (which leads to the determination of the recession velocity of nebulae), and for various other appearances which the geocentric system does not "save"; (*c*) *representativeness:* the heliocentric system is not merely a conventional calculating device but a conceptual reconstruction of facts, as Copernicus and Galileo believed, and as must now be conceded in view of the facts mentioned above; (*d*) it *is fertile:* it has prompted new astronomical discoveries (such as Kepler's laws), new developments in mechanics (e.g., the various theories of gravitation and of tides) and in optics (such as Roemer's measurement of the velocity of light), as well as the conjecturing—now fairly well established—that there is a multiplicity of solar systems (first suggested by Bruno and insisted on by Galileo upon his discovery of Jupiter's satellites); (*e*) *refutability:* it is better refutable by empirical evidence than any conventionalist system, as it does not admit an unending addition of auxiliary hypotheses aimed at saving the central assumptions; moreover, the simple Copernicus-Kepler model was refuted or, rather, improved long ago with the discovery that the real orbits are much more complex than the original ellipses, owing to the perturbations from other planets and to the finite velocity of propagation of the gravitational field; (*f*) *world-view compatibility:* the new astronomy was compatible not only with the new physics but also with the new anthropology and the new ethics, according to which the earth was not the basest place in the universe, and nature was not made to serve man.

What role did simplicity play in the choice among these two rival—but by no means empirically equivalent—theories? Copernicus, referring to the *geometrical* aspect of his theory, did employ the argument from simplicity; but, at the same time, he granted that his theory was "well-nigh contrary to common sense" or, in our terminology, that it was epistemologically more complex than the theory according to which celestial motions are such as they appear to be. But how naively simple the most complex curve Ptolemy could imagine looks as compared with the real orbits of planets as calculated with Newtonian mechanics and perturbation theory! In short, it is false to say that we retain the heliocentric system because it is the simplest: we prefer it, despite its greater complexity, because it is the truest. And simplicity did not intervene in our judgment of its truth value.

3.2. Gravitation Theory

Several theories of gravitation account for roughly the same observed facts as Einstein's theory does, and all of them are *simpler* than it: Whitehead's (1922), Birkhoff's (1945), Belinfante-Swihart's (1957), and others. Thus, for example, Whitehead's theory, conveniently modified, gives the same formula (not merely the same numerical value for some particular case) for the gravitational deflection of light rays as Einstein's theory does.[33] It is true that the recent measurement of the gravitational red-shift of spectral lines (due to the energy lost by photons in escaping from gravitational fields, e.g., in moving upwards in the vicinity of the earth) is in amazing agreement with the value predicted with the help of the theory.[34] Yet, the situation is not as good with regard to other empirical tests of general relativity: (*a*) the deflection of light rays in the vicinity of celestial bodies has been confirmed to within 10–15% accuracy only; (*b*) the advance of the perihelion of planets has been well confirmed in the case of Mercury alone (artificial satellites might provide further evidence); (*c*) the displacement of the perihelion owing to the sun's rotation (another effect predicted by the theory) has not been measured; (*d*) gravitational waves, also predicted by the theory, have not been detected.

Why, then, do most physicists prefer Einstein's theory of gravitation, which is so obviously complex from a syntactical and an epistemological point of view that most astronomers refuse to employ it? The reasons seem to be that, unlike its rivals, Einstein's theory of gravitation (*a*) has a *high serendipic power:* it predicts phenomena which were neither observed before nor predictable on the theories prevailing at the birth of general relativity—like the deflection of light rays and the gravitational red-shift; on the other hand, its rivals have been tailored to fit phenomena *ex post facto:* their degree of inductive character, or ad-hocness, is considerable, whereas Einstein theory's is nil; (*b*) is *extensible:* it provides a framework that might be expanded into a unified field theory; (*c*) *is representational:* it ascribes reality to the gravitational field (or space) and its sources, and contains no adjustable parameters lacking a physical meaning (as does the phenomenological linear theory of gravitation); (*d*) is *deep:* it is relevant to our views on space, time, field, force, and mass; the basic equations may even be regarded, as Schrödinger suggested, as a definition of matter; (*e*) is highly *original:* it is counterintuitive and noncausal enough to deserve attention; (*f*) is *fertile:* it has suggested new observations, some of which are still to be done.

Neither observed facts nor simplicity have played a prominent role in the construction of general relativity (notwithstanding Einstein's own statements about the value of simplicity). Rather, the complexity of the theory has been an obstacle to its acceptance by many,[35] and a major stimulus for the invention of simpler theories. In fact, the theory contains epistemologically complex predicates, such as 'spacetime curvature', and 'gravitational potential', which are not acceptable to empiricist philosophies; and the equations of the theory, though deceivingly "simple" if written in the compact tensor notation, are complex enough to obstruct and even preclude the very formulation of problems, which renders the theory unpractical. This is why many physicists, even granting that Einstein's theory is so far *the truest,* frequently employ or try alternative theories which are syntactically, epistemologically, and pragmatically simpler; but future theories of gravitation will have to be more inclusive and deeper than Einstein's (among other things, they will have to be contiguous with quantum mechanics)—whence they are likely to be even more complex than it.

3.3. Beta-decay theory

The present theory of the beta-decay of neutrons, mesons, hyperons, and other so-called fundamental particles, contains two hypotheses that it was found necessary to complicate in the course of time in order to square the theory with empirical data. One of the hypotheses refers to the existence of the neutrino, the other to certain symmetry properties of the basic equations.

The neutrino hypothesis may conveniently be expounded with reference to the mu-meson decay. If only charge conservation is taken into account, the hypothesis

*H*1 mu-meson → electron

will be sufficient. But it is found that electrons are emitted with a continuous energy spectrum (to the extent to which observation can suggest or test continuity!), which is inconsistent with the assumption that only two bodies are involved (if we further *assume* momentum conservation). *H*1, the simplest hypothesis, was therefore false: a more complex one had to be invented. The next simplest conjecture involves the invention of an unobserved entity, the neutrino:

*H*2 mu-meson → electron + neutrino.

This hypothesis is epistemologically complex; it is also methodologically complicated since the neutrino, owing to its lack of charge and its small (or zero) mass, is remarkably elusive—to the point that many physicists have disbelieved in its existence for years, especially after many independent and elaborate attempts to detect it failed. Still, *H2* is not complex enough: it is consistent with the continuous energy spectrum but inconsistent with the hypothesis of spin conservation, found correct in other fields. This latter hypothesis is respected by introducing a further theoretical entity, namely, the antineutrino:

H3 mu-meson $\rightarrow$ electron + neutrino + antineutrino

This hypothesis is consistent with charge, energy, and spin conservation; but in involves an entity which is empirically indistinguishable from the neutrino. The decay scheme has become more and more complex syntactically, epistemologically, and methodologically.

Of course, *H3* is not the sole hypothesis consistent with the known facts; we may frame a heap of alternative conjectures just by assuming that an arbitrary number n of neutrinos and antineutrinos take part in beta-decay. But there is no point in adopting any one of these more complicated hypotheses as long as we cannot distinguish experimentally among their consequences, and as long as they do not throw new light on the explanation of phenomena. It is here that we appeal to the rule of simplicity. But we do not just choose "the simplest hypothesis compatible with the observed facts", as the inductivist methodology has it: we select the simplest hypothesis of a set of *equally precise* assumptions all compatible with the known *facts* and with the set of *law statements* we regard as relevant and valid. And this is a far cry from unqualified simplicity: it came after a considerable sophistication and when no further complication promised to be fruitful. The rule actually used in scientific research is not just "Choose the simplest", but "Try the simplest first and, if it fails—as it normally should—gradually introduce complications compatible with the bulk of knowledge".

A second hypothesis of the theory is that the laws "governing" this kind of disintegration are not invariant under the reversal of position coordinates (i.e., under the parity transformation $x \rightarrow -x$). Until the work of Lee and Yang (1956), the simplest hypothesis had been made regarding this transformation, namely, that all physical laws are parity-invariant (i.e., do not change under the exchange of left and right). The rejection of this metanomological statement [36] made it possible to identify two kinds of particles (the theta and tau masons)—which involved a taxonomical simplification—and led to predicting previously unsuspected facts, such as the unsymmetry of the angular distribution of decay products.

The theory, as corrected in the above sketched ways, (*a*) had *serendipic power,* (*b*) was *original* to the point of "craziness" (as both the neutrino hypothesis and the parity non-conservation hypothesis have seemed to many), and (*c*) was *deep,* to the point that it dethroned the laboriously acquired belief that no intrinsic differences between left and right could ever be found in nature.

It will not do to count the identification of the theta and tau mesons in favor of the tenet of simplicity: this small simplification introduced in the *systematics* of fundamental particles did not involve a simplification in the basic *theory* but was an assignment of simplicity to nature itself. Besides, it was overcompensated by the introduction of new, less familiar, terms (pseudoscalar and pseudovector contributions to the energy operator), which correspondingly complicated the theorems dependent on them.

Not the abidance by simplicity but the bold inventions of new, complicating, hypotheses was decisive in the building, improvement, and acceptance of the beta-decay theory.

3.4. Theory of evolution

What gave Darwin's theory of evolution through natural selection the victory over its various rivals, notably creationism and Lamarckism? Darwin's theory was in part logically faulty (remember the vicious circle of the "survival of the fittest"); it contained several false or at least unproven assertions ("Each variation is good for the individual", "Acquired characters, if favorable, are inherited", "Sexual selection operates universally"); it had not been tested by observation, let alone by experiment on living species under controlled conditions (the development of antibiotic-resistant strains of bacteria, industrial melanism in butterflies, and a few other processes supporting the theory, were observed one century after *The Origin of Species* appeared); its explanatory power was clearly smaller than that of its rivals (irrefutable theories have the maximum *post factum* explanatory power); it had no inductive basis but was, on the contrary, a bold invention containing high-level unobservables. And, as if these sins were not enough to condemn the theory, Darwin's system was far more complex than any of its rivals: compare the single postulate that states the special creation of every species, or Lamarck's three postulates (stating the immanent tendency to perfection, the law of use and disuse, and the inheritance of acquired characters), with Darwin's system, which included among others the following axioms: "The high rate of population increase leads to population pressure", "Population pressure leads to struggle for life", "In the struggle for life the innately fittest survive", "Favorable differences are inheritable and cumulative", and "Unfavorable characteristics lead to extinction".

The characters which ensured the survival of Darwin's theory despite its complexity and its various genuine shortcomings seem to have been the following: (*a*) *external consistency:* the theory was compatible with evolutionary geology and with the evolutionary theory of the solar system; (*b*) *extensibility and fertility:* the theory was quickly, boldly, and fruitfully expanded to physical anthropology, psychology, and history, and unwarrantedly extrapolated to sociology (social Darwinism) and ontology (Spencerian progressivism); (*c*) *originality:* although the idea of evolution was old, the mechanism proposed by Darwin was new and suggested daring fresh starts in all related fields, as well as the very relation among fields hitherto disconnected; (*d*) *scrutability:* Darwin's theory did not involve inscrutable predicates like 'creation', 'purpose', 'immanent perfection', and the like, and it did not involve unscientific alleged modes of knowing (such as revelation); (*e*) *empirical refutability:* contrary to its rivals, every piece of relevant evidence was conceivably favorable or unfavorable; (*f*) *level-parsimony:* no spiritual entity was invoked to account for lower-level phenomena, and no purely physico-chemical mechanism was resorted to either; (*g*) *metascientific soundness:* in particular, compatibility with the postulate of lawfulness, violated by the hypothesis of creation—but, on the other hand, the theory was inconsistent with the inductivist methodology then dominant, and sounded suspect to some on this score; (*h*) *world-view compatibility:* definite consistency with the naturalist, agonistic, dynamicist, progressivist, and individualist outlook of the *intelligentsia* on which the recent social and cultural changes (notably 1789, Chartism, and 1848) had made a deep impression. These virtues of Darwinism overcompensated its shortcomings and made it worth correcting on several points, until it merged (only in the 1930's) with genetics.

In short, simplicity was not taken into account in the genesis and development of Darwin's theory of evolution.

3.5 Genetic theory

The Mendelian theory of heredity has been under attack from environmentalism, or neo-Lamarckism, since its inception. The theory of the omnipotence of the environment is attractive to many because it is so much closer to common sense, because it is causal, because (if only it were true!) it would enable us to quickly control evolution in a planned way and last, but not least, because it is superficially compatible with an optimistic and progressive view of human life, on which nurture can overcome every shortcoming of nature. On the other hand, Mendelian genetics is formally, semantically, and epistemologically much more complex; it involves theoretical terms such as 'gene'; it calls for the use of statistics; it does not so far afford a precise control of evolution; it suggests rather gloomy prospects and—at least in Weismann's version of the theory—it reinforces the anachronic ontological tenet of the existence of an unchangeable substance (the germ-plasm). Furthermore, the genetic theory does not account satisfactorily for heredity in the case of higher organisms, and many geneticists are beginning to grant a parallel though weaker intervention of the cytoplasm in the transmission of characters.

Why then is Mendelian genetics accepted by most biologists? The main reasons seem to be the following: (*a*) it is *representational:* it locates precisely each hereditary factor in a bit of matter (gene, or gene complex), and it provides a chance mechanism (gene shuffling) that explains the final outcome, whereas environmentalism is a phenomenological theory; (b) it is *consistent* with the theory of evolution through natural selection (as modified to meet precisely this requirement), and with biochemistry (a plausible and precise mechanism of genetic information transmission, and of gene duplication, has recently been invented); (*c*) it has *predictive power:* statistical predictions (not individual ones) are often possible in an accurate manner with the help of its laws; (*d*) it is *refutable* and *confirmable* by experiment (e.g., mutation by direct physical action of X-rays on chromosomes), whereas the environmental theory is confirmable only, since it speaks about vague environmental influences; (*e*) it is *compatible* with some well-grounded and widely accepted philosophical views, such as naturalism (material basis of biological traits), and atomism (existence of discrete units of some sort on every level of organization). Last, but not least, its main enemy, Lysenkoism, has been marred by fraud, dogmatism, and unpleasant associations with curtailments of academic freedom. Yet, who would deny that unacademic and unfair attitudes were apparent in both camps some years ago, just because the whole controversy was regarded as part of a holy war?

At any rate simplicity—which, incidentally, was on the part of Lysenko—played no role in the discussion as compared with ideological and political considerations.

3.6. Test of the tests

Five case histories have ben recalled and analyzed in order to test the tests proposed in sec. 2. (No metascience is scientific if it does not test its hypotheses.) They were not just elementary cases of fitting polynomials to isolated sets of observational data—the favorite example of inductivist treatments of simplicity. The five cases selected for examination consisted of systems of testable hypotheses and were important enough to affect to some extent the modern world-view. In none of them was simplicity found to be a major factor of theory construction or evaluation: quite on the contrary, the theories ultimately chosen were in most respects remarkably more complex than their defeated rivals. Which plainly suggests the objective complexity of reality.

4. CONCLUSION: THE LIGHTNESS OF SIMPLICITIES

4.1. Simplicity neither necessary nor sufficient

While agreement with fact as tested by experience was regarded by metascientists as the sole test of a true theory,[37] simplicity alone seemed to provide the decisive criterion of choice among competing theories. What else could distinguish one theory from another while—in compliance with inductivism—attention was focused on empirical confirmation with neglect of all the remaining factors that as a matter of fact, consciously or unconsciously, intervene in the evaluation of scientific theories?

Gone are those days of *sancta simplicitas:* it is more and more clearly realized that the degree of truth, or degree of sustenance, of scientific theories has never been equated with their degree of confirmation. Many more requisites have always been imposed *de facto* by scientists, and have occasionally been recognized.[38] Twenty requirements—functioning at the pragmatic level as so many assaying criteria—were recognized in sec. 2, and overall simplicity was not included among them for the plain reason that a theory may be simple and false, or complex and approximately true—i.e., for the simple reason that *simplicity is neither a necessary nor a sufficient sign of truth.*

It would be unrealistic to regard any of the twenty requirements, save systematicity, accuracy, and testability, as strictly necessary for calling a set of hypotheses a *scientific theory,* even though they are jointly sufficient for calling it an *approximately true scien-*

tific theory. (The assaying criteria are consequently useful in distinguishing scientific from nonscientific systems and, in particular, in weeding out pseudoscientific theories.) The twenty requirements are rather *desiderata* of theory construction, *means* for attaining truth, and *symptoms* of truth; and, like other desiderata, they are not all mutually compatible, whence a compromise must always be sought.

Now every desideratum—here as elsewhere—can be satisfied in various degrees, and the failure of a theory to rigorously comply with *some* of the above requirements—save systematicity, accuracy, and testability, which are mandatory—should not lead one to reject a theory altogether. Thus, e.g., syntactical correctness and linguistic accuracy are always meager in the beginnings. If a theory is rich in deep transcendent and scrutable concepts, and if it promises to unify wide fields of knowledge, or to be instrumental in the exploration of new territories, it would be short-sighted to dismiss it entirely because of some formal shortcomings; the wisest course will be to work out the theory and to test it: syntactical and semantical neatness will eventually be achieved in this process. Only mature theories fulfil all the requirements in an excellent way. But, then, mature factual theories, like mature persons, are those which are about to be replaced.

What is the place of simplicity in the body of criteria that guide our evaluation of scientific theories? In order to estimate it we should recall, in the first place, that there is a variety of simplicities (sec. 1.1) and, in the second place, that simplicity of some sort is favorable to only a few symptoms of truth, and even so within limits. It was shown in secs. 1.2 and 2.2 that syntactical simplicity is favorably relevant to systematicity—though not necessary to attain it; also, moderate semantical and methodological simplicity were proposed as assaying criteria, mainly for practical reasons. On the other hand, complexity of some sort is associated with eleven other requisites: external consistency, linguistic exactness, empirical interpretability, representativeness, explanatory power, predictive power, depth, extensibility, originality, confirmability, and world-view compatibility. Finally, the rule of simplicity is ambiguous with regard to testability and to level-parsimony and, in the best of cases, it is neutral relatively to the remaining four requirements—syntactical correctness, fertility, scrutability, and metascientific soundness.

It does not seem possible to assign numerical weights to most of the requisites, and it does not seem promising to attempt to quantify the contribution—positive, negative, or nil—of simplicity to those various symptoms. If numbers have to be mentioned in this connection, let us be content with saying that simplicity does not contribute positively to 17 out of 20 major symptoms of truth. As regards most of the symptoms of truth, then, simplicity is similar to phlogiston: it is vague, elusive, and has a negative weight whenever it is not imponderable.

4.2. The role of simplicities in research

The role of simplicities in scientific research—as distinct from its products: data and theories—is, in summary, the following. Simplicities are undesirable in the stage of problem *finding*, since the mere discovery or invention of problems adds to the existing complexity. Simplicities of various sorts, on the other hand, are desirable in the *formulation* of problems and, much less so, in the *solution* of problems—which sometimes demands a complication of the given problem (e.g., the broadening of its setting) or the invention of new, complex concepts, hypotheses, or techniques. Then, some kinds of simplicity—notably syntactical and semantical economy—are *nolens volens* involved in *theory construction,* either because of the forced poverty of every beginning or because an unmanageable complication at a late stage has called for simplification in certain respects (usually syntactical); yet no single theory, if deep and promising, should be sacrificed to simplicity. Finally, syntactical and pragmatical simplicity are, within limits, favorable to the *test* of theories. But, then, simplicity in some regard is usually compensated by complexity in some other respect; suffice to recall the infinite syntactical complexity that must be paid for the epistemological impoverishment of theories brought about by the replacement of transcendent ("auxiliary") expressions by observational ones.[39]

The function of simplicities in scientific investigation is not, at any rate, as important as it had been imagined by conventionalists and empiricists. The main reason for the loss of weight of simplicity is this. The task of the theoretician is not merely to describe experience in the most economical way, but to *build* theoretical models (not necessarily mechanical!) of bits of reality, and to test such images by means of logic, further theoretical constructions, empirical data, and metascientific rules. Such a constructive work certainly *involves* the neglect of complexities but does not *aim* at disregarding them; rather, a desideratum of every new theory is to account for something that had been overlooked in previous views.

This is why we cannot any longer believe in the scholastic maxim *Simplex sigillum veri:* because we know that all our constructions are defective since, deliberately or not, they involve the neglect of an unknown number of factors. Factual theories apply exactly to schematic, impoverished, models or images, and only inexactly to the real referents of these pictures; the simpler the theoretical model, the coarser or more unrealistic it will be. We need not wait for empirical tests in order to discover that *all* our theories are, strictly speaking, false (cf. 1.4). We know this beforehand if only because they all involve *too many simplifications,* as shown by an analysis of the construction and application of factual theories, and by historical experience. Conceptual economy is consequently a sign and a test of transitoriness, i.e., of falsity—to be superseded by a lesser falsity. *Simplex sigillum falsi.*

4.3. Conclusion

The unqualified demand for economy in every respect, or even in some one respect, is definitely incompatible with a number of important desiderata of theory construction—such as, e.g., accuracy, depth, and external consistency—whence simplicity *tout court* should neither be regarded as mandatory nor be counted as an independent criterion on a par with others—let alone above others. The rules of simplicity fall under the general norm *"Do not hold arbitrary (ungrounded) beliefs".*

If framed with all the due precautions to prevent the mutilation of scientific theory, the rule of simplicity will boil down to the norm directing us to minimize *superfluities.* But of course this rule, like every other negative injunction, is insufficient as a theory construction policy; moreover, it does not help us to recognize which elements of a theory are redundant, i.e., which ones discharge neither a logical nor an empirical function. Production is not ensured by specifying what should not be done.

Simplicity is ambiguous as a term and double-edged as a prescription, and it must be controlled by the symptoms of truth rather than be regarded as a factor of truth. To paraphrase Baltasar Gracián—*"Lo bueno, si breve, dos veces bueno"*—let us say that a working theory, if simple, works twice as well—but this is trivial. If a practical advice is wanted as a corollary, let this be: Ockham's razor—like all razors—must be handled with care to prevent beheading science in the attempt to shave off some of its pilosities. In science, as in the barber shop, better alive and bearded than dead and cleanly shaven.

NOTES

1. Cf. Mario Bunge, "The Complexity of Simplicity", presented to the International Congress for Logic, Methodology, and Philosophy of Science (Stanford, (1960), *Jour. Phil.* LIX, 113 (1962). For a fuller analysis, see *The Myth of Simplicity* (Englewood Cliffs, N.J.: Prentice-Hall, 1963).
2. Adolph Lindenbaum, "Sur la simplicité formelle des notions", *Actes de Congrès International de Philosophie Scientifique* (Paris: Hermann, 1936), VII, 28. Nelson Goodman, *The Structure of Appearance* (Cambridge, Mass.: Harvard University Press, 1951), ch. iii, and "Axiomatic Measure of Simplicity", *Jour. Phil.*, 52, 709 (1955). John G. Kemeny, "Two Measures of Complexity", *Jour. Phil.*, 52, 722 (1955). Horst Kiesow, "Anwendung eines Einfachheitsprinzip, auf die Wahrscheinlichkeitstheorie", *Archiv. f. Math. Logik u. Grundlagenforschung, 4, 27* (1958).
3. Dorothy Wrinch and Harold Jeffreys, "On Certain Fundamental Principles of Scientific Inquiry", *Phil. Mag 42, 369* (1921); Harold Jeffreys, *Theory of Probability,* 2nd ed. (Oxford; Clarendon Press, 1948), p. 100. Karl R. Popper, *The Logic of Scientific Discovery* (1935; London; Hutchinson, 1959), secs. 44 to 46, and *Appendix VIII. John G. Kemeny, "The Use of Simplicity in Induction", *Phil. Rev. 62,* 391 (1953). [Ed. note: See page 309 in this volume for the Kemeny article.]
4. Mario Bunge, reference 1.
5. William Craig, "Replacement of Auxiliary Expressions", *Phil. Rev., 65,* 38 (1956).
6. Cf. Wilhelm Ostwald, *Grundriss der Naturphilosophe* (Leipzig: Reclam, 1908), p. 127: simple formulas to express laws of nature can be found only when the conceptual analysis of phenomena is quite advanced.
7. Cf. Ernst Cassirer, *Determinismus und Indeterminismus in der modern Physik* (Göteborg: Elanders 1937), p. 88, being no. 3, vol. XLII, of *Göteborgs Högskolas Arsskrift* (1936).
8. Cf. Patrick Suppes, *Introduction to Logic* (Princeton: Van Nostrand, 1957), p. 169.
9. Nelson Goodman, reference 2, has argued most persuasively in favor of this thesis.
10. Cf. H. J. Eysenck, *Uses and Abuses of Psychology* (London: Penguin, 1953), ch. 12. Ernest Nagel, "Methodological Issues in Psychoanalytic Theory", in S. Hook (ed.), *Psychoanalysis, Scientific Method, and Philosophy* (New York: University Press, 1959), ch. 2.
11. Karl R. Popper, *The Logic of Scientific Discovery* (1935; London: Hutchinson, 1959), sections 44 to 46, and *Appendix VIII.
12. Cf. Hermann Weyl, *Philosophy of Mathematics and Natural Science* (1927; Princeton: Princeton University Press, 1949), p. 156, and Popper, reference 11.
13. Alfred Tarski, "The Semantic Conception of Truth", *Phil. and Phenom. Res., 4,* 341 (1944).
14. Mario Bunge, *Causality* (Cambridge, Mass.: Harvard University Press, 1959), pp. 290–1.
15. Cf. Mario Bunge, "Levels: A Semantical Preliminary", *Rev Metaphys., 13,* 396 (1960), and "On the Connections Among Levels", *Proc. XIIth. Congr. Phil.* (Firenze: Sansoni, 1960), VI 63.
16. The compatibility of astronomy with physics was as essential to Copernicus as was the "saving of appearances", as E. Rosen rightly notes in his Introduction to *Three Copernican Treatises,* 2nd ed. (New York: Dover, 1959) p. 29. "What Copernicus desired was not merely a simpler system, as Burtt thought, but a more reasonable one" *(loc. cit.).* The unification of astronomy and terrestrial mechanics was also an unfulfilled dream of Averroes and the main drive for Galileo and Newton.
17. See, e.g., George R. Price, "Science and the Supernatural", *Science, 122,* 359

(1955). On the other hand C. D. Broad, in "The Relevance of Psychical Research to Philosophy", *Philosophy, 24,* 291 (1949), accepted *ESP* while acknowledging that it would require a radical upheaval in psychology, biology, physics, and philosophy.
18. See F. C. S. Schiller, "Hypothesis", in C. Singer (ed.), *Studies in the History and Method of Science* (Oxford: Clarendon Press, 1921), II, p. 442.
19. The term *serendipity* (lucky accident) was coined by Horace Walpole and revived by Walter Cannon, *The Way of an Investigator* (New York: Norton, 1945), ch. iv, and by Robert K. Merton, *Social Theory and Social Structure,* rev. ed. (Glencoe, Ill.: Free Press, 1957), ch. ii.
20. Further reasons are given in Mario Bunge, reference 14, ch. 12.
21. Henry Margenau, *The Nature of Physical Reality* (New York: McGraw-Hill, 1950), p. 90.
22. Freeman J. Dyson, "Invention in Physics", *Sci. American, 199,* no. 3, p. 80 (1958).
23. Some variables regarded as observable in nonrelativistic quantum mechanics are no longer observable in the relativistic theory, and conditions of observability, such as reality (hermiticity), are open to criticism. It can be shown that a non-hermitian operator may represent, in a number of cases, a pair of observables. Cf. Andrés J. Kálnay, "Sobre los observables cuánticos y el requisito de la hermiticidad" (forthcoming).
24. Karl R. Popper, *The Logic of Scientific Discovery* (1935; London: Hutchinson, 1959), ch. iv.
25. The legitimacy of such irrefutable statements, rejected by Popper, is defended in Mario Bunge, "Kinds and Criteria of Scientific Laws", *Philosophy of Science, 28,* 260 (1961).
26. Mario Bunge, "The Place of Induction in Science", *Phil. Sci., 27,* 262 (1960).
27. Mario Bunge, *Metascientific Queries* (Springfield, Ill.: Charles Thomas, 1959), ch. 4.
28. Cf. Mario Bunge, *Intuition and Science* (Englewood, Cliffs, N. J.: Prentice-Hall, 1962).
29. Carl G. Hempel, "The Concept of Cognitive Significance: A Reconsideration", *Pub. Amer. Acad. Arts and Sciences, 80, 61* (1951). Arthur Pap, "Are Physical Magnitudes Operationally Definable?", in C. West Churchman and P. Ratoosh (eds.), *Measurement: Definitions and Theories* (New York: Wiley, 1959), ch. 9.
30. Mario Bunge, *Causality* (Cambridge, Mass.: Harvard University Press, 1959).
31. According to general relativity, the paths of bodies are determined by the gravitational field, and the latter is in turn determined by the mass distribution. The field strength being proportional to the quantity of matter (as given, e.g., by the nuclear particles), it is not possible to find a coordinate system in which the sun's field turns out to be weaker than the earth's so that that latter might be regarded as stationary and the sun as revolving around it. Gravitational fields are equivalent to accelerations, and the latter can be transformed away by means of suitable coordinate transformations, within differential spacetime volumes alone; e.g., Einstein's elevator must start from a place within the earth's field, and it finally crashes. For a criticism of the erroneous belief that general relativity permits to transform every acceleration away, see V. A. Fock, "Le système de Ptolomée et le système de Copernic á la lumière de la théorie générale de la rélativité", in *Questions scientifiques* (Paris: Ed. de la Nouvelle Critique, 1952), I, 149.
32. For the kinematical and dynamical inequivalence of the geocentric and the Copernican reference axes according to the general theory of relativity, see G. Giorgi and A. Cabras, "Questioni relativistiche sulle prove della rotazione terrestre", *Rendic. Accad. Naz. Lincei,* IX, 313, 1929).
33. J. L. Synge, "Orbits and Rays in the Gravitational Field of a Finite Sphere According to the Theory of A. N. Whitehead", *Proc. Roy. Soc. Lond. A, 211,* 303 (1952).

34. R. V. Pound and G. A. Rebka, Jr., "Apparent Weight of Photons", *Phys. Rev. Letters,* 4, 337 (1960).
35. For a protest against the complexities of general relativity, see P. W. Bridgman, *The Nature of Physical Theory* (New York: Dover, 1936), pp. 89 ff.
36. For an analysis of the logical status of the parity conservation law and other metanomological statements, see Mario Bunge, "Laws of Physical Laws", *American Journal of Physics, 20,* 518 (1961).
37. See, e.g., W. Stanley Jevons, *The Principles of Science,* 2nd. ed. (1877; New York: Dover, 1958), p. 510; Pierre Duhem, *La théorie physique,* 2nd. ed. (Paris: Rivière, 1914), p. 26; see, however, p. 259, where he admits that simplicity is not a sign of certainty.
38. An early recognition of the multiplicity of requirements is found in Heinrich Hertz, *The Principles of Mechanics* (1894; New York: Dover, 1956), Introduction. Hertz listed the following: (1) logical possibility, or compatibility with the "law of thought", (2) predictive power; (3) maximum number of "essental relations of the object" (what I have called depth); (4) "the smaller number of superfluous or empty relations". Half a century elapsed before another scientist-philosopher dared to add non-empirical requirements: Henry Margenau, *The Nature of Physical Reality* (New York: McGraw-Hill, 1950), ch. 5, lists the following "metaphysical requirements on constructs": (1) logical fertility, (2) multiple connections, (3) permanence or stability, (4) extensibility, (5) causality, (6) simplicity and elegance. See also Mario Bunge, *Metascientific Queries* (Springfield, Ill.: Charles Thomas, 1959), pp. 79 ff., and Karl R. Popper, "The Idea of Truth and the Empirical Character of Scientific Theories", presented to the *International Congress for Logic, Methodology, and Philosophy of Science* (Stanford, 1960). In this paper Popper grants that one of the requirements for a good theory is that it "should succeed with at least some of its new predictions" – i.e., that it should be confirmed.
39. William Craig, reference 5.

SUGGESTIONS FOR FURTHER READING

Barker, S. F. *Induction and Hypothesis. A Study of the Logic of Confirmation.* Ithaca, N.Y.: Cornell University Press, 1962.

Carnap, R. *The Logical Foundations of Probability.* Chicago: University of Chicago Press, 1950.

Cohen, L.J. *The Implications of Induction.* London: Methuen and Co. Ltd., 1970.

Hintikka, J. and Suppes, P. (eds.) *Aspects of Inductive Logic.* Amsterdam: North-Holland Publishing Co., 1966.

Kyburg, H.E. *Probability and Inductive Logic.* London: The Macmillan Co., 1970.

Lakatos, I. (ed.) *The Problem of Inductive Logic.* Amsterdam: North-Holland Publishing Co., 1968.

Levi, I. *Gambling with Truth.* New York: Alfred A. Knopf, 1967.

Martin, M.L. and Foster, M.H. (eds.) *Probability, Confirmation, and Simplicity.* New York: The Odyssey Press, Inc., 1966.

Michalos, A.C. *The Popper-Carnap Controversy.* The Hague: Martinus Nijhoff, 1971.

Popper, K.R. *The Logic of Scientific Discovery.* New York: Science Editions, Inc., 1961, pp. 136–214.

Salmon, W.C. *The Foundations of Scientific Inference.* Pittsburgh: University of Pittsburgh Press, 1966.

Skyrms, B. *Choice and Chance.* Belmont, Calif.: Dickenson Publishing Co., Inc., 1966.

SECTION VIII

The Role of Values in Science

In the first article in this section, Nagel attempts to show that "the various reasons . . . [offered] . . . for the intrinsic impossibility of securing objective (i.e., value-free and unbiased) conclusions in the social sciences do not establish what they purport to establish . . ." Thus, just as Kordig had attempted to show that there may be some *theory*-free or neutral facts, Nagel here attempts to show that there may also be some *value*-free or neutral social scientific investigation. Insofar as such attempts are successful, science tends to become the objective, dispassionate, fact/law/theory-finding enterprise that logical positivists and the man on the street, perhaps, would like it to be. I leave it to you to judge exactly how successful these attempts are.

Nagel examines arguments related to " 1) the selection of problems, 2) the determination of the contents of conclusions, 3) the identification of fact, and 4) the assessment of evidence." In the other article in this section, Leach considers an argument that ties value-judgments to scientific investigation at the point at which a scientist must determine whether or not a given hypothesis ought to be accepted. In fact, Leach reopens doors that Nagel had presumably hoped to close in his analysis of arguments related to the assessment of evidence. If there is a moral to this story, it is that one can never be sure that a philosophic door is closed for all time.

READING 19

The Value-Oriented Bias of Social Inquiry

Ernest Nagel

We turn, finally, to the difficulties said to confront the social sciences because the social values to which students of social phenomena are committed not only color the contents of their findings but also control their assessment of the evidence on which they base their conclusions. Since social scientists generally differ in their value commitments, the "value neutrality" that seems to be so pervasive in the natural sciences is therefore often held to be impossible in social inquiry. In the judgment of many thinkers, it is accordingly absurd to expect the social sciences to exhibit the unanimity so common among natural scientists concerning what are the established facts and satisfactory explanations for them. Let us examine some of the reasons that have been advanced for these contentions. It will be convenient to distinguish four groups of such reasons, so that our discussion will deal in turn with the alleged role of value judgments in (1) the selection of problems, (2) the determination of the contents of conclusions, (3) the identification of fact, and (4) the assessment of evidence.

1. The reasons perhaps most frequently cited make much of the fact that the things a social scientist selects for study are determined by his conception of what are the socially important values. According to one influential view, for example, the student of human affairs deals only with materials to which he attributes "cultural significance," so that a "value orientation" is inherent in his choice of material for investigation. Thus, although Max Weber was a vigorous proponent of a "value-free" social science—i.e., he maintained that social scientists must appreciate (or "understand") the values involved in the actions

From *The Structure of Science*, by Ernest Nagel, © 1961 by Harcourt Brace Jovanovich, Inc., and reprinted with their permission.

or institutions they are discussing but that it is not their business as objective scientists to approve or disapprove either those values or those actions and institutions—he nevertheless argued that

> The concept of culture is a *value-concept*. Empirical reality becomes "culture" to us because and insofar as we relate it to value ideas. It includes those segments and only those segments of reality which have become significant to us because of this value-relevance. Only a small portion of existing concrete reality is colored by our value-conditioned interest and it alone is significant to us. It is significant because it reveals relationships which are important to us due to their connection with our values. Only because and to the extent that this is the case is it worthwhile for us to know it in its individual features. We cannot discover, however, what is meaningful to us by means of a "presuppositionless" investigation of empirical data. Rather perception of its meaningfulness to us is the presupposition of its becoming an *object* of investigation.[1]

It is well-nigh truistic to say that students of human affairs, like students in any other area of inquiry, do not investigate everything, but direct their attention to certain selected portions of the inexhaustible content of concrete reality. Moreover, let us accept the claim, if only for the sake of the argument, that a social scientist addresses himself exclusively to matters which he believes are important because of their assumed relevance to his cultural values.[2] It is not clear, however, why the fact that an investigator selects the materials he studies in the light of problems which interest him and which seem to him to bear on matters he regards as important, is of greater moment for the logic of social inquiry than it is for the logic of any other branch of inquiry. For example, a social scientist may believe that a free economic market embodies a cardinal human value, and he may produce evidence to show that certain kinds of human activities are indispensable to the perpetuation of a free market. If he is concerned with processes which maintain this type of economy rather than some other type, how is this fact more pertinent to the question whether he has adequately evaluated the evidence for his conclusion, than is the bearing upon the analogous question of the fact that a physiologist may be concerned with processes which maintain a constant internal temperature in the human body rather than with something else? The things a social scientist *selects for study* with a view to determining the conditions or consequences of their existence may indeed be dependent on the indisputable fact that he is a "cultural being." But similarly, were we not human beings though still capable of conducting scientific inquiry, we might conceivably have an interest neither in the conditions that maintain a free market, nor in the processes involved in the homeostasis of the internal temperature in human bodies, nor for that matter in the mechanisms that regulate the height of tides, the succession of seasons, or the motions of the planets.

In short, there is no difference between any of the sciences with respect to the fact that the interests of the scientist determine what he selects for investigation. But this fact, by itself, represents no obstacle to the successful pursuit of objectively controlled inquiry in any branch of study.

2. A more substantial reason commonly given for the value-oriented character of social inquiry is that, since the social scientist is himself affected by considerations of right and wrong, his own notions of what constitutes a satisfactory social order and his standards of personal and social justice do enter, in point of fact, into his analysis of social phenomena. For example, according to one version of this argument, anthropologists must frequently judge whether the means adopted by some society achieves the intended aim (e.g., whether a religious ritual does produce the increased fertility for the sake of which the ritual is performed); and in many cases the adequacy of the means must be judged by admittedly "relative" standards, i.e., in terms of the ends sought or the standards employed by that society, rather than in terms of the anthropologist's own criteria. Nevertheless, so the argument proceeds, there are also situations in which

> we must apply absolute standards of adequacy, that is evaluate the end-results of behavior in terms of purposes we believe in or postulate. This occurs, first, when we speak of the satisfaction of psycho-physical 'needs' offered by any culture; secondly, when we assess the bearing of social facts upon survival; and thirdly, when we pronounce upon social integration and stability. In each case our statements imply judgments as to the worthwhileness of actions, as to 'good' or 'bad' cultural solutions of the problems of life, and as to 'normal' and 'abnormal' states of affairs. These are basic judgments which we cannot do without in social enquiry and which clearly do not express a purely personal philosophy of the enquirer or values arbitrarily assumed. Rather do they grow out of the history of human thought, from which the anthropologist can seclude himself as little as can anyone else. Yet as the history of human thought has led not to one philosophy but to several, so the value attitudes implicit in our ways of thinking will differ and sometimes conflict.[3]

It has often been noted, moreover, that the study of social phenomena receives much of its impetus from a strong moral and reforming zeal, so that many ostensibly "objective" analyses in the social sciences are in fact disguised recommendations of social policy. As one typical but moderately expressed statement of the point puts it, a social scientist

> cannot wholly detach the unifying social structure that, as a scientist's theory, guides his detailed investigations of human behavior, from the unifying structure which, as a citizen's ideal, he thinks ought to prevail in human affairs and hopes may sometimes be more fully realized. His social theory is thus essentially a program of action along two lines which are kept in some measure of harmony with each other by that theory—action in assimilating social facts for purposes of systematic understanding, and action aiming at progressively molding the social pattern, so far as he can influence it, into what he thinks it ought to be.[4]

It is surely beyond serious dispute that social scientists do in fact often import their own values into their analyses of social phenomena. It is also undoubtedly true that even thinkers who believe human affairs can be studied with the ethical neutrality characterizing modern inquiries into geometrical or physical relations, and who often pride themselves on the absence of value judgments from their own analyses of social phenomena, do in fact sometimes make such judgments in their social inquiries.[5] Nor is it less evident that students of human affairs often hold conflicting values; that their disagreements on value questions are often the source of disagreements concerning ostensibly factual issues; and that, even if value predications are assumed to be inherently capable of proof or disproof by objective evidence, at least some of the differences between social scientists involving value judgments are not in fact resolved by the procedures of controlled inquiry.

In any event, it is not easy in most areas of inquiry to prevent our likes, aversions, hopes, and fears from coloring our conclusions. It has taken centuries of effort to develop habits and techniques of investigation which help safeguard inquiries in the natural sciences against the intrusion of irrelevant personal factors; and even in these disciplines the protection those procedures give is neither infallible nor complete. The problem is undoubtedly more acute in the study of human affairs, and the difficulties it creates for achieving reliable knowledge in the social sciences must be admitted.

However, the problem is intelligible only on the assumption that there is a relatively clear distinction between factual and value judgments, and that however difficult it may sometimes be to decide whether a given statement has a purely factual content, it is in principle possible to do so. Thus, the claim that social scientists are pursuing the twofold program mentioned in the above quotation makes sense, only if it is possible to distinguish between, on the one hand, contributions to theoretical understanding (whose factual validity presumably does not depend on the social ideal to which a social scientist may subscribe), and on the other hand contributions to the dissemination or realization of some social ideal (which may not be accepted by all social scientists). Accordingly, the undeniable difficulties that stand in the way of obtaining reliable knowledge of human affairs because of the fact that social scientists differ in their value orientations are practical difficulties. The difficulties are not necessarily insuperable, for since by hypothesis it is not impossible to distinguish between fact and value, steps can be taken to identify a value bias when it occurs, and to minimize if not to eliminate completely its perturbing effects.

One such countermeasure frequently recommended is that social scientists abandon the pretense that they are free from all bias, and that instead they state their value assumptions as explicitly and fully as they can.[6] The recommendation does not assume that social scientists will come to agree on their social ideals once these ideals are explicitly postulated, or that disagreements over values can be settled by scientific inquiry. Its point is that the question of how a given ideal is to be realized, or the question whether a certain institutional arrangement is an effective way of achieving the ideal, is on the face of it not a value question, but a factual problem—to be resolved by the objective methods of scientific inquiry—concerning the adequacy of proposed means for attaining stipulated ends. Thus, economists may permanently disagree on the desirability of a society in which its members have a guaranteed security against economic want, since the disagreement may have its source in inarbitrable preferences for different social values. But when sufficient evidence is made available by economic inquiry, economists do presumably agree on the factual proposition that, *if* such a society is to be achieved, then a purely competitive economic system will not suffice.

Although the recommendation that social scientists make fully explicit their value commitments is undoubtedly salutary, and can produce excellent fruit, it verges on being a counsel of perfection. For the most part we are unaware of many assumptions that enter into our analyses and actions, so that despite resolute efforts to make our preconceptions explicit some decisive ones may not even occur to us. But in any event, the difficulties generated for scientific inquiry by unconscious bias and tacit value orientations are rarely overcome by devout resolutions to eliminate bias. They are usually overcome, often only gradually, through the self-corrective mechanisms of science as a social enterprise. For modern science encourages the invention, the mutual exchange, and the free but responsible criticisms of ideas; it welcomes competition in the quest for knowledge between independent investigators, even when their intellectual orientations are different; and it progressively diminishes the effects of bias by retaining only those proposed conclusions of its inquiries that survive critical examination by an indefinitely large community of students, whatever be their value preferences or doctrinal commitments. It would be absurd to claim that this institutionalized mechanism for sifting warranted beliefs has operated or is likely to operate in social inquiry as effectively as it has in the natural sciences. But it would be no less absurd to conclude that reliable knowledge of human affairs is unattainable merely because social inquiry is frequently value-oriented.

3. There is a more sophisticated argument for the view that the social sciences cannot be value-free. It maintains that the distinction between fact and value assumed in the preceding discussion is untenable when purposive human behavior is being analyzed, since in this context value judgments enter inextricably into what appear to be "purely descriptive" (or factual) statements. Accordingly, those who subscribe to this thesis claim that an ethically neutral social science is in principle impossible, and not simply that it is difficult to attain. For if fact and value are indeed so fused that they cannot even be distinguished, value judgments cannot be eliminated from the social sciences unless all predications are also eliminated from them, and therefore unless these sciences completely disappear.

For example, it has been argued that the student of human affairs must distinguish between valuable and undesirable forms of social activity, on pain of failing in his "plain duty" to present social phenomena truthfully and faithfully:

> Would one not laugh out of court a man who claimed to have written a sociology of art but who actually had written a sociology of trash? The sociologist of religion must distinguish between phenomena which have a religious character and phenomena which are a-religious. To be able to do this, he must understand what religion is. . . . Such understanding enables and forces him to distinguish between genuine and spurious religion, between higher and lower religions; those religions are higher in which the specifically religious motivations are effective to a higher degree. . . . The sociologist of religion cannot help noting the difference between those who try to gain it by a change of heart. Can he see this difference without seeing at the same time the difference between a mercenary and nonmercenary attitude? . . . The prohibition against value-judgments in social science would lead to the consequence that we are permitted to give a strictly factual description of the overt acts that can be observed in concentration camps, and perhaps an equally factual analysis of the motivations of the actors concerned:

> we would not be permitted to speak of cruelty. Every reader of such a description who is not completely stupid would, of course, see that the actions described are cruel. The factual description would, in truth, be a bitter satire. What claimed to be a straightforward report would be an unusually circumlocutory report.... Can one say anything relevant on public opinion polls ... without realizing the fact that many answers to the questionnaires are give by unintelligent, uninformed, deceitful, and irrational people, and that not a few questions are formulated by people of the same caliber—can one say anything relevant about public opinion polls without committing one value-judgment after another? [7]

Moreover, the assumption implicit in the recommendation discussed above for achieving ethical neutrality is often rejected as hopelessly naive—this is the assumption, it will be recalled, that relations of means to ends can be established without commitment to these ends, so that the conclusions of social inquiry concerning such relations are objective statements which make *conditional* rather than categorical assertions about values. This assumption is said by its critics to rest on the supposition that men attach value only to the ends they seek, and not to the means for realizing their aims. However, the supposition is alleged to be grossly mistaken. For the character of the means one employs to secure some goal affects the nature of the total outcome; and the choice men make between alternative means for obtaining a given end depends on the values they ascribe to those alternatives. In consequence, commitments to specific valuations are said to be involved even in what appears to be purely factual statements about means-ends relations.[8]

We shall not attempt a detailed assessment of this complex argument, for a discussion of the numerous issues it raises would take us far afield. However, three claims made in the course of the argument will be admitted without further comment as indisputably correct: that a large number of characterizations sometimes assumed to be purely factual descriptions of social phenomena do indeed formulate a type of value judgment; that it is often difficult, and in any case usually inconvenient in practice, to distinguish between the purely factual and the "evaluative" contents of many terms employed in the social sciences; and that values are commonly attached to means and not only to ends. However, these admissions do not entail the conclusion that, in a manner unique to the study of purposive human behavior, fact and value are fused beyond the possibility of distinguishing between them. On the contrary, as we shall try to show, the claim that there is such a fusion and that a value-free social science is therefore inherently absurd, confounds two quite different senses of the term "value judgment": the sense in which a value judgment expresses *approval or disapproval* either of some moral (or social) ideal, or of some action (or institution) because of a commitment to such an ideal; and the sense in which a value judgment expresses *an estimate* of the degree to which some commonly recognized (and more or less clearly defined) type of action, object, or institution is embodied in a given instance.

It will be helpful to illustrate these two senses of "value judgment" first with an example from biology. Animals with blood streams sometimes exhibit the condition known as "anemia." An anemic animal has a reduced number of red blood corpuscles, so that, among other things, it is less able to maintain a constant internal temperature than are members of its species with a "normal" supply of such blood cells. However, although the meaning of the term "anemia" can be made quite clear, it is not in fact defined with complete precision; for example, the notion of a "normal" number of red corpuscles that enters into the definition of the term is itself somewhat vague, since this number varies with the individual members of a species as well as with the state of a given individual at different times (such as its age or the altitude of its habitat). But in any case, to decide whether a given animal is anemic, an investigator must judge whether the available evidence *warrants* the conclusion that the specimen is anemic.[9] He may perhaps think of anemia as being of several distinct kinds (as is done in actual medical practice), or he may think of anemia as a condition that is realizable with greater or lesser completeness (just as certain plane curves are sometimes described as better or worse approximations to a circle as defined in geometry); and, depending on which of these conceptions he adopts, he may decide either that his specimen has a certain kind of anemia or that it is anemic only to a certain degree. When the investigator reaches a conclusion, he can therefore be said to be making a "value judgment," in the sense that he has in mind some standardized type of physiological condition designated as "anemia" and that he *assesses* what he knows about his specimen with the measure provided by this assumed standard. For the sake of easy reference, let us call such evaluations of the evidence, which conclude that a given characteristic is in some degree present (or absent) in a given instance, "characterizing value judgments."

On the other hand, the student may also make a quite different sort of value judgment, which asserts that, since an anemic animal has diminished powers of maintaining itself, anemia is an undesirable condition. Moreover, he may apply this general judgment to a particular case, and so come to deplore the fact that a given animal is anemic. Let us label such evaluations, which conclude that some envisaged or actual state of affairs is worthy of approval or disapproval, "appraising value judgments."[10] It is clear, however, that an investigator making a characterizing value judgment is not thereby logically bound to affirm or deny a corresponding appraising evaluation. It is no less evident that he cannot consistently make an appraising value judgment about a given instance (e.g., that it is undesirable for a given animal to continue being anemic), unless he can affirm a characterizing judgment about that instance independently of the appraising one (e.g., that the animal is anemic). Accordingly, although characterizing judgments are necessarily entailed by many appraising judgments, making appraising judgments is not a necessary condition for making characterizing ones.

Let us now apply these distinctions to some of the contentions advanced in the argument quoted above. Consider first the claim that the sociologist of religion must recognize the difference between mercenary and nonmercenary attitudes, and that in consequence he is inevitably committing himself to certain values. It is certainly beyond dispute that these attitudes are commonly distinguished; and it can also be granted that a sociologist of religion needs to understand the difference between them. But the sociologist's obligation is in this respect quite like that of the student of animal physiology, who must also acquaint himself with certain distinctions—even though the physiologist's distinction between, say, anemic and nonanemic may be less familiar to the ordinary layman and is in any case much more precise than is the distinction between mercenary and nonmercenary attitudes. Indeed, because of the vagueness of these latter terms, the scrupulous sociologist may find it extremely difficult to decide whether or not the attitude of some community toward its acknowledged gods is to be characterized as mercenary; and if he should finally decide, he may base his conclusion on some inarticulated "total impression" of that community's manifest behavior, without being able to state exactly the detailed grounds for his decision. But however this may be, the sociologist who claims that a certain attitude manifested by a given religious group is mercenary, just as the physiologist who claims that a certain individual is anemic, is making what is primarily a characterizing value judgment. In making these judgments, neither the sociologist nor the physiologist is necessarily committing himself to any values other than the values of scientific probity; and in this respect, therefore, there appears to be no difference between social and biological (or for that matter, physical) inquiry.

On the other hand, it would be absurd to deny that in characterizing various actions as mercenary, cruel, or deceitful, sociologists are frequently (although perhaps not always wittingly) asserting appraising as well as characterizing value judgments. Terms like 'mercenary,' 'cruel,' or 'deceitful' as commonly used have a widely recognized pejorative overtone. Accordingly, anyone who employs such terms to characterize human behavior can normally be assumed to be stating his disapprobation of that behavior (or his approbation, should he use terms like 'nonmercenary,' 'kindly,' or "truthful'), and not simply characterizing it.

However, although many (but certainly not all) ostensibly characterizing statements asserted by social scientists undoubtedly express commitments to various (not always compatible) values, a number of "purely descriptive" terms as used by natural scientists in certain contexts sometimes also have an unmistakably appraising value connotation. Thus, the claim that a social scientist is making appraising value judgments when he characterizes respondents to questionnaires as uninformed, deceitful, or irrational can be matched by the equally sound claim that a physicist is also making such judgments when he describes a particular chronometer as inaccurate, a pump as inefficient, or a supporting platform as unstable. Like the social scientist in this example, the physicist is characterizing certain objects in his field of research; but, also like the social scientist, he is in addition expressing his disapproval of the characteristics he is ascribing to those objects.

Nevertheless—and this is the main burden of the present discussion—there are no good reasons for thinking that it is inherently impossible to *distinguish* between the characterizing and the appraising judgments implicit in many statements, whether the statements are asserted by students of human affairs or by natural scientists. To be sure, it is not always easy to make the distinction formally explicit in the social sciences—in part because much of the language employed in them is very vague, in part because appraising judgments that many be implicit in a statement tend to be overlooked by us when they are judgments to which we are actually committed though without being aware of our commitments. Nor is it always useful or convenient to perform this task. For many statements implicitly containing both characterizing and appraising evaluations are sometimes sufficiently clear without being reformulated in the manner required by the task; and the reformulations would frequently be too unwieldy for effective communication between members of a large and unequally prepared group of students. But these are essentially practical rather than theoretical problems. The difficulties they raise provide no compelling reasons for the claim that an ethically neutral social science is inherently impossible.

Nor is there any force in the argument that, since values are commonly attached to means and not only to ends, statements about means-ends relations are not value-free. Let us test the argument with a simple example. Suppose that a man with an urgent need for a car but without sufficient funds to buy one can achieve his aim by borrowing a sum either from a commercial bank or from friends who waive payment of any interest. Suppose further that he dislikes becoming beholden to his friends for financial favors, and prefers the impersonality of a commercial loan. Accordingly, the comparative values this individual places upon the alternative means available to him for realizing his aim obviously control the choice he makes between them. Now the *total* outcome that would result from his adoption of one of the alternatives is admittedly different from the *total* outcome that would result from his adoption of the other alternative. Nevertheless, irrespective of the values he may attach to these alternative means, each of them would achieve a result—namely, his purchase of the needed car—that is common to both the total outcomes. In consequence, the validity of the statement that he could buy the car by borrowing money from a bank, as well as of the statement that he could realize this aim by borrowing from friends, is unaffected by the valuations placed upon the means, so that neither statement involves any special appraising evaluations. In short, the statements about means-ends relations are value-free.

4. There remains for consideration the claim that a value-free social science is impossible, because value commitments enter into the very *assessment of evidence* by social scientists, and not simply into the content of the conclusions they advance. This version of the claim itself has a large number of variant forms, but we shall examine only three of them.

The least radical form of the claim maintains that the conceptions held by a social scientist of what constitute cogent evidence or sound intellectual workmanship are the products of his education and his place in society, and are affected by the social values transmitted *by* this training and associated with this social position; accordingly, the values to which the social scientist is thereby committed determine which statements he *accepts* as well-grounded conclusions about human affairs. In this form, the claim is a *factual* thesis, and must be supported by detailed empirical evidence concerning the influences exerted by a man's moral and social values upon what he is ready to acknowledge as sound social analysis. In many instances such evidence is indeed available; and differences between social scientists in respect to what they accept as credible can sometimes be attributed to the influence of national religious, economic, and other kinds of bias. However, this variant of the claim excludes neither the possibility of recognizing assessments of evidence that are prejudiced by special value commitments, nor the possibility of correcting for such prejudice. It therefore raises no issue that has not already been discussed when we examined the second reason for the alleged value-oriented character of social inquiry (pages 449–452).

Another but different form of the claim is based on recent work in theoretical statistics dealing with the assessment of evidence for so-called "statistical hypotheses"—hypotheses concerning the probabilities of random events such as the hypothesis that the probability of a male human birth is one-half. The central idea relevant to the present question that underlies these developments can be sketched in terms of an example. Suppose that, before a fresh batch of medicine is put on sale, tests are performed on experimental animals for its possible toxic effects because of impurities that have not been eliminated in its manufacture, for example, by introducing small quantities of the drug into the diet of one hundred guinea pigs. If no more than a few of the animals show serious after-effects, the medicine is to be regarded as safe, and will be marketed; but if a contrary result is obtained the drug will be destroyed. Suppose now that three of the animals do in fact become gravely ill. Is this outcome significant (i.e., does it indicate that the drug has toxic effects), or is it perhaps an "accident" that happened because of some peculiarity in the affected animals? To answer the question, the experimenter must *decide* on the basis of the evidence between the hypothesis H_2: the drug is toxic, and the hypothesis H: the drug is not toxic. But how is he to decide, if he aims to be "reasonable" rather than arbitrary? Current statistical theory offers him a rule for making a reasonable decision, and bases the rule on the following analysis.

Whatever decision the experimenter may make, he runs the risk of committing either one of two types of errors: he may reject a hypothesis though in fact it is true (i.e., despite the fact that H_1 is actually true, he mistakenly decides against it in the light of the evidence available to him); or he may accept a hypothesis though in fact it is false. His decision would therefore be eminently reasonable, were it based on a rule guaranteeing that no decision ever made in accordance with the rule would commit either type of error. Unhappily, there are no rules of this sort. The next suggestion is to find a rule such that, when decisions are made in accordance with it, the relative frequency of each type of error is quite small. But unfortunately, the risks of committing each type of error are not independent; for example, it is in general logically impossible to find a rule so that decisions based on it will commit each type of error with a relative frequency not greater than one in a thousand. In consequence, before a reasonable rule be proposed, the experimenter must compare the relative importance to himself of the two types of error, and state what risk he is willing to take of commiting the type of error he judges to be the more important one. Thus, were he to reject H_1 *though* it is true (i.e., were he to commit an error of the first type), all the medicine under consideration would be put on sale, and the lives of those using it would be endangered; on the other hand, were he to commit an error of the second type with respect to H_1, the entire batch of medicine would be scrapped, and the manufacturer would incur a financial loss. However, the preservation of human life may be of greater moment to the experimenter than financial gain; and he may perhaps stipulate that he is unwilling to base his decision on a rule for which the risk of committing an error of the first type is greater than one such error in a hundred decisions. If this is assumed, statistical theory can specify a rule satisfying the experimenter's requirement, though how this is done, and how the risk of committing an error of the second type is calculated, are technical questions of no concern to us. The main point to be noted in this analysis is that the rule presupposes certain appraising judgments of value. In short, if this result is generalized, statistical theory appears to support the thesis that value commitments enter decisively into the rules for assessing evidence for statistical hypotheses.[11]

However, the theoretical analysis upon which this thesis rests does not entail the conclusion that the rules actually employed in every social inquiry for assessing evidence necessarily involve some *special* commitments, i.e., commitments such as those mentioned in the above example, as distinct from those generally implicit in science as an enterprise aiming to achieve reliable knowledge. Indeed, the above example illustrating the reasoning in current statistical theory can be misleading, insofar as it suggests that alternative decisions between statistical hypotheses must invariably lead to alternative actions having immediate practical consequences upon which different special values are placed. For example, a theoretical physicist may have to decide between two statistical hypotheses concerning the probability of certain energy exchanges in atoms; and a theoretical sociologist may similarly have to choose between two statistical hypotheses concerning the relative frequency of childless marriages under certain social arrangements. But neither of these men may have any *special* values at stake associated with the alternatives between which he must decide, other than the values, to which he is committed as a member of a scientific community, to conduct his inquiries with probity and responsibility. Accordingly, the question whether any special value commitments enter into assessments of evidence in either the natural or social sciences is not settled one way or the other by theoretical statistics; and the question can be answered only by examining actual inquiries in the various scientific disciplines.

Moreover, nothing in the reasoning of theoretical statistics depends on what particular subject matter is under discussion when a decision between alternative statistical hypotheses is to be made. For the reasoning is entirely general; and reference to some special subject matter becomes relevant only when a definite numerical value is to be assigned to the risk some investigator is prepared to take of making an erroneous decision concerning a given hypothesis. Accordingly, if current statistical theory is used to support the claim that value commitments enter into the assessment of evidence for statistical hypotheses in social inquiry, statistical theory can be used with equal justification to support analogous claims for all other inquiries as well. In short, the claim we have been discussing establishes no difficulty that supposedly occurs in the search for reliable knowledge in the study of human affairs which is not also encountered in the natural sciences.

A third form of this claim is the most radical of all. It differs from the first variant mentioned above in maintaining that there is a necessary *logical* connection, and not merely a contingent or causal one, between the "social perspective" of a student of human affairs and his standards of competent social inquiry, and is in consequence the influence of the special values to which he is committed because of his own social involvements is not eliminable. This version of the claim is implicit in Hegel's account of the "dialectical" nature of human history and is integral to much Marxist as well as non-Marxist philosophy that stresses the "historically relative" character of social thought. In any event, it is commonly based on the assumption that, since social institutions and their cultural products are constantly changing, the intellectual apparatus required for understanding them must also change; and every idea employed for this purpose is therefore adequate only for some particular stage in the development of human affairs. Accordingly, neither the substantive concepts adopted for classifying and interpreting social phenomena, nor the logical canons used for estimating the worth of such concepts, have a "timeless validity"; there is no analysis of social phenomena which is not the expression of some special social standpoint, or which does not reflect the interests and values dominant in some sector of the human scene at a certain stage of its history. In consequence, although a sound distinction can be made in the natural sciences between the origin of a man's views and their factual validity, such a distinction allegedly cannot be made in social inquiry; and prominent exponents of "historical relativism" have therefore challenged the universal adequacy of the thesis that "the genesis of a proposition is under all circumstances irrelevant to its truth." As one influential proponent of this position puts the matter,

> The historical and social genesis of an idea would only be irrelevant to its ultimate validity if the temporal and social conditions of its emergence had no effect on its content and form. If this were the case, any two periods in the history of human knowledge would only be distinguished from one another by the fact that in the earlier period certain things were still unknown and certain errors still existed which, through later knowledge, were completely corrected. This simple relationship between an earlier incomplete and a later complete period of knowledge may to a large extent be appropriate for the exact sciences.... For the history of the cultural sciences, however, the earlier stages are not quite so simply superseded by the later stages, and it is not so easily demonstrable that early errors have subsequently been corrected. Every epoch has its fundamentally new approach and its characteristic point of view, and consequently sees the "same" object from a new perspective.... The very principles, in the light of which knowledge is to be criticized, are themselves found to be socially and historically conditioned. Hence their application appears to be limited to given historical periods and the particular types of knowledge then prevalent.[12]

Historical research into the influence of society upon the beliefs men hold is of undoubted importance for understanding the complex nature of the scientific enterprise; and the sociology of knowledge—as such investigations have come to be called—has produced many clarifying contributions to such an understanding. However, these admittedly valuable services of the sociology of knowledge do not establish the radical claim we have been stating. In the first place, there is no competent evidence to show that the principles employed in social inquiry for assessing the intellectual products are *necessarily* determined by the social perspective of the inquirer. On the contrary, the "facts" usually cited in support of this contention establish at best only a contingent causal relation between a man's social commitments and his canons of cognitive validity. For example, the once fashionable view that the "mentality" or logical operations of primitive societies differ from those typical in Western civilization—a discrepancy that was attributed to differences in the institutions of the societies under comparison—

is now generally recognized to be erroneous, because it seriously misinterprets the intellectual processes of primitive peoples. Moreover, even extreme exponents of the sociology of knowledge admit that most conclusions asserted in mathematics and natural science are neutral to differences in social perspective of those asserting them, so that the genesis of these propositions is irrelevant to their validity. Why cannot propositions about human affairs exhibit a similar neutrality, at least in some cases? Sociologists of knowledge do not appear to doubt that the truth of the statement that two horses can in general pull a heavier load than can either horse alone, is logically independent of the social status of the individual who happens to affirm the statement. But they have not made clear just what are the inescapable considerations that allegedly make such independence inherently impossible for the analogous statement about human behavior, that two laborers can in general dig a ditch of given dimensions more quickly than can either laborer working alone.

In the second place, the claim faces a serious and frequently noted dialectical difficulty—a difficulty that proponents of the claim have succeeded in meeting only by abandoning the substance of the claim. For let us ask what is the cognitive status of the thesis that a social perspective enters essentially into the content as well as the validation of every assertion about human affairs. Is this thesis meaningful and valid only for those who maintain it and who thus subscribe to certain values because of their distinctive social commitments? If so, no one with a different social perspective can properly understand it; its acceptance as valid is strictly limited to those who can do so, and social scientists who subscribe to a different set of social values ought therefore dismiss it as empty talk. Or is the thesis singularly exempt from the class of assertions to which it applies, so that its meaning and truth are not inherently related to the social perspectives of those who assert it? If so, it is not evident why the thesis is so exempt; but in any case, the thesis is then a conclusion of inquiry into human affairs that is presumably "objectively valid" in the usual sense of this phrase—and, if there is one such conclusion, it is not clear why there cannot be others as well.

To meet this difficulty, and to escape the self-defeating skeptical relativism to which the thesis is thus shown to lead, the thesis is sometimes interpreted to say that, though "absolutely objective" knowledge of human affairs is unattainable, a "relational" form of objectivity called "relationism" can nevertheless be achieved. On this interpretation, a social scientist can discover just what his social perspective is; and if he then formulates the conclusions of his inquiries "relationally," so as to indicate that his findings conform to the canons of validity implicit in his perspective, his conclusions will have achieved a "relational" objectivity. Social scientists sharing the same perspective can be expected to agree in their answers to a given problem when the canons of validity characteristic of their common perspective are correctly applied. On the other hand, students of social phenomena who operate within different but incongruous social perspectives can also achieve objectivity, if in no other way than by a "relational" formulation of what must otherwise be incompatible results obtained in their several inquiries. However, they can also achieve it in "a more roundabout fashion," by undertaking "to find a formula for translating the results of one into those of the other and to discover a common denominator for these varying perspectivistic insights."[13]

But it is difficult to see in what way "relational objectivity" differs from "objectivity" without the qualifying adjective and in the customary sense of the word. For example, a physicist who terminates an investigation with the conclusion that the velocity of light in water has a certain numerical value when measured in terms of a stated system of units, by a stated procedure, and under stated experimental conditions, is formulating his conclusion in a manner that is "relational" in the sense intended; and his conclusion is marked by "objectivity," presumably because it mentions the "relational" factors upon which the assigned numerical value of the velocity depends. However, it is fairly standard practice in the natural sciences to formulate certain types of conclusions in this fashion. Accordingly, the proposal that the social sciences formulate their findings in an analogous manner carries with it the admission that it is not in principle impossible for these disciplines to establish conclusions having the objectivity of conclusions reached in other domains of inquiry. Moreover, if the difficulty we are considering is to be resolved by the suggested translation formulas for rendering the "common denominators" of conclusions stemming from divergent social perspectives, those formulas cannot in turn be "situationally determined" in the sense of this phrase under discussion. For if those formulas were so determined, the same difficulty would crop up anew in connection with them. On the other hand, a search for such formulas is a phase in the search for invariant relations in a subject matter, so that formulations of these relations are valid irrespective of the particular perspective one may select from some class of perspectives on that subject matter. In consequence, in acknowledging that the search for such invariants in the social sciences is not inherently bound to fail, proponents of the claim we have been considering abandon what at the outset was its most radical thesis.

In brief, the various reasons we have been examining for the intrinsic impossibility of securing objective (i.e., value-free and unbiased) conclusions in the social sciences do not establish what they purport to establish, even though in some instances they direct attention to undoubtedly important practical difficulties frequently encountered in these disciplines.

NOTES

1. Max Weber, *The Methodology of the Social Sciences,* Glencoe, Ill., 1947, p. 76.
2. This question receives some attention below in the discussion of the fourth difficulty.
3. S. F. Nadel, *The Foundations of Social Anthropology,* Glencoe, Ill., 1951, pp. 53–54. The claim is sometimes also made that the exclusion of value judgments from social science is undesirable as well as impossible. "We cannot disregard all questions of what is socially desirable without missing the significance of many social facts; for since the relation of means to ends is a special form of that between parts and wholes, the contemplation of social ends enables us to see the relations of whole groups of facts to each other and to larger systems of which they are parts."—Morris R. Cohen, *Reason and Nature,* New York, 1931, p. 343.
4. Edwin A. Burtt, *Right Thinking,* New York, 1946, p. 522.
5. For a documented account, see Gunnar Myrdal, *Value in Social Theory,* London, 1958, pp. 134–52.
6. See, e.g., S. F. Nadel, *op. cit.,* p. 54; also Gunnar Myrdal, *op. cit.,* p. 120, as well as his *Political Element in the Development of Economic Theory,* Cambridge, Mass., 1954, esp. Chap. 8.
7. Leo Strauss, "The Social Science of Max Weber," *Measure,* Vol. 2 (1951), pp. 211–14. For a discussion of this issue as it bears upon problems in the philosophy of law, see Lon Fuller, "Human Purpose and Natural Law," *Natural Law Forum,* Vol. 3 (1958), pp. 68-76; Ernest Nagel, "On the Fusion of Fact and Value: A Reply to Professor Fuller," *op. cit.,* pp. 77-82; Lon L. Fuller, "A Rejoinder to Professor Nagel," *op. cit.,* pp. 83-104; Ernest Nagel, "Fact, Value, and Human Purpose," *Natural Law Forum,* Vol. 4 (1959), pp. 26-43.
8. Cf. Gunnar Myrdal, *Value in Social Theory,* London, 1958, pp. xxii, 211-13.
9. The evidence is usually a count of red cells in a sample from the animal's blood. However, it should be noted that "The red cell count gives only an estimate of the *number of cells per unit quantity of blood,"* and does not indicate whether the body's total supply of red cells is increased or diminished.—Charles H. Best and Norman B. Taylor, *The Physiological Basis of Medical Practice,* 6th ed., Baltimore, 1955, pp. 11, 17.
10. It is irrelevant to the present discussion what view is adopted concerning the ground upon which such judgments supposedly rest—whether those grounds are simply arbitrary preferences, alleged intuitions of "objective" values, categorical moral imperatives, or anything else that has been proposed in the history of value theory. For the distinction made in the text is independent of any particular assumption about the foundations of appraising value judgments, "ultimate" or otherwise.
11. The above example is borrowed from the discussion in J. Neyman, *First Course in Probability and Statistics,* New York 1950, Chap. 5, where an elementary technical account of recent developments in statistical theory is presented. For a non-technical account, see Irwin D. J. Bross, *Design for Decision,* New

York, 1953, also R. B. Braithwaite, *Scientific Explanation,* Cambridge, Eng., 1953, Chap. 7.

12. Karl Mannheim, *Ideology and Utopia,* New York, 1950, pp. 271, 288, 292. The essay from which the above excerpts are quoted was first published in 1931, and Mannheim subsequently modified some of the views expressed in it. However, he reaffirmed the thesis stated in the quoted passages as late as 1946, the year before his death. See his letter to Kurt H. Wolff, dated April 15, 1946, quoted in the latter's "Sociology of Knowledge and Sociological Theory," in *Symposium on Sociological Theory* (ed. by Llewellyn Gross), Evanston, Ill., 1959, p. 571.

13. Karl Mannheim, *op. cit.,* pp. 300–01.

READING 20

Explanation and Value Neutrality

James Leach

I

Following the progressive elimination of teleology from the province of the scientist in the seventeenth century, the domains of science and values seemed finally isolated and immune each from the other. Value-free scientific objectivity emerged as the ideal, with the very notion of scientifically supported values exploded as a 'naturalistic fallacy'. Finally, the convergence in value theory of intuitionists, emotivists and oxonians, conjoined with the success and influence of a positivist philosophy of science, consummated the separation of science and values. The issue seemed securely resolved. Particularly enthroned was the value neutrality thesis of science (VN) which prohibits the scientist from making value judgments.

Professor Ernest Nagel, in his illuminating treatise on the logic of explanation,[1] follows the lead of Carnap, Jeffrey and Hempel in opting for this thesis. Most attempts to disunify the empirical sciences, on the grounds that the natural sciences are value free while history or the social sciences are not, he disposes of nicely with an impressive barrage of counter evidence. In particular he correctly rejects the idealist claim that to explain an action one must appraise that action. Hence he rests the unity of the sciences on the exclusion of values.

Nevertheless, there remains one argument against which his barrage clearly misses the mark. And the survival of this argument suffices to revive the issue of value neutrality. For the smoldering cinders of pragmatism, kindled by recent successes in statistical ana-

Reprinted by permission of the author and publisher from the *British Journal of the Philosophy of Science, 19,* 1968, pp. 93–108.

lysis, flare forth to suggest an alternative defence of the unity of science. But now a unity grounded on the inclusion, not the exclusion, of value judgments or decisions.

A survey of recent philosophic literature reveals progressively increased agreement, I think, that the amount of evidence necessary to warrant the acceptance of any particular expanatory hypothesis is not fixed but instead depends on our varied purposes. However, it seems less generally recognised that sufficient evidence also depends on pragmatic costs concerning the significance of making a mistake when acting on the hypothesis in question.

Generally, I shall argue in what follows, the reasonableness or acceptability of probabilistic hypotheses necessitates appraisal according to pragmatic criteria, and hence surrendering the value neutrality thesis even in its weakest form. When such hypotheses occur in fields like history where the data is often meager and where plausible competing hypotheses abound, the requirements for acceptable evidence must be loosened considerably. In other words, the degree of rigidity of such requirements, as a function of these cost factors, helps to distinguish the various empirical disciplines. Hence the goal of such varied disciplines as history, law, and science is not merely truth for its own sake but truth modified by other goals and criteria. Before turning to an elaboration and defence of the argument surviving Nagel's criticism, let me first clarify the VN thesis itself.

II

Though favoured by a long history, the VN thesis remains ambiguous even as formulated in the most recent literature. Generally, the thesis demands the scientist's value scheme to be logically divorced from scientific standards of explanatory validity and reasonable belief. It requires scientific inquiry to be objective both in the sense that the scientist remain evaluatively neutral when validating or appraising the rational correctness of his explanations, and in the sense of yielding the same conclusion for all competent inquirers. In short, the scientist qua scientist must make no value judgments in what has been called the context of justification.

But so formulated, the thesis leaves unspecified what sorts of value the scientist should avoid and hence what type of neutrality to maintain. Since the values one advocates, seeks or avoids are correlated with goals or objectives, the question quickly reduces to a concern with the goals of scientific inquiry. Three different kinds of goal seem relevant. The scientist might seek, first, truth and nothing but the truth; secondly, truth qualified by such other independent theoretical or epistemic values as explanatory power, simplicity, predictive content and relief from doubt; or finally these in addition to other pragmatic values associated with acting on his beliefs.

The VN thesis can, accordingly, take three progressively weaker forms. In its strongest and least defensible version it prohibits the scientist from making any value judgments or decisions at all. The scientist must seek no goals, not even truth. Since he neither accepts nor rejects hypotheses, he has no need, qua scientist, to replace doubt by true belief. At most he assigns degrees of confirmation or probability to hypotheses so that others, decision makers, can decide as to the truth or falsity of the beliefs. Let us call this strong version the "odds maker" view of science and of the VN thesis. Should this position prove too strong, a weaker version of VN might then be advocated to the effect that the scientist, though himself a decision maker, seeks only one goal, has only one overriding value: truth at any cost. He is prohibited only from making any other type of value judgment than appraising hypotheses as to their likelihood to be true. Let us call this the 'truth seeker' view of VN. A still weaker, more liberal, version is also possible. Many people construe science as a quest not only for true hypotheses but also for ones with additional cognitive, theoretical or epistemic values. Thus the scientist, on this 'cognitive decision maker' version of VN, can legitimately gamble with truth so long as the costs involved are purely theoretical. He is proscribed only from making value judgments of a practical sort based on the seriousness of consequences of acting upon his accepted hypotheses.

This liberal interpretation, of course, sets a limit to the VN thesis. For were any further practical values or goals to be countenanced within the methodology of science, little or nothing would remain toward which the scientist must be neutral. The force of the methodological imperative of VN would be negligible. However, one could interpret VN not as related to scientific goals but to subjective biases of the investigator. In this sense the scientist would be prohibited from taking his own values, preferences and attitudes into account when appraising his acceptances. But this seems to me a quite independent issue. For whether the scientist qua scientist has practical goals and hence whether or not he must be neutral toward some values is not in any way a function of whether these goals can be formalized, codified and made objectively uniform for all scientific inquirers. Whether one's personal values can be objectively defended is, indeed, a most important but open question. Nonetheless, it is independent of what kinds of goals scientists ought to pursue, and hence of the VN thesis, in any of the three versions marked off above, which is our concern here.

Thus to challenge VN in its weakest and most cogent form one must defend what we will call the 'practical decision maker' view of science by arguing that the scientist in the logic of his inquiry also makes pragmatic value judgments in pursuit of practical goals. This amounts to arguing that two inquirers, with the same evidence and making the same probability assignments, might nonetheless rationally disagree about the acceptability of a given explanatory hypothesis on the basis of attaching different values to the seriousness in action of making a mistake. The case against VN seems best argued, then, on the grounds of recent work in statistical analysis concerned with the problem of rational decision making in the face of uncertainty. This type of argument, advocated by such experimentalist philosophers as Braithwaite, Churchman and Rudner,[2] builds on the earlier statistical work of Neyman, Savage and Wald in application to problems of quality control and more generally to problems of non-deductive inference.

Applications of statistical procedures are to be found, of course, in most of the natural and social sciences where a major goal is to decide what to believe. But they are also used in determining insurance rates and in operations research where the goal is to determine how to act. Hence, the basic problem of statistics, rational decision making in the face of uncertainty, suggests not so much a structural similarity but a *partial reduction* or dependency relation between these different kinds of problems: how to act and what to believe in the face of uncertainty. Accordingly, Professor R. Rudner has formulated what I take to be the most explicit and compelling attack against the value neutrality thesis by posing his argument in this context.

But since Rudner's case is incomplete as it stands and has subsequently been subjected to sustained counterattack in the recent literature by Professors Carl Hempel and Isaac Levi, it will be instructive to unpack the argument in some detail. We will then be in advantageous position to assess the Hempel-Levi defence of a value free science by locating the form of the thesis and the accompanying issues more clearly. In Rudner's argument four statements are explicitly noted as premises from which the conclusion is claimed to follow. Let us list them as follow:

(1) 'The scientist as scientist accepts or rejects hypotheses.'
(2) 'No scientific hypothesis is ever completely verified.'
(3) Therefore, 'The scientist must make the decision that the evidence is *sufficiently* strong or that the probability is *sufficiently* high to warrant the acceptance of the hypothesis.'

(4) The 'decision regarding the evidence and respecting how strong is "strong enough" is going to be a function of the *importance,* in the typically ethical sense, of making a mistake in accepting or rejecting the hypothesis.'
(5) Therefore, 'the scientist as scientist does make value judgments.'[3]

From the above description of the three interpretations of VN we can see that each version denies the conclusion (5) on the basis of rejecting a different premise. The 'odds maker' view denies (1), the 'truth seeker' view rejects (3), and the 'cognitive decision maker' view denies (4). Hence if all of the premises can be sufficiently supported, the VN thesis will have to be abandoned. Now to some elaboration. Premise (2) states a central tenet of empiricism to which all covering law theorists subscribe. In accord with the requirement of empirical testability, it formulates one version of the fallibilist claim that all scientific explanatory hypotheses are empirical and hence corrigible. No such statement is ever without risk or ever completely confirmed by any amount of evidence. With (2) thus undisputed by covering law theorists, the acceptability of (3) depends only upon that of (1). How then to defend (1)?

The issue here turns on how to interpret the function of the scientist. For to defend the value neutrality thesis by rejecting (1), i.e. to deny (5) by denying that the scientist accepts or rejects hypotheses, commits one to the 'odds maker' or 'guidance-counselor' view of science. On this rendering, espoused by Carnap[4] and earlier by Hempel[5], the scientist simply assigns degrees of confirmation to hypotheses relative to the total available evidence. He thus serves only as an advisor to policy makers who might want to apply such information to practical affairs. Carnap of course defines logical or inductive inference in this way, so that all probabilistic explanatory inferences in science consist not in an attempt to replace doubt by true or reasonable belief, but only in analytically assigning degrees of confirmation to hypotheses relative to the given evidence. Hence, if this strong construal of the VN thesis is convincing, the objection to (1), and thus to (5), concedes that the scientist qua policy maker must make the decision required by (3), yet insists that the scientist qua scientist, i.e. qua odds maker, merely determines the degree of probability for an hypothesis. The VN thesis would therefore dictate giving up (1). Determining how costly this price is and whether wisdom demands such a price then become our immediate problems.

No doubt the issue as to the aim and function of science is embarrassingly complex and difficult. At the present time we have no completely adequate answer. But Hempel's more recent analysis of explanation itself entails that the scientist do more than assign degrees of confirmation. Suppose we ask whether the covering law theory is intended as an explication of 'an explanation' or of a 'rationally acceptable explanation'? Clearly covering law theorists intend the latter. They have never been satisfied to analyse the notion of explanation merely in terms of a valid argument containing empirical laws among its explanans. Instead, they correctly insist that the explanans also be rationally acceptable in order to provide good grounds for the explanandum, and hence to be a 'correct explanation'.

Further, when attempting to eliminate the inconsistency of inductive inference, Hempel clearly recognises the inadequacy of his earlier requirements for deductive systematisations as sufficient conditions of rationally acceptable explanations generally. He thus appeals for instance to the principle of total evidence. This immediately transcends the scope of inductive logic proper to the application or methodology of induction and to general pragmatic (as opposed to syntactic and semantical) considerations. Now, if the task were simply to explicate the notion of a formally sound (deductively or inductively) potential explanation, then pragmatic and methodological matters would be irrelevant: the scientist might escape as an odds maker by simply assigning degrees of confirmation to explanatory hypotheses. But since the covering law theory is designed to explicate the notion of an acceptable or reasonable explanation, which often requires acceptable statistical hypotheses as premises, the principle of total evidence and other methodological principles must be invoked. And this means that the total evidence, at least, must already have been accepted. In this case we must abandon the strong version of the VN thesis whereby the scientist pursues no values or utilities at all, not even truth. The scientist is at least a truth seeker and must decide when an hypothesis is to be accepted as true or rejected as false.

This suffices for the claim in premise (1) that an adequate explication or rational reconstruction of scientific procedures must include a statement to the effect that the scientist as scientist accepts or rejects hypotheses. Hence, if these considerations are cogent, (3) follows as a consequence. That is, the scientist must be seen not just as a counsellor of policy makers, as an odds maker, but as himself a decision maker. One of his major objectives is to replace doubt by true or reasonable belief, and to do this by deciding in the face of uncertainty when the evidence for an explanatory hypothesis suffices to warrant his belief in the hypothesis.

Moreover, if corrigibility or the chance of making an error

were the only relevant factor for the decision required (3), the scientist would never reach any decision. He would merely keep increasing the amount of evidence indefinitely. But if (1) is true, if the scientist must decide in the face of uncertainty, then, contrary to the 'truth seeker' view, some additional criterion or rules of acceptance needs be invoked in order to make a reasonable decision. Even in the pursuit of truth one must assess or attach utilities to the degree true answers are desirable and mistakes undesirable. To weight all mistakes equally comes to refusing to accept an hypothesis as warranted until all the evidence is both in and affirmative. Thus the second version of VN, that the scientist makes utility or value judgments regarding truth only and pursues truth at any cost, must also be rejected. Further, since (5) clearly follows from (4) and with (3) now established, the tenability of Rudner's argument rests upon the move from (3) to (4), from the fact that scientists must make decisions to the fact that these decisions are a function of the seriousness of error relative to pragmatic goals.

IV

Unfortunately, Rudner offers no argument in the article cited for this cardinal connection. Instead, he fills the gap by appealing to examples from quality control and to some guiding theories of statistical inference. 'If', to use one of his cases:

> the hypothesis under consideration were to the effect that a toxic ingredient of a drug was not present in lethal quantity, we would require a relatively high degree of confirmation or confidence before accepting the hypothesis—for the consequences of making a mistake here are exceedingly grave by our moral standards. On the other hand, if say, our hypothesis states that, on the basis of a sample, a certain lot of machine-stamped belt buckles was not defective, the degree of confidence we should require would be relatively not so high.[6]

Taking such an example as a paradigm case of scientific inquiry, he follows the statisticians by assimilating cases of deciding what to believe to cases of deciding how to act in the face of uncertainty. He concludes that the seriousness of making a mistake, reflected in the cost or value factor, must be weighed in any scientific assessment of statistical explanatory hypotheses.

What is unfortunate in the appeal to examples at this pivotal juncture in the argument, however, is not the lack of better grounds for moving from (3) to (4), for there are indeed better grounds. Rather, the suppression of the needed assumptions has led some defenders of the VN thesis to serious misunderstandings. As a result they reconstruct Rudner's argument by appealing to an altogether unnecessary premise to fill the gap, and then proceed to challenge this very premise. Professor Isaac Levi, for instance, clearly sees the required logical connection between believing or accepting an hypothesis and acting on the basis of the hypothesis relative to some specific objective or goal. This stresses again the need to interpret explanatory hypotheses, at least in part, methodologically as means or instruments adopted to achieve some specified objectives. Rudner's argument assumes, with Hempel and Levi concurring, that all scientific problem solving is a species of purposive behavior. The assessment of the rational acceptability of explanatory hypotheses is consequently relativised to goals which require weighing cost or utility factors in addition to degrees of probability or confirmation. The relevance of decision theory to the logic of inductive inference therefore rests in part at least on its applicability to inductive behavior.

But when Levi states explicitly the assumptions necessary to move from premise (3) to (4) and hence to the denial of the VN thesis (5), he switches issues. He commits Rudner and Churchman to two additional premises, one of which is unnecessarily strong. He nevertheless makes it the centre of controversy. The two additional assumptions cited by Levi are as follows:

(6) 'To choose to accept a hypothesis H as true (or to believe that H is true) is equivalent to choosing to act on the basis of H relative to some specific objective P.'

(7) 'The degree of confirmation an hypothesis H must have before one is warranted in choosing to act on the basis of H relative to an objective P is a function of the seriousness of the error relative to P resulting from basing the action on the wrong hypothesis.'[7]

Now (7) he accepts on the authority of the statistical theories of Pearson, Neyman and Wald. But since (7) without (6) obviously fails to yield (4), the tenability of (4) is made to rest on (6), on some sort of identity or equivalence between believing an hypothesis and acting on the basis of it relative to some specific objective P. For, as Levi queries, 'Otherwise how can the role of value judgments in determining minimum probabilities be explained?[8]

But surely a weaker commitment suffices in response to this question. Whether or not Rudner and Churchman would accept (6) as an adequate formulation of their suppressed premise seems simply irrelevant to the issue. Hence Levi and others raise a false issue and thereby confuse what I think are the important ones. Let me clarify. Levi assumes that the desired conclusion presupposes a 'behaviouralist' construal of belief whereby 'accepting a proposition H as true' is synonymous with, equivalent to, equated with or reducible to 'acting on the basis of H relative to a practical objective P'.

However such an assumption is not essential to yield (4). A behaviouralist view of belief, in other words, is unnecessarily strong. The only conditions required for the argument are that belief be logically tied to action in such a way that beliefs can be *partially justified* by practical considerations of action, and that different mistakes in action be taken with different degrees of seriousness. And these conditions, satisfied by a critical cognitivist view as well as by a behavioural account of belief, assure that the acceptance of an hypothesis is not merely the result of a quest for a single value or goal, truth and nothing but the truth. The logical connection between belief and action would, on the critical cognitivist view, be considerably weaker than identity, deductive entailment, or even material equivalence. It might in fact be as weak as nomological implication even in its defeasible version, i.e. with a *ceteris paribus* clause. Hence critical cognitivism contrasts with behaviouralism by viewing action or behaviour merely as evidential manifestations, signs, criteria or contingently necessary conditions of belief. Thus in place of the equivalence condition in (6) two nomological implications suffice:

(6′) To choose to accept an hypothesis H as true implies as a contingently necessary condition the disposition to act on the basis of H relative to some specific objective P.

(6″) To be disposed to act on the basis of H relative to some specific objective P implies as a contingently necessary condition acting on the basis of H relative to P in the appropriate circumstances C.

Once these replace (6), the issue no longer turns on a behaviouralist reduction of belief to action but clearly focuses on the questions posed by Hempel's attempt to reinforce the principle of total evidence by formulating acceptance rules of application for rational belief. The issue concerns, as we have seen, whether the scientist pursues one or more goals and what kind of values the assigned utilities are to represent, e.g. pragmatic or just cognitive utilities.

Once again, however, confusion threatens. For Rudner's premise (4), invoking the notion of the seriousness or importance of error, does not commit one to a purely instrumentalist view. It does not make science merely a handmaiden of practice. Nor does it require pragmatic costs or utilities to replace truth, high confirmation, simplicity, or even explanatory power as appropriate criteria for appraising the reasonableness of hypotheses. Nor does it entail defining any of these criteria in terms of pragmatic utility. For these can be interpreted as independent weights, each important for an adequate theory of inductive inference and for the assessment of rational choice among explanatory hypotheses. Yet no criterion is either a necessary or sufficient condition of the reasonableness of a belief or, more precisely, of a decision to accept a belief. The proper relative weights to be assigned these criteria, moreover, remains a difficult and open question.

The burden of Rudner's argument, then, is to show that pragmatic costs enter into the scientist's decision to accept or reject hypotheses. Thus the move from (3) to (4) requires the suppressed premises (6′), (6″) and (7) to provide the defeasible link between the decision to accept H and actions performed upon H in pursuit of pragmatic goals. This amounts to resting the reasonableness or rational acceptability of H relative to P not only on whether H is true or has high epistemic utility, but also upon its pragmatic value, upon whether or not action based on H is warranted.

One further caution. The point is not that if the appropriate act is unwarranted, the disposition and belief leading to the act are thereby unreasonable. For the epistemic interdependence of acts, dispositions and beliefs makes the appropriate logical connection between these notions much more complex than this stronger claim allows. Rather we argue the weaker claim: the unwarrantedness of the action is but one determining factor in appraising the disposition and belief. No doubt the conceptual meanings of 'belief' and 'action' and hence the adequacy of (6′) and (6″) still await clearer analysis. Almost inconceivable to me, however, is that an adequate analysis of 'belief' could be provided which is logically *un*related to 'action'. For there would then seem to be no way to account for the important influence of beliefs on action. Hence, if 'X believes H, relative to P', in conjunction with other statements, defeasibly implies 'X performs act A, relative to P' then it seems reasonable that the appraisal of A should bear on the appraisal of the decision to believe H. This I take to be the crux of Rudner's case.

V

Having warded off the false issues concerning the move from (3) to (4), a behaviouralist reduction of belief to action and an instrumentalist reduction of truth to utility, let us return to the main issue at hand. Suppose, then, an inquirer to be seeking the truth and nothing but the truth, to be concerned only with adding true statements into the body of scientific knowledge K. To be rational, in the second version of VN, the inquirer must weight each possible mistake he might make as equally serious. But such a condition cannot be satisfied on Rudner's argument since different mistakes have different degrees of seriousness or importance for scientists pursuing varied goals. This conflict serves as a basis to continue the defence of (4). It will also be helpful at this juncture to see how Hempel develops his notion of 'epistemic utility' in defence of the 'cognitive decision maker' view of science.

The problem of establishing acceptance rules for scientific explanatory hypotheses, of deciding which scientific hypotheses to accept and thus add to the body of scientific knowledge, turns on what kind of values the utilities assigned to outcomes are to represent. Hempel maintains that:

> the utilities should reflect the value or disvalue which the different outcomes have from the point of view of pure scientific research rather than the practical advantages or disadvantages that might result from the application of an accepted hypothesis, according as the latter is true or false. Let me refer to the kind of utilities thus vaguely characterized as *purely scientific, or epistemic, utilities.*[9]

Beyond the need to clarify this distinction between pragmatic and epistemic values, the problem concerns finding a measure of the epistemic utility of adding an hypothesis h to the previously established system of knowledge K. Carnap's rule of maximising estimated utility warrants the acceptance or rejection of h as epistemically rational in accordance with the following rule:

'Accept or reject h, given K, according as $c(h,\mathrm{K}) > \frac{1}{2}$; or $c(h, \mathrm{K}) < \frac{1}{2}$; when $c(h, \mathrm{K}) = \frac{1}{2}$, h may be accepted, rejected, or left in suspense.'[10]

Since on this account a scientist is interested only in accepting h when it is true, the possible correct answers of accepting h when true and rejecting h when false are considered equally desirable and the corresponding mistakes equally undesirable. Hence, we are led to the unsettling recommendation of accepting, and hence of acting upon the basis of, h when its degree of confirmation or probability is merely 0·51.

The obvious difficulty with Carnap's account, beyond leading to paradox, is that the rule is much too liberal to be suitable for even pure scientific procedure. Further, Levi notes the same problem when other methods are used instead of Carnap's maximising utility (e.g. Bayes's method and the method of significance testing proposed by Neyman and Pearson). He also acknowledges that accepting h 'when its degree of confirmation is low does not seem reasonable unless this acceptance is reduced to action undertaken to realise some objective other than seeking the truth and nothing but the truth'.[11] Levi, however, pursues the third option, suspended judgment of remaining in doubt, as a way of avoiding the difficulty and hence of defending the view that scientists accept or reject hypotheses in quest of truth and nothing but the truth. In this sense, the scientist suspends judgment on h when he feels the total available evidence warrants neither the acceptance nor the rejection of h.

Consider one of Levi's examples of the Bayes method applied to the problem of replacing doubt by true belief. Suppose an experimental psychologist wants to determine whether or not some person has extrasensory perception. By assuming that the subject has ESP if he guesses correctly the colours of cards drawn randomly with frequency greater than 0·6, and otherwise not, the psychologist decides on the basis of a sample of guesses whether the long range frequency of correct guesses is 0·6 or less (H_1), or is greater than 0·6 (H_2), when A_1 is the act of accepting hypothesis H_1.

> Since the experimenter presumably is interested only in accepting the hypothesis as true which is true, he should consider the possible correct answer O_{11} (accepting H_1 when it is true) and O_{22} (accepting H_2 when it is true) as equally desirable and the corresponding 'mistakes' or 'errors' O_{12} and O_{21} as equally undesirable. . . . Consequently, the matrix for this problem would be as follows:
>
> | | H_1 | H_2 |
> |---|---|---|
> | A_1 | 1 | 0 |
> | A_2 | 0 | 1 |
>
> The Bayes method recommends adopting A_1 if the probability of H_1 is greater than $\frac{1}{2}$ and A_2 if the probability of H_1 is less than $\frac{1}{2}$. Consequently, if the probability of H_2 . . . is 0·51, we would be warranted in accepting H_2 (adopting A_2).[12]

Now, to avoid this unsettling and unreasonable consequence, yet to preserve application of the method to cases where the goal is to seek the truth and nothing but the truth, Levi revises the problem by adding a third option: the act S of suspending judgment or remaining in doubt. In this case the matrix changes to the following:

	H_1	H_2
A_1	1	0
A_2	0	1
S	k	k

Here the utility k is the value of act S when the hypothesis H_1 is true. The strategy consists in assigning a sufficiently high value to *k* so that a high degree of confirmation is required to warrant the acceptance of any hypothesis. But the price of thus avoiding the unsettling consequence amounts to selecting arbitrarily the value assigned to *k*.

VI

However, this arbitrary assignment brings us back to the crux of Rudner's move from premise (3) to (4), from the fact that scientists must make decisions (which Levi and Hempel concede) to the fact that these decisions are a function of the seriousness of error. For Rudner's argument hinges on the assumption that any arbitrary selection of a value assigned to *k* in such cases as the above would be unreasonable. Particularly so if there are grounds in action for making such assignments of values. This is why premises (6′) and (6″), which connect beliefs or the acceptance of hypotheses with acting upon the basis of them relative to some goal, provide the crucial link between the decision to accept in (3) and the different degrees of seriousness or error in (4). Instead of assigning a value to *k* arbitrarily or on the basis of sheer convention, we take our cue from (6′), from the fact that belief implies a disposition to act. Our choice of a value for *k* then becomes a function of how serious a mistake would be if we acted upon the hypothesis, i.e. if we accepted a false hypothesis or rejected a true one. This in turn depends upon how useful the hypothesis in question is to us relative to our goals. So, while on purely logical grounds the value assigned to *k* is arbitrary, still a rational choice of assignments can and must be made on extra-logical grounds, viz. on the basis of purposiveness and pragmatic utilities. Relevant to this choice are the purposes for which we intend to use the hypothesis, and the relative advantages and disadvantages of acting upon it relative to these objectives.

It should be noted that Levi is not unaware of this type of consideration. After noting the unhappy result, the arbitrary assignment of a value for *k*, he offers further revision of his non-behavioural decision-theoretic approach to scientific inference by introducing the notion of 'degrees of caution'. In the case of his truth-seeking experimenter, we have a scientist who 'takes mistakes more seriously in relation to eliminating doubt' than might some others. Unlike a more tenacious scientist, he exercises a higher degree of caution, indicating 'that to seek truth and nothing but the truth is to aim at a complex objective. It is an attempt to eliminate doubt tempered by an interest in finding truth and avoiding error.' [13] But this of course opens the door to the move from premise (3) to (4). It means, as Levi himself acknowledges, 'that the choice of a problem whose goal is to replace doubt by true belief and, as a consequence, the choice of a degree of caution in realizing such a goal can be both influenced and justified by practical considerations'.[14]

Moreover, the fact that this conclusion implies nothing about any kind of behavioural reduction constitutes less solace for the value neutrality theses than Levi recognises. Nevertheless, in concentrating his effort against (6) Levi fails to see that by granting that the scientist must reckon with the seriousness of mistakes, though not a reductionism of belief to action, he has nonetheless conceded that different mistakes need to be taken with different degrees of seriousness. In other words, he totally neglects (6′) and (6″) which together suffice to get Rudner's argument to the conclusion (5), that scientists qua scientists make value judgments regarding the seriousness of errors in acting upon accepted or rejected hypotheses. For, the different purposes for which one acts clearly require different cost factors and, as a result, different degrees of seriousness. Hence Levi in effect concedes (5), the denial of VN even in its weakest form.

Clearly then, if the preceding considerations are cogent, the scientist must take into account in his decision to accept or reject hypotheses not only epistemic but also pragmatic utilities. Consequently the decision will vary with the kind of action to be based upon the hypothesis. Consider again the example from applied science cited earlier regarding the toxic ingredient of a drug. Because of the different utilities involved, a decision rule might warrant, on the same evidence, that the hypothesis in question should be accepted if applied to experimental animals, but rejected if applied to humans where an error would obviously be more serious. Hence the decision to accept or reject hypotheses is in part instrumental to action; it is partially a decision to adopt one of alternative courses of action. And this decision can be rationally defended only by considering the pragmatic gains and losses attached to the possible outcomes of the actions. This amounts to resting the acceptance of an hypothesis partially upon whether or not it leads to warranted action.

Having thus unpacked and defended the experimentalist argument opposing the varied versions of the value neutrality thesis of covering law theorists, let me now present it in full array:

(1) The scientist as scientist accepts or rejects hypotheses.
(2) No scientific hypothesis is ever completely verified, but all are corrigible.
(3) Therefore, the scientist must make the decision that the evi-

dence is sufficiently strong to warrant the acceptance of the hypothesis.

(6′) To choose to accept an hypothesis H as true (or to believe that H is true) implies as a contingently necessary condition the disposition to act on the basis of H relative to some specific objective P.

(6″) To be disposed to act on the basis of H relative to some objective P implies as a contingently necessary condition acting on the basis of H in the appropriate circumstances C.

(7) The degree of confirmation that a hypothesis H must have before one is warranted in choosing to act on the basis of H relative to an objective P is a function of the seriousness of the error relative to P resulting from basing the action on the wrong hypothesis.

(4) Therefore, the decision in (3) regarding the evidence and respecting how strong is 'strong enough' to rationally accept H is going to be a function of the importance or seriousness, in the typically ethical sense, of making a mistake in accepting or rejecting the hypothesis.

(5) Hence, the scientist as scientist does make pragmatic value judgments.

VII

Let me consider briefly one final but important objection to the foregoing argument, an objection posed most recently by Hempel, Nagel and Levi, but the roots of which lie deeply embedded in Aristotle's distinction between the theoretical and practical sciences. The experimentalist argument, it would seem, forces a reconsideration both of the meaning of scientific and historical objectivity, and of the relations between theoretical, technological and policy making aspects of rational inquiry. The issue of historical objectivity will be considered on another occasion.[15] Let me accordingly limit discussion to one related point.

Though Hempel is unable to find any 'satisfactory general way of resolving the issue between the two conceptions of science', the cognitive and pragmatic, he still claims 'that it would be pointless to formulate criteria of applicability by reference to pragmatic utilities; for we are concerned here with purely theoretical (in contrast to applied) explanatory and predictive statistical arguments.'[16] Nagel follows suit by objecting that the experimentalist argument misleadingly 'suggests that alternative decisions between statistical hypotheses must invariably lead to alternative actions having immediate practical consequences upon which different special values are placed'.[17]

This kind of objection forces one additional clarification. For the argument against the VN theses in no way depends on considering in the weighting of pragmatic utilities, only the 'alternative actions having *immediate* practical consequences'. No doubt some applied historical and scientific hypotheses will be closely linked logically to such actions. But others, of a higher degree of 'theoretical purity' will be more tenuously linked to such actions via the former ones. Despite the tenuousness of this linkage, it seems as ill-conceived to disregard the purposive and practical dimension of science as to disregard its empirical dimension. Surely there is an important sense in which the experimental data garnered from laboratory tests are logically related, as confirmation or disconfirmation, to 'purely theoretical' hypotheses. But then the practical considerations involved in experimental design of tests would bear heavily on the criteria of acceptance for even such theoretical hypotheses. Moreover, this linkage need be no more tenuous than the linkage of empirical significance between hypotheses containing theoretical constructs and relevant observational statements. And as it might be argued in Quinean fashion both that no observational statements are quite free of theory, and that no theoretical statements are quite free of experimental considerations, so it might also be argued in extension that no hypotheses are quite free of practical consequences.

In other words, it seems to make as much sense to speak of degrees of practical consequences and associated pragmatic utilities, with a system or network of hypotheses linked at its edges to immediate practical consequences, as it does for Hempel to speak of degrees of testability or empirical significance with a network of hypotheses linked at its edges to immediate observations. Surely nothing in either linkage precludes the other. It should also come as no surprise that the goals or objectives used to weight the consequences of wrong estimations in the 'purely theoretical' sciences will be of a more complex nature and will require more general criteria than those used in the more practical sciences such as quality control and engineering. Hence, even if there are cases where the choice of policy for deciding to accept or reject hypotheses does not depend directly upon immediate practical consequences, still the costs or pragmatic utilities will be relevant in a more indirect manner. This should serve to dethrone the traditional VN thesis even in its weakest form.

NOTES

1. E. Nagel, *The Structure of Science* (New York: Harcourt & Brace, 1961), pp. 485–502.
2. R. Braithwaite, *Scientific Explanation* (New York: Harpers, 1953); C.W. Churchman, 'Statistics, Pragmatics, Induction', *Philosophy of Science, 15* (1948); R. Rudner, 'The Scientist Qua Scientist Makes Value Judgments', *Philosophy of Science, 20* (1953.).
3. Rudner, p. 2.
4. R. Carnap, *Logical Foundations of Probability* (Univ. Chicago Press, 1950).
5. C. Hempel, 'Review of Churchman's Theory of Experimental Inference', *Journal of Philosophy, 46* (1949).
6. Rudner, p. 2.
7. I. Levi, 'Must the Scientist Make Value Judgments?' *Journal of Philosophy, 67* (1961), 348. For additional instances of invoking (6) in a similar context see the discussion by Kyburg, Black, Sellars, Jeffrey, Levi and Bar-Hillel of a paper by Braithwaite in Kyburg and Nagel (eds.), *Induction: Some Current Issues.* (Conn: Wesleyan Univ. Press, 1963), pp. 196–204.
8. Levi, 'On the Seriousness of Mistakes', *Philosophy of Science, 29* (1962), 50.
9. C. Hempel, 'Inductive Inconsistencies', *Logic and Language* (Dordrecht-Holland: Reidel Publishing Co., 1962), p. 154. Cf. also Hempel's most recent statement in 'Recent Problems in Induction', R. Colodny (Ed.) *Mind and Cosmos* (Pittsburgh: University of Pittsburgh Press, 1966).
10. C. Hempel, 'Deductive Nomological *vs.* Statistical Explanation', *Minnesota Studies in the Philosophy of Science,* iii (Minnesota: University of Minnesota Press, 1962) 155.
11. I. Levi, 'On the Seriousness of Mistakes', p. 55.
12. *Ibid.,* p. 54.
13. *Ibid.,* p. 56.
14. *Ibid.,* p. 57.
15. J. Leach, 'Historical Objectivity and Value Neutrality', submitted to *Inquiry.*
16. Hempel, 'Deductive-Nomological *vs.* Statistical Explanation', p. 162.
17. Nagel, *The Structure of Science,* p. 497.

SUGGESTIONS FOR FURTHER READING

Brodbeck, M. (ed.) *Readings in the Philosophy of the Social Sciences.* New York: The Macmillan Co., 1968, pp. 79–138.

Brody, B.A. (ed.) *Readings in the Philosophy of Science.* Englewood Cliffs, N.J.: Prentice-Hall, Inc., 1970, pp. 540–570.

Fried, C. *An Anatomy of Values.* Cambridge: Harvard University Press, 1970.

Handy, R. *The Measurement of Values.* St. Louis: Warren H. Green, Inc., 1970.

———. *Value Theory and the Behavioral Sciences.* Springfield, Ill.: C.C. Thomas, 1969.

Kaplan, A. *The Conduct of Inquiry.* San Francisco: Chandler Publishing Co., 1964, pp. 370–405.

Kemeny, J.G. *A Philosopher Looks at Science.* Princeton, N.J.: D. Van Nostrand Co., Inc., 1959, pp. 230–243.

Krimerman, L.I. (ed.) *The Nature and Scope of Social Science.* New York: Appleton-Century-Crofts, 1969, pp. 704–758.

Laszlo, E. and Wilbur, J.B. (eds.) *Human Values and Natural Science.* New York: Gordon and Breach Science Publishers, 1970.

Margenau, H. *Ethics and Science.* Princeton, N.J.: D. Van Nostrand Co., Inc., 1964.

Meehan, E.J. *Value Judgment and Social Sciences.* Homewood, Ill.: Dorsey Press, 1970.

SECTION IX

The Social Responsibilities and Control of Science

The primary aim of the first article in this section is to explain the nature of the problem of determining the social and/or political responsibilities of scientists *as* scientists, and to indicate some of the historical and possible future consequences of perceiving this problem in different ways. Haberer is not particularly flattering in his assessment of most scientists' appreciation of the problem. "Scientists," he claims, "have almost always been pliant partners, willing under almost any conditions to accommodate to a given political order." Moreover, "The failure of scientists has lain in their moral obtuseness, in their incapacity to define, delineate, or even to recognize the nature of the problem of responsibility." Nevertheless, it is not entirely clear that Haberer has been able to elucidate the problem with much more precision than those whom he criticizes.

We are told that a scientist's commitment to science is not an "*ultimate* commitment," but it is not clear what such a commitment would imply. It seems to imply, for Haberer, a willingness to die for one's convictions, and although some scientists have had such a willingness, many others have not. Does that mean that a scientist's commitment is ultimate, or not? All scientists, or some? Does it mean that a scientist's commitments *ought* to involve such willingness?

Haberer contrasts the intrinsic value of truth and knowledge with its instrumental value, i.e., with knowledge as power, and he points out that scientists have typically been willing to downplay the former for the sake of the latter. Generally, the purest of pure science tends to be recommended on the grounds of its "eventual usefulness." What must be noticed about this distinction is that it is *not* the same distinction discussed by Leach in the preceding article. Leach contrasted truth with usefulness, but Haberer is contrasting useless with useful truth, or useless with useful scientific knowledge. Hence, the questions that were raised in our appraisal of Bunge's first article (Reading 2), when it touched upon the problem of totally useless scientific truth, return again. This time, however, we are concerned with the possibility of such truth from the point of view of the responsibility of its producer and the impact of the pursuit of such truth on society as a whole. If there is such a thing as useless truth, what would constitute a responsible attitude toward it, for scientists and for ordinary people?

The question that has just been raised drives us straight to the problem of constructing a science policy that will guide people who are responsible for the allocation of public funds and other resources in science and technology. In Linder's contribution to this section, he suggests another path to the same problem, namely, the fact that "progress has progressed in an unbalanced way." Then he outlines four aspects of a solution to the problem of devising a science policy. He goes a bit further than Haberer down the road to a constructive proposal, and he is also concerned more explicitly with technological developments.

Finally, in the third article, Baram presents a lucid discussion of the need for and problems in the identification of a regulating agency or office for science and technology. Law enforcement agencies, he claims, are generally empowered to act only after the harm has been done. Furthermore, "Crusaders are in short supply, and citizens' groups lack funds, technical expertise, and national political strength." On top of these problems, there is also a short supply of scientific manpower, so that when we look around for unbiased regulators, we find ourselves dipping into the same pool of resources. In response to these problems, Baram recommends the fostering of "independent adversaries" and interdisciplinary education. The latter suggestion has been proposed so often by so many people that one begins to feel that the triumph of interdisciplinary education is virtually assured. Unfortunately, however, this seems to be far from the truth.

READING 21

Scientific Leadership and Social Responsibility

Joseph Haberer

POLITICS AND THE LEADING INSTITUTIONS

That politics is an inherent component of the scientific enterprise has been the underlying premise of this work. For the most part, an awareness by scientists of this ubiquitousness has been missing. Throughout modern times relationships with the political order were peripheral, which partly explains this indifference. But also it may be attributed to a pronounced aversion to politics itself as a form of human activity. In a very profound sense, scientists have shown an ambivalence towards politics which has characterized other major vocational communities, such as the religious and military, whose concerns are not primarily political. Such communities are nonetheless infused with problems which require political modes of thought and political instrumentalities. In their initial development into leading institutions they devote their energies almost exclusively to the imperatives derived from the central concern—be it salvation of the soul or physical defense of one's society—around whose loci their respective activities are organized.

Even at the beginning, an incursion of political questions and practices always occurs, though in a covert and implicit fashion. The founders of modern science assumed that the kingdom of science would dispense with the concerns and the practices that come under the rubric of policies. Their expectation was that science would eventually make possible the total elimination of politics within society, excluding it both as a mode of activity and as a form of institutional organization. Very much like the founders of the early Christian church, they saw no need for a developed social and political theory, since the imminent culmination of his-

From *Politics and the Community of Science,* by J. Haberer, © 1969 by Litton Educational Publishing Inc. Reprinted by permission of Van Nostrand Reinhold Company.

tory would make such theories superfluous. Moreover, the two realms of politics and science were considered antithetical, inherently at odds with each other. This rejection of politics as a primary activity derived precisely from the attitude which the founders of modern science shared with the early church fathers. Both considered politics derivative and inextricably corrupting. Unlike the Greek view, which considered politics a primary and enhancing form of human experience, Christianity saw in it the very embodiment of a sinfulness and lack of grace which had to be overcome. During the nascent period of the Christian church, expectations of the imminent arrival of the promised second coming were subjected to a continuous disappointment. As decade followed decade, the growing church had to begin to cope more systematically with political and social questions. Thus the early church fathers, St. Augustine and others, were compelled to face up to the task of coping with a world in which choices of a political sort were necessary.

Modern science also began with an aversion toward politics, rejecting it as a fundamentally alien activity. The prophetic strain in the founders of modern science posited a kingdom of science in which politics would become unnecessary, would be replaced by mere administration. It may be this visionary impulse, which became part of the scientific outlook, which partly explains the a-historical and a-political posture of modern science. Yet the anticipatory tranformation of society into a scientific Utopia was not fulfilled. As science turned into the secular church of modern man, at the same time its phenomenal growth turned it into a major social institution, one requiring an increasingly heavy involvement of its institutional leaders with the political order. This tranformation resulted in the gradual recognition of the importance of the political order to the growth and well-being of the scientific enterprise. It has also forced upon it an intensified internal politicalization. The changed scale of scientific activity alone, reflected, for example, in the construction of large research centers and expanding scientific bureaucracies, makes visible patterns of thought and action characterized by the term "the politics of science."

What is being suggested is that the aversion to politics has historic parallels, that these result from conflicting basic values, and contrary definitions of priorities which distinguish major social institutions from each other. Initial rejection of the political order becomes eventually a highly ambivalent *modus operandi* with it, one which rejects the spirit of politics, while coming to terms with the realities which undergird its practices. As the major institution acquires large-scale influence and garners to itself increased power, it tends more and more to become a politically charged form of social organization, wherein institutional and organizational problems require political approaches for their solution. Matters of public policy become increasingly important and require more and more attention. Thus the early church turns into the Church of Rome, a highly politicized religious order, and the protestant reformation produces Calvin's Genevan theocracy. As this metamorphosis takes place, spokesmen develop an ideology for their vocational community. Considerable effort goes into promulgating this ideology throughout the wider social order, for the inner logic of any potentially leading institution is to move toward transforming its central values into the dominant values of society. At the same time, the emerging leaders are also impelled—almost against their will—to cope with the growing public problems, both internal and external, of their enterprise. This requires prior decisions on fundamental questions pertaining to the political constitution of the vocational community. For example: Who is properly a member of the community and on the basis of what criteria? Who speak authoritatively for the community? How are binding decisions to be arrived at? All this requires the formulation of political theory which will *inter alia* provide an appropriate framework through which to make judgments and decisions of a public nature.

Science appears to be at that juncture. Its leaders increasingly devote their energies to questions of public policy, in two ways: First, in establishing a *policy for science,* that is, for making decisions within the community of science, which affect the work and direction of those within a field, specialty, science organization; second, in their involvement in a *science policy,* the decision making process and its outcome in the broader political system. Both are evidently very much related and intertwined. Both are largely the outcome of a growth in scale: an immense growth in size of the membership which makes up the community and an immense growth in material resources allocated to it by society. Sound investments of the order that are now common in all highly developed countries, are made because of the expectation that basic research and development are the sinews which make possible the very existence and the well-being of a modern industrial society. Such large-scale institutions, and investments, also require a significant measure of planning, hierarchical leadership and bureaucracy.

The modern scientific enterprise cannot be run on laissez-faire principles, where individual scientific entrepreneurs operate in the market place of ideas, where, so to speak, a scientific "invisible hand" automatically regulates its affairs. In large-scale institutional situations, the Cartesian model becomes increasingly disfunctional and irrelevant. At best, it pertains only in select and special enclaves such as an institute for advanced studies. Even here, the eminent scientist is likely to become involved in other capacities; as a consultant, or as the principal investigator of a government-sponsored research project—roles which very likely place him in an equivocal position. In any case, to the extent that he functions in a large university, one which is increasingly dependent upon public funds, he becomes enmeshed not only in the politics of science, but also in an institutional network which conforms more and more closely to the Baconian model.

SCIENCE, LEADERSHIP, AND POLITICS

Scientists, particularly their leaders, have acted rather consistently in their involvement with the political order. This pattern may be described in two ways: First, the scientific leadership has tended, almost without exception, to acquiesce in any fundamental confrontation with the state, especially when opposition was likely to evoke serious sanctions. Second, the concomitant rise of modern science and the modern nation-state has placed the focus upon power as the crucial variable. Statesmen and scientists have tended to become craftsmen of power, even when their purpose or symbolic expressions vary. One danger in this emphasis is that practitioners in both enterprises may come to treat power as its own justification, and act as if its extension were a goal in itself.

Before about 1850, science needed only the state's goodwill to work. It now very frequently depends upon the state for the resources required to pursue large-scale investigations. At the same time, the modern state depends upon the power provided by science through technology so much that, in a very real sense, it could not exist in its present form without science. State and science, or politics and scientific knowledge, have thus become increasingly interdependent. As a result, scientists are increasingly involved in the affairs of government. Moreover, the large-scale institutionalization of science and the great number of scientists that now make up the community of scientists, have also created certain professional interests which its leadership seeks to protect.[1]

When involved in politics, the scientist-leader has usually preferred to exercise personal influence with political leaders, to operate in or near the inner councils of government.[2] Scientists, however, now appear more prepared to engage in elective or grass-roots politics.[3] However, the scientific leader, if he belongs to the theoretical elite, if he is a Cartesian in spirit (Planck, Oppenheimer, Szilard, Einstein, for example), usually enters the political arena with the conviction or hope that he can manipulate statesmen for his own purposes. In the end, he is likely to find himself used, or discarded when he is no longer useful.

Moreover, the political acumen of these leaders has tended to be poor. Bacon, Descartes, Planck, Oppenheimer and others have demonstated poor judgment at critical periods of their public careers. Bacon, Oppenheimer and Planck fell from positions of great influence—two ended their public careers in a form of open disgrace. Even Descartes, who because of his disengagement from politics avoided the onus of public downfall, nevertheless suffered a (personally felt) humiliation associated with the failure of his primary mission. Even so, Descartes' disengagement was probably more apparent than real since he, too, followed the pattern of attempting to exercise personal influence with rulers.

These scientist-leaders all tried to acquire some influence (and power) within science and with government. This effort was likely to be engaged in under a euphemistic guise, to disassociate their activity from the stigma attached to politics. Less noticeable than their desire for influence, but no less significant, has been their unwillingness to accept the burdens and obligations of political involvement. As advisors, consultants, or administrators, scientific leaders have been prepared to accept the duties of office, but they are usually unwilling to examine the larger meanings of such responsibility , or to recognize that political involvement, contributing to policy making, carry an accountability beyond questions of technical competence and judgment. Whatever the reasons for this—perhaps the disposition to confuse responsibility with guilt—as scientific leaders have moved toward the centers of political power, they have faced the burdens of politics. Although often apparently fascinated by political power—close connection with the state is usually welcomed rather than shunned—they have not readily recognized or accepted the ambiguities associated with the political. The practitioner in politics must deal with the price of his engagement—including the moral burden created by the tension between means and ends, theory and practice. Refusal or inability to recognize this is likely to endanger other men.

It is of the greatest significance that, from about 1850, the professionalization of science[4] began to affect the traditional dominance of the Cartesian model of science. Today, the Baconian model dominates most of science, at least in highly industrialized societies where the community, as a whole, can no longer be institutionalized along Cartesian lines. To the extent that basic science and applied science (technology) can be distinguished, the former tends to be Cartesian and the latter Baconian. However, such distinctions are becoming increasingly tenuous, for the two have converged on many levels. Science still tends to speak with the voice of Descartes and act with the hand of Bacon—its self-image leans toward a Cartesian definition, its institutional and public posture toward a Baconian expression. Internally, the preferred terms involve the ideal of basic science; in external affairs (vis-à-vis the political order) utility is the common justification. The theologians of science are Cartesian, even as their church is being transformed from a small body of the elect to a large order organized on Baconian principles.

SCIENCE AND ULTIMATE COMMITMENT

Scientists generally believe that their community is first and foremost dedicated to the pursuit of scientific truth or knowledge for its own sake. The history of modern science is filled with examples of great scientists who have worked without utilitarian motives: curiosity, a passion for finding answers or resolving the mystery of nature. Nevertheless, though we may agree that a commitment exists, the philosophy of science and the conduct of scientists have consistently failed to state or to demonstrate that it is in any sense an *ultimate* commitment. The founders explicitly disavowed it. The scientist tends to accept, in practice, a priority of values imposed by other communities. He will not engage in conflict with church or state when either can seriously jeopardize his safety. If his conscience or his scientific integrity make a situation intolerable, or if outside demands become irreconcilable with his work, he may attempt to escape by retreating within himself, perhaps ceasing to work. In most cases, however, he will prefer to compartmentalize the two realms and continue his work, so long as the state does not actually persecute him. He is thus not prepared to rally to the defense of colleagues where it is truly dangerous to do so,[5] nor does he understand the strength derived from unified action in defense of one's community.

In essence, then, the scientist tends to believe in some sort of ultimate commitment, but he is usually unaware of or unwilling to face up to the real meaning of such a commitment. This attitude emerged from the propensity to view science instrumentally. Its leading ideologists have equated science with power, and judged its worth in terms of instrumental value—not mere material utility, but as a means of resolving the fundamental problem of man's stature. Power, however, tends to become its own justification as well as its own end. For science, power becomes the "effecting of all things possible." By invoking the imperatives of progress, the scientific enterprise tends to be directed toward testing the limits of that which is possible.[6]

It may be argued, as evidence to counter this interpretation, that many scientists today appear more cognizant of the value of basic science than before, and frequently urge intensified work in basic science. But an examination of such pleas, especially those made in public, shows that almost without exception they are joined with an argument of necessity, of eventual usefulness. Basic science is increasingly viewed as a "national resource," which must be nourished and developed to avoid stagnation in a highly industrialized society.[7] This is not a position taken merely as the price of gaining the general community's support. To be sure, advocacy of science as an intrinsic good carries little weight with that community, especially when it is asked to make enormous investments for the development of that basic science. Even scientists themselves do not seem to consider this a very persuasive argument. They tend to be torn between a desire to believe that science has some intrinsic value and a need to justify it to themselves and to others as a useful and indispensable enterprise within society. Although intrinsic values and utility need not necessarily conflict, scientists increasingly think of basic science in instrumental terms.

A community must be prepared to demand sacrifice of its members. Such sacrifice may never be necessary, but on those rare occasions when a challenge arises, it must, to be meaningful, transcend mere words. Precisely because science lacks this characteristic—because scientists speak of their ultimate commitment without being prepared to put it to the test—it remains an embryonic community. Lacking the anchor of ultimate commitment, the community has shown a marked predisposition to fill this hiatus by emphasizing instrumentalism. Truth and Knowledge become equated with their manifestation as Power, so that Power has tended to become the key value within science. The very successes of the scientific method are likely to be judged in terms of their eventual tranformation into means of manipulating and controlling nature, or as demonstrating the extension of man's power over his environment—even over other men. Bacon's equation was prophetic. Science has become power incarnate.

Cartesian and Baconian philosophers may well be so deeply imbedded in the modern scientific consciousness, almost platitudinous, that their underlying assumptions are accepted unquestioningly. Indeed, science itself has almost never been profoundly examined by its leading practitioners. As Whitehead put it, "science repudiates philosophy... it has never cared to justify its faith."[8] For example, Hume's devastating and apparently unanswerable epistemological critique of scientific method was simply ignored—it was sufficient that the method brought success.[9] Socratic wisdom, the philosophic spirit, have not the manifest utility of the scientific method.

For the individual, the unexamined life may be safe, even if hardly worth living. For a community, it is likely to become very dangerous. Loss of a sense of history, indifference to the perennial problems which confront a community, can lead only to the development of a fragile sense of identity. Nor is commitment strengthened by facile optimism or by faith in the inevitable conjuncture of the scientific method with humanistic ends. Such a community will remain too fragmented to defend itself adequately against the incursions of communities with a more securely anchored sense of identity. If strong identity and strong commitment are somehow related, then weakness in either is bound to have serious consequences not only for science itself, but also for the polity.

The crucible which most profoundly tests an ultimate commitment is a fundamental attack on its social and communal embodiments. Whether such a threat manifests itself in a campaign to suppress its ideology, destroy its institutions, or persecute its leaders, the strength of the commitment will be reflected in the response by members of the community, acting both individually and collectively, to the threat. In the final analysis, adherence to an ultimate commitment may require a specific kind of courage: the readiness, should this be necessary, to sacrifice one's life. Such a testing is arduous indeed. Meeting it successfully calls for fortitude and heroism. Heroism is usually defined as a willingness to involve oneself in the fate of others, to risk one's very life in defense of members of the community or of the values embodied in that community.

There is, however, a peculiar quality to the figure of the scientist-hero. His designation derives from a confrontation which differs significantly from that of other kinds of heroes in other communities. The scientist-hero is usually pictured in the isolation of his laboratory, or, as Newton, in the solitude of Nature. The adversary he confronts and eventually defeats is a hiatus in knowledge, or an unresolved problem which his persistence and scientific imagination master. His heroism lies precisely in the victory of his determination and creative achievement over the scientific opinion of his day. This victory involves a certain courage—the danger of ostracism is always present—and may well be characterized as heroic, yet it differs from the phenomena more generally so designated. For the scientist *qua* scientist focuses on a non-human universe of matter and energy. The community of science is inclined to ignore continuing conflicts as alien and reprehensible. Since the scientist's heroism is not viewed in terms of an involvement and relationship with other scientists, he does not become a hero by defending other scientists or even the values of the enterprise itself. The very faith that science itself will survive, that the method of science will insure its triumph, provided there are scientists to practice, tends to give the enterprise a sense that it will prevail whatever the fate of its discrete individual members. Scientists may show great courage in their individual capacities and in other roles. A medical scientist may knowingly expose himself to great danger in his attempt to find the cure for a disease, an atomic scientist may engage in hazardous work so as to strengthen the security of his country. But scientists to date have not been prepared to take similar risks on behalf of the values, institutions, and members of their vocational community.

In a confrontation with the state, victims within the community of science have almost always faced destruction alone. The scientist, Baconian or Cartesian, has been unwilling (or unable) to think of sacrificing himself for the good of the community of science or in defense of his colleagues. Again, this is not to say that scientists have not been prepared to sacrifice their lives for another community of which they are members—the nation-state, for example—but only that they have not demonstrated a willingness to do so on behalf of their vocational community. Indeed, heroism or martyrdom has been viewed as senseless and meaningless.[10]

That this attitude reflects not only the founders' thinking but also that of the community's leadership throughout its modern history is demonstrated both by its conduct and also by the words of its wise men. Such a theme recurs in the writings of Einstein, for example. The absurdity of Galileo's attempt to convince his opponents evoked these comments:

> Undoubtedly more than almost anyone else he strove passionately to get at the truth. But I find it hard to understand that as a mature individual he should have thought it worth the effort to try against such great odds to convince that crowd in Rome, which was so entangled in their own superficial and trivial interests, of the validity of the new scientific discoveries. Was this really so important to him

> that he should have wasted his last years on it? The revocation that had been forced from him was really insignificant. After all, Galileo's arguments were accessible to anyone searching for knowledge. Those who were somewhat enlightened would know that his official disavowal had been forced from him. Certainly I cannot conceive that old Galileo's obstinate inner autonomy necessitated in any urgent way that he go into the lion's den at Rome to fight with priests and (hack) politicians. In any case, I can't imagine that I would have undertaken this in order to defend the theory of relativity. I would think: the truth is incomparably stronger than I am, and it would seem ridiculous and quixotic to try to protect it with sword and Rosinante.[11]

Max Born makes very much the same point:

> When Galileo's trial was discussed and somebody blamed him for having failed to stand up for his convictions, [David] Hilbert [a famous mathematician at Göttingen] answered quite heatedly: "But he was not an idiot. Only an idiot could believe that scientific truth needs martyrdom—that may be necessary in religion, but scientific results prove themselves in due time." This kind of teaching has directed my way in life and in science.[12]

The significance of this attitude is likely to be overlooked because the community of modern science has only rarely had to confront questions on the nature of its ultimate commitment. Leading scientists have encouraged either a disengagement from politics or have viewed science as a natural ally and servant of the state. This relatively peaceful history–which began to be threatened only during this century–has been conducive to the belief that the community is proof against any and all ill winds. When such confrontation has become unavoidable, the community of politics has invariably prevailed. Of course, this is by and large also true of confrontations between the modern state and the religious community. What is remarkable, however, is that when the interests and values of science confront those of politics, the community of science has ultimately tended to acquiesce with astonishing ease permitting, with only token opposition, the destruction of colleagues and scientific institutions, and the perpetration of obvious assaults upon the integrity of science itself.

Scientists as individuals have often been as eager to demonstrate their willingness to sacrifice themselves as have other citizens. But the scientist lacks, indeed seems unable to envisage, ultimate commitment of the sort that other institutional communities, such as church, army, political party, tend to encourage and consider indispensable. To be sure, practice often fails to conform to such ideals; still, significantly, these institutional communities call for, on one level or another, what may be called an "internal civic courage."

The community of science, from its inception, dispensed with such an ideal. It has not been prepared to urge its members toward any ultimate commitment to the life of the community. Yet when basic values, the institutions, and the integrity of science are threatened, it is such sacrifice and possible "martyrdom" that the members of a true community must be prepared to undergo in its defense. One may then wonder whether the community of science has any ultimate or intrinsic values at all—or if any are expressed in the institutional life of the enterprise. The philosophy of the founders of modern science, and the subsequent history of the community, leave room for doubt.

SCIENCE AND RESPONSIBILITY

If any "lesson" can be derived from the preceding studies, it is that scientists and particularly their leaders, failed to recognize and exercise a responsibility which was appropriate to their community. A galaxy of problems and questions now revolve around the theme of the "social responsibility of science." These have become a matter of great concern only within the last few decades, for historically this subject was at best only of peripheral concern. When science moved to the center of the stage, its salience to the crucial social, economic, and political problems of the time—that is, to public questions—became increasingly apparent.

To consider the issue of responsibility a simple one would be a serious error. Moralists may take comfort from their stark either/or response to problems involving moral choices. Human situations involving such choices rarely permit simple alternatives. Political problems and the decisions affecting public issues often require excruciating choices and demand a willingness to choose between alternatives none of which are without their flawed moral consequences. The best intentioned acts of the good man may result in disastrous consequences for others. As Machiavelli noted, "a [states]man who wishes to make a profession of goodness in everything must necessarily come to grief among so many who are not good."[13]

At issue then is not that the responses of scientists to the crises which confronted them were less than morally perfect. That scientists are human, perhaps all too human, should not occasion any surprise nor should it subject them to any special reproach. Rather the failure of scientists has lain in their moral obtuseness, in their incapacity to define, delineate, or even to recognize the nature of the problem of responsibility. Characteristically, responsibility has been recognized only in its narrower sense. Scientists have been willing to be held responsible for the calibre of their scientific work; or when acting in administrative positions for their performance in terms of the formal responsibilities attached to their positions. Beyond this methodological and bureaucratic responsibility scientists have not, at least until very recently, ventured.

There exists a need to consider and clarify the problem of social responsibility as it applies to science. In this discussion, the term social responsibility subsumes decisions and choices affecting the public domain. In this sense, almost all matters in the area of social responsibility have some political implications, so that social and political responsibility are very closely intertwined and may, for our purposes, be considered as almost synonymous. One may well begin with the generally accepted meaning attributed to the concept.[14] Simply put, responsibility suggests accountability. It means the act of being held answerable for what one does or does not do, the willingness or unwillingness to be so held accountable.[15] While the question of responsibility and that of "guilt" are often connected, they are not the same, and need to be distinguished. An irresponsible act entails a measure of guilt, of moral culpability. To raise the questions of responsibility should not, however, predispose one toward a premature adverse judgment. Only the evidence, the actual conduct in specific situations, based on appropriate criteria provide the grounds for making legal, ethical or moral judgments.

In more specific terms, what then does it mean to be responsible, to be willing to hold oneself accountable in science? Does it mean, for example, that scientists ought to withhold their knowledge, if they consider that there is a great risk that it will be misused, even endanger human life and welfare? Yet almost every discovery and invention has within it a potential for misuse. How would scientists decide whether the probabilities for misuse were greater than those for constructive application? Indeed, discoveries in pure science often have led to unanticipated utilization. Since the scientist is not a seer, he has no way of knowing whether or not the long-range social consequences will be beneficial rather than detrimental. Even if he can make an educated guess, it is still necessary to clarify the criteria by which such decisions are made. For example, is it to be an individual one based primarily on his own judgment and conscience; or is it to be a collective action in which the combined will predominates? Under most conditions such questions and the choices which are entailed in them are clearly extraordinarily difficult ones to apply in practice. Only in the more extreme cases, for example, in scientific experimentation on unwilling human "subjects" or in the area of chemical-biological warfare research might explicit guidelines and restraints be clearly delineated. But even here, choices are not clear-cut. For if it can be argued that chemical-biological research is an area in which scientific research would be immoral, then why not include all military type research or all knowledge that could potentially be useful to a military establishment? But given the nature of the contemporary nation-state system, to refuse to assist the national defense machinery—except in the unlikely event scientists were universally to refuse to do so for their respective nations—the consequences for the country in which such restraint was exerted might be most harmful for its citizens in their power confrontations with other states. This is only to reiterate that no easy answers are available. At the same time it becomes important not to evade the whole issue. What is required, on the one hand, is a willingness to confront the problem and to make decisions based on carefully thought out criteria, and, on the other hand, to be fully cognizant of the tenuousness of judgment and the fallibility of man in his existential situation. Men must choose and act for, paradoxically, not to do so is in effect but a form of choice and action. The cardinal error is either to remove oneself from accepting the responsibility of making choices or else in their making to do so with an unveering *hubris*.

APPROACHES TO SOCIAL RESPONSIBILITY

Since the end of World War II, as the problem of social responsibility surfaced, different approaches to it have been taken. Some scientists simply do not see this as a real problem for their vocational community. Their position is that the scientist's cardinal duty is to do good science. In contrast to the technician, the work of the basic scientist is seen essentially as a private activity. Social consequences which his discoveries may harbor, the misuse to which they may be put are, insofar as he is a scientist, outside his domain.[16] P. W. Bridgman, the distinguished physicist and philosopher, has been the most prominent spokesman for this point of view. Bridgman interpreted the new concern with the social responsibility of science as reflecting a rising demand by society and some fellow scientists that every scientist was morally obligated to "see to it that the uses society makes of scientific discoveries are beneficent." If this notion of responsibility were accepted, the consequences for science would be ominous. Society would blame the scientist for any failures in the effort to control the effects of his work, would apply legal sanctions against him, and most likely deprive him of his scientific freedom. But it is society's responsibility to control the consequences of science and not the scientist's concern in his capacity as scientist.[17] Moreover, for Bridgman, this insistence upon assuming responsibility would put unreasonable demands upon the scientist's time:

> For if I personally had to see to it that only beneficent uses were made of my discoveries, I should have to spend my life oscillating between some kind of forecasting bureau, to find what might be the uses made of my discoveries, and lobbying in Washington to procure the passage of special legislation to control the uses. In neither of these activities do I have any competence, so that my life would be embittered and my scientific productivity cease.[18]

Bridgman confuses the issue by creating a straw man. In his argument he sets up a position which does not reflect that of the advocates for greater social responsibility.[19] Moreover, he equates moral responsibility with legal responsibility when the two are not the same. Neither in society at large, nor among scientists, has there been any voice which has suggested that scientists be held legally responsible for the eventual social consequences of their discoveries, especially so, when these are so often impossible to predict. In any case, advocates for a greater social responsibility have usually done little more than urge that scientists both individually and collectively show a greater concern for the implications and social consequences of their work.

To be sure, the Cartesian scientist in his search for scientific knowledge has generally not been explicitly concerned with practical applications and may thus have found grounds for not considering the social consequences of his work. Nonetheless, for several reasons, the persuasiveness of this position has been seriously eroded. For one thing, the role differentiation between the basic and applied scientist has become less sharp. It becomes more difficult to know where basic and applied science begin and end, for the two are more and more interdependent. Not only do basic discoveries frequently find some application, but important theoretical discoveries increasingly derive from work in applied fields. Moreover, the same scientist may find himself in the role of basic and applied scientist at various points in his career. In the past, when the pursuit of basic science was predominantly supported by private means or subsidized by private patrons, the question of public accountability, of social responsibility hardly arose. Now basic science has become in many fields a costly endeavor, one which cannot be adequately supported by private means, but only by governments through the use of public funds. In such a situation, the issue of responsibility is bound to arise.[20] As a result, it becomes more and more difficult to maintain the posture of science as an essentially private pursuit of knowledge, made public by the scientist at his discretion. At a time when science has so clearly become a valuable societal resource, when its enormous and decisive consequences for society have become rather apparent, Bridgman's position that for the scientist *qua* scientist there is no social responsibility lacks persuasiveness and appears rather more an effort to simply evade the problem by asserting that it does not really exist.

One further development vitiates the force of Bridgman's position. Science has become professionalized.[21] Every profession—be it law, medicine, education. or architecture—is concerned with the effects of its practice on the society in which it operates. While particular professions may hold to a narrow or broad definition of that responsibility, none can evade coming to terms with some acceptance of that responsibility, since a profession is fundamentally service oriented. What it provides to individuals, organizations or governments may be esoteric, practical, or theoretical, manual or symbolic. Its activities, however, take place in the public domain, and are therefore held accountable on that level. A profession recognizes this when it takes upon itself the task of setting standards, codes of ethics, in short, when it regulates members or potential members.[22]

An influential and widely prevailing approach is that of scientists who take a pluralist position. They see themselves essentially as a profession in a society of many divergent and competing groups, each of which has interests which it seeks to protect or to further. Indeed, most science policy controversies probably involve just such professional interests. At the same time, pluralists recognize that there are some special tasks and responsibilities which they can and ought to perform. Principally, they see this in terms of an educative function; to provide and interpret for those in government as well as for the concerned lay public the key scientific information necessary for making rational and wise policy decisions. The main thrust of their informational activities is to create an awareness of the political economic and political consequences of those policies which are critically affected by scientific and technological factors. Efforts in this direction are varied. These include informal or formal attempts (e.g., before legislative bodies) either in an individual capacity or as representatives of science organizations to persuade officials on specific issues, whether it be the test ban treaty, the ecological effects of pesticides, the brain drain, population control or any number of other issues. Organizations such as the American Federation of Scientists, the Scientists' Institute for Public Information, the Society for the Social Responsibility of Science (a very small, Quaker influenced organization of scientists) in the United States; The Atomic Scientists Association in Great Britain; and the International World Federation of Scientific Workers are each concerned with science public policy and seek through their newsletters and journals to provide information and influence opinion. Other journals, notably *The Bulletin of the Atomic Scientists, Science,* and *Minerva* provide the major forum for the pluralists.

Most scientists follow the pluralist approach in which their involvement on broad issues is limited primarily to an educative and advisory role. Some scientists, however, have gone beyond this in advocating that scientists ought not to work in areas of research or in institutional settings where this is detrimental to the integrity of science and to human welfare. Norbert Wiener, in an open letter published in 1946, put this position most forcefully:

> When . . . you turn to me for information concerning controlled missiles, there are several considerations which determine my reply. In the past, the comity of scholars has made it a custom to furnish scientific information to any person seriously seeking it. However, we must face these facts: The policy of the government itself during and after the war, say in the bombing of Hiroshima and Nagasaki, has made it clear that to provide scientific information is not a necessarily innocent act, and may entail the gravest consequences. One therefore cannot escape reconsidering the established custom of the scientist to give information to every person who may inquire of him. The interchange of ideas, one of the great traditions of science, must of course receive certain limitations when the scientist becomes an arbiter of life and death. . . .
>
> . . . The experience of the scientists who have worked on the atomic bomb has indicated that in any investigation of this kind the scientist ends by putting unlimited powers in the hands of the people whom he is least inclined to trust with their use. It is perfectly clear also that to disseminate information about a weapon in the present state of our civilization is to make it practically certain that that weapon will be used. In that respect the controlled missile represents the still imperfect supplement to the atom bomb and to bacterial warfare.
>
> The practical use of guided missiles can only be to kill foreign civilians indiscriminately, and it furnishes no protection whatsoever to civilians in this country. . . .
>
> If therefore I do not desire to participate in the bombing or poisoning of defenseless peoples—and I most certainly do not—I must take a serious responsibility to those to whom I disclose my

> scientific ideas. . . . I do not expect to publish any future work of mine which may do damage in the hands of irresponsible militarists.
>
> I am taking the liberty of calling this letter to the attention of other people in scientific work. I believe it only proper that they should know of it in order to make their own independent decisions, if similar situations would confront them[23]

In 1957, 18 prominent German atomic scientists, including Hahn, Heisenberg and von Weizsäcker, signed a statement expressing alarm at the possibility that the German Army would be equipped with tactical atomic weapons. They felt it was their responsibility to provide the public with information about the destructive effects and potentialities of these strategic atomic weapons. After citing such facts, they concluded:

> We know that it is very difficult to draw political consequences from these facts. Since we are not politicians, one might deny us the right to judge these questions; however, our activity in pure science and its applications, which brings us into contact with many young people in this field, has bestowed upon us a *responsibility* for the possible consequences of this activity. This is why we cannot keep silent in these political matters. We support wholeheartedly the idea of freedom as represented today by the Western World against the ideas of communism. We do not deny that the mutual fear of the hydrogen bomb represents today an essential contribution to the maintenance of peace in the whole world and of freedom in part of the world. However, we consider, in the long run, this way to maintain peace and freedom as completely unreliable and the danger in the case of its failure as deadly.
>
> We do not feel competent to make concrete propositions for the policies of the great powers. We think that today a small country such as the Federal Republic can protect itself best and promote world peace by renouncing explicitly and voluntarily the possession of atomic weapons of any kind. Be that as it may, *none of the undersigned would be ready in any way to take part in the production, the tests, or the application of atomic weapons.*[24]

It is in any case quite likely that there are scientists who refuse to work in a scientific setting—be it industrial, governmental, or university—which involved them in an area of research they considered reprehensible (such as chemical-biological warfare); or one in which security restrictions became onerous and loyalty oaths demeaning. Others may simply reject working for any scientific organization or on any scientific project supported by funds from the military. It is, however, difficult to know how widespread such decisions may have been. Employment choices which are determined on the basis of moral criteria are usually private ones. After he has made this decision, the scientist rarely will declare publicly the ethical ground on which he rejected the professional opportunity before him. Nonetheless, his decision and others like it will have public implications. For example, if enough scientists responded in this way, governments with large-scale research and development programs would experience serious difficulties in recruiting a sufficient number of high caliber scientists on which such programs largely depend for their success.

CONCLUSION

There has been a growing recognition by scientists that the large-scale impact in the twentieth century has created for them an added responsibility for the social consequences of their work. Yet relatively scant attention has been given to the responsibility which has accrued to science through its professionalization. Any major vocational group when it turns into a profession needs first of all to acquire a considerable degree of internal autonomy. It must be able to protect the essential values embodied in their practice, the institutions on which it depends, and the members which make up its community from attacks which would weaken or destroy them. For example, the medical and legal practitioners engage in their work through institutions (e.g., hospitals, medical schools, private practice), and depend upon relationships essential to the performance of their tasks (e.g., confidentiality between practitioner and client, accountability to one's peers on matters of professional conduct, and so on). While professional autonomy is certainly not unlimited, and public law will restrict or regulate their practice, the intrusion of public authorities into internal professional matters usually is a reaction against a profession's abuse of its privileged position or to the neglect of its more obvious social responsibilities. Nonetheless, every profession properly concerns itself with the protection of its basic values and institutions. Responsibility here entails coming to the defense of one's community when it is under attack, supporting colleagues when their professional, or personal, life is jeopardized by arbitrary action. Moreover, under such conditions one would expect members of the vocational community, particularly its leaders, to be prepared to take some risks to defend their enterprise. A doctor, for example, who accepts the values implicit in the Hippocratean oath is expected to treat the sick regardless of their political persuasion. Put in a position where, let us say, the state insists that he no longer provide services to a politically undesirable group, the logic of his professional values would dictate a refusal to let himself be bound in this way. It would require him to act on the basis of professional imperatives even though this might place him in considerable danger. Lawyers have similar cardinal obligations. At stake here is a necessary commitment which in a very real sense is so intrinsic that it must be considered an ultimate one. To go against it is to betray one's vocation, and by implication also oneself. Still, the effectiveness and strength of such a commitment will depend very much upon its internalization in members of a profession. This internalization is developed through the socialization which members of a profession undergo, primarily through their training and apprenticeship, and in professional associations. Its strength will rest upon the continuing common effort to maintain an awareness of what their professional commitment specifies and an unceasing vigilance to protect and enhance the practice which reflects it. A great deal here depends upon the relationship of leaders to the members of their vocational community. More specifically, the task of leaders is to reflect appropriate models for others to follow, alert against danger, and mobilize the community against attacks on the integrity of their vocational domain. This always requires a judgment as to when such an attack has taken or is likely to take place.

Only in extreme cases are such judgments relatively easy to make, for most problems of professional ethics and responsibility are not ones which threaten the profession in a fundamental sort of way. Situations perceived as threatening more likely involve a possible diminution of a profession's influence and power within a society, which is not quite the same thing, even though it may be interpreted that way. For example, enactment of a medicare program or of socialized medicine does not per se pose a fundamental threat to the basic commitment of the medical profession. Nor does it follow that those who attack the science "establishment" or those who insist on more effective budgetary controls over tax-funded science organizations are hostile toward science or scientists as such. Not infrequently criticism of a profession is misinterpreted as an attack aimed at crippling it.

There are situations where the integrity of a profession is unquestionably under attack. This work has explored three situations where conflicts centered on the ultimate commitment of science, and examined the response to each crisis. We have suggested that the appropriate responsibility of leaders entails two things. First, it necessitates an effort to make one's constituency aware of the danger that confronts it. Although the extent of the danger may not be fully recognized, usually such an awareness is already present in a considerable portion of the community. Since the community will most likely be divided itself—unanimity is unlikely even in extreme situations—the task is to lead those who have not gone over to the other side into a recognition of the danger, encourage them through persuasion and example to defend their community. Second, it requires that leaders organize their constituency, and mobilize their members in the defense of their constituency. To do this entails obvious risks.

Situations in extremis have been rather rare for the community of science. Only in this century have some situations approximating this intensity occurred. Until the rise of totalitarian systems a threat to the integrity of science never seemed a problem. Even so, no government, no major ideology, no effective social or political power has been anti-science per se; none aimed at eliminating or destroying science or weakening the institutions that made it possible. Attacks on science, scientific institutions, or scientists have not been aimed at crippling the enterprise, but rather at controlling and manipulating it for the advancement of a political regime or in support of an ideology. Science in the modern world has been perceived instrumentally not only by statesmen, but by scientists themselves. This does not mean that scientific leaders made no claims for science other than instrumental ones. Scientific spokesmen have articulated an ideology of science which at one level focuses on the intrinsic value of knowledge. While the theologians of science have stressed the intrinsic values of science, the history of modern science, in effect, suggests that the movement has from the beginning leaned very heavily in the direction of an instrumentalism, characterized by an inversion of priorities whereby knowledge as power incarnate became its primary impetus and the disinterested search for knowledge became of secondary importance.

A strong *methodological ethic* and a weak *institutional ethic* have infused the scientific enterprise. The honesty and integrity of science about which so much is said refers to the factor of self-correctiveness inherent in the method of basic science—a method which makes it highly probable that the cheat, the sloppy worker will be found out. It does not pay to falsify scientific work, for the risk of discovery is very high. In addition to this self-correctiveness, the methodological work ethic is psychologically internalized. Few scientists would even consider falsifying their data, committing plagiarism or otherwise going counter to the methodological norms. Such internalization means that the policing function in science as it pertains to its knowledge output need be only minimal. For the methodological ethic is part of the inner logic, the intrinsic nature of science as a technique, as an approach to the discovery of knowledge. Eliminate this methodological ethic and science as we know it ceases to exist. However, it is an ethic which is applicable only to the way in which scientists go about doing their scientific work. In all other matters it is inapplicable, for it can make valid judgments only about those aspects to which its own procedure pertains.

There is no significant carryover of the methodological ethic into the institutional realm. Members of the community have in the past related to each other for the most part only in accordance with the canons of the methodological ethic. But this ethic has little bearing on the political, social, and interpersonal relations within science nor does it provide any guidelines for dealing with the external social and political order. Almost no guild, no community ethic has been developed. Until very recently, scientists simply did not concern themselves with the task of developing a binding institutional ethic for no major issues seemed to arise which would call for it. Moreover, the leaders of the modern state recognized very early the importance of science to the state, encouraged its growth (or at least showed no antipathy) and quite correctly perceived no likely danger from the possible collective action of scientists.

Conflicts when they arose between the political order and the scientific community were not the outcome of a basic animus by the state to science, but rather revolved around specific issues over the influence of leading scientists (Einstein in Germany, for example,); over the leadership of scientists whose political views were considered either unsympathetic or favorable to the political or ideological line of the regime in power (Planck in Germany, Lysenko in the Soviet Union); [25] or over the allocation of resources for science and within science. In recent decades the impingement of specific areas of research or fields of investigation into sensitive ideological positions has sometimes posed a problem. Both in National Socialist Germany and Stalinist Russia some of the biological sciences suffered from the impact of ideological restrictions. Since the beginning of modern science the degree of restriction has depended on the scope of intellectual freedom which was possible in a given social order. It does not appear to have been very much affected by an institutional ethic of science which would have obligated members to defend the community's autonomy or push for the extension of the boundaries of free inquiry in a hostile environment. Moreover, whether democracy and science are especially compatible is in any case an open question. Modern science, it should be noted, arose and flourished in autocratic and aristocratic political and social milieus; scientists in France and England were very creative before the advent of modern democracy.

Germany under the Bismarckian and Prussian hegemony became the leading science community in the 19th century. Many fields of science have flourished in the Soviet Union.

One further indication of the absence of an institutional ethic has been the pattern of prudential acquiescence. Scientists have almost always been pliant partners, willing under almost any conditions to accommodate to a given political order. All this suggests a major factor which contributed to the hiatus in perceived responsibility, the stunted feeling of community, the pronounced acquiescence which has characterized modern science. This is not to suggest that the problem of responsibility has simply been absent, but rather that on a conscious level it has until recently rarely been given serious attention. Most of the time consideration of it seems to have been driven underground, repressed, rationalized away. When the issue came up at all, it was projected onto the social order and defined away as of no concern for the scientist *qua* scientist. Moreover, since its beginning and throughout most of the history of modern science, the issue of responsibility simply did not arise in a fundamental sort of way. Large-scale assumption infused the social enterprise of science: science was considered intrinsically a boon; all scientific knowledge was thought in the long run to be desirable, hence extending the boundaries of knowledge was praiseworthy; scientific knowledge and its applications were expected to prove beneficial to mankind, therefore the more knowledge was accumulated, the greater the blessings that would follow. Reflected in the scientific temperament was the optimism of the Renaissance and the Enlightenment, which took for granted that progress was linear, that man was perfectable, that science would make possible a man-made Garden of Eden. Given these overriding assumptions, this deeply infused faith and optimism, there was no need to break one's head over the question of the social, political, or moral responsibility of science. An invisible hand was at work. Each scientist striving on his own, through following the norms of the methodological ethic, would become a partner in this historic fulfillment. If knowledge was intrinsically good, and knowledge was also power, it followed that the power derived from this knowledge would also be intrinsically good.

Experience has shown otherwise. As faith in inevitable progress eroded under the impact of twentieth century events—including two world wars and the rise of totalitarian regimes which used a science-based technology for morally abhorrent purposes—a facile optimism gave way to increasing pessimism. It hardly needs belaboring that the results of science have been a double-edged matter. Whether science is put to constructive or destructive, moral or immoral uses, it can now be clearly seen depends greatly upon the conscious choice of those human participants who are in a position to influence the outcome. Even so, the element of choice is still a limited one. Because science and its applications entail unforeseen consequences, these always present a problematic component and leave open the possibility of some undesirable effects which no one could anticipate. A reckless manipulation and despoiling of nature through the introduction of large-scale technological and scientific experimentation has begun to have visibly destructive effects. In some cases the vital ecological balance upon which man and the animal kingdom depend for their survival has begun to unhinge. The long-range consequences and impact of *any* scientific or technological innovation—be it atomic weapons testing, creation of a Van Allen belt, heart transplants—cannot be predicted with certainty. What has become apparent is that the very technology which modern science engendered, appears at this juncture of history more and more to threaten rather than enhance human welfare. It is almost as if the very Nature which Bacon and Descartes sought to conquer, has turned upon those who attempted to dominate and master her. Part of the difficulty undoubtedly arose from the hubris which characterized the Baconian and Cartesian scientist, his pronounced disinterest in first-order philosophical questions, and the superficial awareness of the social implications of his work. Only after the explosions at Hiroshima and Nagasaki did scientists begin to reconsider their traditional attitudes.

The metamorphosis of science into a professional, large-scale enterprise brought with it significant changes which demanded attention. Society provided the resources through which science could grow and flourish. It supported the educational facilities which trained its members, and allocated funds to scientific institutions which undergirded it. It made possible a fuller utilization of available scientific resources and the development of projected potentialities. Scientists functioned with the autonomy and prerogatives attached to a profession. But public support entailed reciprocal obligations. Science was seen as a national resource to be harnessed to national goals and policies. Massive governmental financing brought with it problems derived from the dependence upon this support.

As the scientist became a leading figure in society, some reservations about his role began to surface, especially toward atomic scientists who most clearly symbolized the heightened prestige of scientists in the post-Hiroshima age. Among segments of the public and public officials an ambivalent attitude toward the scientist was noted.[26] To be sure, his work enhanced human welfare. Yet much of it suggested the image of a latter-day magician who had unleashed forces in a world unprepared for their proper use, while he himself, like the sorcerer's apprentice, was incapable of controlling these forces adequately. Others saw in science the last best hope for saving man from impending disaster. Scientists would have to make men see the error of following the irrational and deranged ways of power politics and convert them to the rational and sane spirit embodied in the scientific approach.[27] For the most part, the prevalent attitude of the science community and its leaders has been entrepreneurial.[28] Society was making a heavy investment in science. Scientists would be held accountable for the way they used the resources put at their disposal. Their work would be judged primarily on the "pay-offs" which were attributable to the investment in science. Wherever governments and scientists are involved with issues of a national policy for science similar considerations prevail.[29]

Big science—large-scale institutionalization, a vastly growing membership, mushrooming specialization—makes accelerating demands upon limited resources. Not all the expectations of scientists for financial support can be satisfied. Nor can all meritorious research proposals receive it. Conflicts engendered by relative scarcity require criteria by and through which choices over the allocation of resources can be arrived at. Such choices occur in a context which involves the interests of both those who make the decisions and those who are affected by them. Scientists who actually make such decisions or who influence them will be thought responsible and accountable by those affected and concerned with the particular decision—by individual scientists, leaders of scientific organizations, public officials and others who believe they have a stake in the decision under scrutiny.

In the institutional life of science on matters of public policy the question of who is to be held responsible for a decision, by whom, and on what grounds is one that is usually left unexplored. Scientific influentials are usually co-opted by an incumbent leadership, chosen by a small group of influential scientists who themselves are not elected by the larger constituency which they represent. Who speaks and acts for science will not usually depend upon nor be greatly influenced by the mass of scientists. The leaders within the science establishment for the most part determine and make these choices unencumbered by the demands of scientists not in the elite group that they have a voice in these decisions. For the profound structural changes within science have had a negligible impact on the constitutions of scientific organizations or on the processes by which its leaders are chosen. These processes remain deeply elitist and oligarchic. They are based on a conception of the representative-leader as a *trustee*. As such he is not to be chosen by those whose interests he represents, but obligated to defend and further those interests as his conscience and judgment dictate. There is a distrust of the democratic electoral, more egalitarian process by which all members may participate in decisions which affect the whole community.

The oligarchic and elitist constitutional structure characteristic of most scientific organizations was defensible and very likely appropriate when science was small and involved only minimally in public affairs. However, in an age when scientific organizations and institutions have become very much involved in public affairs, the question of what properly constitutes representativeness, of who really speaks for science or for particular constituencies of science, is an important though largely neglected problem. It may well be that leaders within various science establishments often express the thinking and work on behalf of those whom they represent. Nonetheless, in the area of making decisions on public questions affecting the internal and external affairs of science, the scientific enterprise has not become more democratic. At a time when their leaders frequently speak for them on controversial public issues, these very leaders and spokesmen cannot in any direct way be held accountable by the great majority of scientists who make up their constituency. If responsibility entails accountability, then ways of making the institutional life of science more truly representative must be found.

If science has become or does become a (or *the*) leading institution of society, one may anticipate some concern with developing more articulate political theory which is applicable to the concerns of that enterprise. As the number of scientists, and the resources at their command, increases, so does the need to establish more elaborate and formal political practices and institutional devices to cope with the internal and external affairs of science.

There is cause for grave concern in all this. For, as this study has attempted to demonstrate, the history of modern science is predicated on a disjuncture between power and responsibility, between a potent method (praxis) and ultimate ends. It may be, as C. P. Snow put it, that scientists have had "the future in their bones." [30] But the question remains: What kind of future has been implicit in their work and in their conduct? Given the historical evidence, is increasing dependence on science and scientists likely to be an omen or ominous? Is that future to be welcomed?

To raise these questions, or to examine critically the response of scientists to the crises which have faced them, is not to express hostility toward science or scientists. There is no reason to think that the pattern described is fixed, or that the historical developments through which it was expressed are the outcome of any inexorable movement. If any "iron laws" govern the historical and institutional development of science, they have yet to be discovered. The past is influential and conditions the life of a community, but it is not determining; the future is never entirely open, choices are limited. It is precisely this possibility of choice—of avoiding the illusion that we may become complete masters of our fate or the conclusion that we are, in the last analysis, only victims—that accentuates one of the imperative tasks before us. Scientists and non-scientists must seek to understand the enterprise of science in more comprehensive ways, to recognize the larger problems that confront it (and us), to dare to examine these in terms of fundamental political categories and problems, and to relate it to the ultimate concerns of men.

NOTES

1. The news reports in *Science, Nature,* and *Minerva* over the last several years provide examples of this effort to protect interests or to change policy so as to enlarge the institutional resources available to science. Within limits, what is operative here is a phenomenon common to the rise of professionalism, namely the development of a guild mentality.
2. This was by and large true of scientists (like Szilard) who were active after 1945 in the scientists' movement against military dominance in the AEC, and of the Federation of American Scientists since. The grassroots politics of 1945–1947 tended to be directed at changing the views or votes of political influentials, mainly in Washington, and was in any case an *ad hoc* venture.
3. Note the 1964 Presidential election campaign, in which networks of scientists and engineers formed to campaign and use their influence on behalf of President Johnson. Cf. *Science,* CXLVI, 232f, 380f, 1440f, and 1561f.
4. There are several indications of this congealing professionalism. In disciplines which became professionalized (like chemistry, physics) the role of the amateur declined and in fact almost disappeared. Professionalism also created more structured training criteria, and a form of educational certification.
5. This is not the same as saying that scientists were callous toward their persecuted German colleagues. On the contrary, in England and the United States particularly, they—and the world of the universities in general—provided great assistance, and, indeed, made it possible for most who escaped the National Socialists to re-establish themselves. These impressive achievements are described in Norman Bentwich, *The Rescue and Achievement of Refugee Scholars* (The Hague: Mortimus Nijhoff, 1953); Helga Press, *Die Deutsche Akademische Emigration, op. cit.;* and Radio Bremen, *Auszug des Geistes.*
6. Bacon, "New Atlantis," *Works,* III, p. 288 (Classics Club Ed.). In this connection, we may recall Oppenheimer's attitude about having no choice but to proceed with scientific work which promises to be successful or "technically sweet."

7. This view, which began to gather momentum during and after World War II, is now almost universally held, particularly in highly industrialized, technological societies and in underdeveloped countries which seek to change their situation. The literature which supports this argument is now massive; representative of such thinking in the United States are: Bush, *Science, The Endless Frontier,* the 1960 reprint (Washington, D.C.: The National Science Foundation) contains a stimulating introduction by Alan T. Waterman; House Committee on Select Research, *Hearings, Science Research and Development Programs;* and House Committee on Science and Astronautics, *Hearings, Government and Science,* 88th Congress, first and second sessions, 1963–64, see esp Vol. III, Summary of Hearings, pp. 1199ff.; National Academy of Sciences, *Federal Support of Basic Research in Institutions of Higher Learning* (Washington, D.C.: 1964); *Daedalus,* XCI (Spring, 1962), contains a series of stimulating and pertinent essays on the theme "Science and Technology in Contemporary Society." A perusal of the content and editorials of *Science* during the last decade, or of any other publication directed at a general public of scientists *(Nature, The New Scientist, Bulletin of the Atomic Scientists, Minerva),* will also demonstrate this.
8. *Science and the Modern World,* p. 17. Pascal is a notable exception.
9. *Ibid.* see also pp. 141–43 and *passim.*
10. Oppenheimer's behavior during and after his hearing, as well as the statements and activities of German scientists under National Socialism, were based (in part) upon this conviction.
11. In a letter dated July 4, 1949 to Max Brod, author of the historical novel, *Galileo in Gefangenheit* (Galileo's Imprisonment), quoted in Carl Seelig, *Albert Einstein,* p. 210. Since Einstein made this point on several other occasions, we may assume that it was not merely an afterthought. Cf. Reichinstein, *Albert Einstein,* p. 33 and Vallentin, *The Drama of Albert Einstein,* p. 15 Einstein wrote this after his experiences in Germany. In view of his bitter reproaches of German intellectuals in the 1930's for not defending their scientific colleagues, it is hard to fathom. Significantly, he pointed to intellectuals, and in his public statements, including those directed at the Prussian and Bavarian Academies of Science, never used the word scientists.
12. Max Born, "Recollections of Max Born: How I Became a Physicist," *Bulletin of the Atomic Scientists,* Vol. XXI, No. 7 (September 1965), p. 4.
13. Niccolò Machiavelli, *The Prince,* Chapter XV, in *The Prince and the Discourses* (New York: Random House, Modern College Library Edition, 1950), p. 56. For amplification of the view that ethics and political action are inherently in conflict cf. Hans J. Morgenthau, *Scientific Man vs. Power Politics* (Chicago: The University of Chicago Press, 1946), Reinhold Niebuhr, *Moral Man and Immoral Society* (New York: Charles Scribner's Sons, 1932), and Sheldon S. Wolin, *Politics and Vision,* (Boston: Little, Brown and Company, 1960), pp. 224–228.
14. Contemporary American and British philosophy has treated the problem of responsibility primarily in terms of a linguistic and positivist analysis, which confines the subject to very narrow boundaries, making it irrelevant to all but a small group of professional philosophers. For examples of this approach see T. D. Weldon, *States and Morals* (London: J. Murry, 1946), R. M. Hare, *The Language of Morals* (New York: Oxford University Press, 1952), and Sidney Hook (editor) *Determinism and Freedom in the Age of Modern Science* (New York: New York University Press, 1958).
15. The cluster of meanings is illustrated by the definition of any standard dictionary. See, for example, that of *The American College Dictionary* (New York: Random House, 1966), p. 1034, which provides the following meanings for the concept "responsible":

1. answerable or accountable as for something within one's power
2. involving accountability or responsibility: *a responsible position*
3. chargeable with being the author, cause of occasion of something (followed by *for*)

4. having a capacity for moral decisions and therefore accountable; capable of rational thought or action
5. able to discharge obligations or pay debts
6. reliable in business or other dealings, showing reliability.

16. Eric Hutchinson, "Science and Responsibility," *American Scientist,* Vol. 52, No. 1 (March 1964), p. 42A.
17. P. W. Bridgman, "Scientists and Social Responsibility," *The Bulletin of the Atomic Scientists,* Vol. 4, No. 3 (March 1948), pp. 69–72.
18. *Ibid.,* p. 70.
19. See the rejoinder by Harold Urey, Lee A. DuBridge, and Eugene Rabinowitch in "Comments on Dr. Bridgman's Article," *Bulletin of the Atomic Scientists,* Vol. 4, No. 3 (March 1948), pp. 72–75. I. I. Rabi agreed with Bridgman and concluded: "Our real social responsibility, and it is enough to load any man to his full capacity, is to do good, sound, honest science and to publish the results as clearly and objectively as we know how. Beyond that we enter the shadow land of dilletantism." *Ibid.,* p. 72.
20. One indication of the rapidly growing interest in the impact of science and technology on human affairs and on public policy is the number of publications which now deal on a regular basis with one aspect or another of this interaction. Over 100 such English language serials are listed in *A Guide to Serial Publications Reporting on Science Affairs,* published by The Institute for the Study of Science in Human Affairs, Columbia University [ISHA Bulletin No. 2, October 1968]. For some perceptive studies on the policy aspects of this development, see Daniel S. Greenberg, *The Politics of Pure Science* (New York: The American Library, 1967); Alvin M. Weinberg, *Reflections on Big Science* (Cambridge, Mass.: The M.I.T. Press, 1967); Ralph E. Lapp, *The New Priesthood* (New York: Harper and Row, 1965).
21. John J. Beer and W. David Lewis, "Aspects of the Professionalization of Science," *Daedalus,* Vol. 92, No. 4 (Fall 1963), pp. 764–784.
22. Everett C. Hughes, "Professions," *Daedalus,* Vol. 92, No. 4 (Fall 1963), p. 655.
23. Norbert Wiener, 'A Scientist Rebels," *Bulletin of the Atomic Scientists,* Vol. 3, No. 1 (January 1947), p. 31.
24. "Declaration of the German Nuclear Physicists," *Bulletin of the Atomic Scientists,* Vol. 13, No. 6 (June 1957), p. 228. Italics added. For a discussion of the background to this manifesto, see Carl Friedrich von Weizäcker, *Die Verantwortung der Wissenschaft im Atomalter* (Göttingen: Vanderhoeck & Ruprecht, 1957), pp. 16–30.
25. On the Lysenko controversy see *The Situation in Biological Sciences* Verbatim Report of the Proceedings of the Lenin Academy of Agricultural Sciences of the U.S.S.R., July 31–August 7, 1948 (Moscow: Foreign Languages Publishing House, 1949); Julian Sorell Huxley, *East and West: Lysenko and World Science* (New York: H. Schuman, 1949); Conway Zirkle (ed.), *Death of a Science in Russia* (Philadelphia: University of Pennsylvania Press, 1949); Pamela N. Wrinch, "Science and Politics in the U.S.S.R.: The Genetics Debate," *World Politics,* Vol. 3, No. 4 (July 1951), pp. 486–519.
26. See Harry S. Hall, "Scientists and Politicians," *Bulletin of the Atomic Scientists,* Vol. 12 (1956), pp. 46–52; Walter Hirsch, "The Image of the Scientist in Science Fiction: A Content Analysis," *American Journal of Sociology,* Vol. 63 (1958), pp. 506–512; Stephen B. Withey, "Public Opinion about Science and Scientists," *Public Opinion Quarterly,* Vol. 23 (1959), pp. 382–388.
27. *Supra,* Chapter 10, pp. 255.
28. See, for example, the papers of the Panel on Basic Research and National Goals, National Academy of Science, *Basic Research and National Goals* (Washington, D.C.: U.S.G.P.O., March 1965). This was the Academy's Report to the House Committee on Science and Astronautics.
29. Belief in the utility of science does not exclude a belief in its intrinsic value. What I am suggesting is that massive social support for science flows preponderantly from the expectation that many of its findings will eventually result in social and technological applications.

30. C. P. Snow, *The Two Cultures and Second Look* (Mentor ed.; New York: New American Library, 1964), p. 16. See also his *Science and Government.* Examination and anlysis of Snow's thesis (which would require too extensive a treatment to be undertaken here) would reinforce these arguments. That Snow's work was favorably received by the scientific community seems more a response to his obvious preference for the scientific over the literary culture than to his profound analysis. To be sure, he raises significant questions, but their treatment appears to me superficial in almost every respect. He avoids fundamental problems, except as transformed into primarily technological concerns (e.g., how to bridge the economic, scientific, and technological gap between the have and have-not countries in the world). Moreover, the thrust of the history related in *Science and Government* goes counter to the message with which it ends and which is the heart of *The Two Cultures.*

READING 22

The Role of Science Policy in Solving Social Problems

Staffan Burenstam Linder

THE UNBALANCED PROGRESS OF PROGRESS

The concept of science policy may be new and fashionable but the problem is as old as science itself. To allocate resources between competing projects and to determine the total size of the scientific effort is a science policy problem which those interested in scientific work have always been exposed to. Columbus, in trying to get funds to pay his way across the ocean, encountered this problem of science policy, if not explicitly so at least implicitly.

But, in the old days there was a more deep-going problem in science policy than the money matter. This was the problem of how to choose between truth and dogma when the two collided. If Columbus encountered the menial aspect of science policy Copernicus met the old, fundamental problem of science policy. At that time the struggle between the two sets of ideals must have caused severe social strains and the way the conflict was resolved has wide social consequence. The results of science have transformed society but already the acceptance of the method of science made deep changes in religious and political life. Once the right of scientists of pursuing the truth has been accepted, the freedoms of thought and speech which we consider basic to our culture are more secure. The strength behind the scientific ideals is shown by the fact that in countries where the liberal freedoms are sometimes denied, few efforts are nonetheless made to tamper with academic freedom. The difficulty of doing the latter probably restrains sinister wishes to do the former. However, even if the threat to the ideals of scientific method have not completely disap-

Reprinted by permission of the U.S. Government Printing Office from *International Science Policy,* Committee on Science and Astronautics, U.S. House of Representatives, 1971, pp. 133–141.

peared and may present themselves again, this is not what we think of when we nowadays discuss science policy.

Once scientific efforts were not only accepted but unhesitatingly promoted as an engine of progress, science policy did not cause much attention. The money matter, of course, existed but at a time when thought was speculative rather than empirical science was also inexpensive. It is only during the last ten years that science policy has begun to attract wide attention again. The reason for this was first that the costs of research, at least within the natural sciences, rose rapidly at the same time as it became more widely recognized that technical progress was the force behind the spectacular economic growth. Options became wider and costs higher. Thus, the problem of promoting research without incurring prohibitive costs and of allocating resources between competing projects became more important.

However, even if money problems form a part of science policy more now than before, the chief reason for the present attention given science policy lies elsewhere. It is a fundamental problem that again has risen although now in a different shape.

Now it is not truth versus dogma but truth versus truth or, more clearly, truth versus abuse of truth.

The problem is not that truth may destroy dogmas but that truth may destroy itself.

That science is no guarantee of happiness may be accepted. And that economic growth built on the results of science and technology is no guarantee of happiness may also be accepted although perhaps more reluctantly. But, what some have come to suspect during recent years is that, actually, the causation runs the opposite way. Science, technology, and economic growth may, instead of not guaranteeing happiness, actually guarantee unhappiness.

This suspicion is of great social consequence. The idea of progress is the basis of our civilization since a number of centuries. It is since the Enlightenment the foundation of our social contract. It offers the hope which is required for the acceptance, the temporary acceptance, of the unavoidable ills of an unfinished society that is built on a principle that does not accept any ills. This is the principle of egalitarianism in the sense of happiness for all. I think that everyone who has been involved in practical politics is aware of the importance of the idea of progress as the escape when, unavoidably, hard choices present themselves.

If the current suspicions are well-founded it may prove that their causes, through the abuse of truth and science, will lead to the destruction of our civilization. The suspicions themselves may cause defeatism and even a reaction against science and technology that may completely change the direction and position of our civilization. There may be a Luddite reaction against research.

These problems are social problems which call for adequate science policies. They require measures of an economic and social kind, too, but for continued progress engineered through further scientific and technological advances these advances must be shaped through science policy in a form which prevents social strains.

In order to see more clearly what is required it is interesting to analyze the reasons for the present disbelief in science. I think that it is possible to arrange such an analysis around the theme that *the problem of progress is that progress has progressed in an unbalanced way.* I shall point at five different kinds of such unbalances.

To begin with I shall repeat the often made observation that there is an unbalance between the progress of science and of man. There is no particular reason why the scientific efforts should have resulted in any moral advancement. However, for the results of science not to be abused, such an advancement is called for. Presently, it is not magnanimity and greatness but fear that provides uncertain checks. Even if that fear is needed for restraint, it is a grinding feeling to live with.

Secondly, there is an unbalance between technological progress and the capacity of man to adjust to the demands of the new technology. New techniques of production are intended to be our servant but sometimes they become our master. For the utilization of new techniques certain capacities not available to all are required. For instance various statistics indicate that life in the big urban centers exposes people to unexpected hardships.

In modern society there seem to be growing minority groups which in one way or another have failed to adjust to the new demands. To this category belong criminals, narcotics, alcoholics and people otherwise socially and mentally handicapped. These minorities constitute serious social problems, class problems of a non-Marxian kind. Their situation is aggravated by the fact that costs for rehabilitation and care are rising with rising wages, these services being highly labor intensive requiring skilled labor. The minorities of this kind are exposed to a new kind of poverty the background of which is not a low level of productivity but inability both to adjust to the demands of a high productivity system and to obtain adequate services in such an economy.

Against the background of rising service costs we can also see the increasing difficulties for medical care and old age care.

The third and fourth type of unbalance also concerns new types of poverty caused by economic growth. There is a dangerous misunderstanding that economic growth leads to general affluence. It only leads to partial affluence. It is the quantity of goods that increases but many other ingredients in a good life do not become more plentiful. One important example of this is that time at our disposal for various activities together with our consumption goods does not increase. We only have our 24 hours per day at our disposal. When, as a result of economic growth, the quantity of goods increases whereas the amount of time is fixed, time becomes more scarce in relation to goods. Time becomes more expensive. It must be more carefully husbanded. We all know how this is done in production where more capital goods are combined with each hour of labor to yield a higher productivity. The same happens in consumption. Greater quantities of goods are combined with each hour of leisure to yield a higher level of satisfaction. Since activities differ as to the ease with which it is possible in this way to increase the pleasure associated with them, there will be a reallocation of time among the various activities. Some will expand, others will decline. Activities which it is easy to make more pleasurable with an additional amount of consumption goods per unit of time will expand, and vice versa. This reallocation will rise to important social changes, changes which at least some commentors would call problematic. Most of those activities which the original economic philosophers like John Stuart Mill considered especially desirable and which they hoped and thought that economic growth would stimulate are activities, the pleasures of which cannot easily be expanded with bigger and bigger injections of consumption goods. To these activities belong reflection and meditation, most of the cultural activities, friendship and love, peace and quiet.

It is already pointed out the increasing scarcity of time which makes it difficult to solve the rehabilitation problems of the minorities which have been unable to adjust to the high productivity economy. It may also be that the strains of the hecticness of modern life is one reason why some people have difficulties of adjusting and fall out into the minorities.

The fourth unbalance is perhaps the most dangerous of the new poverty. Economic growth leads to an influence of goods but to a scarcity of natural resources like air and water. The supply of natural resources decreases and the demand for natural resources increases as a result of production and consumption. When, through economic growth, production and consumption actually increase, the scarcity of natural resources becomes even more pronounced. As a result of this unbalance we face all the environmental problems, the threat of which causes serious social problems.

There may be an erosion not only of the soul but also of the spirit.

A fifth unbalance is caused by the uneven geographic spread of the benefits of modern science and technology. They have not reached the underdeveloped countries and the depressed areas and poor regions in the rich countries. Thus, it is not only the new forms of poverty but also traditional poverty that causes social strains.

This is no complete catalogue of the social problems caused by the unbalanced progress of progress. It is an analysis of some of the mechanisms and the resulting troubles, which, in the judgment of many, have lowered the prestige that science deserves for its great contributions. To help to correct the unbalances we require a science policy. Thus, science policy again takes on the importance it originally had but which it lost when the conflict between truth and dogma had been settled. There are three main strategies that may be pursued:

(*a*) Continuation of present policies.
(*b*) Prevention of scientific and technological progress.
(c) Adoption of a cautious, but well financed, program of technological development.

A continuation of present policies is dangerous. The unbalances are not—as could be hoped—self-correcting in a way that unbalances in the market system are. At least the self-correcting forces are not working fast enough to be relied on. There are no supply-demand forces operating smoothly through a price mechanism to correct unbalances as had an invisible hand been active. In science policy the hand must be visible.

The second strategy—to let the hand stop the clock of progress —has its advocates at least within economics where there has been some talk of the "no-growth-economy". It may also be noted that funds for science during the last few years have stagnated or even declined. However, to accept a science policy which would yield a "no-science-society" would not be acceptable. It would leave uncorrected some of the unbalances which exist. It would mean that other unbalances would, actually, increase. The dangerous new poverty in the form of environmental problems would deteriorate as even a "no-growth-economy" would lead to a continued wearing down of natural resources. Finally, to accept suddenly a no-progress-society would represent a dramatic change in our basic attitudes and aspirations, a change for which we are not prepared. It would probably cause severe political strain as so many political conflicts are solved through promises of betterment and outward signs that the promises can be at least partly fulfilled.

Instead, our conclusions must be that we require science, and require it badly, to solve the problems that the application of science has caused. However, this science must not result in new problems bigger than those solved. There is in the progress of science the inner elements of tragedy: science has created forces that can be controlled only through more science but this new science may impose its own threats. To avoid a final tragedy we need a science policy and that science policy must follow the third strategy of a cautious, but well-financed, progress of technological development. Such a science policy can be divided into four parts.

(A) Determination of the Total R. & D. Effort and Allocation of Funds between Competing Projects

This is the traditional resource problem which has become more pressing due to the wider options and higher costs. There is a need for policies to manage science as such both within a country and through international cooperation to get the highest possible returns. However, in the final analysis other considerations must now enter. There must, as we have seen, be an evaluation of the wider social consequences. We require what we can call a "management of insights" on the following three levels influencing the final allocation of resources.

(B) Regulation

As a result of the dangerous side effects of many of the results of science it is important, as part of a science policy, to determine if and how these results can be controlled. The control of atomic power both in a military and civilian context may be mentioned as an example. The possibilities of exerting such control must be taken into account in determining whether to promote or not promote technological development in particular fields.

(C) Forecasting

Through technological forecasting which is part of the modern future studies it is possible to make informed guesses of prospects and problems. Such forecasting can improve the quality of the allocation of funds between competing projects. It can do this especially by serving as the starting point for technological assessment.

(D) Assessment

This is the part of science policy which is the most discussed recently. Assessment goes beyond mere forecasting in trying to evaluate systematically the advantages and the risks of new technologies. The assessment can be either of a certain technique or of the possibilities of solving a certain problem through alternative strategies using different technologies.

The analysis of the various unbalances of progress serves not only to illustrate the need for a science policy but also to facilitate comment both on the possibilities and on the limitations of science policy.

There are certainly possibilities for science policy to help us to remain within a context of real progress. It should be possible to find ways of relieving the problem of traditional poverty, underdevelopment, due to an uneven spread of technology. Through a determined effort technologies suitable for the conditions of underdeveloped countries could certainly be promoted. The underdeveloped countries form what has been called a "research desert" and since our economies have blossomed through technology, their economies are unlikely to grow sufficiently fast in such a desert.

Similarly, it would be surprising if important results could not be reached through a careful assessment both of new technology from the point of view of its ecological consequences and of environmental problems and alternative technologies to handle them. It would be surprising since so little has been done in this direction and since in this case technologies interact with nature rather than with the human soul. Thus, it should be possible to advance through a utilization of the methods and arguments of the natural sciences themselves. In view of the seriousness of the environmental problems science policy would be called for even if it could not do anything more than to relieve these pressures and the social strains that follow from a powerless watching of the thickening clouds.

But that there are limitations of science policy should not be overlooked in the present enthusiasm for such policies. Science policy cannot do much to correct the unbalance and pressure which arise when belligerent and antagonistic man is better and better equipped for annihilation and only grinding fear can provide checks. Hopefully, peace and conflict research in an assessment could be found to provide insights into how a more enjoyable foundation for peace than fear could be found. Some consolation could also be found from the fact that the superpowers in an evaluation have reached the conclusion that some defense research is associated with dangers for all parties concerned. Thus, they have shown signs of stopping such research. Such a sign is the U.S. declaration of not wanting to utilize biological weapons even for defense. Research on such weapons carries the risk that methods for their production could become simple enough for a multitude of countries to learn them. However, on the whole, it is rather a change of heart than a change in science policy that is required to correct the unbalance between technology and the inner qualities of man.

A change in values rather than in science policy would also be required for more people to avoid a stressing pattern of life where economic growth is permitted to result in a hectic utilization of time in commodity intensive activities. No science policy can correct this kind of imbalance even if scientific explorations of modern ways of living to more people may reveal some of its futilities.

Opponents to science policy may argue that not only does such a policy leave many problems unsolved but it creates new problems. There may, in particular, be a fear that science policy would mean a manipulation for political purposes. However, modern science policy is not concerned with the old problem of the conflict between truth and dogma. In fact, science policy may be required to avoid political manipulation through the results of technology. The risk presently is not so much that certain projects will be stopped for partisan purposes but that certain projects will be promoted for such ends. A science policy could permit a more open debate on the disadvantages and merits of the policies chosen.

To reduce the risks for political abuse basic research should not be exposed to more than unavoidable allocation process of scarce resources. Furthermore, technological assessment need not be applied to the social sciences. Social research hardly results in new technologies of a dangerous sort. Thus, there is no need for control and assessment of its result.

Writings within the social sciences may instead be politically controversial and a system for interference could be abused. Furthermore, research within the social sciences is inexpensive which means that careful evaluation for appropriate allocation of money is not as necessary.

Finally, a limitation on science policy is that it is easy to say but difficult to make. Dr. Harvey Brooks has rightly pointed out that "many of the current demands for better scientific planning are probably as naive as the early demands for economic planning".

One limitation exists for small countries. As in the economic context, they will be highly dependent on what happens elsewhere. They have difficulties in making a wise allocation of resources in isolation. They do not have resources to make all around technological assessments. Technologies developed elsewhere become, as a rule, part of their system without any particular evaluation. For the small countries to pursue a science policy it is necessary to engage in international cooperation. Only the U.S. has resources wide enough to rely on a national science policy. However, even the U.S. must in many instances fall back on international cooperation and will be influenced by actions taken in other countries. One field where world wide cooperation is required is in the control of the environmental hazards.

A limitation of a different kind arises out of uncertainties. No matter how carefully science policy evaluations are made they must be carried out against the background of great uncertainties about the future. But, there are also uncertainties in the present. Through improved methods for cost-benefit analysis it may be possible to base decisions on firmer ground. Yet, it is extremely difficult to attach weights to the advantages of competing technologies and strategies.

The same observation could perhaps be made with respect to most political decisions. However, correct decisions in science policy requires more expertise than is generally found among parliamentarians who must have many other strings on their bows. The more complicated the world and the more important the science policies, the more important that politicians shift some of their interest over to an understanding of the technical complexities and that they are assisted in so doing through institutional arrangements facilitating their contact with technological problems.

The uncertain situation of politicians in the formation of wise science policy is a difficulty. An even greater difficulty arises when and if attempts are made to introduce a greater element of direct democracy in the formation of science policy. This is a dilemma. The difficulty of the layman to participate in science policy discussions is equal to the necessity of widening this debate. Otherwise it may be impossible to obtain a political basis for many decisions that must be taken. Immediate small benefits may in the popular debate loom larger than long run imperatives. In various ways—for instance, through an improvement of general education—such difficulties may be reduced. However, their unavoidable existence puts an even greater responsibility on politicians to show leadership and demonstrate political will in the search for wise science policies.

Against this background the Committee on Science and Astronautics is to be praised for its many important initiatives in stimulating debate of science policy matters.

READING 23

Social Control of Science and Technology

Michael S. Baram

Science and technology increasingly work changes in the complex matrix of society. These changes pervade our ecological systems and our physical and psychic health. Less perceptibly, they pervade our culture, values, and value-based institutions, such as the law. In turn, our values and institutions shape the progress and utilization of science and technology.

Science and technology have provided society with enormous material benefits and a higher standard of living and health. Yet these benefits have been accompanied by alarming rates of resource consumption and new hazards to ecological systems and health. Social response to these unexpected problems has been of a remedial nature—that is, how to diminish pollution through regulation and technology. However, since our values and institutions shape the progress and use of science and technology, the fundamental social response must come from change in these values and institutions. To the extent possible, this change should yield preventive or a priori controls.

This important task can be described as the need to formulate coherent and humane social controls on science and technology.

Of course science and technology are not discrete activities: They describe a process that ranges from basic research through applied research and development technology to application and use technologies. Most social change occurs during the latter stages, in which technology is manifested either in specific acts, such as organ transplantation techniques, or as part of a major public or private system, such as nuclear energy or computer applications.

Reprinted by permission of the author and *Science,* Vol. 172, 7 May 1971, copyright 1971 by the American Association for the Advancement of Science, pp. 535–539.

Events throughout this process have become highly dependent on federal funds since World War II. In 1969, approximately 65 percent of the funds spent in the United States on basic and applied research and development technology was provided by federal agencies. This reliance on federal support provides even further justification for public interest in the social control of science and technology.

The most substantial expenditures and investments occur during the development technology stage, after a number of important decisions have been made about prototypes, production, application, and use. Of the approximately $17 billion of federal support for research and development in 1969, it is estimated that $5 billion went for research and $12 billion for development. Production and application activities undoubtedly involved billions more. Similar ratios prevail in the private sector.

These investments must be considered in human as well as economic terms, for it is during the development and subsequent stages that large numbers of engineers, administrators, managers, production and shop personnel, salesmen, and subcontractors commit their careers, personal values, and families—and ultimately their communities—to the specific technological activity or system. Therefore, all subsequent social controls must consider the political, economic, and human factors that have been developed.

Numerous social controls on institutions and individuals generating or utilizing science and technology have been developed over the years. Table 1 suggests, in general terms, what these controls are and where they function in the various stages of science and technology. The legal doctrines in the table all operate during the advanced technology stages—after decisions committing economic and human resources have been made and, normally, after injury has occurred. By this time, fully developed systems and practices are in use, without coherent controls.

This has led to condemnation of law as a modern system of control. As Jacques Ellul has said (1):

> The judicial regime is simply not adapted to technical civilization, and this is one of the causes of its inefficiency and of the ever greater contempt felt toward it.
>
> Law is conceived as a function of a traditional society. It has not registered the essential transformation of the times. Its content is exactly what it was three centuries ago. It has experienced only a few fragmentary transformations (such as the corporation)—no other attempts at modernization have been made. Nor have form and methods varied any more than content. Judicial technique has been little affected by the techniques that surround us today; had it been, it might have gained much in speed and flexibility.
>
> Faced with this importance of the law, society passes to the oppo-

> site extreme and burdens administration with everything that is the product of the times in the judicial sphere. Administration, because it is better adapted from the technical point of view, continually enlarges its sphere at the expense of the judicial, which remains centered on vanishing problems such as codicils, community reversions, and the like. These last, and all similar problems that are the exclusive concern of our law, are problems that relate to an individualistic society of private property, political stability, and judicial subtlety.

In specific terms, the legal system has not been responsive to new social conditions. For example, it has not functioned as an effective control on science and technology because it does not operate early enough in the process. Harold Green, in discussing this issue, has said (2):

> The basic question is whether our legal system is capable of imposing effective social control over new technologies before they inflict very substantial, or even irreparable, injury upon society. It seems clear that we cannot rely on the courts alone to protect society against fast-moving technological developments. Judge-made rules of law always come after, and usually long after, the potential for injury has been demonstrated. . . .

This characteristic of retroactivity limits the ability of the legal system to respond to a number of modern social problems, in particular the harmful effects of science and technology and the problem of environmental deterioration. Retroactivity is inherent in a legal system based on the values and conflicts of the private sector of society. The courts have not been designed to serve as oracles or social planners, but to grapple with actual conflict manifested in specific acts or injuries. They lack the technical, astrological, or other expertise needed for the difficult task of evaluating the present, diffuse effects or the future effects of science and technology. Consequently, the courts are reluctant to impose controls and have rarely intruded on the substantive aspects of decisions of public agencies, which presumably are technically expert.

Judicial procedures that have reinforced concepts of justice and due process, such as statutes of limitations and rules of evidence and standing, have also brought an immobility to the law to the extent that it cannot respond easily to such issues as deleterious damage or public health.

Recent developments in environmental litigation have ameliorated some of these procedural obstacles, particularly the issue of standing for citizens' groups alleging other than economic injuries. But some feel that this brief honeymoon is already over. In Sierra Club v. Hickel (3), the Ninth Circuit Court of Appeals denied that the Sierra Club had standing, since it had not alleged

TABLE 1. *Where social controls occur in science and technology.*

Sources of control	Basic science	Applied Science	Development technology	Production, application, and use technology
Scientific peer groups	X	X		
Professional associations				X
Federal government				X
Executive action		X	X	X
Agency programs	X	X	X	X
Agency regulation	X	X	X	X
Agency security classification	X	X	X	X
Congressional hearings		X	X	X
Congressional legislation and funding	X	X	X	X
Industry-consumer markets			X	X
Industrial associations and labor unions			X	X
Insurance				X
Crusaders and citizens' groups			X	X
Law				
Patents, copyrights, trade secrets			X	X
Torts				X
Constitutional rights				X
Land use			X	X
Consumer protection				X
Experimentation		X	X	
Education-ethics	X	X	X	X

that its members would be affected, beyond displeasure, by the scheduled action of the Department of the Interior. (This action was the approval of a commercial and recreational development, in the heart of Yosemite National Park, to be carried out by the Walt Disney Corporation.) This may indicate that the bounds of procedural flexibility have been reached.

The list of problems is incomplete, but sufficient to justify the conclusion of a recent law review note: "The passive nature of the courts and the difficulties encountered in their use make it clear that they cannot serve as society's primary instrument for technology assessment" (4).

To return to Table 1, the controls of the private sector are similarly clustered in the advanced technology and use stages. For obvious reasons, industrial decisions and insurance controls are implemented without full consideration for the public interest.

Decisions are made on market or profit considerations, based on what the consumer wants or can be manipulated to want, and do not consider larger public interests in the preservation of natural resources or public health, for example. Advertisements boost the sales of items that are attractive to individual consumers, but that collectively erode environmental quality, other public interests, and, ultimately, private interests. Sales of snowmobiles to the new breed of armchair sportsmen have climbed to 500,000 annually and provide a noisome case in point.

The automobile represents the ultimate absurdity. The automobile birthrate is now treble the human birthrate in the United States: 10 million automobiles are produced for every 3 million human beings. Death rates occur in a similar ratio. Automobiles produce most of our air pollution, are dangerously designed, and are not economically recycled. How much longer can these absurd ratios and harmful effects be tolerated, despite the importance of the industry to the economy? Obviously, many of our problems labeled technology-induced or environmental are, in reality, the behavioral problems of a materialistic society. As such, we cannot expect effective private sector controls to emerge, nor can we expect courts to alter such "normal" behavior.

Crusaders and citizens' groups have recently proven somewhat effective as technology-curbers, but they have not provided coherent, a priori controls. Crusaders are in short supply, and citizens' groups lack funds, technical expertise, and national political strength. They can only attack discrete problems, often on a local scale, and must ultimately resort to the legal system with its shortcomings. Their task is made extremely difficult by the fact that, once again, substantial economic and human commitments have already been made in support of harmful developments, on a scale far larger than the immediate interests represented by such groups. Without substantial evidence of harm to public health, such groups appear to represent merely their own esthetic or otherwise elitist values, or a Luddite revival. This is not said to disparage such activities: They have served to educate and involve citizens, and they represent an exciting and valuable development.

The public agencies have actual and potential social control functions that cover the complete spectrum of scientific and technological activities. But this role is inextricably wound up with their several other functions, which include the promotion of certain activities for national purposes such as defense or the balance of payments. Reasons for their failure to exert social control have often been cited and are true to varying degrees: bureaucracy and inertia, ignorance and lack of sensitivity to noneconomic interests, fragmentation of authority by congressional design or by subsequent developments.

Legislation has proven to be no guarantee of implementation. The Refuse Act section of the 1899 Rivers and Harbors Act is a potentially powerful source of authority for combating most forms of water pollution as they occur. Yet for 70 years it had been ignored by the Corps of Engineers and the Department of Justice.

The idea of reorganizing the federal agencies or creating new administrative bodies to better control science and technology has been under discussion for some time. Under this rational measure, one or several new and prescient groups would function as long-range planners with coherent control authority. For example, a single agency could, perhaps, determine national and regional energy needs and then plan, license for construction, and regulate in the public interest more effectively than the present multiple-agency condition. Reliance on teams of technical experts and experts from such other fields as law, health, and economics could be built into reorganization plans of this kind.

These are certainly steps in the right direction. Of our present array of social controls, perhaps the public agencies, which support most research and development, could effectively perform assessment and control functions when they are most important—before large-scale development and the commitment of economic and human resources.

Hugh Folk, in considering present and future social control by public agencies, has already discerned some pragmatic problems (5). Experts will once again be drawn from the same pool. Many of them will actually have contributed, in industry or government, to the problems they will be called upon to solve. Few experts will be able to apply their disciplinary background to a wider range of social issues. And experts will need extraordinary courage to function in a truly critical sense, since their careers will still be rooted in the same industrial-governmental university milieu. What will happen to the expert who tries to serve the public interest by calling for a halt to a particular line of research? A test case is now before us involving radiation safety standards. John Goffman and Arthur Tamplin have challenged the Atomic Energy Commission and its affiliates in industry and the universities.

Folk's central thesis must be repeated here: assessment and control are essentially policy-making processes and, as such, will be embroiled in political controversy. He fears the repetition of the nonrational policy-making that occurs in our present agency framework and that results in agency establishment of "standards at levels slightly below that at which people complain vigorously

. . . thus keep[ing] the public sullen but not mutinous." Designs for central assessment and control authorities must meet these issues squarely if real change is to occur.

Finally, let us briefly consider peer groups, well positioned to assess and control early in the basic and applied science stages.

Based on personal observation, in part, I do not think scientific peer groups presently have the objectivity or capability to function as coherent and humane social controls. The members of a peer group share the narrow confines of their discipline, and individual success is measured by the degree to which one plunges more deeply into and more narrowly draws the bounds of his research. There are no peer group rewards for activities or perceptions that extend beyond the discipline or relate it to social problems. Members are therefore neither motivated nor trained to relate their peer group activity to broader social concerns. Probably because of their closeness and commitment to their work, they are unable to objectively assess implications and recommend controls.

Genetic research today provides us with a case in point. It is proceeding rapidly in the United States and England, and, periodically, significant breakthroughs are announced. Members of the peer group and others have frequently discussed the potential applications of their work, and it has become a fashionable topic. Despite the potential for genetic engineering and its misuse for political and social goals repugnant to our professed values, this work continues at an urgent pace. I would think that the historical evidence of the political misuse of science and technology in this century would at least bring about a slight pause or slowdown in activities until our legal and other control systems had time to prepare principles regarding experimentation, as well as other public and private safeguards.

It is a disturbing experience to discuss these issues with biologists. Their responses avoid the central issue of slowing or suspending work to formulate controls and include the following:

- "If we don't do it, somebody else will";
- "Don't worry about secret and horrible developments—all work is done in large, expensive labs funded by the government";
- "Further work will improve the health of society and upgrade the gene pool";
- "Cloning of humans is at least 5 [or 10] years away";
- "Science is intrinsically valuable in its contribution to man's collective knowledge, and it must not be controlled for social purposes of any sort."

Self-enclosed peer groups cannot be entrusted with self-control, perhaps because of their narrow disciplinary backgrounds or self-interest, and perhaps because our educational system does not foster ethical and interdisciplinary values in professional training (6).

The social control of science and technology will be a troublesome and never wholly successful undertaking. It bears the potential to politicize and regiment intellectual activity, which has been realized in Russian genetics. Nor will the task lend itself to a specific solution—there are no administrative, legislative, or judicial panaceas.

Of course, it must also be recognized that future impact assessment and derivative control will always be limited, as man's intellectual and imaginative resources are limited. Even our measuring devices are still too crude to discern pernicious impacts in many cases. The earlier the assessment takes place in the process of science and technology, the more speculative it is. But the practice must begin, and develop, and pervade all the social control mechanisms we now have and may devise.

To begin, there are a number of reforms that can be introduced in our present array of social controls. Administrative agencies must be reorganized sensibly in light of new national objectives and available scientific and technological resources.

Legislation must be generated to provide guidelines for the administrative agencies similar to those provided by the National Environmental Policy Act (NEPA). Substantive and procedural duties are imposed by NEPA on all federal agencies to implement a broad policy of preventing and eliminating environmental damage. Section 102(2) of NEPA requires that the federal agencies, in their policies, recommendations, and other major federal actions affecting environmental quality, shall (7):

> A) utilize a systematic, interdisciplinary approach . . . in decision making which may have an impact on man's environment;
>
> B) . . . insure that presently unquantified *environmental* amenities and values . . . be given appropriate consideration in decision making along with economic and technical considerations;
>
> C) include in every major recommendation . . . and other major federal actions . . . [a] detailed statement . . . on (i) the environmental impact of the proposed action, (ii) any adverse environmental effects which cannot be avoided should the proposal be implemented, (iii) alternatives to the proposed action, (iv) the relationship between. . . short term uses. . . and long term productivity, (v) any irreversible and irretrievable commitments of resources which would be involved . . .
>
> D) study, develop and describe appropriate alternatives. . . .

We can only speculate about what impact NEPA will have on environmental quality. Perhaps its primary significance will be to instill certain *habits* and *values* in federal officials and the experts they consult: the habits of interdisciplinary assessment and consideration of alternatives, and a value system that would include health and ecological considerations.

NEPA will probably slow down the agency decision-making process, and this will help matters. Finally, NEPA will bring about the generation of information by federal agencies. This information should become available in useful form to concerned citizens who invoke the Freedom of Information Act (8). The agencies' broad-based studies of harmful effects and alternatives will be helpful, either because of contents or omissions, to environmental action groups. Hopefully, executive privilege and other exceptions to the Freedom of Information Act will not be invoked to the detriment of congressional purpose as expressed in NEPA. Unfortunately, this has already occurred in Soucie v. DuBridge (9), where the Office of Science and Technology report to the President on the SST was successfully withheld from conservationists.

Obviously, NEPA will also bring about some assessment and agency control of science and technology when environmental effects are predicted. However, there is a need for legislation, similarly grounded in a multiple-value system and the habit of assessment, that will more directly confront the need for a priori control of science and technology. This legislation should be directed at the substantial agency sponsorship of research and development, thereby regulating federal procurement and government contractor activities.

Independent adversaries must be fostered. A tax-exempt status ruling by the Internal Revenue Service would be a helpful first step for citizens' groups pursuing activities in the public interest—for example, groups that have demonstrated their concern for public health. Multiple-year grants to interdisciplinary groups, perhaps based at universities, could foster independent adversaries by establishing new career patterns. Congress, through its committee structure and reference service, should assist in this process.

Citizens should continue to press for responses from the legal system. Environmental litigation to date has been marked by ingenuity, but it lacks a coherent rationale. If Sierra Club v. Hickel is an omen of anything, it may be that the mere displeasure or the aggravation of elitist values of a citizens' group will not be sufficient to challenge agency and industrial actions that serve economic or public recreational interests, even though on a crass and commercial basis. Perhaps this is as it should be. Litigation to control environmental quality and science and technology should seek a coherent and important *raison d'être*—public health.

Public health—in both physical and psychic terms—includes esthetic and recreational values and the importance of ecosystems. It can therefore provide the nexus between citizen group social action or litigation and the public interest. The federal agencies, under NEPA, must now consider health effects. Establishing public health as the nexus does not simplify decision-making, but it can reduce subjective value clashes and will cause science and technology to be used in a self-evaluative and beneficial manner.

Finally, the most important social control must be discussed—education. The training and values of our professionals in law, engineering, and other fields must be responsive to the problems that beset society. The intense specialization that marks graduate education fosters narrow professionalism. Peer groups have not rewarded members who apply their training to problems that extend beyond disciplinary confines.

Our graduate schools and departments represent artificial divisions of knowledge and experience, and they deprive students of important opportunities and professional qualities. Substantive specialization and procedural barriers prevent students from working with colleagues in other disciplines and, often, from doing clinical work that is related to social issues. As a result, they are unfamiliar with the values, attitudes, and methods of other disciplines and unable to synthesize and apply them to social issues. These limitations in training are then reflected in careers and social problems.

No new degree programs will provide us with the answers. Rather, every degree program we now have must be enriched with interdisciplinary, clinical, and. preferably, problem-oriented components. Many exciting educational experiments, such as Cornell's "Science, Technology, and Society" program, are being conducted in institutions across the country.

Several innovative developments are taking place in the Boston area. At M.I.T., the school of engineering is moving to confront problems of biomedical engineering, public systems, and environmental quality. The civil engineering department has brought into its faculty and academic structure an interdisciplinary team made up of a political scientist, a lawyer, and an economist to work with the engineering faculty on water resources, transportation systems, systems engineering, and environmental quality. Engineering students can now enrich their academic programs with courses and research that relate their engineering disciplines to the full complexity of the social context in which they will eventually work.

A number of engineering students have joined members of the Harvard and Boston University law schools' environmental law societies on projects confronting local and national environmental issues.

Professor Jerrold Zacharias is now working on adapting M.I.T.'s advanced degree programs to specifically train students for college teaching careers in science and engineering. The mastery of a discipline, educational methods and technology, ethical and legal materials, and interdisciplinary research are now considered to be important features of this development. Graduates will be expected to bring breadth and innovative qualities to their teaching careers and relate their discipline to the social context.

Finally, at Boston University Law School, the new Center for Law and the Health Sciences has established a program that enables law students to work with graduate students from other disciplines on health-related social problems. Student and faculty participants are drawn from different disciplines and institutions, and students receive academic credit through ad hoc institutional arrangements.

David Bazelon, chief judge of the Washington, D.C., Federal Court of Appeals, has played a major role in this undertaking, as chairman of the center. In a summer pilot program, 15 graduate students from Boston University, Brandeis, Harvard, and M.I.T. were divided into four interdisciplinary teams. Each team confronted a complex health problem: genetic counseling, health insurance reform, multiple-service health centers, and the training of mental health professionals. Each team contained a law student, medical student, economist, or urban planner and a student from a discipline particularly relevant to the problem—for example, bioengineering. Twelve faculty members, representing a number of disciplines, served as a general resource to the students at scheduled meetings and informal sessions.

Interdisciplinary education presents a number of organizational problems and a number of unique educational benefits. Much was learned from the summer pilot program, and the academic year program is now being implemented. Problem orientation has proven to be an important aspect of the interdisciplinary program, in that it forces students to learn, synthesize, and then apply their knowledge. At the same time, students are able to exercise considerable initiative in defining and working on problems in a context of competing values. The center hopes thereby to enrich the graduate education of a number of students and enable them to function effectively in health-related careers.

The social control of science and technology, through the training of new kinds of professionals, is one of the most important tasks at hand for law schools, schools of science and engineering, and other programs of higher education. This task must become an ongoing process, and it needs interdisciplinary cooperation and public support. Faculty in schools of professional training in medicine, law, and other fields, are needed to help build and implement these new programs of public service and must rejoin the university. In addition, these new programs must be related to the social system and values, for only through individual and collective wisdom and temperance, induced by an appreciation of the values of others, will we control science and technology in a coherent and humane fashion (10).

REFERENCES AND NOTES

1. J. Ellul, *The Technological Society* (Random House, New York, 1964), p. 251.
2. H. Green, *The New Technological Era: A View From the Law* (Monograph 1, Program of Policy Studies, George Washington Univ., Washington, D.C., 1967).
3. Sierra Club *v.* Hickel, U.S. Court of Appeals, Ninth Circuit (1970).
4. B. Portnoy, "The Role of the Courts in Technology Assessment," *Cornell Law Rev.* 55, 861 (1970).
5. H. Folk, "The role of technology assessment in public policy," paper given at the Boston meeting of the AAAS, 29 December 1969.
6. See H. Morgenthau, "Modern Science and Political Power," *Columbia Law Rev.* 64, 1386 (1964).
7. 42 U.S.C. 4331 *et seq*. A full review of the National Environment Policy Act is presented in an article by R. C. Peterson of Yale Law School *(Title I of the National Environmental Policy Act of 1969)*. It is available from the Environmental Law Institute, 1346 Connecticut Ave., NW, Washington, D.C.
8. 5 U.S.C. 552
9. Soucie *v.* DuBridge, U.S. District Court, District of Columbia (1970).
10. Further information on the programs discussed above is available from the author, Room 1-376, M.I.T., Cambridge, Mass. 02139.

SUGGESTIONS FOR FURTHER READING

Canada, Senate Special Committee on Science Policy. *A Science Policy for Canada, Vol. 2.* Ottawa: Information Canada, 1972.

Horowitz, I.L. (ed.) *The Use and Abuse of Social Science.* New Brunswick, N.J.: Transaction Books, 1971.

Morison, R.S. "Science and Social Attitudes." *Science,* 165 (1969), pp. 150–156.

Ravetz, J.R. *Scientific Knowledge and Its Social Problems.* Oxford: Clarendon Press, 1971, pp. 273–364.

Schooler, D. *Science, Scientists and Public Policy.* New York: The Free Press, 1971.

Shils, E. (ed.) *Criteria for Scientific Development, Public Policy and National Goals.* Cambridge: M.I.T. Press, 1968.

United States, Congress, House, Subcommittee on Science, Research and Development of the Committee on Science and Astronautics. *National Science Policy.* 91st Cong., 2d sess., 1970. Washington, D.C. U.S. Government Printing Office, 1970.

York, C.M. "Steps toward a National Policy for Academic Science." *Science,* 172 (1971), pp. 643–651.

Yovits, M.C. *et al.* (eds.) *Research Program Effectiveness.* New York: Gordon and Breach, Science Publishers, Inc., 1966.

SECTION X

The Impact of Science and Technology on Society

The first article in this section, by Callahan, examines the ethical norms that "should be brought to bear in controlling population growth." It is, like the three articles following it, a case study of the interaction of scientific knowledge, technological expertise and morality. By considering these individual cases in some detail, we should be able to put a sharper focus on the interface between science, technology and philosophy than we have been doing in the preceding two sections.

With respect to population growth, we are supposed to have some scientific evidence that excess growth "poses critical dangers to the future of the species, the ecosystem, individual liberty and welfare, and the structure of social life," according to Callahan. Moreover, there are certain technological means of reducing growth, not all of which are consistent with our moral principles and social customs. Hence, in order to put our scientific and technological knowledge to work, we must clarify the benefits and costs of using them in one way rather than another. We must examine the likely consequences from the point of view of principles, values and aims that are not scientific or technological, but primarily ethical.

In the Morison-Kass exchange, we have a conflict over proposed definitions of "death" that have been put forward in the light of new knowledge and medical technology. Near the end of Kass's article, he presents an argument in defense of "loyal agents" which seems to be generalizable for any employee—doctor, lawyer, mechanic, plumber, policeman, Green Beret and so on. When you assess his argument, you should consider it in this light, and try to decide whether or not it is possible to purge his proposal of its apparently egoistic bias. Does Kass see the responsibility of physicians (and other "loyal agents") properly, or does Morison? Or, can we have it both ways?

The primary aim of the last article in this book is to show that the application of a useful methodology for economics "is inappropriate to the analysis of decision making in military strategy and foreign policy." In other words, Wolff tries to show that what is good reasoning in economics is poor reasoning in military strategy and foreign policy because certain conditions present in economics are not present in the other areas. What we have then, if Wolff is right, is the misapplication of technology or technique, a misapplication which is revealed not by technology itself but by other considerations.

READING 24

Ethics and Population Limitation

What ethical norms should be brought to bear in controlling population growth?

Daniel Callahan

Throughout its history, the human species has been preoccupied with the conquest of nature and the control of death. Human beings have struggled to survive, as individuals, families, tribes, communities, and nations. Procreation has been an essential part of survival. Food could not have been grown, families sustained, individuals supported, or industry developed without an unceasing supply of new human beings. The result was the assigning of a high value to fertility. It was thought good to have children: good for the children themselves, for the parents, for the society, and for the species. While it may always have been granted that extenuating circumstances could create temporary contraindications to childbearing, the premise on which the value was based endured intact. There remained a presumptive right of individual procreation, a right thought to sustain the high value ascribed to the outcome: more human beings.

That the premise may now have to be changed, the value shifted, can only seem confounding. As Erik Erikson has emphasized, it is a risky venture to play with the "fire of creation," especially when the playing has implications for almost every aspect of individual and collective life (*1*). The reasons for doing so would have to be grave. Yet excessive population growth presents such reasons—it poses critical dangers to the future of the species, the ecosystem, individual liberty and welfare, and the structure of social life. These hazards are serious enough to warrant a reexamination

Reprinted by permission of the author and *Science*, Vol. 175, 4 February 1972, copyright 1972 by the American Association for the Advancement of Science, pp. 487–494.

and, ultimately, a revision of the traditional value of unrestricted procreation and increase in population.

The main question is the way in which the revision is to proceed. If the old premise—the unlimited right of and need for procreation—is to be rejected or amended, what alternative premises are available? By what morally legitimate social and political processes, and in light of what values, are the possible alternatives to be evaluated and action taken? These are ethical questions, bearing on what is taken to constitute the good life, the range and source of human rights and obligations, the requirements of human justice and welfare. If the ethical problems of population limitation could be reduced to one overriding issue, matters would be simplified. They cannot. Procreation is so fundamental a human activity, so wide-ranging in its personal and social impact, that controlling it poses a wide range of ethical issues. My aim here is primarily to see what some of the different ethical issues are, to determine how an approach to them might be structured, and to propose some solutions.

With a subject so ill-defined as "ethics and population limitation," very little by way of common agreement can be taken for granted. One needs to start at the "beginning," with some basic assertions.

FACTS AND VALUES

There would be no concern about population limitation if there did not exist evidence that excessive population growth jeopardizes present and future welfare. Yet the way the evidence is evaluated will be the result of the values and interests brought to bear on the data. Every definition of the "population problem" or of "excessive population growth" will be value-laden, expressive of the ethical orientations of those who do the defining. While everyone might agree that widespread starvation and malnutrition are bad, not everyone will agree that crowding, widespread urbanization, and a loss of primitive forest areas are equally bad. Human beings differ in their assessments of relative good and evil. To say that excessive population growth is bad is to imply that some other state of population growth would be good or better—for example, an "optimum level of population." But as the demographic discussion of an optimum has made clear, so many variables come into play that it may be possible to do no more than specify a direction: "the desirability of a lower *rate* [italics added] of growth" (*2*).

If the ways in which the population problem is defined will reflect value orientations, these same definitions will have direct implications for the ways in which the ethical issues are posed. An apocalyptic reading of the demographic data and projections can, not surprisingly, lead to coercive proposals. Desperate problems are seen to require desperate and otherwise distasteful solutions (*3*). Moreover, how the problem is defined, and how the different values perceived to be at stake are weighted, will have direct implications for the priority given to population problems in relation to other social problems. People might well agree that population growth is a serious issue, but they might (and often do) say that other issues are comparatively more serious (*4*). If low priority is given to population problems, this is likely to affect the perception of the ethical issues at stake.

WHY ETHICAL QUESTIONS ARISE

Excessive population growth raises ethical questions because it threatens existing or desired human values and ideas of what is good. In addition, all or some of the possible solutions to the problem have the potential for creating difficult ethical dilemmas. The decision to act or not to act in the face of the threats is an ethical decision. It is a way of affirming where the human good lies and the kinds of obligations individuals and societies have toward themselves and others. A choice in favor of action will, however, mean the weighing of different options, and most of the available options present ethical dilemmas.

In making ethical choices, decisions will need to be made on (i) the human good and values that need to be served or promoted—the ends; (ii) the range of methods and actions consistent and coherent with those ends—the means; and (iii) the procedure and rationale to be used in trying to decide both upon ends and means and upon their relation to each other in specific situations—the ethical criteria for decision-making. A failure to determine the ends, both ultimate and proximate, can make it difficult or impossible to choose the appropriate means. A failure to determine the range of possible means can make it difficult to serve the ends. A failure to specify or articulate the ethical criteria for decision-making can lead to capricious or self-serving choices, as well as to the placing of obstacles in the way of rational resolution of ethical conflicts.

In the case of ethics and the population problem, both the possibilities and the limitations of ethics become apparent. In the face of a variety of proposals to solve the population problem, some of them highly coercive, a sensitivity to the ethical issues and some greater rigor in dealing with them is imperative. The most fundamental matters of human life and welfare are at stake. Yet because of the complexity of the problem, including its variability from one nation or geographical region to the next, few hard and fast rules can be laid down about what to do in a given place at a given time.

Still, since some choices must be made (and not to choose is to make a choice as well), the practical ethical task will be that of deciding upon the available options. While I will focus on some of the proposed options for reducing birthrates, they are not the only ones possible. Ralph Potter has discussed some others (5).

> It has generally been assumed that policy must be primarily, if not exclusively, concerned with bringing about a decline in the rate of population increase through a reduction in the birthrate. But there are other choices. It is generally considered desirable but impossible to increase resources at a sufficient pace and through an adequate duration to preserve the present level of living for all within an expanding population. It is generally considered possible but undesirable to omit the requirement that all persons have access to that which is necessary for a good life. There is still the option of redefining what is to be considered necessary for a good life or of foregoing some things necessary for a good life in order to obtain an equitable distribution in a society that preserves the autonomy of parents to determine the size of their families.

A useful way of posing the issue practically is to envision the ethical options ranked on a preferential scale, from the most desirable to the least desirable. For working purposes, I will adopt as my own the formulation of Kenneth E. Boulding: "A moral, or ethical, proposition is a statement about a rank order of preferences among alternatives, which is intended to apply to more than one person" (6). Ethics enters at that point when the preferences are postulated to have a value that transcends individual tastes or inclinations. Implicitly or explicitly, a decision among alternatives becomes an ethical one when it is claimed that one or another alternative *ought* to be chosen—not just by me, but by others as well. This is where ethics differs from tastes or personal likings, which, by definition, imply nonobligatory preferences that are applicable to no more than one person (even if the tastes are shared).

GENERAL ETHICAL ISSUES

I will assume at the outset that there is a problem of excessive population growth, a problem serious for the world as a whole (with a 2 percent annual growth rate), grave for many developing nations (where the growth rates approaches 3 percent per annum), and possibly harmful for the developed nations as well (with an average 1 percent growth rate). The threats posed by excessive population growth are numerous: economic, environmental, agricultural, political, and sociopsychological. There is considerable agreement that something must be done to meet these threats. For the purpose of ethical analysis, the first question to be asked is, "In trying to meet these threats, what human ends are we seeking to serve?" Two kinds of human ends can be distinguished—proximate and ultimate.

Among the important proximate ends being sought in attempts to reduce birthrates in the developing countries are a raising of literacy rates, a reduction in dependency ratios, the elimination of starvation and malnutrition, more rapid economic development, and an improvement in health and welfare services; among these ends in the developed countries are a maintenance or improvement of the quality of life, the protection of nonrenewable resources, and the control of environmental pollution. For most purposes, it will be sufficient to cite goals of this sort. But for ethical purposes, it is critical to consider not just proximate, but ultimate ends as well. For it is legitimate to ask of the specified proximate ends what ultimate human ends they are meant to serve. Why is it important to raise literacy rates? Why is it necessary to protect nonrenewable resources? Why ought the elimination of starvation and malnutrition to be sought? For the most part, these are questions that need not be asked or that require no elaborate answers. The ethical importance of such questions is that they force us to confront the goals of human life. Unless these goals are confronted at some point, ethics cannot start or finish.

Philosophically, solving the population problem can be viewed as determining at the outset what final values should be pursued. The reason, presumably, that a reduction in illiteracy rates is sought is that it is thought valuable for human beings to possess the means of achieving knowledge. The elimination of starvation and malnutrition is sought because of the self-evident fact that human beings must eat to survive. The preservation of nonrenewable resources is necessary in order that human life may continue through future generations. There is little argument about the validity of these propositions, because they all presuppose some important human values: knowledge, life, and survival of the species, for instance. Historically, philosophers have attempted to specify what, in the sense of "the good," human beings essentially seek. What do they, in the end, finally value? The historical list of values is long: life, pleasure, happiness, knowledge, freedom, justice, and self-expression, among others.

This is not the place to enter into a discussion of all of these values and the philosophical history of attempts to specify and rank them. Suffice it to say that three values have had a predominant role, at least in the West: freedom, justice, and security-survival. Many of the major ethical dilemmas posed by the need for population limitation can be reduced to ranking and interpreting these three values. Freedom is prized because it is a condition for self-determination and the achievement of knowledge. Justice, particularly distributive justice, is prized because it entails equality of treatment and opportunity and an equitable access to those resources and opportunities necessary for human development. Security-survival is prized because it constitutes a fundamental ground for all human activities.

Excessive population growth poses ethical dilemmas because it forces us to weight and rank these values in trying to find solutions. How much procreative freedom, if any, should be given up in order to insure the security-survival of a nation or a community? How much security-survival can be risked in order to promote distributive justice? How much procreative freedom can be tolerated if it jeopardizes distributive justice?

Ethical dilemmas might be minimized if there were a fixed agreement on the way the three values ought to be ranked. One could say that freedom is so supreme a value that both justice and security-survival should be sacrified to maintain it. But there are inherent difficulties in taking such a position. It is easily possible to imagine situations in which a failure to give due weight to the other values could result in an undermining of the possibility of freedom itself. If people cannot survive at the physical level, it becomes impossible for them to exercise freedom of choice, procreative or otherwise. If the freedom of some is unjustly achieved at the expense of the freedom of others, then the overall benefits of freedom are not maximized. If security-survival were given the place of supremacy, situations could arise in which this value was used to justify the suppression of freedom or the perpetuation of social injustice. In that case, those suppressed might well ask, "Why live if one cannot have freedom and justice?"

For all of these reasons it is difficult and perhaps unwise to specify a fixed and abstract rank order of preference among the three values. In some circumstances, each can enter a valid claim against the others. In the end, at the level of abstractions, one is forced to say that all three values are critical; none can permanently be set aside.

THE PRIMACY OF FREEDOM

In the area of family planning and population limitation, a number of national and international declarations have given primacy to individual freedom. The Declaration of the 1968 United Nations International Conference on Human Rights is representative (*7, 8*): ". . . couples have a basic human right to decide freely and responsibly on the number and spacing of their children and a right to adequate education and information in this respect." While this primacy of individual freedom has been challenged (*9*), it retains its position, serving as the ethical and political foundation of both domestic and foreign family planning and population policies. Accordingly, it will be argued here that (i) the burden of proof for proposals to limit freedom of choice (whether on the grounds of justice or security-survival) rests with those who make the proposals, but that (ii) this burden can, under specified conditions, be discharged if it can be shown that a limitation of freedom of choice in the name of justice or security-survival would tend to maximize human welfare and human values. This is only to say that, while the present international rank order of preference gives individual freedom primacy, it is possible to imagine circumstances that would require a revision of the ranking.

One way of approaching the normative issues of ranking preferences in population limitation programs and proposals is by locating the key ethical actors, those who can be said to have obligations. Three groups of actors can be identified: individuals (persons, couples, families), the officers and agents of voluntary (private-external) organizations, and the government officials responsible for population and family planning programs. I will limit my discussion here to individuals and governments. What are the ethical obligations of each of the actors? What is the right or correct course of conduct for them? I will approach these questions by first trying to define some general rights and obligations for each set of actors and then by offering some suggested resolutions of a number of specific issues.

I begin with individuals (persons, couples, families) because, in the ranking of values, individual freedom of choice has been accorded primacy by some international forums—and it is individuals who procreate. What are the rights and obligations of individuals with regard to procreation?

Individuals have the right voluntarily to control their own fertility in accordance with their personal preferences and convictions (7, p. 15). This right logically extends to a choice of methods to achieve the desired control and the right to the fullest possible knowledge of available methods and their consequences (medical, social, economic, and demographic, among others).

Individuals are obligated to care for the need and respect the rights of their existing children (intellectual, emotional, and physical); in their decision to have a child (or another child), they must determine if they will be able to care for the needs and respect the rights of the child-to-be. Since individuals are obliged to respect the rights of others, they are obliged to act in such a way that these rights are not jeopardized. In determining family size, this means that they must exercise their own freedom of choice in such a way that they do not curtail the freedom of others. They are obliged, in short, to respect the requirements of the common good in their exercise of free choice (*10*). The source of these obligations is the rights of others.

The role of governments in promoting the welfare of their citizens has long been recognized. It is only fairly recently, however, that governments have taken a leading role in an antinatalist control of fertility (*11*). This has come about by the establishment, in a number of countries, of national family planning programs and national population policies. While many countries still do not have such policies, few international objections have been raised against the right of nations to develop them. So far, most government population policies have rested upon and been justified in terms of an extension of freedom of choice. Increasingly, though, it is being recognized that, since demographic trends can significantly affect national welfare, it is within the right of nations to adopt policies designed to reduce birthrates and slow population growth.

A preliminary question must, therefore, be asked. Is there any special reason to presume or suspect that governmental intervention in the area of individual procreation and national fertility patterns raises problems which, in *kind,* are significantly different from other kinds of interventions? To put the question another way, can the ethicopolitical problems that arise in this area be handled by historical and traditional principles of political ethics, or must an entirely new ethic be devised?

I can see no special reason to think that the formation of interventionist, antinatalist, national population policies poses any unique *theoretical* difficulties. To be sure, the perceived need to reduce population growth is historically new; there exists no developed political or ethicopolitical tradition dealing with this specific problem. Yet the principle of governmental intervention in procreation-related behavior has a long historical precedent: in earlier, pronatalist population policies, in the legal regulation of marriage, and in laws designed to regulate sexual behavior. It seems a safe generalization to say that governments have felt (and generally have been given) as much right to intervene in this area as in any other where individual and collective welfare appears to be at stake. That new forms of intervention may seem to be called for or may be proposed (that is, in an anti- rather than pronatalist direction) does not mean that a new ethical principle is at issue. At least, no such principle is immediately evident.

Yet, if it is possible to agree that no new principles are involved, it is still possible to argue that a further extension of an old principle—the right of government intervention into procreation-related behavior—would be wrong. Indeed, it is a historical irony that, after a long international struggle to establish individuals' freedom of choice in controlling their own fertility, that freedom should immediately be challenged in the name of the population crisis. Irony or not, there is no cause to be surprised by such a course of events. The history of human liberty is studded with instances in which, for a variety of reasons, it has been possible to say that liberty is a vital human good and yet that, for the sake of other goods, restriction of liberty seems required. A classical argument for the need of a government is that a formal and public apparatus is necessary to regulate the exercise of individual liberty for the sake of the common good.

In any case, the premise of my discussion will be that governments have as much right to intervene in procreation-related behavior as in other areas of behavior affecting the general welfare. This right extends to the control of fertility in general and to the control of individual fertility in particular. The critical issue is the way in which this right is to be exercised—its conditions and limits—and that issue can only be approached by first noting some general issues bearing on the restriction of individual freedom of choice by governments.

Governments have the right to take those steps necessary to insure the preservation and promotion of the common good—the protection and advancement of the right to life, liberty, and property. The maintenance of an orderly and just political and legal system, the maintenance of internal and external security, and an equitable distribution of goods and resources are also encompassed within its rights. Its obligations are to act in the interests of the people, to observe human rights, to respect national values and traditions, and to guarantee justice and equality. Since excessive population growth can touch upon all of these elements of national life, responses to population problems will encompass both the rights and the obligations of governments. However, governmental acts should represent collective national decisions and be subject to a number of stipulations.

I now recapitulate the points made so far and summarize some propositions, which I then use to suggest solutions to some specific ethical issues.

1) General moral rules: (i) individuals have the right to freedom of procreative choice, and they have the obligation to respect the freedom of others and the requirements of the common good; (ii) governments have the right to take those steps necessary to secure a maximization of freedom, justice, and security-survival, and they have the obligation to act in such a way that freedom and justice are protected and security-survival enhanced.

2) Criteria for ethical decision-making: (i) one (individual, government, organization) is obliged to act in such a way that the fundamental values of freedom, justice, and security-survival are respected; (ii) in cases of conflict, one is obliged to act in such a way that any limitation of one or more of the three fundamental values—a making of exceptions to the rules concerning these values—continues to respect the values and can be justified by the promise of increasing the balance of good over evil.

3) Rank order of preference: (i) those choices of action that ought to be preferred are those that accord primacy to freedom of choice; (ii) if conditions appear to require a limitation of freedom, this should be done in such a way that the direct and indirect harmful consequences are minimized and the chosen means of limitation are just—the less the harm, the higher the ranking.

SOME SPECIFIC ETHICAL ISSUES

Since it has already been contended that individual freedom of choice has primacy, the ethical issues to be specified here will concentrate on those posed for governments. This focus will, in any event, serve to test the limits of individual freedom.

Faced with an excessive population growth, a variety of courses are open to governments. They can do nothing at all. They can institute, develop, or expand voluntary family planning programs. They can attempt to implement proposals that go "beyond family planning" (*12*).

Would it be right for governments to go beyond family planning if excessive population growth could be shown to be a grave problem? This question conceals a great range of issues. Who would decide if governments have this right? Of all the possible ways of going beyond family planning, which could be most easily justified and which would be the hardest to justify? To what extent would the problem have to be shown to be grave? As a general proposition, it is possible ethically to say that governments would have the right to go beyond family planning. The obligation of governments to protect fundamental values could require that they set aside the primacy of individual freedom in order to protect justice and security-survival. But everything would depend on the way they proposed to do so.

Would it be right for governments to establish involuntary fertility controls? These might include (if technically feasible) the use of a mass "fertility control agent," the licensing of the right to have children, compulsory temporary or permanent sterilization, or compulsory abortion (*12*). Proposals of this kind have been put forth primarily as "last resort" methods, often in the context that human survival may be at stake. "Compulsory control of family size is an unpalatable idea to many," the Ehrlichs have written, "but the alternatives may be much more horrifying . . . human survival seems certain to require population control programs. . . ." (*3*, p. 256). Their own suggestion is manifestly coercive: "If . . . relatively uncoercive laws should fail to bring the birthrate under control, laws could be written that would make the bearing of a third child illegal and that would require an abortion to terminate all such pregnancies" (*3*, p. 274).

That last suggestion requires examination. Let us assume for the moment that the scientific case has been made that survival itself is at stake and that the administrative and enforcement problems admit of a solution. Even so, some basic ethical issues would remain. "No one," the United Nations has declared, "shall be subjected to torture or to cruel, inhuman, or degrading treatment or punishment" (*13*, Article 5). It is hard to see how compulsory abortion, requiring governmental invasion of a woman's body, could fail to qualify as inhuman or degrading punishment. Moreover, it is difficult to see how this kind of suggestion can be said to respect in any way the values of freedom and justice. It removes free choice altogether, and in its provision for an abortion of the third child makes no room for distributive justice at all; its burden would probably fall upon the poorest and least educated. It makes security-survival the prime value, but to such an extent and in such a way that the other values are ignored altogether. But could not one say, when survival itself is at stake, that this method would increase the balance of good over evil? The case would not be easy to make (i) because survival is not the only human value at stake; (ii) because the social consequences of such a law could be highly destructive (for example, the inevitably massive fear and anxiety about third pregnancies that would result from such a law); and (iii) because it would be almost impossible to show that this is the *only* method that would or could work to achieve the desired reduction in birthrates.

Would it be right for governments to develop "positive" incentive programs, designed to provide people with money or goods in return for a regulation of their fertility? These programs might include financial rewards for sterilization, for the use of contraceptives, for periods of nonpregnancy or nonbirth, and for family planning bonds or "responsibility prizes" (*12,* p. 2). In principle, incentive schemes are noncoercive; that is, people are not forced to take advantage of the incentive. Instead, the point of an incentive is to give them a choice they did not previously have.

Yet there are a number of ethical questions about incentive plans. To whom would they appeal most? Presumably, their greatest appeal would be to the poor, those who want or need the money or goods offered by an incentive program; they would hold little appeal for the affluent, who already have these things. Yet if the poor desperately need the money or goods offered by the incentive plan, it is questionable whether, in any real sense, they have a free choice. Their material needs may make the incentive seem coercive to them. Thus, if it is only or mainly the poor who would find the inducements of an incentive plan attractive, a question of distributive justice is raised. Because of their needs, the poor have less choice than the rich about accepting or rejecting the incentive; this could be seen as a form of exploitation of poverty. In sum, one can ask whether incentive schemes are or could be covertly coercive, and whether they are or could be unjust (*14*). If so, then while they may serve the need for security-survival, they may do so at the expense of freedom and justice.

At least three responses seem possible. First, if the need for security-survival is desperate, incentive schemes might well appear to be the lesser evil, compared with more overtly coercive alternatives. Second, the possible objections to incentive schemes could be reduced if, in addition to reducing births, they provided other benefits as well. For instance, a "family planning bond" program would provide the additional benefit of old-age security (*15*). Any one of the programs might be defended on the grounds that those who take advantage of it actually want to control births in any case (if this can be shown). Third, much could depend upon the size of the incentive benefits. At present, most incentive programs offer comparatively small rewards; one may doubt that they offer great dilemmas for individuals or put them in psychological straits.

The objection to such programs on the grounds of coercion would become most pertinent if it can be shown that the recipients of an incentive benefit believe they have no real choice in the matter (because of their desperate poverty or the size of the benefit); so far, this does not appear to have been the case (*16*).

While ethical objections have been leveled at incentive programs because of some experienced corrupt practices in their implementation, this seems to raise less serious theoretical issues. Every program run by governments is subject to corruption; but there are usually ways of minimizing it (by laws and review procedures, for instance). Corruption, I would suggest, becomes a serious theoretical issue only when and if it can be shown that a government program is *inherently* likely to create a serious, inescapable, and socially damaging system of corruption. This does not appear to be the case with those incentive programs so far employed or proposed.

Would it be right for governments to institute "negative" incentive programs? These could take the form of a withdrawal of child or family allowances after a given number of children, a withdrawal of maternity benefits after a given number, or a reversal of tax benefits, to favor those with small families (*12,* p. 2). A number of objections to such programs have been raised. They are directly coercive in that they deprive people of free choice about how many children they will have by imposing a penalty on excess procreation; thus they do not attach primary importance to freedom of choice. They can also violate the demands of justice, especially in those cases where the burden of the penalties would fall upon those children who would lose benefits available to their siblings. And the penalties would probably be more onerous to the poor than to the rich, further increasing the injustice. Finally, from quite a different perspective, the social consequences of such programs could be most undesirable. They could, for instance, worsen the health and welfare of those mothers, families, and children who would lose needed social and welfare benefits. Moreover, such programs could be patently unjust in those places where effective contraceptives do not exist (most places at present). In such cases, people would be penalized for having children whom they could not prevent with the available birth control methods.

It is possible to imagine ways of reducing the force of these objections. If the penalties were quite mild, more symbolic than actual [as Garrett Hardin has proposed (*17*)], the objection from the viewpoint of free choice would be less; the same would apply to the objection from the viewpoint of justice. Moreover, if the penalty system were devised in such a way that the welfare of children and families would not be harmed, the dangerous social consequences would be mitigated. Much would depend, in short, upon the actual provisions of the penalty plan and the extent to which it could minimize injustice and harmful social consequences. Nonetheless, penalty schemes raise serious ethical problems. It seems that they would be justifiable only if it could be shown that security-survival was at stake and that, in their application, they would give due respect to freedom and justice. Finally, it would have to be shown that, despite their disadvantages, they promised to increase the balance of good over evil—which would include a calculation of the harm done to freedom and justice and a weighing of other, possibly harmful, social consequences.

An additional problem should be noted. Any penalty or benefit scheme would require some method of governmental surveillance and enforcement. Penalty plans, in particular, would invite evasion—for example, hiding the birth of children to avoid the sanctions of the scheme. This likelihood would be enhanced among those who objected to the plan on moral or other grounds, or who believed that the extra children were necessary for their own welfare. One does not have to be an ideological opponent of "big government" to imagine the difficulties of trying to ferret out violators or the lengths to which some couples might go to conceal pregnancies and births. Major invasions of privacy, implemented by a system of undercover agents, informants, and the like, would probably be required to make the scheme work. To be sure, there are precedents for activities of this kind (as in the enforcement of income tax laws), but the introduction of further governmental interventions of this kind would raise serious ethical problems, creating additional strains on the relationship between the government and the people. The ethical cost of an effective penalty system would have to be a key consideration in the development of any penalty program.

Would it be right for governments to introduce antinatalist shifts in social and economic institutions? Among such shifts might be a raising of marriage ages, manipulation of the family structure away from nuclear families, and bonuses for delayed marriage (*12*, pp. 2–3). The premise of these proposals is that fertility patterns are influenced by the context in which choices are made and that some contexts (for example, higher female employment) are anti- rather than pronatalist. Thus instead of intervening directly into the choices women make, these proposals would alter the environment of choice; freedom of individual choice would remain. The attractiveness of these proposals lies in their noninterference with choice; they do not seem to involve coercion. But they are not without their ethical problems, at least in some circumstances. A too-heavy weighting of the environment of choice in an antinatalist direction would be tantamount to an interference with freedom of choice—even if, technically, a woman could make a free choice. In some situations, a manipulation of the institution of marriage (for example, raising the marriage age) could be unjust, especially if there exist no other social options for women.

The most serious problems, however, lie in the potential social consequences of changes in basic social institutions. What would be the long-term consequences of a radical manipulation of family structure for male-female relationships, for the welfare of children, for the family? One might say that the consequences would be good or bad, but the important point is that they would have to be weighed. Should some of them appear bad, they would then have to be justified as entailing a lesser evil than the continuation of high birthrates. If some of the changes promised to be all but irreversible once introduced, the justification would have to be all the greater. However, if the introduction of shifts in social institutions had some advantages in addition to antinatalism—for instance, greater freedom for women, a value in its own right—these could be taken as offsetting some other, possibly harmful, consequences.

Would it be right for the government of a developed nation to make the establishment of a population control program in a developing nation a condition for extending food aid (*18, 19*)? This would be extremely difficult to justify on ethical grounds. At the very least, it would constitute an interference in a nation's right to self-determination (*20*). Even more serious, it would be a direct exploitation of one nation's poverty in the interests of another nation's concept of what is good for it; and that would be unjust. Finally, I would argue that, on the basis of Article 3 of the "Universal Declaration of Human Rights" (*21*), a failure to provide needed food aid would be a fundamental violation of the right to life (when that aid could, without great cost to the benefactor nation, be given). The argument that such aid, without an attendant population control program, would only make the problem worse in the long run, is defective. Those already alive, and in need of food, have a right to security-survival. To willfully allow them to die, or to deprive them of the necessities of life, in the name of saving even more lives at a later date cannot be justified in the name of a greater preponderance of good over evil. There could be no guarantee that those future lives would be saved, and there would be such a violation of the rights of the living (including the right to life) that fundamental human values would be sacrificed.

Would it be right for a government to institute programs that go beyond family planning—particularly in a coercive direction—

for the sake of future generations? This is a particularly difficult question, in great part because the rights of unborn generations have never been philosophically, legally, or ethically analyzed in any great depth (*22*). On the one hand, it is evident that the actions of one generation can have profound effects on the options available to future generations. And just as those living owe much of their own welfare to those who preceded them (beginning with their parents), so, too, the living would seem to have obligations to the unborn. On the other hand though, the living themselves do have rights—not just potential, but actual. To set aside these rights, necessary for the dignity of the living, in favor of those not yet living would, I think, be to act arbitrarily.

A general solution might, however, be suggested. While the rights of the living should take precedence over the rights of unborn generations, the living have an obligation to refrain from actions that would endanger future generations' enjoyment of the same rights that the living now enjoy. This means, for instance, that the present generation should not exhaust nonrenewable resources, irrevocably pollute the environment, or procreate to such an extent that future generations will be left with an unmanageably large number of people. All of these obligations imply a restriction of freedom. However, since the present generation does have the right to make use of natural resources and to procreate, it must be demonstrated (not just asserted) that the conduct of the present generation poses a direct threat to the rights of future generations. In a word, the present generation cannot be deprived of rights on the basis of vague speculations about the future or uncertain projections into the future.

Do governments have the right unilaterally to introduce programs that go beyond family planning? It is doubtful that they do. Article 21 of the "Universal Declaration of Human Rights" (*13*) asserts that "Everyone has the right to take part in the government of his country, directly or through freely chosen representatives. . . . The will of the people shall be the basis of the authority of government." There is no evident reason that matters pertaining to fertility control should be exempt from the requirements of this right. By implication, not only measures that go beyond family planning, but family planning programs as well require the sanctions of the will of the people and the participation of the people in important decisions.

A RANKING OF PREFERENCES

The preceding list of specific issues by no means exhausts the range of possible ethical issues pertaining to governmental action; it is meant only to be illustrative of some of the major issues. Moreover, the suggested solutions are only illustrative. The complexities of specific situations could well lead to modifications of them. That is why ethical analysis can rarely ever say exactly what ought to be done in *x* place at *y* time by *z* people. It can suggest general guidelines only.

I want now to propose some general ethical guidelines for governmental action, ranking from the most preferable to the least preferable.

1) Given the primacy accorded freedom of choice, governments have an obligation to do everything within their power to protect, enhance, and implement freedom of choice in family planning. This means the establishment, as the first order of business, of effective voluntary family planning programs.

2) If it turns out that voluntary programs are not effective in reducing excessive population growth, then governments have the right, as the next step, to introduce programs that go beyond family planning. However, in order to justify the introduction of such programs, it must be shown that voluntary methods have been adequately and fairly tried, and have nonetheless failed and promise to continue to fail. It is highly doubtful that, at present, such programs have "failed"; they have not been tried in any massive and systematic way (*23*).

3) In choosing among possible programs that go beyond family planning, governments must first try those which, comparatively, most respect freedom of choice (that is, are least coercive). For instance, they should try "positive" incentive programs and manipulation of social structures before resorting to "negative" incentive programs and involuntary fertility controls.

4) Further, if circumstances force a government to choose programs that are quasi- or wholly coercive, they can justify such programs if, and only if, a number of prior conditions have been met: (i) if, in the light of the primacy of free choice, a government has discharge the burden of proof necessary to justify a limitation of free choice—and the burden of proof is on the government (this burden may be discharged by a demonstration that continued unrestricted liberty poses a direct threat to distributive justice or security-survival; and (ii) if, in light of the right of citizens to take part in the government of their country, the proposed limitations

on freedom promise, in the long run, to increase the options of free choice, decisions to limit freedom are collective decisions, the limitations on freedom are legally regulated and the burden falls upon all equally, and the chosen means of limitations respect human dignity, which will here be defined as respecting those rights specified in the United Nations' "Universal Declaration of Human Rights" (*13*). The end—even security-survival—does not justify the means when the means violate human dignity and logically contradict the end.

As a general rule, the more coercive the proposed plan, the more stringent should be the conditions necessary to justify and regulate the coercion. In addition, one must take account of the possible social consequences of different programs, consequences over and above their impact on freedom, justice, and security-survival. Thus, if it appears that some degree of coercion is required, that policy or program should be chosen which (i) entails the least amount of coercion, (ii) limits the coercion to the fewest possible cases, (iii) is most problem-specific, (iv) allows the most room for dissent of conscience, (v) limits the coercion to the narrowest possible range of human rights, (vi) threatens human dignity least, (vii) establishes the fewest precedents for other forms of coercion, and (viii) is most quickly reversible if conditions change.

While it is true to say that social, cultural, and political life requires, and has always required, some degree of limitation of individual liberty—and thus some coercion—that precedent does not, in itself, automatically justify the introduction of new limitations (*24*). Every proposal for a new limitation must be justified in its own terms—the specific form of the proposed limitation must be specifically justified. It must be proved that it represents the least possible coercion, that it minimizes injustice to the greatest extent possible, that it gives the greatest promise of enhancing security-survival, and that it has the fewest possible harmful consequences (both short- and long-term).

FREEDOM AND RISK-TAKING

The approach I have taken to the ethics of population limitation has been cautionary. I have accepted the primacy of freedom of choice as a given not only because of its primacy in United Nations and other declarations, but also because it is a primary human value. I have suggested that the burden of proof must lie with those proposals, policies, or programs that would place the primacy elsewhere.

At the same time, I have laid down numerous conditions necessary to discharge the burden of proof. Indeed, these conditions are so numerous, and the process of ethical justification so difficult, that the possibility of undertaking decisive action may seem to have been excluded. This is a reasonable concern, particularly if time is short. Is it reasonable to give the ethical advantage to freedom of choice (*25*)? Does this not mean that a great chance is being taken? Is it not unethical to take risks of that sort, and all the more so since others, rather than ourselves, will have to bear the burden if the risk-taking turns out disastrously? In particular, would it not be irresponsible for governments to take risks of this magnitude?

Three kinds of responses to these questions are possible. First, as mentioned, it can and has been argued that freedom of choice has not been adequately tested. The absence of a safe, effective, and inexpensive contraceptive has been one hindrance, particularly in developing countries; it is reasonable to expect that such a contraceptive will eventually be developed. The weakness of existing family planning programs (and population policies dependent upon them) has, in great part, been the result of inadequate financing, poor administration, and scanty research and survey data. These are correctable deficiencies, assuming that nations give population limitation a high priority. If they do not give population limitation a high priority, it is unlikely that more drastic population policies could be successfully introduced or implemented. Very little effort has been expended anywhere in the world to educate people and persuade them to change their procreation habits. Until a full-scale effort has been made, there are few good grounds for asserting that voluntary limitation will be ineffective.

Second, while the question of scientific-medical-technological readiness, political viability, administrative feasibility, economic capability, and assumed effectiveness of proposals that would go beyond family planning is not directly ethical in nature, it has important ethical implications. If all of these categories seem to militate against the practical possibility of instituting very strong, immediate, or effective coercive measures, then it could become irresponsible to press for or support such measures. This would especially be the case if attention were diverted away from what could be done, for example, an intensification of family planning programs.

Third, primacy has been given to freedom of choice for ethical reasons. Whether this freedom will work as a means of population limitation is a separate question. A strong indication that freedom of choice will be ineffective does not establish grounds for rejecting it. Only if it can be shown that the failure of this freedom to reduce population growth threatens other important human values, thus establishing a genuine conflict of values, would the way be open to remove it from the place of primacy. This is only another way of asserting that freedom of choice is a right, grounded in a commitment to human dignity. The concept of a "right" becomes meaningless if rights are wholly subject to tests of economic, social, or demographic utility, to be given or withheld depending upon their effectiveness in serving social goals.

In this sense, to predicate human rights at all is to take a risk. It is to assert that the respect to be accorded human beings ought not to be dependent upon majority opinion, cost-benefit analysis, social utility, governmental magnanimity, or popular opinion. While it is obviously necessary to adjudicate conflicts among rights, and often to limit one right in order to do justice to another, the pertinent calculus is that of rights, not of utility. A claim can be entered against the primacy of one right only in the name of one or more other important rights. The proper route to a limitation of rights is not directly from social facts (demographic, economic, and so on) to rights, as if these facts were enough in themselves to prove the case against a right. The proper route is from showing that the social facts threaten rights, and in what way, to showing that a limitation of one right may be necessary to safeguard or enhance other rights. To give primacy to the right of free choice is to take a risk. The justification for the risk is the high value assigned to the right, a value that transcends simply utilitarian considerations.

REFERENCES AND NOTES

1. E.H. Erikson, *Insight and Responsibility* (Norton, New York, 1964), p. 132.
2. B. Berelson, in *Is There an Optimum Level of Population?*, S. F. Singer, Ed. (McGraw-Hill, New York, 1971), p. 305.
3. See, for instance, P. R. Ehrlich and A. H. Ehrlich, *Population, Resources, Environment Issues in Human Ecology* (Freeman, San Francisco, 1970), pp. 321–324.
4. A 1967 Gallup Poll, for example, revealed that, while 54 percent of those surveyed felt that the rate of American population growth posed a serious problem, crime, racial discrimination, and poverty were thought to be comparatively more serious social problems. J. F. Kanther, *Stud. Fam. Plann.* No. 30 (May 1968), p. 6.
5. R. B. Potter, Jr., in *Freedom, Coercion and the Life Sciences,* L. Kass and D. Callahan, Eds., in press.
6. K. E. Boulding, *Amer. Econ. Rev. 59,* 1 (March 1969).
7. *Final Act of the International Conference on Human Rights* (United Nations, New York, 1968), p. 15.
8. "Declaration on Population: The World Leaders Statement," *Stud. Fam. Plann.* No. 26 (January 1968), p. 1.
9. For instance, not only has Garrett Hardin, in response to the "The World Leaders' Statement" *(8),* denied the right of the family to choose family size, he

has also said that "If we love the truth we must openly deny the validity of the Universal Declaration of Human Rights, even though it is promoted by the United Nations" [*Science 162,* 1246 (1968)]. How literally is one to take this statement? The declaration, after all, affirms such rights as life, liberty, dignity, equality, education, privacy, and freedom of thought. Are none of these rights valid? Or, if those rights are to remain valid, why is only the freedom to control family size to be removed from the list?

10. See A. S. Parkes, in *Biology and Ethics,* F. J. Ebling, Ed. (Academic Press, New York, 1969), pp. 109–116.

11. In general, "antinatalist" means "attitudes or policies directed toward a reduction of births," and "pronatalist" means "attitudes or policies directed toward an increase in births."

12. See B. Berelson, *Stud. Fam. Plann.* No. 38 (February 1969), p. 1.

13. Universal Declaration of Human Rights," in *Human Rights: A Compilation of International Instruments of the United Nations* (United Nations, New York, 1967).

14. See E. Pohlman and K. G. Rao, *Licentiate 17,* 236 (1967).

15. See, for instance, R. G. Ridker, *Stud. Fam. Plann.* No. 43 (June 1969), p. 11.

16. The payments made in six different family planning programs are listed in *Incentive Payments in Family Planning Programmes* (International Planned Parenthood Federation, London, 1969), pp. 8–9.

17. G. Hardin, *Fam. Plann. Perspect. 2,* 26 (June 1970).

18. See, for example, W. H. Davis, *New Repub.* (20 June 1970), p. 19.

19. P. R. Ehrlich, *The Population Bomb* (Ballantine Books, New York, 1968), pp. 158–173.

20. See the "International Covenant on Economic, Social and Cultural Rights," Article 1, section 1, paragraph 1, in *Human Rights: A Compilation of International Instruments of the United Nations* (United Nations, New York, 1967), p. 4: "All people have the right to self-determination. By virtue of that right they freely pursue their economic, social and cultural development."

21. "Everyone has the right to life, liberty and the security of person" (13).

22. One of the few recent discussions on the obligation to future generations is in M. P. Golding [*UCLA Law Review 15,* 457 (February 1968)].

23. See D. Nortman, in *Reports on Population/Family Planning* (Population Council, New York, December 1969), pp. 1–48. Judith Blake is pessimistic about the possibilities of family planning programs [*J. Chronic Dis. 18,* 1181 (1965)]. See also J. Blake [*Science 164,* 522 (1969)] and the reply of O. Harkavy, F. S. Jaffe, S. M. Wishik [*Ibid. 165,* 367 (1969)].

24. See E. Pohlman, *Eugen. Quart. 13,* 122 (June 1966): "The spectre of 'experts' monkeying around with such private matters as family size desires frightens many people as being too 'Big Brotherish.' But those involved in eugenics, or psychotherapy, or child psychology, or almost any aspect of family planning are constantly open to the charge of interfering in private lives, so that the charge would not be new. . . . Of course, many injustices have been done with the rationale of being 'for their own good.' But the population avalanche may be used to justify—perhaps rationalize—contemplation of large-scale attempts to manipulate family size desires, even rather stealthily." This mode of reasoning may explain how some people will think and act, but it does not constitute anything approaching an ethical justification.

25. P. R. Ehrlich (*19,* pp. 197–198) argues that the taking of strong steps now to curb population growth is the wiser and safer gamble than doing nothing or too little. This seems to me a reasonable enough position, up to a point. That point would come when the proposed steps would seriously endanger human dignity; an ethic of survival, at the cost of other basic human values, is not worth the cost.

26. This article is an abridgment of an "Occasional Paper" [*Ethics and Population Limitation* (Population Council, New York, 1971)] and was written while the author was a staff associate at the Population Council in 1969–70. I would particularly like to thank Bernard Berelson for his suggestions and criticisms.

READING 25

Death: Process or Event?

Robert S. Morison

Most discussions of death and dying shift uneasily, and often more or less unconsciously, from one point of view to another. On the one hand, the common noun "death" is thought of as standing for a clearly defined event, a step function that puts a sharp end to life. On the other, dying is seen as a long-drawn-out process that begins when life itself begins and is not completed in any given organism until the last cell ceases to convert energy.

The first view is certainly the more traditional one. Indeed, it is so deeply embedded, not only in literature and art, but also in the law, that it is hard to free ourselves from it and from various associated attitudes that greatly influence our behavior. This article analyzes how the traditional or literary conception of death may have originated and how this conception is influencing the way in which we deal with the problem of dying under modern conditions. In part, I contend that some of our uses of the term "death" fall close to, if not actually within, the definition of what Whitehead called the "fallacy of misplaced concreteness" (1). As he warned, "This fallacy is the occasion of great confusion in philosophy," and it may also confuse our handling of various important practical matters.

Nevertheless, there is evidence that the fallacy may be welcomed by some physicians because it frees them from the necessity of looking certain unsettling facts in the face.

In its simplest terms, the fallacy of misplaced concreteness consists in regarding or using an abstraction as if it were a thing, or, as Whitehead puts it, as a "simple instantaneous material configuration." Examples of a relatively simple kind can be found

Reprinted by permission of the author and *Science,* Vol. 173, 20 August 1971, copyright 1971 by the American Association for the Advancement of Science, pp. 694–698.

throughout science to illustrate the kinds of confusion to which the fallacy leads. Thus, our ancestors who observed the behavior of bodies at different temperatures found it convenient to explain some of their observations by inventing an abstraction they called heat. All too quickly the abstract concept turned into an actual fluid that flowed from one body to another. No doubt these conceptions helped to develop the early stages of thermodynamics. On the other hand, the satisfaction these conceptions gave their inventors may also have slowed down the development of the more sophisticated kinetic theory.

It should be quite clear that, just as we do not observe a fluid heat, but only differences in temperature, we do not observe "life" as such. Life is not a thing or a fluid any more than heat is. What we observe are some unusual sets of objects separated from the rest of the world by certain peculiar properties such as growth, reproduction, and special ways of handling energy. These objects we elect to call "living things." From here, it is but a short step to the invention of a hypothetical entity that is possessed by all living things and that is supposed to account for the difference between living and nonliving things. We might call this entity "livingness," following the usual rule for making abstract nouns out of participles and adjectives. This sounds rather awkward, so we use the word "life" instead. This apparently tiny change in the shape of the noun helps us on our way to philosophical error. The very cumbersomeness of the word "livingness" reminds us that we have abstracted the quality for which it stands from an array of living things. The word "life," however, seems much more substantial in its own right. Indeed, it is all too easy to believe that the word, like so many other nouns, stands for something that must have an existence of its own and must be definable in general terms, quite apart from the particular objects it characterizes. Men thus find themselves thinking more and more about life as a thing in itself, capable of entering inanimate aggregations of material and turning them into living things. It is then but a short step to believing that, once life is there, it can leave or be destroyed, thereby turning living things into dead things.

Now that we have brought ourselves to mention dead things, we can observe that we have invented the abstract idea of death by observing dead things, in just the same way that we have invented the idea of life by observing living things. Again, in the same way that we come to regard life as a thing, capable of entering and leaving bodies, we come to regard death as a thing, capable of moving about on its own in order to take away life. Thus, we have become accustomed to hearing that "death comes for the archbishop," or, alternatively, that one may meet death by "appointment in Samara." Only a very few, very sophisticated old generals simply fade away.

In many cases then, Death is not only reified, it is personified, and graduates from a mere thing to a jostling woman in the marketplace of Baghdad or an old man, complete with beard, scythe, and hourglass, ready to mow down those whose time has come. In pointing to some of the dangers of personification, it is not my purpose to abolish poetry. Figures of speech certainly have their place in the enrichment of esthetic experience, perhaps even as means for justifying the ways of God to man. Nevertheless, reification and personification of abstractions do tend to make it difficult to think clearly about important problems.

ABSTRACTIONS CAN LEAD TO ARTIFICIAL DISCONTINUITY

A particularly frequent hazard is the use of abstractions to introduce artificial discontinuities into what are essentially continuous processes. For example, although it is convenient to think of human development as a series of stages, such periods as childhood and adolescence are not "instantaneous configurations" that impose totally different types of behavior on persons of different ages. The infant does not suddenly leave off "mewling and puking" to pick up a satchel and go to school. Nor at the other end of life does "the justice, . . with eyes severe and beard of formal cut" instantly turn into "the lean and slipper'd pantaloon." The changes are gradual; finally, the pantaloon slips through second childishness into "mere oblivion, sans teeth, sans eyes, sans taste, sans everything" (2). Clearly we are dealing here with a continuous process of growth and decay. There is no magic moment at which "everything" disappears. Death is no more a single, clearly delimited, momentary phenomenon than is infancy, adolescence, or middle age. The gradualness of the process of dying is even clearer than it was in Shakespeare's time, for we now know that various parts of the body can go on living for months after its central organization has disintegrated. Some cell lines, in fact, can be continued indefinitely.

The difficulty of identifying a moment of death has always been recognized when dealing with primitive organisms, and the conventional concept has usually not been applied to organisms that reproduce themselves by simple fission. Death as we know it, so to speak, is characteristic only of differentiated and integrated organisms, and is most typically observed in the land-living vertebrates in which everything that makes life worth living depends on continuous respiratory movements. These, in turn, depend on an intact brain, which itself is dependent on the continuing circulation of properly aerated blood. Under natural conditions, this tripartite, interdependent system fails essentially at one and the same time. Indeed, the moment of failure seems often to be dramatically marked by a singularly violent last gasping breath. Observers of such a climactic agony have found it easy to believe that a special event of some consequence has taken place, that indeed Death has come and Life has gone away. Possibly even some spirit or essence associated with Life has left the body and gone to a better world. In the circumstances surrounding the traditional deathbed, it is scarcely to be wondered at that many of the observers found comfort in personifying the dying process in this way, nor can it be said that the consequences were in any way unfortunate.

Now, however, the constant tinkering of man with his own machinery has made it obvious that death is not really a very easily identifiable event or "configuration." The integrated physiological system does not inevitably fail all at once. Substitutes can be devised for each of the major components, and the necessary integration can be provided by a computer. All the traditional vital signs are still there—provided in large part by the machines. Death does not come by inevitable appointment, in Samarra or anywhere else. He must sit patiently in the waiting room until summoned by the doctor or nurse.

Perhaps we should pause before being completely carried away by the metaphor. Has death really been kept waiting by the machines? If so, the doctor must be actively causing death when he turns the machines off. Some doctors, at least, would prefer to avoid the responsibility, and they have therefore proposed a different view of the process (3). They would like to believe that Death has already come for the patient whose vital signs are maintained by machine and that the doctor merely reveals the results of his visit. But if Death has already come, he has certainly come without making his presence known in the usual way. None of the outward and visible signs have occurred—no last gasp, no stopping of the heart, no cooling and stiffening of the limbs. On the other hand, it seems fairly obvious to most people that life under the conditions described (if it really is life) falls seriously short of being worth living.

IS A "REDEFINITION" OF DEATH ENOUGH?

We must now ask ourselves how much sense it makes to try to deal with this complex set of physiological, social, and ethical variables simply by "redefining" death or by developing new criteria for pronouncing an organism dead. Aside from the esoteric philosophical concerns discussed so far, it must be recognized that practical matters of great moment are at stake. Fewer and fewer people die quietly in their beds while relatives and friends live on, unable to stay the inevitable course. More and more patients are subject to long, continued intervention; antibiotics, intravenous feeding, artificial respiration, and even artificially induced heartbeats sustain an increasingly fictional existence. All this costs money—so much money, in fact, that the retirement income of a surviving spouse may disappear in a few months. There are other costs, less tangible but perhaps more important—for example, the diversion of scarce medical resources from younger people temporarily threatened by acute but potentially curable illness. Worst of all is the strain on a family that may have to live for years in close association with a mute, but apparently living, corpse.

An even more disturbing parameter has recently been added to the equation. It appears that parts of the dying body may acquire values greater than the whole. A heart, a kidney, someday even a lung or a liver, can mean all of life for some much younger, more potentially vigorous and happy "donee."

Indeed, it appears that it is primarily this latter set of facts which has led to recent proposals for redefining death. The most prominent proposals place more emphasis on the information-processing capacity of the brain and rather less on the purely mechanical and metabolic activities of the body as a whole than do the present practices. The great practical merit of these proposals is that they place the moment of death somewhat earlier in the continuum of life than the earlier definition did. By so doing, they make it easier for the physician to discontinue therapy while some of what used to be considered "signs of life" are still present, thus sparing relatives, friends, and professional attendants the anguish and the effort of caring for a "person" who has lost most of the attributes of personality. Furthermore, parts of the body which survive death, as newly defined, may be put to other, presumably more important uses, since procedures such as autopsies or removal of organs can be undertaken without being regarded as assaults.

In considering the propriety of developing these new criteria, one may begin by admitting that there is nothing particularly unusual about redefining either a material fact or a nebulous abstraction. Physical scientists are almost continuously engaged in redefining facts by making more and more precise measurements. Taxonomists spend much of their time redefining abstract categories, such as "species," in order to take into account new data or new prejudices. At somewhat rarer intervals, even such great concepts as force, mass, honor, and justice may come up for review.

Nevertheless, in spite of the obvious practical advantages and certain theoretical justifications, redefinition of abstractions can raise some very serious doubts. In the present instance, for example, we are brought face to face with the paradox that the new definitions of death are proposed, at least in part, because they provide that certain parts of the newly defined dead body will be *less dead* than they would have been if the conventional definition were still used. Looked at in this light, the proposed procedure raises serious ethical questions (4). The supporters of the new proposal are, however, confronted every day by the even more serious practical problems raised by trying to make old rules fit new situations. Faced with a dilemma, they find it easier to urge a redefinition of death than to recognize that life may reach a state such that there is no longer an ethical imperative to preserve it. While one may give his support to the first of these alternatives as a temporary path through a frightening and increasingly complicated wilderness, it might be wise not to congratulate ourselves prematurely.

As our skill in simulating the physiological processes underlying life continues to increase in disproportion to our capacity to maintain its psychological, emotional, or spiritual equality, the difficulty of regarding death as a single, more or less coherent event, resulting in the instantaneous dissolution of the organism as a whole, is likely to become more and more apparent. It may not be premature, therefore, to anticipate some of the questions that will then increasingly press upon us. Some of the consequences of adopting the attitude that death is part of a continuous process that is coextensive (almost) with living may be tentatively outlined as follows.

An unprejudiced look at the biological facts suggests, indeed, that the "life" of a complex vertebrate like man is not a clearly defined entity with sharp discontinuities at both ends. On the contrary, the living human being starts inconspicuously, unconsciously, and at an unknown time, with the conjugation of two haploid cells. In a matter of some hours, this new cell begins to divide. The net number of living cells in the organism continues to increase for perhaps 20 years, then begins slowly to decrease. Looked at in this way, life is certainly not an all-or-none phenomenon. Clearly the amount of living matter follows a long trajectory of growth and decline with no very clear beginning and a notably indeterminate end. A similar trajectory can be traced for total energy turnover.

A human life is, of course, far more than a metabolizing mass of organic matter, slavishly obeying the laws of conservation of mass and energy. Particularly interesting are the complex interactions among the individual cells and between the totality and the environment. It is, in fact, this complexity of interaction that gives rise to the concept of human personality or soul.

Whatever metaphors are used to describe the situation, it is clear that it is the complex interactions that make the characteristic human being. The appropriate integration of these interactions is only loosely coupled to the physiological functions of circulation and respiration. The latter continue for a long time after the integrated "personality" has disappeared. Conversely, the natural rhythms of heart and respiration can fail, while the personality remains intact. The complex human organism does not often fail as a unit. The nervous system is, of course, more closely coupled to personality than are the heart and lungs (a fact that is utilized in developing the new definitions of death), but there is clearly something arbitrary in tying the sanctity of life to our ability to detect the electrical potential charges that managed to traverse the impedance of the skull.

If there is no infallible physiological index to what we value about human personality, are we not ultimately forced to make judgments about the intactness and value of the complex interactions themselves?

"VALUE" OF A LIFE CHANGES WITH VALUE OF COMPLEX INTERACTIONS

As the complexity and richness of the interactions of an individual human being wax and wane, his "value" can be seen to change in relation to other values. For various reasons it is easier to recognize the process at the beginning than at the end of life. The growing fetus is said to become steadily more valuable with the passage of time (5): its organization becomes increasingly complex and its potential for continued life increases. Furthermore, its mother invests more in it every day and becomes increasingly aware of and pleased by its presence. Simultaneous with these increases in "value" is the increased "cost" of terminating the existence of the fetus. As a corollary, the longer a pregnancy proceeds, the more reasons are required to justify its termination. Although it may be possible to admire the intellectual ingenuity of Saint Thomas and others who sought to break this continuous process with a series of discontinuous stages and to identify the moment at which the fetus becomes a human being, modern knowledge of the biological process involved renders all such efforts simply picturesque. The essential novelty resides in the formation of the chromosomal pattern—the rest of the development is best regarded as the working out of a complicated tautology.

At the other end of life the process is reversed: the life of the dying patient becomes steadily less complicated and rich, and, as a result, less worth living or preserving. The pain and suffering involved in maintaining what is left are inexorably mounting, while the benefits enjoyed by the patient himself, or that he can in any way confer on those around him, are just as inexorably declining. As the costs mount higher and higher and the benefits become smaller and smaller, one may well begin to wonder what the point of it all is. These are the unhappy facts of the matter, and we will have to face them sooner or later. Indeed, attempts to face the facts are already being made, but usually in a gingerly and incomplete fashion. As we have seen, one way to protect ourselves is to introduce imaginary discontinuities into what are, in fact, continuous processes.

A similar kind of self-deception may be involved in attempts to find some crucial differences among the three following possibilities that are open to the physician attending the manifestly dying patient.

1) Use all possible means (including the "extraordinary measures" noted by the Pope) to keep the patient alive.
2) Discontinue the extraordinary measures but continue "ordinary therapy."
3) Take some "positive" step to hasten the termination of life or speed its downward trajectory.

Almost everyone now admits that there comes a time when it is proper to abandon procedure 1 and shift to procedure 2 although there is a good deal of disagreement about determining the moment itself. There is much less agreement about moving to procedure 3, although the weight of opinion seems to be against ever doing so.

The more one thinks of actual situations, however, the more

one wonders if there is a valid distinction between allowing a person to die and hastening the downward course of life. Sometimes the words "positive" and "negative" are used, with the implication that it is all right to take away from the patient something that would help him to live but wrong to give him something that will help him to die.

The intent appears to be the same in the two cases, and it is the intent that would seem to be significant. Furthermore, one wonders if the dividing point between positive and negative in this domain is any more significant than the position of zero on the Fahrenheit scale. In practice, a physician may find it easier not to turn on a respirator or a cardiac pacemaker than to turn them off once they have been connected, but both the intents and the results are identical in the two cases. To use an analogy with mathematics, subtracting one from one would seem to be the same as not adding one to zero.

Squirm as we may to avoid the inevitable, it seems time to admit to ourselves that there is simply no hiding place and that we must shoulder the responsibility of deciding to act in such a way as to hasten the declining trajectories of some lives, while doing our best to slow down the decline of others. And we have to do this on the basis of some judgment on the quality of the lives in question.

Clearly the calculations cannot be made exclusively or even primarily on crude monetary or economic criteria. Substantial value must be put on intangibles of various kinds—the love, affection, and respect of those who once knew the fully living individual will bulk large in the equation. Another significant parameter will be the sanctity accorded to any human life, however attenuated and degraded it may have become. Respect for human life as such is fundamental to our society, and this respect must be preserved. But this respect need not be based on some concept of absolute value. Just as we recognize that an individual human life is not infinite in duration, we should now face the fact that its value varies with time and circumstance. It is a heavy responsibility that our advancing command over life has placed on us.

It has already been noted that in many nations, and increasingly in the United States, men and women have shouldered much the same kind of responsibility—but apparently with considerably less horror and dismay—at the beginning of the life-span. In spite of some theological misgivings and medical scruples, most societies now condone the destruction of a living fetus in order to protect the life of the mother. Recent developments have greatly broadened the "indications" to include what is essentially the convenience of the mother and the protection of society against the dangers of over-population.

A relatively new, but very interesting, development is basing the decision of whether or not to abort purely on an assessment of the quality of the life likely to be lived by the human organism in question. This development has been greatly enhanced by advances in the technique of amniocentesis, with its associated methods for determining the chromosomal pattern and biochemical competence of the unborn baby. Decisions made on such grounds are difficult, if not impossible, to differentiate, in principle, from decisions made by the Spartans and other earlier societies to expose to nature those infants born with manifest anatomical defects. We are being driven toward the ethics of an earlier period by the inexorable logic of the situation, and it may only increase our discomfort without changing our views to reflect that historians (6) and moralists (7) both agree that the abolition of infanticide was perhaps the greatest ethical achievement of early Christianity.

ISSUE CANNOT BE SETTLED BY ABSOLUTE STANDARDS

Callahan (5) has recently reviewed all the biological, social, legal, and moral issues that bear on decisions to terminate life in its early stages and argues convincingly that the issue cannot be settled by appeals to absolute rights or standards. Of particular importance for our purpose, perhaps, is his discussion of the principle of the "sanctity of life," since opposition to liberalizing the abortion laws is so largely based on the fear of weakening respect for the dignity of life in general. It is particularly reassuring, therefore, that Callahan finds no objective evidence to support this contention. Indeed, in several countries agitation for the liberalization of abortion laws has proceeded simultaneously with efforts to strengthen respect for life in other areas—the abolition of capital punishment, for example. Indeed, Callahan's major thesis is that modern moral decisions can seldom rest on a single, paramount principle; they must be made individually, after a careful weighing of the facts and all the nuances in each particular case.

The same considerations that apply to abortion would appear to apply, in principle, to decisions at the other end of the life-span. In practice, however, it has proven difficult to approach the latter decisions with quite the same degree of detachment as those involving the life and death of an unborn embryo. It is not easy to overlook the fact that the dying patient possesses at least the remnants of a personality that developed over many decades and that involved a complicated set of interrelationships with other human beings. In the case of the embryo, such relationships are only potential, and it is easier to ignore the future than to overlook the past. It can be argued, however, that it should be easier to terminate a life whose potentialities have all been realized than to interrupt a pregnancy the future of which remains to be unfolded.

Once it is recognized that the process of dying under modern conditions is at least partially controlled by the decisions made by individual human beings, it becomes necessary to think rather more fully and carefully about what human beings should be involved and what kinds of considerations should be taken into account in making the decisions.

Traditionally it has been the physician who has made the decisions, and he has made them almost exclusively on his own view of what is best for the patient. Only under conditions of special stress, where available medical resources have been clearly inadequate to meet current needs, has the physician taken the welfare of third parties or "society" into account in deciding whether to give or withhold therapy. Until recently, such conditions were only encountered on the battlefield or in times of civilian catastrophe such as great fires, floods, or shipwrecks. Increasingly, however, the availability of new forms of therapy that depend on inherently scarce resources demands that decisions be made about distribution. In other words, the physician who is considering putting a patient on an artificial kidney may sometimes be forced to consider the needs of other potiential users of the same device. The situation is even more difficult when the therapeutic device is an organ from another human being. In some communities, the burden of such decisions is shifted from a single physician to a group or committee that may contain nonmedical members.

These dramatic instances are often thought of as being special cases without much relationship to ordinary life and death. On the other hand, one may look upon them as simply more brilliantly colored examples of what is generally true but is not always so easy to discern. Any dying patient whose life is unduly prolonged imposes serious costs on those immediately around him and, in many cases, on a larger, less clearly defined "society." It seems probable that, as these complex interrelationships are increasingly recognized, society will develop procedures for sharing the necessary decisions more widely, following the examples of the committee structure now being developed to deal with the dramatic cases.

It is not only probable, but highly desirable that society should proceed with the greatest caution and deliberation in proposing procedures that in any serious way threaten the traditional sanctity of the individual life. As a consequence, society will certainly move very slowly in developing formal arrangements for taking into account the interests of others in life-and-death decisions. It may not be improper, however, to suggest one step that could be taken right now. Such a step might ease the way for many dying patients without impairing the sanctity or dignity of the individual life: instead, it should be enhanced. I refer here to the possibility of changing social attitudes and laws that now restrain the individual from taking an intelligent interest in his own death.

The Judeo-Christian tradition has made suicide a sin of much the same character as murder. The decline of orthodox theology has tended to reduce the sinfulness of the act, but the feeling still persists that there must be something wrong with somebody who wants to end his own life. As a result, suicide, when it is not recognized as a sin, is regarded as a symptom of serious mental illness. In this kind of atmosphere, it is almost impossible for a patient to work out with his doctor a rational and esthetically satisfactory plan for conducting the terminating phase of his life. Only rarely can a great individualist like George Eastman or Percy Bridgman (8) transcend the prevailing mores to show us a rational way out of current prejudice. Far from injuring the natural rights of the individual, such a move can be regarded as simply a restoration of a right once greatly valued by our Roman ancestors, who contributed so much to the "natural law" view of human rights. Seneca (9), perhaps the most articulate advocate of the Roman view that death should remain under the individual's control, put the matter this way: "To death alone it is due that life is not a punishment, that erect beneath the frowns of fortune, I can preserve my mind unshaken and be master of myself."

REFERENCES

1. A. N. Whitehead, *Science and the Modern World* (Macmillan, New York, 1967), pp. 51–55.
2. Shakespeare, *As You Like It,* Act II, Scene vii, 11. 144–166.
3. H. K. Beecher, *J. Am. Med. Assoc. 205,* 337 (1968).
4. P. Ramsey, in *Updating Life and Death,* D. R. Cutler, Ed. (Beacon, Boston, 1969), p. 46; H. Jonas, in *Experimentation with Human Subjects,* P. A. Freund, Ed. (Braziller, New York, 1970), pp. 10–11.

5. D. Callahan, *Abortions: Law, Choice and Morality* (Macmillan, New York, 1970).
6. W. E. H. Lecky, *History of European Morals from Augustus to Charlemagne* (Appleton, New York, 1870), vol. 1.
7. H. Sidgwick, *Outlines of the History of Ethics* (Macmillan, London, 1886; Beacon, Boston, 1960).
8. G. Holton, *Bull. At. Sci. 18* (No. 2), 22 (February 1962).
9. Seneca, *Ad Marciam, de Consolatione,* XX, translated by W. E. H. Lecky, in *History of European Morals from Augustus to Charlemagne* (Appleton, New York, 1870), vol. 1, p. 228.

READING 26

Death as an Event: A Commentary on Robert Morison

Attempts to blur the distinction between a man alive and a man dead are both unsound and dangerous.

Leon R. Kass

As I understand R. S. Morison's argument, it consists of these parts, although presented in different order. First: He notes that we face serious practical problems as a result of our unswerving adherence to the principle, "Always prolong life." Second: Although *some* of these problems could be solved by updating the "definition of death," such revisions are scientifically and philosophically unsound. Third: The reason for this is that life and death are part of a continuum; it will prove impossible, in practice, to identify any border between them because theory tells us that no such border exists. Thus: We need to abandon both the idea of death as a concrete event and the search for its definition; instead, we must face the fact that our practical problems can only be solved by difficult judgments, based upon a complex cost-benefit analysis, concerning the value of the lives that might or might not be prolonged.

I am in agreement with Morison only on the first point. I think he leads us into philosophical, scientific, moral, and political error. Let me try to show how.

SOME BASIC DISTINCTIONS

The difficulties begin in Morison's beginning, in his failure to distinguish clearly among aging, dying, and dead. His statement that

Reprinted by permission of the author and *Science*, Vol. 173, 20 August 1971, copyright 1971 by the American Association for the Advancement of Science, pp. 698–703.

"dying is seen as a long-drawn-out process that begins when life itself begins" would be remarkable, if true, since it would render dying synonymous with living. One consequence would be that murder could be considered merely a farsighted form of euthanasia, a gift to the dying of an early exit from their miseries (1). But we need not ponder these riddles, because what Morison has done is to confuse dying with aging. Aging (or senescence) apparently does begin early in life (though probably not at conception), but there is no clear evidence that it is ever the cause of death. As Sir Peter Medawar has pointed out (2):

> Senescence, then, may be defined as that change of the bodily faculties and sensibilities and energies which accompanies aging, and which renders the individual progressively more likely to die from accidental causes of random incidence. Strictly speaking, the word "accidental" is redundant, for all deaths are in some degree accidental. No death is wholly "natural"; no one dies *merely* of the burden of the years.

As distinguished from aging, dying would be the process leading from the incidence of the "accidental" cause of death to and beyond some border, however ill-defined, after which the organism (or its body) may be said to be dead.

Morison observes, correctly, that death and life are abstractions, not things. But to hold that "livingness" or "life" is the property shared by living things, and thus to abstract this property *in thought,* does not necessarily lead one to hold that "life or "livingness" is a thing in itself with an existence apart from the objects said to "possess" it. For reification and personification of life and death, I present no argument. For the adequacy of the abstractions themselves, we must look to the objects described.

What about these objects: living, nonliving, and dead things? A person who believes that living things and nonliving things do not differ in kind would readily dismiss "death" as a meaningless concept. It is hard to be sure that this is not Morison's view. When he says, "These objects *we elect to call* 'living things' [emphasis added]," is he merely being overly formal in his presentation, or is he deliberately intimating that the distinction between living and nonliving is simply a convention of human speech, and not inherent in the nature of things? My suspicions are increased by his suggestion that "substitutes can be devised for each of the major components [of a man], and the necessary integration can be provided by a computer." A living organism comprising mechanical parts with computerized "integration"? Morison should be asked to clarify this point: Does he hold that there is or is not a *natural* distinction between living and nonliving things? Are his arguments about the fallacy of misplaced concreteness of "death" and "life" merely secondary and derivative from his belief that living and nonliving or dead objects do not differ in kind (3)?

If there is a natural distinction between living and nonliving things, what is the proper way of stating the nature of that difference? What is the real difference between something alive and that "same" something dead? To this crucial question, I shall return later. For the present, it is sufficient to point out that the real source of our confusion about death is probably our confusion about living things. The death of an organism is not understandable because its "aliveness" is not understood except in terms of nonliving matter and motion (4).

One further important distinction must be observed. We must keep separate two distinct and crucial questions facing the physician: (i) When, if ever, is a person's life no longer worth prolonging? and (ii) When is a person in fact dead? The first question translates, in practice, to: When is it permissible or desirable for a physician to withhold or withdraw treatment so that a patient (still alive) may be allowed to die? The second question translates, in practice, to: When does the physician pronounce the (ex)patient fit for burial? Morison is concerned only with the first question. He commendably condemns attempts to evade this moral issue by definitional wizardry. But regardless of how one settles the question of whether and what kind of life should be prolonged, one will still need criteria for recognizing the end. The determination of death may not be a very interesting question, but it is an extremely important one. At stake are matters of homicide and inheritance, of burial and religious observance, and many others.

In considering the definition and determination of death, we note that there is a difference between the meaning of an abstract concept such as death (or mass or gravity or time) and the operations used to determine or measure it. There are two "definitions" that should not be confused. There is the conceptual "definition" or meaning and the operational "definition" or meaning. I think it would be desirable to use "definition of death" only with respect to the first, and to speak of "criteria for determining that a death has occurred" for the second. Thus, the various proposals for updating the definition of death (5), their own language to the contrary, are not offering a new definition of death but merely refining the procedure stating that a man has died. Although there is much that could be said about these proposals, my focus here is on Morison's challenge to the concept of death as an event, and to the possibility of determining it.

THE CONCEPT OF DEATH

There is no need to abandon the traditional understanding of the concept of death: Death is the transition from the state of being alive to the state of being dead. Rather than emphasize the opposition between death and life, an opposition that invites Morison to see the evils wrought by personification, we should concentrate, for our purposes, on the opposition between death and birth (or conception). Both are transitions, however fraught with ambiguities. Notice that the notion of transition leaves open the question of whether the change is abrupt or gradual and whether it is continuous or discontinuous. But these questions about *when* and *how* cannot be adequately discussed without some substantive understanding of *what* it is that dies.

What dies is the organism as a whole. It is this death, the death of the individual human being, that is important for physicians and for the community, not the "death" of organs or cells, which are mere parts.

> The ultimate, most serious effect of injury is death. Necrosis is death but with this limitation; it is death of cells or tissue WITHIN A LIVING ORGANISM. Thus we differentiate between *somatic death,* which is death of the whole, and *necrosis,* which is death of the part.
>
> From a tissue viewpoint, even when the whole individual dies, he dies part by part and at different times, For instance, nerve cells die within a few minutes after circulation stops, whereas cartilage cells may remain alive for several days. Because of this variation in cellular susceptibility to injury, it is virtually impossible to say just when all the component parts of the body have died. Death of composite whole, the organism as an INTEGRATED functional unit, is a different matter. Within three or four minutes after the heart stops beating, hypoxia ordinarily leads to irreversible changes of certain vital tissues, particularly those of the central nervous system, and this causes the INDIVIDUAL to die (6).

The same point may perhaps be made clear by means of an anecdote. A recent discussion on the subject of death touched on the postmortem perpetuation of cell lines in tissue culture. Someone commented, "For all I know, I myself might wind up in one of those tissue-culture flasks." The speaker was asked to reconsider whether he really meant "I myself" or merely some of his cells.

IS DEATH A DISCRETE EVENT?

A proof that death is not a discrete event (7)—that life and death are part of a continuum—would thus require evidence that the organism as a whole died progressively and continuously. This evidence Morison does not provide. Instead he calls attention to the continuity of the different ages of man and to growth and decay, but he does not show that any of these changes are analogous to the transition of death. The continuity between childhood and adolescence says nothing about whether the transition between life and death is continuous. He also mentions the "postmortem" viability of cells and organs. He says that "various parts of the body can go on living for months after its central organization has disintegrated." It should be clear by now that the viability of *parts* has no necessary bearing on the question of the whole. His claim that the beginning of life is not a discrete event ("the living human being starts inconspicuously, unconsciously, and at an unknown time, with the conjugation of two haploid cells"), even granting the relevance of the analogy with death, is really only a claim that we do not see and hence cannot note the time of the event. Morison himself more than once identifies the beginning as the discrete event in which egg and sperm unite to form the zygote, with its unique chromosomal pattern.

Only in a few places does Morison even approach the question of the death of the organism as a whole. But his treatment only serves to discredit the question. ["The nervous system is, of course, more closely coupled to personality than are the heart and lungs (a fact that is utilized in developing the new definitions of death), but there is clearly something arbitrary in tying the sanctity of life to our ability to detect the electrical potential charges that managed to traverse the impedance of the skull."] Lacking a concept of the organism as a whole, and confusing the concept of death with the criteria for determining it, Morison errs by trying to identify the whole with one of its parts and by seeking a single "infallible physiological index" to human personhood. One might as well try to identify a watch with either its mainspring or its hands; the watch is neither of these, yet it is "dead" without either. Why is the concept of the organism as a whole so difficult to grasp? Is it because we have lost or discarded, in our reductionist biology, all notions of organism, of whole? (8).

Morison also attempts to discredit the "last gasp" as indicative of death as a discrete event: "Observers of such a climactic agony have found it easy to believe that a special event of some consequence has taken place, that indeed Death has come and Life has gone away." But if we forget about reification, personification, spirits fleeing, Death coming, Life leaving—is this not a visible sign of the death of the organism as a whole? This is surely a reasonable belief, and one which, if it now seems unreasonable, seems so only because of our tinkering.

Morison credits "the constant tinkering of man with his own machinery" for making it "obvious that death is not really a very easily identifiable event. . . ." To be sure, our tinkering has, in some cases, made it difficult to decide when the moment of death occurs, but does it really reveal that no such moment exists? Tinkering can often obscure rather than clarify reality, and I think this is one such instance. I agree that we are now in doubt about some borderline cases. But is the confusion ours or nature's? This is a crucial question. If the indeterminacy lies in nature, as Morison believes, then all criteria for determining death are arbitrary and all moments of death a fiction. If, however, the indeterminacy lies in *our* confusion and ignorance, then we must simply do the best we can in approximating the time of transition.

We are likely to remain ignorant of the true source of the indeterminacy. If so, then there is absolutely no good reason for insisting that it is nature's, and at least two good reasons for blaming ourselves. (i) It is foolish to abandon or discredit nature as a standard in matters of fundamental human importance: birth, death, health, sickness, origin. In the absence of this standard, we are left to our tastes and our prejudices about the most important human matters; we can never have knowledge, but, at best, only social policy developed out of a welter of opinion. (ii) We might thereby be permitted to see how we are responsible for confusing ourselves about crucial matters, how technological intervention (with all its blessings) can destroy the visible manifestations and signs of natural phenomena, the recognition of which is indispensible to human community. Death was once recognizable by any ordinary observer who could see (or feel or hear). Today, in some difficult cases, we require further technological manipulation (from testing of reflexes to the electroencephalogram) to make manifest latent signs of a phenomenon, the visible signs of which an earlier intervention has obscured.

In the light of these remarks, I would argue that we should not take our bearings from the small number of unusual cases in which there is doubt. In most cases, there is no doubt. There is no real need to blur the distinction between a man alive and a man dead or to undermine the concept of death as an event. Rather, we should ask, in the light of our traditional concepts (though not necessarily with traditional criteria), whether the persons in the twilight zone are alive or not, and find criteria on the far side of the twilight zone in order to remove any suspicion that a man may be pronounced dead while he is yet alive.

DETERMINING WHETHER A MAN HAS DIED

In my opinion, the question, "Is he dead?" can still be treated as a question of fact, albeit one with great moral and social consequences. I hold it to be a medical-scientific question in itself, not only in that physicians answer it for us. Morison treats it largely as a social-moral question. This is because, as I indicated above, he does not distinguish the question of when a man is dead from the question of when his life is not worth prolonging. Thus, there is a conjoined issue: Is the determination of death a matter of the true, or a matter of the useful or good?

The answer to this difficult question turns, in part, on whether or not medicine and science are in fact capable of determining death. Therefore, the question of the true versus the good (or useful) will be influenced by what is in fact true and knowable about death as a medical "fact." The question of the true versus the good (or useful) will also be influenced by the truth about what is good or useful, and by what people think to be good and useful. But we can and should also ask, "What is the truth?" about which one of these concerns—scientific truth or social good—is uppermost in the minds of people who write and speak about the determination of death.

To turn to Morison's paper in the light of the last question, it seems clear that his major concern is with utility. He abandons what he calls "esoteric philosophical concerns," his own characterization of his scientific discussion about death, to turn to "practical matters of great moment." Despite his vigorous scientific criticisms of the proposals for "redefining" death, he thinks they have "great practical merit," and thus he does not really oppose them as he would any other wrong idea. Am I unfair in thinking that his philosophical and scientific criticism of the concept of death is really animated by a desire to solve certain practical problems? Would the sweeping away of the whole concept of death for the unstated purpose of forcing a cost-benefit analysis of the value of prolonging lives be any less disingenuous than a redefinition of death for the sake of obtaining organs?

Morison properly criticizes those who would seek to define a man out of existence for the purpose of getting at his organs or of saving on scarce resources (9). He points out that the redefiners take unfair advantage of the commonly shared belief that a body, once declared dead, can be buried or otherwise used. His stand here is certainly courageous. But does he not show an excess of courage, indeed rashness, when he would decree death itself out of existence for the sake of similar social goods? Just how rash will be seen when his specific principles of social good are examined.

THE ETHICS OF PROLONGING LIFE

We are all in Morison's debt for inviting us to consider the suffering that often results from slavish and limitless attempts to prolong life. But there is no need to abandon traditional ethics to deal with this problem. The Judeo-Christian tradition, which teaches us the duty of preserving life, does not itself hold life to be the absolute value. The medical tradition, until very recently, shared this view. Indeed, medicine's purpose was originally *health,* not simply the unlimited prolongation of life or the conquest of disease and death. Both traditions looked upon death as a natural part of life, not as an unmitigated evil or as a sign of the physician's failure. We sorely need to recover this more accepting attitude toward death (10) and, with it, a greater concern for the human needs of the dying patient. We need to keep company with the dying and to help them cope with terminal illness (11). We must learn to desist from those useless technological interventions and institutional practices that deny to the dying what we most owe them—a good end. These purposes could be accomplished in large measure by restoring to medical practice the ethic of allowing a person to die (12).

But the ethic of allowing a person to die is based on a consideration of the welfare of the dying patient himself, rather than on a consideration of benefits that accure to others. This is a crucial point. It is one thing to take one's bearings from the patient and his interests and attitudes, to protect his dignity and his right to a good death against the onslaught of machinery and institutionalized loneliness; it is quite a different thing to take one's bearings from the interests of, or costs and benefits to, relatives or society. The first is in keeping with the physician's duty to act as the loyal agent of his patient; the second is a perversion of that duty, because it renders the physician in this decisive test of his loyalty, merely an agent of society, and ultimately, her executioner. The first upholds and preserves the respect for human life and personal dignity; the second sacrifices these on the ever-shifting altar of public opinion.

To be sure, the physician always operates within the boundaries set by the community—by its allocation of resources, by its laws, by its values. Each physician, as well as the profession as a whole, should perhaps work to improve these boundaries and especially to see that adequate resources are made available to better the public health. But in his relations with individual patients, the physician must serve the interest of the patient. Medicine cannot retain trustworthiness or trust if it does otherwise (13).

On this crucial matter, Morison seems to want to have it both ways. On the one hand, he upholds the interest of the deteriorating individual himself. Morison wants him to exercise a greater control over his own death, "to work out with his doctor a rational and esthetically satisfactory plan for conducting the terminating phase of his life." On the other hand, there are hints that Morison would like to see other interests served as well. For example, he says: "It appears that parts of the dying body may acquire values greater than the whole." Greater to whom? Certainly not to the patient. We are asked to consider that "Any dying patient whose life is unduly prolonged imposes serious costs on those immediately around him and, in many cases, on a larger, less clearly defined 'society.'" But cannot the same be said for any patient whose life is prolonged? Or is Morison suggesting that the "unduliness" of "undue" prolongation is to be defined in terms of social costs? In a strictly patient-centered ethic of allowing a person to die, these costs to others would not enter—except perhaps as they might influence the patient's own judgment about prolonging his own life.

In perhaps the most revealing passage, in which he merges both the interests of patient and society, Morison notes:

> ...the life of the dying patient becomes steadily less complicated and rich, and, as a result, less worth living or preserving. The pain and suffering involved in maintaining what is left are inexorably mounting, while the benefits enjoyed by the patient himself, or that he can in any way confer on those around him, are just as inexorably declining. As the costs mount higher and higher and the benefits become smaller and smaller, one may well begin to wonder what the point of it all is. These are the unhappy facts of the matter, and we have to face them sooner or later.

What are the implications of this analysis of costs and benefits? What should we do when we face these "unhappy facts?" The implication is clear: We must take, as the new "moment," the point at which the rising cost and declining benefit curves intersect, the time when the costs of keeping someone alive outweigh the value of his life. I suggest that it is impossible, both in principle and in practice, to locate such a moment, dangerous to try, and dangerously misleading to suggest otherwise. One simply cannot write an equation for the value of a person's life, let alone for comparing two or more lives. Life is incommensurable with the cost of maintaining it, despite Morison's suggestion that each be entered as one term in an equation (14).

Morison's own analogy—abortion—provides the best clue as to the likely consequences of a strict adoption of his suggestions. I know he would find these consequences as abhorrent as I. No matter what one can say in favor of abortion, one can't say that it is done for the benefit of the fetus. His interests are sacrificed to those of his mother or of society at large. The analogous approach to the problem of the dying, the chronically ill, the elderly, the vegetating, the hopelessly psychotic, the weak, the infirm, the retarded—and all others whose lives might be deemed "no longer worth preserving"—points not toward suicide, but toward murder. Our age has witnessed the result of one such social effort to dispense with "useless lives."

To be fair, in the end, Morison explicitly suggests only that we make acceptable the practices of suicide and assisted suicide, or euthanasia. But in offering this patient-centered suggestion for reform, he challenges the ethics of medical practice, which has always distinguished between allowing to die and deliberately killing. Morison questions the validity of this distinction: "The intent appears to be the same in the two cases, and it is the intent that would seem to be significant." But the intent is not the same, although the outcome may be. In the one case, the intent is to desist from engaging in useless "treatments" precisely because they are no longer treatments, and to engage instead in the positive acts of giving comfort to and keeping company with the dying patient. In the other case, the intent is indeed to directly hasten the patient's death. The agent of the death in the first case is the patient's disease; in the second case, his physician. The distinction seems to me to be valuable and worth preserving.

Nevertheless, it may be true that the notion of a death with dignity encompasses, under such unusual conditions as protracted, untreatable pain, the right to have one's death directly hastened. It may be an extreme act of love on the part of a spouse or a friend to administer a death-dealing drug to a loved one in such agony. In time, such acts of mercy killing may be legalized (15).

But when and if this happens, we should insist upon at least this qualification: The hastening of the end should never be undertaken for anyone's benefit but the dying patient's. Indeed, we should insist that he spontaneously demand such assistance while of sound mind, or, if he were incapable of communication at the terminal stage, that he have made previous and very explicit arrangements for such contingencies. But we might also wish to insist upon a second qualification—that the physician not participate in the hastening. Such a qualification would uphold a cardinal principle of medical ethics: Doctors must not kill.

SUMMARY

1) We have no need to abandon either the concept of death as an event or the efforts to set forth reasonable criteria for determining that a man has indeed died.

2) We need to recover both an attitude that is more accepting of death and a greater concern for the human needs of the dying patient. But we should not contaminate these concerns with the interests of relatives, potential transplant recipients, or "society." To do so would be both wrong and dangerous.

3) We should pause to note some of the heavy costs of technological progress in medicine: the dehumanization of the end of life, both for those who die and for those who live on; and the befogging of the minds of intelligent and moral men with respect to the most important human matters.

REFERENCES

1. This calls to mind the following exchange from Shakespeare's *Julius Caesar,* immediately following Caesar's assassination (Act III, scene i, ll. 101–105): "Casca: Why, he that cuts off twenty years of life/Cuts off so many years of fearing death. Brutus: Grant that, and then is death a benefit. So are we Caesar's friends, that have abridg'd/His time of fearing death."

2. P. B. Medawar, *The Uniqueness of the Individual* (Basic Books, New York, 1957), p. 55.

3. Would A. N. Whitehead himself have considered life and death as exemplifying his "fallacy of misplaced concreteness"? I seriously doubt it. See, for example, two of his essays, "Nature Lifeless" and "Nature Alive" [in *Modes of Thought* (Free Press, New York, 1968), pp. 127–147, 148–169]; and *Science and the Modern World* (Mentor, New York, 1948).

4. For an excellent discussion of the problematic status of "life" in modern scientific thought, see H. Jonas [*The Phenomenon of Life: Toward a Philosophical Biology* (Dell, New York, 1968)], especially the first essay (pp. 7–37).

5. The most prominent proposal is contained in the report of the ad hoc committee of the Harvard Medical School, H. K. Beecher, chairman [*J. Am. Med. Assoc. 205,* 337 (1968)].
6. H. C. Hopps, *Principles of Pathology* (Appleton-Century-Crofts, New York, 1959), p. 78. This passage also suggests, in opposition to Morison, that the notion of death as a discrete event has a distinguished medical and scientific history and is not simply an artistic, literary, or legal fiction. The dead body that was lately alive is a concrete fact, a fact understood to some extent even by animals. One must wonder about the sort of scientific understanding of the world which tells us that the apparent change in state from a man alive to a man dead is but an illusion. If this is an illusion, then what is not?
7. To say that something is a discrete event does not mean that it need be instantaneous. Moreover, even instantaneous events take time, for how long is an instant?
8. See works by Whitehead (*3*) and by Jonas (*4*) for consideration of the problem of organism.
9. Such second-party benefits are, without embarrassment, admitted to be a major (if not *the* major) reason for updating the criteria for pronouncing a man dead [See especially the opening paragraph of the Harvard committee report (*5*)]. In support of the new criteria, Beecher has written: "[I]t is within our power to take a giant step forward in relieving the shortages of donor material.... The crucial point is agreement that brain death is death indeed, even though the heart continues to beat." And again: "Thus, if these new criteria of brain death are accepted, the tissues and organs now consigned to the grave can be utilized to restore those who, although critically ill, can still be saved" [*Daedalus* (Spring 1969), p. 291 and p. 294]. Indeed, the new criteria have been so linked with transplantation that one physician has publicly referred to them as a "new definition of heart donor eligibility" [D. D. Rutstein, *Daedalus* (Spring 1969), p. 526]. It can be only regarded as unsavory and dangerous, both for medicine and for the community at large, to permit the determination of one person's death to be contaminated by a consideration of the needs of others. Having said this, however, I hasten to add that the redefiners also think that their criteria do happen to fit the fact of death. The authors claim that they are true criteria, capable of scientific, and not simply utilitarian, justification. All the experience to date in using these criteria for pronouncing a patient dead supports the validity of this claim.
10. More generally, modern biomedical science needs to come to terms with human mortality. With the President making the conquest of cancer a national goal, and with others proposing crash programs to conquer genetic disease, heart disease, stroke, and aging (to each his favorite malady), medicine will soon be called upon to do battle with death itself, as if death were just one more disease. Fortunately, such a battle will not succeed, for death is not only inevitable, but also biologically, psychologically, and spiritually desirable.
11. E. Kübler-Ross, *On Death and Dying* (Macmillan, New York, 1969).
12. See P. Ramsey [*The Patient as Person: Explorations in Medical Ethics* (Yale Univ. Press, New Haven, Conn., 1970), pp. 113–164] for an excellent account of this ethic.
13. The exceptional cases cited by Morison (battlefield or civilian catastrophes) do not provide a precedent for allowing considerations of "the welfare of third parties or 'society'" to intrude upon the doctor's treatment of his patient under ordinary circumstances. What is special about these cases is that the survival of the entire group or community, as a group or community, is in jeopardy, not simply that they represent "conditions of special stress where available medical resources are clearly inadequate to meet current needs." There is an overriding, acknowledged single principle, the survival of the group, which justifies the practice of "triage" or "disaster medicine" under conditions of battle, great fires, floods, or shipwrecks. Those who are most able to be returned to function and most able, when functioning again, to save others are treated first. No such danger to community safety or survival is entailed by the ordinary (though by

no means simple or trivial) problems that result from the usual scarcity of medical resources. See Ramsey *(12,* pp. 256–259) and P. A. Freund, *Daedalus* (Spring 1969), pp. xiii–xiv.

14. Morison writes: "Another significant parameter will be the sanctity *accorded to* any human life [emphasis added]." Life either has sanctity or it does not. Sanctity cannot be given or taken away by human accord (indeed, "sanctity" implies and requires "the sacred" and the divine), although men can, of course, choose to deny or ignore that human life possesses it. The difficulties and dangers of the cost-benefit approach to matters of life and death would not be lessened by placing the decisions in the hands of public committees. A widely-discussed citizens' committee in Seattle, which selected, on grounds including "social worth," from many medically fit candidates those who could use the few artificial kidney machines, has been disbanded. Its members felt incapable of judging the comparative value of individual lives when life and death are at stake. The problem resides not in any deficiencies of the Seattle citizens, but in the human impossibility of their task.

15. Strictly speaking, I doubt if we could establish the *right* to be mercifully killed. Rights imply duties, and I doubt that we can make killing the *duty* of a friend or loved one.

16. I am genuinely grateful to R. S. Morison for his stimulating and provocative paper. He has helped me begin to see more clearly what some of the serious and important question are.

READING 27

Maximization of Expected Utility as a Criterion of Rationality in Military Strategy and Foreign Policy

Robert P. Wolff

I

The nuclear age has given rise to many problems of policy formation, and not the least of these relate to the formation of grand military strategy. In the past, peacetime military strategies have been simple conceptions, like the Maginot Line, which seized upon some dominant feature of the geography and elaborated a theory around it. As has frequently been noted, these theories had a way of going wrong, but since the resulting wars were won by the masses of material and men rather than by superior sand-table deductions, the failures of the theoreticians were not fatal. There was time to test new weapons, learn anew the capacities of old ones, correct errors of prediction, and regroup forces in a winning combination. The consequences of error might be costly, as in the Second World War where our failure to heed the clear warnings of impending Japanese attack laid upon us the terrible burden of endless Pacific island-conquests. Nevertheless, even the Second World War as a whole can, I think, be judged a rational price to pay for the goal of destroying the Axis powers.

It is now generally recognized that the invention of nuclear weapons, in particular of missiles with thermonuclear warheads, has completely altered the character of war and of military strategy

Reprinted by permission of the author and *Social Theory and Practice, 1,* Spring 1970, pp. 99–111.

as well. Nuclear conflicts are projected as ranging from short, half-hour wars to what Herman Kahn, wryly but accurately, calls a long-war—two to thirty days. In such a conflict quite obviously the weaponry and troops would be those in existence at its inception. There would be time neither for mobilization nor for the conversion of factories to war production. Indeed, it might even prove impossible to activate stockpiled war material which was not in a ready-to-use condition. What may not be quite so obvious is that the non-material components of the war—the judgments, decisions, strategies—would also be those of the pre-war period. As the first great attack approached there would be time only to put into effect the contingency plans already formulated. After the blow had struck, disorder and fragmentation of command and control would be inevitable. Not merely grand strategy, but even detailed tactical responses, would have to be pre-arranged. The war might effectively be over before the top military leaders were able to contact one another and concert their activities.

As the conditions under which a future war could be fought are different from those of our past experience, so too are the weapons and their consequences. I shall not here review the dimensions of potential destruction, for they are familiar to all. Allow me only to remark that in an age of thermonuclear missiles, defense has given way to deterrence as the central concept of military strategy. We have given up any serious hope of denying the enemy access to our territory—the traditional objective of defense policy. Instead, we aim at a military force which can threaten sufficient punishment to dissuade an enemy from contemplating attack. [3, 9 chapter 1].

The nuclear revolution has thus brought three significant changes in the context and aims of military planning: First, the necessity of having, in actual existence, a set of complete strategies for a spectrum of possible wars, with little hope of *ad hoc* adjustments after hostilities have started; Second, a massive increase in the level and significance of potential destruction, necessitating policy planning with regard to situations of which we have little or no practical experience; and Third, a fundamental alteration in the central concepts of military planning, thereby rendering unhelpful, or even harmful, the long experience and instinctive wisdom of our military leaders.

Our planners have been cut loose from their moorings and set adrift on a sea of speculation. They have been forced to make weighty decisions on scanty evidence; they have had to reason their way along proliferating networks of future possibilities without even a method of decision-making to guide them. Glenn Snyder describes the situation vividly in the opening lines of the preface of his book. *Deterrence and Defense.* [9, v.]

> This book was motivated largely by the apparent lack of any definite agreed criteria of making rational decisions in national security policy. Many of the "great debates" in this field seem to be characterized, even partially caused, by the lack of such criteria. We have no scarcity of prescriptive arguments for or against some policy alternative, and many of these are very well reasoned and powerfully stated. Such prescriptive analyses should certainly continue, both in the scholarly world and in the government. But there is an urgent need for something else—a systematic theory, or method, which can be used by the governmental generalist, or the interested layman, to compare and weigh the arguments for particular policies, which often sharply conflict.

Faced with this problem, certain students of military policy have turned to the disciplines of economics and game theory in an attempt to adapt their concepts and methods to the analysis of grand strategy. The aim has not been to solve problems or deduce conclusions—the theory is too primitive and the facts of war too complex for that. Rather, the more modest hope has been entertained of imposing a measure of conceptual order on the chaos of apocalyptic forebodings and rule-of-thumb guesswork into which our post-war military policy was degenerating. The key to the new methodology is a concept of rational decision-making which made its first appearance two centuries ago, and has since undergone many transformations at the hands of economists and mathematicians: namely, *maximization of expected utility*.

The notion of expected utility was first proposed by Daniel Bernoulli in 1730 as a solution to an apparent paradox known as the problem of the St. Petersburg Gamble. [2] Suppose that you are offered an opportunity to bet a sum of money on the toss of a coin, with the understanding that if it lands heads you will be paid one dollar. If it lands tails, it will be tossed again, and this time you will win *two* dollars on heads. The coin will be tossed again and again until it finally lands heads, and for each additional toss required your winnings are doubled. In brief, you are paid 2^{n-1} dollars for the first heads, n being the toss on which it appears. Now it was a commonplace of Bernoulli's time that the rational way to estimate the value of a bet was by calculating its so-called mathematical expectation. This is the sum of the values of the possible outcomes, discounted by their probabilities. Thus:

$$\text{Mathematical Expectation} = \sum_{i=1}^{n} p_i v_i$$

$$\text{where} \quad \sum_{i=1}^{n} p_i = 1.$$

Employing this formula, the mathematical expectation, or expected value, of the St. Petersburg gamble is

$$\sum_{i=1}^{\infty} \frac{1}{2}n \times 2^{n-1} = \frac{1}{2} + \frac{1}{2} + \frac{1}{2} + \frac{1}{2} + \ldots\ldots$$

As this sum does not converge, *any* finite sum of money is less than the expected value, and hence is a reasonable amount to pay for entering the game. But, Bernoulli pointed out, this is clearly absurd. No sensible man would wager so much as one hundred dollars on this game, let alone one million or one hundred million. Hence, he concluded, men must not value equal increments of money equally, irrespective of the state of their pocketbooks. They must attach a decreasing utility to additional increments, in such a way that the value of a sum is inversely proportioned to the size of their fortune. When men calculated, he supposed, they were really attempting to maximize a quantity known as subjective utility, and not monetary wealth. By assuming that the relation between this utility and money was given by the formula $u(x) = \log x$, Bernoulli thought himself able to resolve the paradox of the gamble.

Since the time of Bernoulli, the notion of subjective utility has undergone considerable refinement. Economists no longer speak of utility as a quantity or perceived attribute of objects. Instead, they adopt the neutral position that men can be treated *as if* they were attempting to maximize such a quantity. More precisely, it is said that if men can order their preferences for a variety of possible states of affairs, and if they can consistently express preferences for probability combinations of pairs of states, then a function can be defined which maps their preference onto the real numbers, and is invariant up to a linear transformation. In deference to the history of the concept, this function is called a utility function.

Bernoulli himself was not quite clear whether he was offering a *description* of the way men actually make choices, or a *proposal* as to how they ought to make choices, and this ambiguity has continued in utility theory to the present day. Maurice Allais, in a critical analysis of what he calls the "American School" suggests that utility theory may be interpreted in at least three different ways: [1]

1. First, the theory of cardinal utility may be taken as an empirically verifiable description of the manner in which all or most men actually choose among risky alternatives. Friedman and Savage and Mosteller and Nogee appear to adopt this interpretation. [4, 7]
2. Second, the theory may be deduced *a priori* from a set of axioms of rationality which have an independent plausibility, and are sufficiently powerful to yield an axiomatic theory of utility. The implication is that the theory offers a model of rational decision-making, and hence an analysis of the way in which men ought, in their own interest, to choose. Von Neumann and Morgenstern, Marschak, and many others have followed this line, refining and perfecting the model by various mathematical techniques. [8, 6]
3. Third, one may observe men who are generally acknowledged to be prudent, and attempt inductively to elict from their behavior a set of rules of rationality which can then be worked up into a formal system. Allais himself adopts this technique, criticizing the work of the American School for ignoring certain rules of decision-making which he claims to be central to the behavior of prudent men.

How are we to classify the employment of the concept of expected utility by strategical theories such as Thomas Schelling and Glenn Snyder? Are they describing ordinary human behavior, deducting criteria of rationality from *a priori* premises, or codifying the common sense of reasonable men? The answer, I think, is that they are in fact doing no one of these things alone. Let me try to make clear the way in which their enterprise differs from that of the economists and game theories whose insights they have borrowed.

As I have suggested above, the military strategists are faced by three problems: they must formulate complete plans in advance, they must deal with a wholly new range of possible consequences, and they must build their theories around the unfamiliar concept of deterrence rather than the familiar concept of defense. In other words, their task is to decide what constitutes a rational choice in contexts and with respect to outcomes which men have never before confronted. They are not describing how nations actually formulate deterrence strategies; nor are they deducting rationality-models from general axioms; nor are they merely codifying the decision-making rules of prudent men. It seems to me that they are engaged in an enterprise which combines elements of the second and third approaches, and also has certain distinctive features of its own. That is, they lay down certain axioms of rationality, guided in part by their awareness of the customs of rational men, and they then make a *proposal* that the resulting method of decision-making be extended to the new and unexperienced realm of nuclear war.

I want to emphasize very strongly the fact that this extension of the method of expected-value maximization is a recommendation and not a demonstration. Snyder, Schelling, and the others are offering a suggestion as to the method of decision-making we ought to use in this new area of human choice. They base their proposal upon analogies between the bargaining and other economic situations to which utility theory has been applied in the past, and the military situations for which a method of decision-making is sought. The sophistication and elegance of the mathematics must not divert *us* from the central question, which is whether the proposed extension is wise and useful.

Such a question cannot be decisively settled by argument. The most that the supporters of utility-maximization can do is to exhibit the method in its original economic context, and then point to the similarities with strategic problems. Those, like myself, who are skeptical of the fruitfulness of the proposed application can only present evidence of dissimilarities, and try to show why these cast doubt upon the adequacy of utility-maximization as a measure of rationality in deterrence contexts.

I am not attempting to question the formal adequacy of game theory nor am I concerned here with its applicability to the domain of economic choice. My aim is to call in question the fruitfulness and wisdom of extending the game-theoretic criteria of rationality to situations of grand deterrence strategy. (I include the adjective "grand" in order to avoid consideration of the applicability of game theory to tactical problems. My impression is that in such situations it can be of great value.)

I shall present four arguments against the use of expected-value maximization as a criterion of rationality in military strategy. In each case I will point to some feature of strategic decisions which makes them unlike ordinary economic decisions, and I will then suggest that this difference is critical for the methodology in question.

1. The first limitation on the use of expected value maximization is its assumption that the reserves of the player in a game situation are sufficient to avoid the risk of exhaustion in the short run. If some one outcome or not unlikely series of outcomes of a strategy will bankrupt the player, then it may be a poor strategy for him to adopt, even though its expected value is quite high. For example, suppose a gambling game offers the opportunity to bet a dollar, with a ten percent probability of winning $20 and a ninety percent probability of losing the dollar. The

expected value of the game (or of the strategy of playing the game, as opposed to not playing it) is, according to the formula, (1/10 x 20) – (9/10 x 1), or $1.10. Over the long run, this is obviously a very good game to play as it promises better than 100 percent return on the investment. What if you only have a five dollar stake, however? Should you play? The danger is that five straight losses may occur before one of the big $20 payoffs. Since we are interested in a five-play run, we can transfer this iterated game into a new one-shot game with six possible outcomes. The bet is now five dollars, and the payoffs are a total loss (equal to five straight losses of the old game), $16 (or one win), $37 (or two wins), and so forth. The schedule of probabilities and payoffs, together with the expected values, is given in Table 1.

The expected value of the new game is $5.50, or (5 x 1.10), for it is formally equivalent to five rounds of the one-play game. But the telescoped form reveals that there is almost a 60 percent chance of being wiped out before at least one win comes along to swell the stake. Hence this is a much less attractive game to a man with limited funds than might at first appear. If a second game presents itself with a less spectacular expected value but a negligible probability of total loss, it would surely be rational to play it instead.

TABLE I

Outcome	Payoff (in $)	Probability	Expected Value (in $)
1. No wins	–5	.59	–2.95
2. One win, four losses	16	.328	5.25
3. Two wins, three losses	37	.0729	2.70
4. Three wins, two losses	58	.0081	.46
5. Four wins, one loss	79	.00045	.04
6. Five wins	100	.00001	. . .
		.99946=1.0	$5.50

The relevance of this limitation to national policy is clear. If we believe that nuclear war will mean the destruction of the United States as a world power, or even its demise as an organized society, then even the temptation of considerable gain (such as the permanent defeat of the Soviet Union) may be insufficient to lure us into a strategy which threatens war. Limited actions, like limited losses in gambling, are of an entirely different order than total war, for there is always the possibility of recouping losses in a later limited action. Total war, on the other hand, closes off all future "plays." It is like going bankrupt. Hence we must be exceedingly wary of policies which are urged on the grounds that they offer a high expected value, but which also threaten too high a probability of total loss. A methodology which fails to take this consideration into account is clearly inadequate.

2. A second restriction, related to that of limited resources, is the assumption that the values of the several outcomes are of roughly comparable orders of magnitude. This is necessary in order that no single outcome dominate every expected value which it is a component. For example, consider a situation in which there are five possible outcomes, rated as follows: A = 10, B = 15, C = −20, D = 50, and E = −100,000. It is obvious that any strategy which offers the slightest meaningful probability of outcome E will have a lower expected value than any other strategy which omits E entirely from among its possible outcomes. For example, if strategy S has a 1/10 of 1 percent chance of yielding E, and a 99 and 9/10 percent chance of D, then its expected value will be (.001 x − 100,000) plus (.000 x 50), or −50.05. This is more than two and one-half as bad as the worst possible strategy which does *not* threaten E at all (i.e., the strategy, if it exists, which *ensures* C). Thus the overriding consideration in the situation will be to avoid E. Any alternative will be preferable to even a minute chance of E. The method of expected value maximization will simply be irrelevant to decision making in such a situation (or rather, it will be limited by the condition that prior to expected value calculations, all strategies be eliminated from consideration which threaten E with any probability whatsoever).

In view of this simple arithmetical truth, it is significant that when students of military strategy employ the concepts of game theory for analytical purposes, they usually fill their matrices with "orienting numbers" which are in fact of comparable orders of magnitude. For instance Glenn Snyder offers a mathematical example to illustrate certain principles of deterrence. He considers the consequences of an all-out Soviet nuclear attack on the U.S. or alternatively of a Soviet ground attack on Western Europe, to both of which U.S. can make no response or a massive retaliatory response or—in the case of the ground attack—a ground force response. The pay-offs matrix which he presents for the United States is given in Table II [9, p. 270]

The nuclear attack on the United States is less costly in the case of massive retaliation because of the assumption that certain Soviet weapons are thereby destroyed which were not used in the first strike. (Presumably, Snyder has made an error in setting the cost of the ground force response to nuclear attack equal to 0. Since in that case a ground force response is impossible, it would come under the heading of "no response" and hence cost −500.)

TABLE II

	Soviet Union		
United States	No Attack	Nuclear Attack on U.S.	Ground Attack on Western Europe
No response	0	−500	−100
Massive response	0	−400	−400
Ground force response	0	0 [n. b.]	−150

What these figures state is that the difference to the United States between an obliterating nuclear exchange and the uncontested loss of Europe is three times as serious as the difference between nuclear exchange and an unanswered knock-out blow from the Soviet Union (i.e.), (400 − 100) / (500 − 400) or (3/1.) Also, since the *status quo ante* is set equal to zero, the figures tell us that the difference between the unanswered nuclear strike and the uncontested loss of Europe is four times as great as the difference between the loss of Europe and the *status quo*. Whether Americans really evaluate these various possibilities in this manner is, of course, left open by Snyder. He specifically disavows any significance for his numbers. However, by choosing commensurable quantities he conveys the impression that his calculations of expected values represent a believable and usable model of decision-making. Because of the comparability of the values assigned to the various outcomes, he can realistically discuss the changes in probability of outcomes which must be effected in order to change the relative desirability of various strategies. Had he assigned a value of −5,000 or −50,000 to the knock-out nuclear attack, he would have so over-balanced his calculations that no realistic alternative defense postures would alter them significantly. In other words, in order to make the methodology of expected value calculation appear *relevant* to the making of strategic decisions, it is necessary to choose illustrative numbers which place the cost of total destruction in the same range as the cost and gains of other policy objectives.

Many of the debates over American foreign policy can be looked at as if they were attempts to support the claims of varying payoff figures in a deterrence game. One group of authors emphasizes the absolute horror of a nuclear holocaust and argues that the result of a war would be the cessation of American society as we know it. They are in effect proposing that we assign to nuclear war a negative value so great that it dominates every expected value calculation and makes even a small chance of total war sufficient to condemn a strategy.

In opposition to this view of nuclear war, there are several possible lines of argument, each of which attempts in a different way to demonstrate the commensurability of nuclear war with other outcomes. The most direct is simply to deny that nuclear war would be so overwhelmingly destructive, or at least that its destructive effects could be cut to acceptable limits by appropriate measures of passive defense. This, in effect, is what Herman Kahn has tried to do in the first part of his book, *On Thermonuclear War.* [5, 3–116]. For reasons which I cannot detail here, I think that Mr. Kahn's estimates are overly optimistic even for the quite small-scale nuclear war which he postulates. For larger wars, he himself would have grave doubts.

A second possibility is to insist on the even greater negative value of some other outcome, thereby once again giving point to the technique of expected value calculation. We are familiar with this line of counter-argument in the form of the slogan, "Better dead than Red!"

Alternatively, one can reintroduce the calculation of expected value by insisting on the extremely large *positive* value of some outcome, whose possibility balances the possibility of nuclear war. In a sense, this is the logic behind the argument that the opportunity of destroying Russia or inflicting defeats upon her is worth the risk of being destroyed ourselves. Unfortunately, the proponents of this view have failed to comprehend the revolution in military relations. Today, the choice is not between obliterating Russia or being obliterated; it is between mutual destruction and mutual survival. Hence this response, although formerly it would have served to defend the methodology of expected value calculation, is now irrelevant to the problem of deterrence.

There is still another response to the argument that a nuclear war would be obliterating, and hence must be avoided at all costs. This is the argument that no course of action—*not even unilateral disarmament*—can safeguard us against a nuclear attack, and that our wisest course is therefore to face the possibility of war, include it in our calculations and work always to diminish the possibility without sacrificing our fundamental values or national interest. This, I take it, is the position of every president since John Kennedy, and of a considerable number of foreign policy experts. I am not concerned here with its soundness or rationality; I only wish to point out that it, in common with the apocalyptic view described above, has no use for expected-value calculation. If I understand this foreign policy correctly, it does *not* assume that the prospect of greater gains justifies a larger risk of nuclear war. Instead, it lays down certain minimal national interests for which we are willing to risk total destruction. Within the context of these primary interests, it strives to minimize the threat of war, while pursuing our secondary interests by non-nuclear means. While the technique of expected-value calculation might conceivably be relevant to choices among secondary goals, it has no role to play in the formulation of deterrence policy. To propose it as a methodology of decision-making is actually to make a substantive recommendation of far-reaching implications. Such a method presupposes a willingness to run greater risks of war in order to achieve more valuable foreign policy victories. Hence it is by no means a neutral methodology, as its proponents claim.

3. We have come upon a third objection to the methodology of utility theory, namely its failure to take account of aversion to risk. We can give this point a mathematical expression by means of a brief hypothetical example. Let us suppose, contrary to what has been urged above, that the negative value of a nuclear war is sufficiently comparable to other possible outcomes to allow of meaningful expected value calculations. Let us further suppose that when faced with a choice between two strategies, both of which threaten a measurable probability of nuclear war, we are willing to base our decision upon a consideration of expected values. Our attitude is that if we must run some risk of war, and if the two probabilities are not too dissimilar, we are prepared to take a slightly greater risk for a chance at a greater gain. Stated symbolically, if a and b are possible outcomes of considerable positive value, $a > b$, and if d is nuclear war, then:

(1) $ra + (1 - r)d > rb + (1 - r)d$ where $o < r < 1$

and furthermore, there is some $\epsilon > o$ such that

(2) $(r - \epsilon)a + (1 - r + \epsilon)d > rb + (1 - r)d$

However, assume that we are extremely anxious to eliminate any risk of nuclear war, and therefore place an especially high value on no-risk strategies. This is the converse of what is called a "love of danger" or "love of gambling." There will now be outcomes which are worth less than either a or b, but which are preferred to any probability combination of a or b with d because they exclude the threat of war. That is to say, given c, $a > c > d$,

(3) $c > ra + (1 - r)d$ for any $r < 1$.

But this contradicts the postulates of the axiomatic theory of utility, as stated for example by Marschak. According to Marschak's Postulate II, [6,] $a > c > d$ implies

(4) $c = ra + (1 - r)d$ for some $r, o < r$
This contradicts (3).

The "aversion to risk," as we may label this phenomenon, seems to me an eminently reasonable attitude for a policy planner to adopt. A private individual who places his life in jeopardy in order to increase his gain may not be irrational. Man's time is short, and death comes soon enough, no matter how cautiously one lives. But for a statesman to adopt a similar viewpoint toward the life of his nation would be grossly irresponsible and unreasonable. Consequently, in planning the deterrence policy of a nation, it is wise to avoid the implications of the methodology of expected value calculation.

(4.) My last argument raises a problem of a somewhat different nature.

As I have already indicated, utility theory is explicitly designed to take account of the divergence between objective payoffs and our subjective preferences for them. A given sum of money may be worth more to me under some conditions than under others. Furthermore, I may place value upon subjective elements which have no objective monetary equivalents. In a friendly poker game, for example, the pleasure of the evening may lead me to play in a less than ruthless style, in order not to clean out a novice who has agreed to sit in. Such behavior, while economically irrational in a narrow sense, is nonetheless an attempt to maximize overall subjective utility, and hence can be viewed as perfectly rational. In the formal presentations of utility theory, it is assumed that a set of outcomes, $p_1, p_2, \ldots, p_i$, can be ordered according to preferability. No limits are placed upon what may be treated as an outcome, and of course I am permitted to arrange the outcomes in any way which strikes my fancy, subject only to certain consistency conditions of well-ordering.

In order to see the complications which this produces, let us revert to the case of the gamble discussed above. The bet, it will be recalled, was one dollar, with a 10 percent chance of a $20 win, and a 90 percent chance of loss. The problem, as we saw, was that with a limited stake of $5, the probability was too high of being cleaned out before a win. In order to make the case more personal, let me suppose that I am on my way to the store to buy my wife a birthday present. The use of a subjective utility measure is designed precisely to register my particular concern over losing the money for the present. I can express this concern by assigning the loss of the entire five dollars a negative value *more than five times* as great as that of the loss of one dollar. If my aversion to losing all my money is great enough, and if the temptation of possible gain is not disproportionately seductive, then for me the expected value of the strategy will be negative, and I will choose not to play. Specifically, suppose that the various outcomes of the five-shot game are assigned the values listed in Table III, as measured in some arbitrarily chosen unit of subjective utility.

In this revised calculation, the expected value of the game is less than the value of not playing at all and so I decide not to play. There is nothing mysterious about this new calculation. It is merely a mathematical expression of the fact that in the stated circumstances, five dollars has a greater than usual value to me.

TABLE III

Outcome	Subjective Utility	Probability	Expected Value
1. No wins	−15	.59	−8.85
2. One win, four losses	16	.328	5.25
3. Two wins, three losses	37	.0729	2.70
4. Three wins, two losses	58	.0081	.46
5. Four wins, one loss	79	.00045	.04
6. Five wins	100	.00001	...
		.99946=1.0	− .40

Now, however, suppose that the gambler offers me a chance to play the game one step at a time, instead of in its five-shot portmanteau form. He knows that I have a strong gambling streak and hopes to woo me away from my husbandly resolve. At this point, I figure that four dollars will still buy a nice present, so I take the plunge, setting loss of one dollar equal to precisely minus one utile. I lose, and consider whether to play again. The game is looking slightly less attractive, for I have only four dollars and I have begun to worry about the present. At the same time, my gambling spirit is aroused. I am hooked, as gamblers say, and so I put down my money. I lose again and in a desperate attempt to recoup slap down another dollar and lose again. Now I have only two dollars left. This is too little to buy any sort of respectable present, so I play twice more and lose everything.

We can assume that my utility function changes from play to play, as the conditions of my pocketbook and of my frame of mind alter. But we have no reason at all to suppose that this change, summed over the five plays, exactly equals the subjective utility of the five-shot game. For me, the prospect of losing five dollars all at once may seem better or worse than the sum of the gains and losses at each stage of the iterated series of one-shot games. Of course it is *possible* that the two will be equal, but it is certainly not necessary, as game theory must assume.

The dilemma can be heightened by the addition of a few more fanciful psychological details. Suppose that I have been through such temptations before and know how I tend to react to them. I very much want to reform and hence, when presented with the five-shot game, place a high negative value on total loss, making the expected value of it negative. However, I also know that my feelings will change in the heat of the game, and that if I play the game one shot at a time, losing as I go, the overall sum of expected values will be positive. What is it rational for me to do?

The source of the trouble is the obscurity surrounding the original expressions of preference on which the definition of subjective utility was based. When I try to decide whether I would rather lose five dollars all at once or one at a time, am I supposed to imagine myself in the midst of play or in a pregame condition? If there is a contradiction between the two methods of estimating, which should I adopt as more rational? Should I strive to adjust my mid-game attitudes to the calmly considered preferences of the pre-game period, or should I ignore my pre-game sobriety as unrepresentative of my true gambling spirit? Obviously, utility theory can give no answer. The dilemma sounds somewhat like a debate between a puritan and a romantic. It is a very important matter, but one which falls under the heading of substantive policy, not procedure or methodology.

Very striking and significant analogies to this curious dilemma can be found in the game theoretic approach to nuclear strategy. Glenn Snyder seems particularly prone to confusion of this sort. Near the beginning of his book, he defines "rationality" as:

> ... choosing to act in the manner which gives best promise of maximizing one's value position, on the basis of a sober calculation of potential gains and losses, and probabilities of enemy actions.

"This definition," he goes on, "is broad enough to allow the inclusion of such 'emotional' values as honor, prestige, and revenge as legitimate ends of policy. It may be perfectly rational, in other words, to be willing to accept some costs solely to satisfy such emotions, but of course if the emotions inhibit a clear-eyed view of the consequences of an act, they may lead to irrational behavior." [9, 25]

Suppose that in conformity with this definition of rationality, our policy-makers in the present (pre-war) state perform the following expected value calculation. Taking into account objective factors like loss of life and property, change in power balance, etc., *and also* such emotional factors as outrage over these losses, they conclude that massive retaliation with the remnants of a nuclear striking force would be irrational if there were no chance of counter-force effects, and if the enemy could respond by totally wiping out the remaining parts of the United States. In other words, all things considered, they conclude that should the Russians succeed in getting in a crippling first-strike, the only rational response would be surrender. Now suppose that the attack actually occurs, and that by some miracle these policy planners are still alive. They feel an overwhelming anger and desire revenge even at the cost of self-annihilation. Their world has crumbled around them; they are willing to see it disappear completely if only they can inflict some terrible blow to the enemy. According to Snyder, it is now perfectly rational for them to unleash their remaining nuclear striking force, for their subjective value structure has changed enough to make the expected value of this suicidal retaliation positive rather than negative. Note that it is not their *calculation* of values, but their *values themselves* which have changed, just as in the gambling example above I imagined my feelings to change from play to play, and hence also my utility estimates.

If Snyder's definition is taken literally, virtually any series of contradictory decisions is rational, so long as each in turn accords with one's feelings of the moment. It becomes impossible to distinguish between rational and irrational policies: should we act on the basis of pre-attack or post-attack utility estimates? In the pre-attack situation, should we plan on the basis of the way we feel then, or the way we expect to feel *after* the attack?

There are two alternatives: Either we interpret the methodology as neutrally as possible, in which case it seems to reduce to an unhelpful tautology which fails to sort strategies into the categories of rational and irrational; or else we can adopt some meaningful interpretation, in which case, as noted above, we build very powerful policy assumptions into our methodology, and thereby conceal them from the scrutiny to which they should increasingly be subjected.

This, in fact, seems to me to be the lesson which we are taught by all four of the cases which I have analysed. Expected value maximization is urged as a substantively neutral method which makes no presuppositions about what is preferable but merely tells us how to maximize whatever it is we decide to place value upon. Upon examination, we discover that when the method is extended to the realm of military strategy, it carries with it very powerful value-assumptions, which are in need of critical examination. In summary, the assumptions which I have attempted to bring to light are:

1. That nuclear war will not bankrupt us, or remove us altogether from the game of deterrence.
2. That the consequences of nuclear war are of roughly the same magnitude of value as other possible outcomes of deterrence strategies.
3. That no allowance in our calculations should be made for our aversion to the risk of nuclear war, over and above our negative evaluation of its consequences.
4. That there is some one frame of mind or condition in which value estimation of future events should be made, and that all variant value estimates resulting from changed objective or subjective conditions should be ignored.

Since the application of utility maximization to military strategy involves these four presuppositions, it is clearly not a value-neutral methodology as its proponents claim. I have tried to suggest my reasons for believing these presuppositions to be false. If I am correct, then the methodology, although useful in economic contexts, is inappropriate to the analysis of decision making in military strategy and foreign policy.

REFERENCES

1. Allais, Maurice, "Le comportement de l'homme rationnel devant le risque: Critique des postulats et axioms de l'école Americaine," *Econometrica, XXI,* 503–546.
2. Bernoulli, Daniel, "Exposition of a New Theory on the Measurement of Risk," *Econometrica, XXI,* 23–36.
3. Brodie, Bernard, *Strategy in the Missile Age,* Princeton, 1959.
4. Friedman, M. and Savage, L. J., "The Unity Analysis of Choices involving Risk," *J. of Political Economy, LVI,* 279–304.
5. Kahn, Herman, *On Thermonuclear War,* Princeton, 1960.
6. Marschak, J., "Rational Behavior, Uncertain Prospects, and Measurable Utility," *Econometrica, XVIII,* 111–141.
7. Mosteller, F., and Nogee, P., "An Experimental Measurement of Utility," *J. of Political Economy, LIX,* 371–404.
8. vonNeumann, J., and Morgenstern, O., *Theory of Games and Economic Behaviour,* second ed., Princeton, 1947.
9. Snyder, G., *Deterrence and Defence,* Princeton, 1961.

SUGGESTIONS FOR FURTHER READING

Baier, K. and Rescher, N. (eds.) *Values and the Future.* New York: The Free Press, 1969.

Burke, J.G. (ed.) *The New Technology and Human Values.* Belmont, Calif.: Wadsworth Publishing Co., 1966.

Callahan, D. *Abortion: Law, Choice and Morality.* London: The Mac millan Co., 1970.

Ellul, J. *The Technological Society.* New York: Random House, 1964.

Ford, A.B. "Casualties of Our Time." *Science,* 167 (1970), pp. 256–263.

Horowitz, I.L. (ed.) *The Use and Abuse of Social Science.* New Brunswick, N.J.: Transaction Books, 1971.

Kübler-Ross, E. *On Death and Dying.* London: The Macmillan Co., 1969.

Rescher, N. "The Allocation of Exotic Medical Lifesaving Therapy." *Ethics,* 79 (1969), pp. 173–186.

Schon, D.A. *Technology and Change.* New York: Dell Publishing Co., Inc., 1967.

Smith, H.L. *Ethics and the New Medicine.* Nashville, Tenn.: Abingdon Press, 1970.

Starr, C. "Social Benefit Versus Technological Risk." *Science,* 165 (1969), pp. 1232–1238.

Index

INDEX